PHYSICS LABORATORY EXPERIMENTS

PHYSICS LABORATORY EXPERIMENTS

SECOND EDITION

Jerry D. Wilson
Lander College

D. C. Heath and Company
Lexington, Massachusetts Toronto

Published simultaneously in Canada.

Printed in the United States of America.

International Standard Book Number: 0–669–09909–0

PREFACE

Physics Laboratory Experiments was written for students of introductory physics. The major purpose of laboratory experiments is to augment and supplement the learning of basic physical principles while introducing laboratory procedures and equipment. Toward this end, each experiment (except Experiment 3) is organized into the following sections:

- Advance Study Assignment
- Introduction
- Equipment Needed
- Theory
- Experimental Procedure
- Laboratory Report
- Questions

In this second edition I have substantially revised material as a result of feedback from the many instructors and students who used the first edition. Retained in this edition are those features that proved to be successful. A summary of pedagogical and practical features is given below.

- **Advance Study Assignment.** Students often come to the laboratory unprepared, even though they should have read the experiment to become familiar with it before the lab period. To address this problem, I have written an Advance Study Assignment preceding each experiment. The assignment consists of questions that can be answered from the Theory and Experimental Procedures sections of the experiment. Students are lead through the theory, calculations, and procedures pertaining to the experiment and consequently perform the experiment with greater confidence and gain better conceptual understanding. I recommend that the Advance Study Assignment be collected at the beginning of the laboratory period.
- **New Experiments.** Three new experiments in this Second Edition give the instructor more selection in experiments on electricity and light:

  Experiment 34 *Multiloop Circuits: Kirchhoff's Rules*
  Experiment 43 *Transistor Characteristics*
  Experiment 47 *Polarized Light*

  I have refined procedures for greater clarity and understanding.
- **New to This Edition: an Instructor's Resource Manual.** Each chapter contains a discussion of the experiment, teaching hints, answers to selected questions, and a post-laboratory quiz with short-answer and essay questions. Graph masters are provided for the instructor to supply students with graph paper. A list of scientific equipment suppliers and a summary of major equipment needs for all experiments are also included, along with a list of references for laboratory safety procedures.
- **Illustrations.** Over 200 photographs and diagrams illustrate equipment and experimental setups, in order to minimize error and to ensure lab safety.
- **Maximum application to available equipment.** Laboratory equipment at many institutions is often limited, and available equipment is usually of the standard variety purchased from scientific supply companies. The experimental procedures in this manual are described for different types of common laboratory apparatus, thus maximizing the application of the manual. For example, in Experiment 5, Uniformly Accelerated Motion, I describe three procedures for measuring. The Instructor's Resource Manual indicates those experiments that can be performed using a common set of equipment.

- **Economical experiments.** Many of the experiments may be done with readily available, economical equipment and materials. This is particularly important in modern physics experiments, where cost is a limiting factor. See, for example,

Experiment 17	*Hooke's Law and Simple Harmonic Motion* (includes elongation of a rubber band)
Experiment 42	*Rectification: Semiconductor Diodes*
Experiment 51	*The Mass of an Electron: e/m Measurement* (does not require expensive Helmholtz coils)
Experiment 52	*Exponential Functions*
Experiment 53	*Chart of Nuclides*
Experiment 57	*The Pi-Mu-e Decay Process*

- **Data analysis.** Experiment 1, Experimental Error and Data Analysis, serves as an introduction to laboratory investigation and discusses data uncertainty and analysis as well as proper graphing procedures. In subsequent experiments, students are generally asked to analyze a set of data before proceeding to the next part of the experiment, thus gaining a better understanding of the purpose of the various procedures and of the total experiment. If students delay analysis until they have taken all the experimental data, they may rush through experiments, simply generating numbers. In doing so, they can fail to grasp the purpose of a particular procedure or find too late that a particular set of data is invalid due to experimental error, which may affect the experimental results as a whole.
- **Example calculations.** In the Theory section of some experiments, I have included, where appropriate, example calculations involving the formulas and mathematics used in the experiment.
- **Laboratory reports.** Since a standardized laboratory report greatly facilitates grading by the instructor, I have provided a laboratory report for each experiment, for recording data, calculations, and experimental results. Only the laboratory report and the questions pertaining to the experiment need to be handed in for grading. Many of the laboratory report tables have been reorganized in this edition for easier data recording and analysis.
- **Thorough coverage.** Topics often omitted in other lab manuals are included, for example, experimental procedures using the air track, an introduction to the oscilloscope, transistor characteristics, and polarized light.

Acknowledgments

I wish to express my sincere thanks to all of the people and companies whose helpful suggestions and assistance were so important in the preparation of this manual. In particular, grateful acknowledgment goes to reviewers of the second edition manuscript: Thomas Damon, Pikes Peak Community College; Walerian Majewski, Northern Virginia Community College; and Peter Skoner, St. Francis College. Their many helpful suggestions have been incorporated in this edition. Special thanks go to Robert Neff for the idea of the Chart of Nuclides experiment and the Ealing Corporation for permission to use their Pi-Mu-*e* Decay Process experiment. Also, I greatly appreciate the photographs provided by Central Scientific Co., Inc., Sargent-Welch Scientific Company, and The Ealing Corporation.

My special recognition is given to I. L. Fischer of Bergen Community College, Paramus, N.J., who in serving as a consultant made many constructive suggestions for procedure and experiment revisions. His contributions also include the new experiments, Multiloop Circuits (Experiment 34), and Transistor Characteristics (Experiment 43), which draw on his background as an electronic engineer. He is now a physics professor, and students receive the benefits of his being a fine experimentalist and teacher.

Jerry D. Wilson

CONTENTS

PHYSICS LABORATORY EXPERIMENTS

Introduction

A. WHY WE MAKE EXPERIMENTAL MEASUREMENTS

When you can measure what you are speaking about and express it in numbers, you know something about it; but when you cannot measure it, when you cannot express it in numbers, your knowledge is of meager and unsatisfactory kind.

LORD KELVIN

As Lord Kelvin so aptly expressed, we measure things to know about them—so that we can describe objects and understand phenomena. Experimental measurement is the cornerstone of the scientific method, which holds that no theory or model of nature is tenable unless the results it predicts are in accord with experiment.

The main purpose of an introductory physics laboratory is to provide "hands-on" experiences of various physical principles. In so doing, one becomes familiar with laboratory equipment and procedures and with the scientific method.

In general, the theory of a physical principle will be presented in an experiment, and the predicted results will be tested by experimental measurements. Of course, these well-known principles have been tested many times before, and there are accepted values for certain physical quantities. Basically you will be comparing your experimentally measured values to accepted theoretical or measured values. Even so, you will experience the excitement of the scientific method. Imagine that you are the first person to perform an experiment to test a scientific theory.

B. GENERAL LABORATORY PROCEDURES

Safety

The most important thing in the laboratory is your safety and that of others in the lab. Experiments are designed to be done safely, but proper caution should always be exercised.

The chief danger comes from a lack of knowledge of the equipment and procedures. Upon entering the physics lab at the beginning of the lab period, you will probably find the equipment for an experiment on the laboratory table. Restrain your curiosity and do not play with the equipment. You may hurt yourself and/or the equipment. A good general rule is:

Do not touch or turn on laboratory equipment until it has been explained and permission has been given by the instructor.

Also, certain items used in various experiments can be particularly dangerous, for example, hot objects, electricity, mercury lamps, and radioactive sources. In some instances, such as with hot objects and electricity, basic common sense and knowledge are required. However, in other instances, such as with mercury lamps and radioactive sources, you may not be aware of the possible dangers. Mercury lamps may emit ultraviolet radiation that can be harmful to your eyes. Consequently, some sources need to be properly shielded. Some radioactive sources are solids and are encapsulated to prevent contact, whereas others are in liquid form and are transferred during an experiment. Some general safety precautions for the use of radioactive materials are:

1. Radioactive materials should be used only by or under the supervision of a person properly informed about the nature of the material.
2. Care should be taken to avoid unnecessary handling or contact with the skin.

3. Mouth pipetting should be prohibited where material is in liquid form.
4. Eating, drinking, and smoking should not be permitted in the area where radioactive materials are being used.
5. Protective gloves or forceps should be used when the material is handled or transferred.
6. All persons working with radioactive material should thoroughly wash their hands afterward.
7. When not in use, materials should be stored in an appropriately labeled container and in a place of limited access.
8. Should an accident occur (particularly if it involves radioactive materials), it should be reported immediately to the laboratory instructor.

Equipment Care

The equipment provided for the laboratory experiment is expensive and in some instances quite delicate. If used improperly, certain pieces of apparatus can be damaged. The general rule given above concerning personal safety also applies to equipment care.

Even after familiarizing one's self with the equipment, it is often advisable or required to have an experimental setup checked and approved by the instructor before putting it into operation. This is particularly true for electrical experiments. Applying power to improperly wired circuits can cause serious damage to meters and other pieces of apparatus.

If a piece of equipment is broken or does not function properly, it should be reported to the laboratory instructor. Also, after you complete an experiment, the experimental setup should be disassembled and left neatly as found, unless you are otherwise instructed.

Laboratory Reports

A Laboratory Report is provided for each experiment in which experimental data are recorded. This should be done *neatly*. Calculations of experimental results should be included. Remember, the neatness, organization, and explanations in the Laboratory Report represent the quality of your work.

Name .... Section .... Date ....

Lab Partner(s) ....

EXPERIMENT 1 *Experimental Error and Data Analysis*

ADVANCE STUDY ASSIGNMENT

Read the experiment and answer the following questions.

1. Do experimental measurements give the true value of a physical quantity? Explain.

2. Distinguish among the types and causes of experimental errors. Give a specific example of each.

3. How can each type of experimental error be reduced or minimized?

4. Why does the accuracy of an experiment depend in general on systematic errors and the precision on random errors?

5. What is the effect of reporting an experimental value with more figures or digits than are significant?

6. Distinguish between percent error and percent difference.

7. Explain why the average of the deviations of a set of measurements is always zero.

(continued)

8. How is the scatter or dispersion of a set of measurements reported, and what does this indicate?

9. What is the statistical significance of one standard deviation? Two standard deviations?

10. What is the slope of a straight line and how is it determined from a graph?

EXPERIMENT **1**

Experimental Error and Data Analysis

I. INTRODUCTION

Laboratory investigations involve taking measurements of physical quantities, and the process of taking any measurement always involves some uncertainty or experimental error. Suppose that you and another person independently took several measurements of a physical quantity (e.g., the length of an object). It is unlikely that you both would come up with exactly the same results. Therefore, questions such as the following arise:

1. How do you compare your experimental result to the accepted value of a physical quantity?
2. Whose data are better, or how does one express the degree of uncertainty or error in experimental measurements?
3. How does one graphically analyze and report experimental data?

In this introductory study experiment we examine the types of experimental errors and some methods of error and data analysis that will be used in subsequent experiments in which measurements are actually made.

II. EQUIPMENT NEEDED

- Pencil and ruler
- 3 sheets of Cartesian graph paper

(Optional)

- Hand calculator
- French curve

III. THEORY

A. *Types of Errors*

Experimental errors can be generally classified as being of three types: personal, systematic, and random.

PERSONAL ERROR

Personal error arises from personal bias or carelessness in reading an instrument, in recording observations, or in mathematical calculations. Examples of personal errors (sometimes called illegitimate errors) include:

1. In performing a series of measurements an observer may become biased in favor of the first observation. Falsely assuming this observation to be correct, the observer attempts to make other measurements agree with it, for example, through biased estimations of fractional scale divisions, and rejects measurements that greatly deviate. This incorrectly gives more significance to one reading than to succeeding ones. All observations taken under the same experimental condition are equally valid and should be retained for analysis.
2. Errors in reading a scale. Reading a value from a scale involves lining up an object with the marks on the scale. The apparent distance between two objects, and hence the value of the reading, depends on the position of the eye. A reading may appear to be different when viewed with one eye

or the other, or when the head is moved from side to side (horizontal scale) or up and down (vertical scale).

This apparent change in position due to a change in the position of the eye is called *parallax.* For example, the position of the mercury meniscus on a thermometer scale may appear different if viewed from above or below a line of sight perpendicular to the scale (Fig. 1-1). Also, when measuring length with a meter stick placed flat against the object, the thickness of the meter stick holds the scale about 0.7 cm from the object, and readings may vary considerably due to parallax (Fig. 1-1). Such errors can be minimized by using a line of sight perpendicular to the scale and placing the meter stick edgewise against the object.

3. Not observing significant figures in calculations. (This is discussed below.)

SYSTEMATIC ERROR

Systematic errors are associated with particular measurement instruments or techniques, such as an improperly calibrated instrument or bias on the part of the observer. Conditions from which systematic errors can result include:

1. An improperly "zeroed" instrument (e.g., a micrometer or ammeter) (Fig. 1-2).
2. A thermometer that reads 101°C when immersed in boiling water at standard atmospheric pressure. The thermometer is improperly calibrated since the reading should be 100°C.
3. Personal bias of an observer, who, for example, always takes a low reading of a scale division. Thus, a personal error may be a systematic error.

Avoiding systematic errors depends on the skill of the observer to detect them and to prevent or correct them.

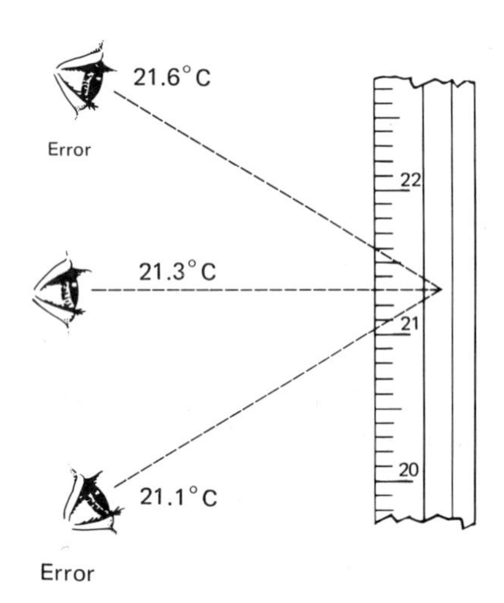

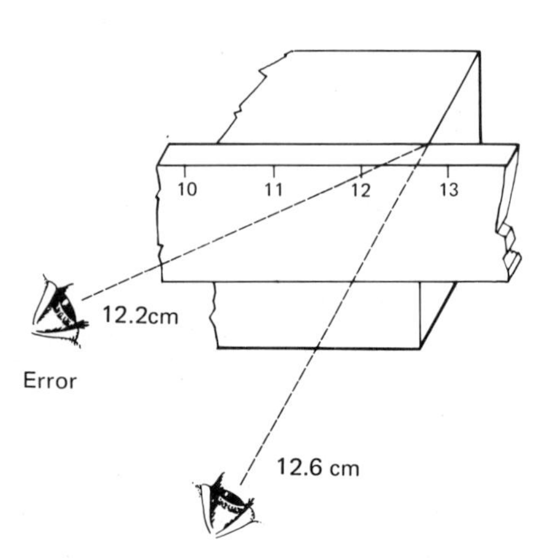

Fig. 1-1 Examples of personal error in reading a scale due to parallax.

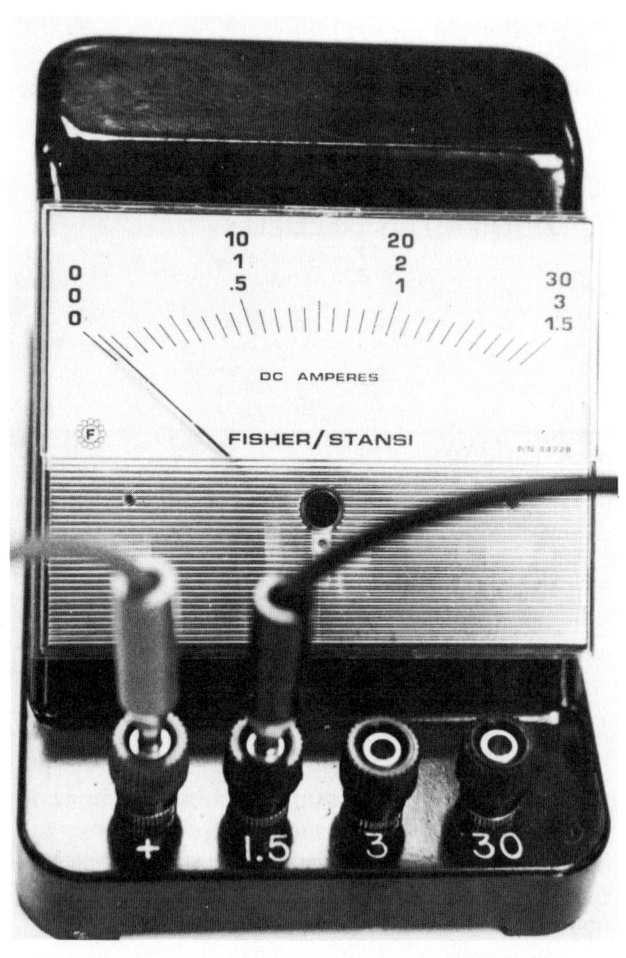

Fig. 1-2 An improperly zeroed instrument gives rise to systematic error. In this case, the ammeter would systematically give a incorrect reading larger than the true value.

RANDOM ERROR

Random errors result from unknown and unpredictable variations in experimental situations. Random errors are also referred to as accidental errors and are sometimes beyond the control of the observer. Conditions in which random errors can result include:

1. Unpredictable fluctuations in temperature or line voltage.
2. Mechanical vibrations of the experimental setup.
3. Unbiased estimates of measurement readings by the observer.

The effect of random errors can be reduced and minimized by improving and refining experimental techniques and repeating the measurement a sufficient number of times so that the erroneous readings become statistically insignificant.

B. *Accuracy and Precision*

The **accuracy** of an experiment is a measure of how close the experimental result comes to the true value. That is, it is a measure of the correctness of the result.

☐ **Example 1.1** Two independent experiments result in the determination of the value of π to be 3.140 and 3.143, respectively. The second result is more accurate than the first because the true value of π is 3.142 (to four significant figures). ☐

The **precision** of an experiment is a measure of its reliability, or how reproducible the result is. That is, it is a measure of the magnitude of uncertainty of the result without reference to what the result means.

☐ **Example 1.2** Two independent experiments give two sets of data with the expressed results and uncertainties of 2.5 ± 0.1 cm and 2.5 ± 0.2 cm, respectively. The first result is more precise than the second because the spread in the first measurements is between 2.4 and 2.6 cm, whereas the spread in the second measurements is between 2.3 and 2.7 cm. That is, the measurements of the first experiment are less uncertain than those of the second. ☐

The accuracy of an experimental value depends in general on systematic errors. The precision of an experimental value depends on random errors.

C. *Significant Figures*

The degree of certainty of a measurement is implied by the way the result is written or reported. When reading the value of an experimental measurement from a calibrated scale, only a certain number of digits can be obtained or read.

The **significant figures** (sometimes called "**significant digits**") in an experimental measurement include all the numbers that can be read directly from the instrument scale plus one doubtful or estimated number (Fig. 1-3). The number of significant figures that are included in a result is determined as follows:*

1. The leftmost nonzero digit is the most significant.
2. If there is no decimal point, the rightmost nonzero digit is the least significant.
3. If there is a decimal point, the rightmost digit is the least significant, even if it is zero.
4. All digits between the last and most significant digits are considered to be significant.

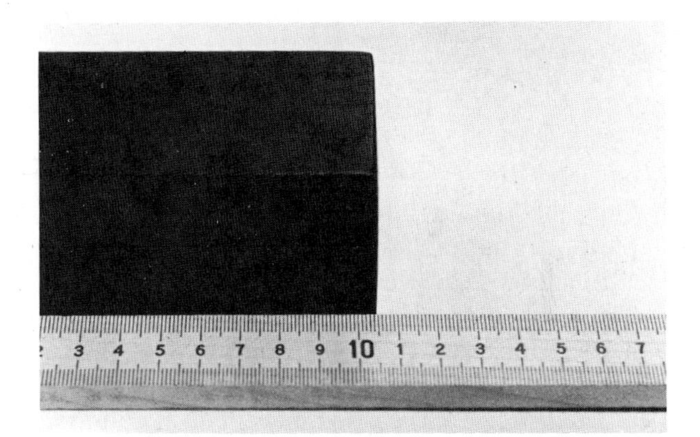

Fig. 1-3 The edge of the object is at the 10.45 cm position of the meter stick. This reading has four significant figures, where the 5 is the estimated or doubtful (least significant) figure.

☐ **Example 1.3** All of the following numbers have four significant figures: 1000.; 11.10; 2783; 278,300; 0.001230; 10.01; 10.00. ☐

In practice, a problem sometimes arises if the decimal point is omitted and the rightmost digit is zero. For example, by the rules given above, the number 4820 has only three significant figures, but the last zero digit may be physically significant in a measurement (the estimated or doubtful figure).

*Later, in calculations using experimental values, the number of significant figures included must be determined.

This problem is resolved by writing the number in powers-of-10 or scientific notation, 4.820×10^3, which shows explicitly that the rightmost zero is significant.

This procedure is also helpful in expressing the significant figures in large numbers. For example, suppose that the average distance from the earth to the sun, 93,000,000 miles, is known to only four significant figures. This is easily expressed in powers-of-10 notation: 9.300×10^7 mi.

MULTIPLICATION AND DIVISION

In the multiplication or division of two or more numerical measurements, the number of significant figures in the final answer can be no greater than the number of significant figures in the measurement with the least number of significant figures.

☐ **Example 1.4** (a) To show how doubtful figures are carried through a multiplication process, the doubtful figures are underlined in the following example:

$$\begin{array}{r} 6.2\underline{7} \\ 5.\underline{5} \\ \hline \underline{3135} \\ 31\underline{35} \\ \hline 3\underline{4.485} \rightarrow 3\underline{4}. \end{array}$$

Only the first doubtful figure from the left is reported. The result is rounded off to this figure. (Rounding off procedures are discussed below.) The result of any operation with a doubtful figure is doubtful. Notice in the multiplication of the $\underline{5} \times \underline{7}$ for the second 5 in the example that when a doubtful figure is carried into the next column, the figure in that column is doubtful, i.e., $[5 \times \underline{7} = \underline{35}$ and $(5 \times 2) + \underline{3} = 1\underline{3}]$.

(b) The division of $374/2\underline{9} = 1\underline{3}$ is shown in Fig. 1-4 as done on a hand calculator. The result must be rounded off to two significant figures. (Why?) Reporting more figures would imply greater significance than given by the measurements and **a result cannot be made more significant by a mathematical operation.** ☐

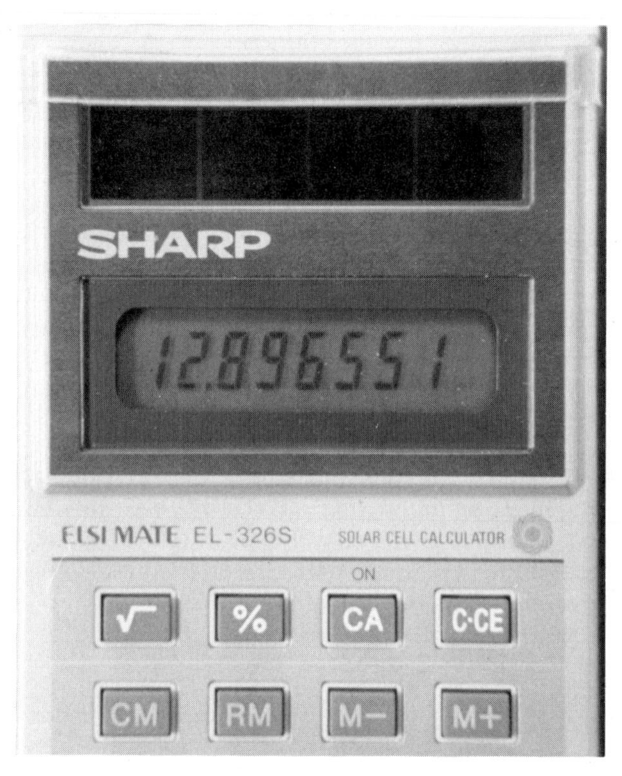

Fig. 1-4 Insignificant figures on a calculator. The calculator shows the result of the division operation 374/29. Since there are only two significant figures in the 29, the result can have no more and must be rounded off to 13.

ROUNDING OFF

The nonsignificant figures are dropped from a result if they are to the right of the decimal point (as in Example 1.4) and are replaced by zeros if they are to the left of the decimal point. The last significant figure retained should be rounded off. The procedure for doing this is as follows. Treat the excess digits to the right of the desired number of significant figures as a decimal fraction and round off according to the following rules:

1. If the fraction is less than 0.5, leave the least significant figure unchanged.
2. If the fraction is greater than or equal to 0.5, increase the least significant figure by one.*

☐ **Example 1.5**

Rule 1. In Example 1.4(a) the result can have only two significant figures, which are 3$\underline{4}$ since the number to the right of the $\underline{4}$ expressed as a decimal fraction ($.\underline{485}$) is less than 0.5.

24.4762 → 24.48 (to four significant figures)

1.014 → 1.01 (to three significant

Rule 2. 1.025 → 1.03 figures)

Rounding off that results in zeros:

487.56 → 490 (to two significant figures)

*In this method, five digits, 0, 1, 2, 3, and 4 are rounded down; and five digits, 5, 6, 7, 8, and 9 are rounded up.

The zero is insignificant, so to avoid confusion scientific notation should be used:

$$487.56 \rightarrow 4.9 \times 10^2$$

ADDITION AND SUBTRACTION

In addition and subtraction, adding or subtracting begins with the first column from the left that contains a doubtful figure. The numbers are rounded off to this column and all digits to the right are dropped.

Example 1.6 Adding a column of numbers:

42.31		42.3
0.0621		0.1
512.4	→	512.4
2.57		2.6
		557.4

since the 4 in the 512.4 defines the first column from the left with a doubtful figure.

In general in computations, figures that are not significant should be dropped continually so as to save labor and to avoid false conclusions.

D. *Expressing Experimental Error and Uncertainty*

PERCENT ERROR

The object of some experiments is to determine the value of a well-known physical quantity, for example, the value of π.

The **accepted or "true" value** of such a quantity found in textbooks and physics handbooks is the most accurate value (usually rounded off to a certain number of significant figures) obtained through sophisticated experiments or mathematical methods.

The **absolute difference** between the experimental value E and the accepted value A, written $|E - A|$, is the positive difference in the values (e.g., $|2 - 4| = |-2| = 2$ and $|4 - 2| = 2$). Simply subtract the smaller value from the larger, even though the symbols may be written in reverse. For a set of measurements, E is taken as the average value of the experimental measurements.

The **fractional error** is the ratio of the absolute difference to the accepted value:

$$\text{fractional error} = \frac{\text{absolute difference}}{\text{accepted value}} = \frac{|E - A|}{A} \qquad \textbf{(1-1)}$$

The fractional error is commonly expressed as a percentage to give the **percent error** of an experimental value*

$$\text{percent error} = \frac{\text{absolute difference}}{\text{accepted value}} \times 100\% = \frac{|E - A|}{A} \times 100\% \qquad \textbf{(1-2)}$$

Example 1.7 A cylindrical object is measured to have a diameter d of 5.25 cm and a circumference c of 16.38 cm. What is the experimental value of π and the percent error of the experimental value if the accepted value of π is 3.14?

Solution $d = 5.25$ cm, $c = 16.38$ cm:

$$c = \pi d \quad \text{or} \quad \pi = \frac{c}{d} = \frac{16.38}{5.25} = 3.12$$

Then $E = 3.12$ and $A = 3.14$, so

$$\text{percent error} = \frac{|E - A|}{A} \times 100\% = \frac{|3.12 - 3.14|}{3.14} \times 100\% = \frac{0.02}{3.14} \times 100\% = 0.64\%$$

PERCENT DIFFERENCE

It is sometimes instructive to compare the results of two equally reliable measurements when an accepted value is not known. The comparison is expressed as a **percent difference**, which is the ratio of the absolute difference between the experimental values E_2 and E_1 to the average or mean value of the two results, expressed as a percent.

$$\text{percent difference} = \frac{\text{abs. difference}}{\text{average}} \times 100\% \qquad \textbf{(1-3)}$$

or

$$\text{percent difference} = \frac{|E_2 - E_1|}{(E_2 + E_1)/2} \times 100\%$$

Dividing by the average or mean value of the experimental values is a logical choice, because there is no way of deciding which of the two results is better.

*Although generally defined with the absolute difference $|E - A|$, some instructors prefer to use $(E - A)$, which results in (+) or minus (−) percent errors, e.g., −0.64% in Example 1.7. In the case of a series of measurements and computed percent errors, this gives an indication of systematic error.

☐ **Example 1.8** What is the percent difference between two measured values of 4.6 cm and 5.0 cm?

Solution With $E_1 = 4.6$ cm and $E_2 = 5.0$ cm.

$$\text{percent difference} = \frac{|E_2 - E_1|}{(E_2 + E_1)/2} \times 100\%$$

$$\text{percent difference} = \frac{5.0 - 4.6}{(5.0 + 4.6)/2} \times 100\%$$

$$= \frac{0.4}{4.8} \times 100\% = 8.3\%$$ ☐

When there are three or more measurements, the percent difference is found by dividing the absolute value of the difference of the extreme values (i.e., values with greatest difference) by the average or mean value of the measurements.

AVERAGE (MEAN) VALUE

Most experimental measurements are repeated several times, and it is very unlikely that identical results will be obtained for all trials. For a set of measurements with predominantly random errors (i.e., the measurements are all equally trustworthy or probable), it can be shown mathematically that the true value is most probably given by the average or mean value.

The **average *or* mean value** $\bar{x}$ of a set of N measurements is

$$\bar{x} = \frac{x_1 + x_2 + x_3 + \cdots + x_N}{N} = \frac{1}{N}\sum_{i=1}^{N} x_i \quad \textbf{(1-4)}$$

where the summation sign Σ is a shorthand notation indicating the sum of N measurements from x_1 to x_N. ($\bar{x}$ is commonly referred to simply as the mean.)

☐ **Example 1.9** What is the average or mean value of the set of numbers 5.42, 6.18, 5.70, 6.01, and 6.32?

$$\bar{x} = \frac{1}{N}\sum_i x_i$$

$$= \frac{5.42 + 6.18 + 5.70 + 6.01 + 6.32}{5}$$

$$= 5.93 \quad \textbf{(1-5)}$$ ☐

DEVIATION FROM THE MEAN

Having obtained a set of measurements and determined the mean value, it is helpful to state how much the individual measurements are scattered from the mean. A quantitative description of this scatter or dispersion of measurements will give an idea of the precision of the experiment.

The **deviation** d_i from the mean of any measurement with a mean value $\bar{x}$ is

$$d_i = x_i - \bar{x}$$

(d_i is sometimes referred to as the **residual** rather than the deviation.)

As defined, the deviation may be positive or negative, since some measurements are larger than the mean and some are smaller. The average of the deviations of a set of measurements is always zero, so the mean of the deviations is not a useful way of characterizing the dispersion.

MEAN DEVIATION

To obtain what is called the **mean *or* average deviation** of a set of N measurements, the absolute deviations $|d_i|$ are determined; that is,

$$|d_i| = |x_i - \bar{x}| \quad \textbf{(1-6)}$$

The *mean deviation* $\bar{d}$ is then

$$\bar{d} = \frac{|d_1| + |d_2| + |d_3| + \cdots + |d_N|}{N}$$

$$= \frac{1}{N}\sum_{i=1}^{N} |d_i| \quad \textbf{(1-7)}$$

(Although commonly called the mean deviation, a more appropriate term would be the **mean absolute deviation**.)

☐ **Example 1.10** What is the mean deviation of the set of numbers given in Example 1.9?

Solution First find the absolute deviation of each of the numbers using the determined mean of 5.93.

$$d_1 = |5.42 - 5.93| = 0.51$$
$$d_2 = |6.18 - 5.93| = 0.25$$
$$d_3 = |5.70 - 5.93| = 0.23$$
$$d_4 = |6.01 - 5.93| = 0.08$$
$$d_5 = |6.32 - 5.93| = 0.39$$

Then

$$\bar{d} = \frac{1}{N}\sum_{i=1}^{N} |d_i|$$

$$= \frac{0.51 + 0.25 + 0.23 + 0.08 + 0.39}{5} = 0.29$$ ☐

The mean deviation is a measure of the dispersion of experimental measurements about the mean (i.e., a measure of precision). It is common practice to report the experimental value E_v of a quantity in the form

$$E_v = \bar{x} \pm \bar{d}$$

In Example 1.10, this would be $E_v = 5.93 \pm 0.29$. It is also common practice to express the dispersion of the mean deviation as a percent of the mean:

$$E_v = \bar{x} \pm \frac{\bar{d}}{\bar{x}} \times 100\%$$

In this case, for Example 1.10, we have

$$E_v = 5.93 \pm \frac{0.29}{5.93} \times 100\% = 5.93 \pm 4.9\%$$

The $\pm$ term gives a measure of the precision of the experimental value. The accuracy of the mean value of a set of experimental measurements (5.93 in the above example) may be expressed in terms of percent error or percent difference.

STANDARD DEVIATION (optional)

To avoid the problem of negative deviations and absolute values, it is statistically convenient to use the square of the deviation.

The **variance** σ^2 of set of measurements is the average of the square of the deviations:

$$\sigma^2 = \frac{(x_1 - \bar{x})^2 + (x_2 - \bar{x})^2 + (x_3 - \bar{x})^2 + \cdots + (x_N - \bar{x})^2}{N}$$

$$= \frac{d_1^2 + d_2^2 + d_3^2 + \cdots + d_N^2}{N}$$

$$= \frac{1}{N} \sum_{i=1}^{N} (x_i - \bar{x})^2 = \frac{1}{N} \sum_{i=1}^{N} d_i^2 \qquad \textbf{(1-8)}$$

The square root of the variance σ is called the **standard deviation,***

$$\sigma = \sqrt{\frac{1}{N} \sum_{i=1}^{N} (x_i - \bar{x})^2} = \sqrt{\frac{1}{N} \sum_{i=1}^{N} d_i^2} \qquad \textbf{(1-9)}$$

Because we take the average of the squares of the deviations and then the square root, the standard deviation is sometimes called the **root-mean-square deviation,** or simply the **root-mean square**. Notice that σ always has the same units as x_i and that it is always positive.

☐ **Example 1.11** What is the standard deviation of the set of numbers given in Example 1.9?

Solution First find the square of the deviation of each of the numbers

$$d_1^2 = (5.42 - 5.93)^2 = 0.26$$
$$d_2^2 = (6.18 - 5.93)^2 = 0.06$$
$$d_3^2 = (5.70 - 5.93)^2 = 0.05$$
$$d_4^2 = (6.01 - 5.93)^2 = 0.01$$
$$d_5^2 = (6.32 - 5.93)^2 = 0.15$$

Then

$$\sigma = \sqrt{\frac{1}{N} \sum_{i=1}^{5} d_i^2}$$

$$= \left(\frac{0.26 + 0.06 + 0.05 + 0.01 + 0.15}{5}\right)^{1/2}$$

$$= 0.33$$

The experimental value E_v is then commonly reported as

$$E_v = \bar{x} \pm \sigma = 5.93 \pm 0.33$$ ☐

The standard deviation is used to describe the precision of the mean of a set of measurements. For a normal distribution of random errors,* it can be statistically shown that the probability that an individual measurement will fall within one standard deviation of the mean, which is assumed to be the true value, is 68 percent (Fig. 1-5). The probability of a

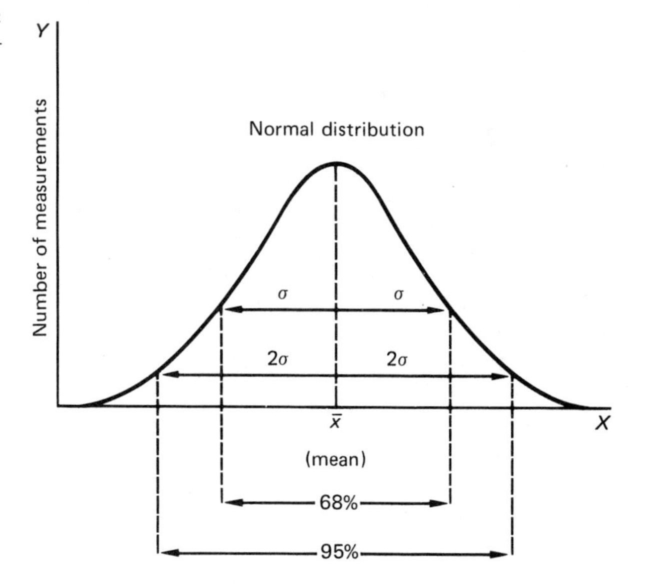

Fig. 1-5 A normal or Gaussian distribution "bell-shaped" curve. If random errors follow this distribution, there is a 68 percent probability that a measurement will be within one standard deviation of the mean value, $\bar{x} \pm \sigma$. The probability of a measurement falling within two standard deviations of the mean, $\bar{x} \pm 2\sigma$, is 95 percent.

*For a small number of measurements, it can be statistically shown that a better value of the standard deviation is given by $\sigma = \sqrt{[1/(N-1)]\Sigma d_i^2}$ where N is replaced by $N - 1$. Your instructor may wish to use this form of the standard deviation.

*This *normal*, or *Gaussian, distribution* is represented by a "bell-shaped" curve (Fig. 1-5). That is, the scatter or dispersion of the measurements is assumed to be symmetric about the true value of a quantity.

measurement falling within two standard deviations is 95 percent.

E. *Graphical Representation of Data*

It is often convenient to represent experimental data in graphical form, not only for reporting, but also to obtain information.

GRAPHING PROCEDURES

Quantities are commonly plotted using rectangular Cartesian axes (X and Y). The horizontal axis (X) is called the abscissa, and the vertical axis (Y), the ordinate. The location of a point on the graph is defined by its coordinates x and y, written (x, y), referenced to the origin O, the intersection of the X and Y axes.

When plotting data, choose axis scales that are easy to plot and read. Graph A in Fig. 1-6 shows an example of scales that are too small. This "bunches up" the data making the graph too small and the major horizontal scale values make it difficult to read intermediate values. Choose scales so that the major portion of the graph paper is used. Graph B in Fig. 1-6 shows the data plotted with more appropriate scales.*

When the data points are plotted, draw a smooth line connecting the points. "Smooth" suggests that the line does not have to pass exactly through each point but connects the general areas of significance of the data points in Graph B with approximately an equal number of points on each side of the line, that is, the "curve of best fit."

In cases where several determinations of each experimental quantity are made, the average value is plotted and the mean deviation or the standard deviation may be plotted as error bars. For example, the following data are plotted in Fig. 1-7. A smooth line is drawn within the error bars. (Your instructor may want to explain the use of a French curve at this point.)

Data Table

Mass (kg)	Period	±	$\bar{d}$ (s)
0.01	0.63	±	0.17
0.05	1.40	±	0.21
0.10	1.99	±	0.25
0.20	2.81	±	0.20
0.40	3.97	±	0.31
0.60	4.87	±	0.13
0.75	5.44	±	0.20
0.90	5.96	±	0.18
1.00	6.28	±	0.14
1.20	6.88	±	0.29

Graphs should have (see Fig. 1-7):

1. Each axis labeled with the quantity plotted.
2. The units of the quantities plotted.
3. The title of the graph on the graph paper (commonly listed as the y coordinate versus the x coordinate).
4. Your name and date on the graph.

*As a general rule, it is convenient to choose the unit of the first major scale division to the right or above the origin or zero point as 1, 2, or 5 (or multiples or submultiples thereof, e.g., 10 or 0.1) so the minor (intermediate) scale divisions can be easily interpolated and read.

STRAIGHT-LINE GRAPHS

Two quantities (x and y) are often linearly related; that is, they have an algebraic relationship of the form $y = ax + b$, where a and b are constants. When the values of such quantities are plotted, the graph is a straight line, as shown in Fig. 1-8.

The a in the algebraic relationship is called the **slope** of the line and is equal to the ratio of the intervals $\Delta y/\Delta x$. Any set of intervals may be used to determine the slope of a straight-line graph; for example, in Fig. 1-8:

$$a = \frac{\Delta y_1}{\Delta x_1} = \frac{15\text{ cm}}{2.0\text{ s}} = 7.5\text{ cm/s}$$

$$a = \frac{\Delta y_2}{\Delta x_2} = \frac{45\text{ cm}}{6.0\text{ s}} = 7.5\text{ cm/s}$$

However in practice, points should be chosen relatively far apart. For best results, points corresponding to data points should not be chosen, even if they appear to lie on the line.

The b in the algebraic relationship is called the **intercept** and is equal to the value of the y coordinate where the graph line intercepts the Y axis. In Fig. 1-8, $b = 3$ cm. Notice from the relationship that $y = ax + b$, when $x = 0$ and $y = b$. If the graph line intercepts at the origin (0, 0), then $b = 0$.

The equation of the line in the graph in Fig. 1-8 is then $d = 7.5t + 3$. The general equation for uniform motion has the form $d = d_0 + vt$; hence, the initial displacement $d_0 = 3$ cm and the speed is $v = 7.5$ cm/s.

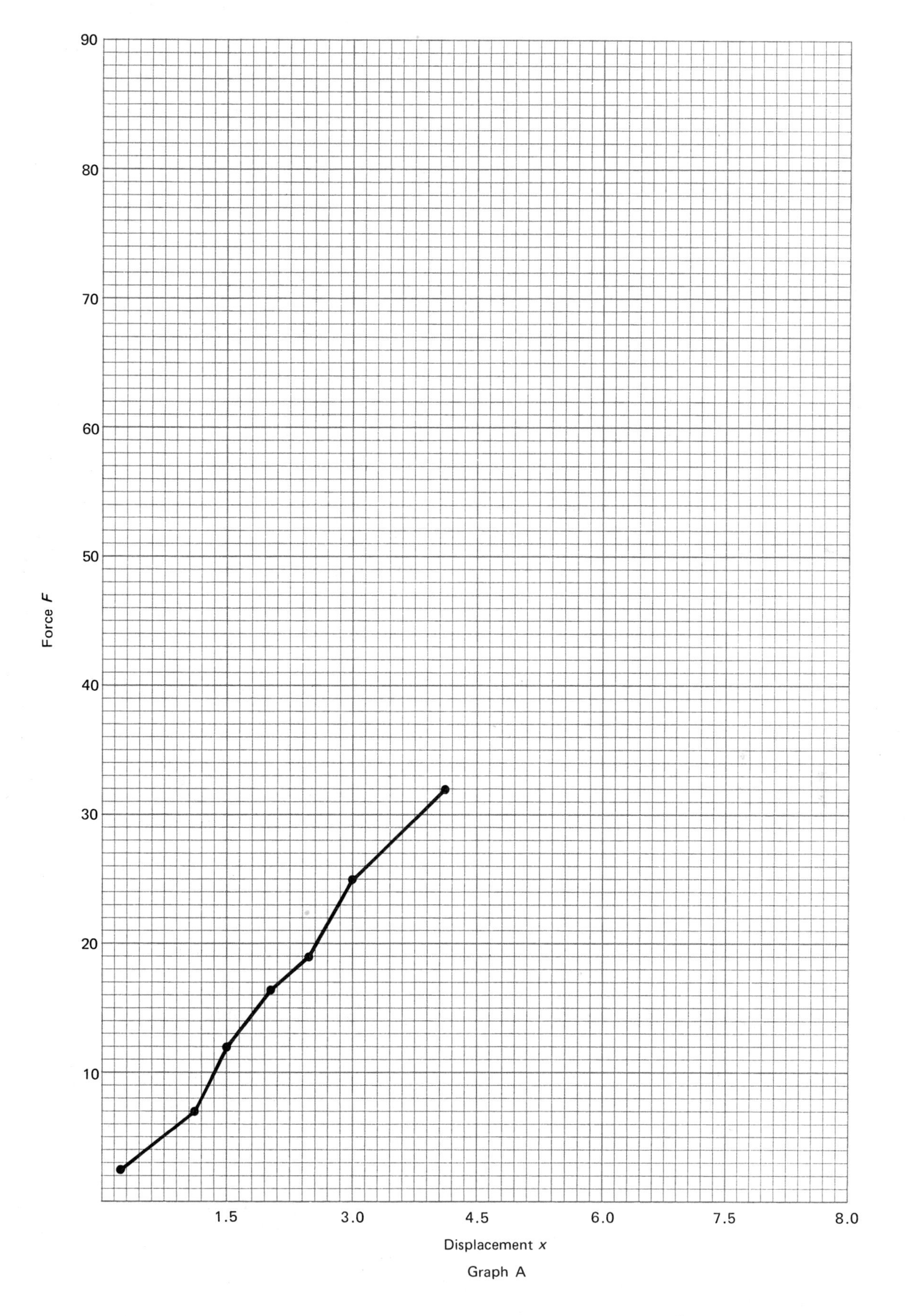

Fig. 1-6 Example of improperly (Graph A) and properly (Graph B) plotted and labeled graphs. See the text for a description.

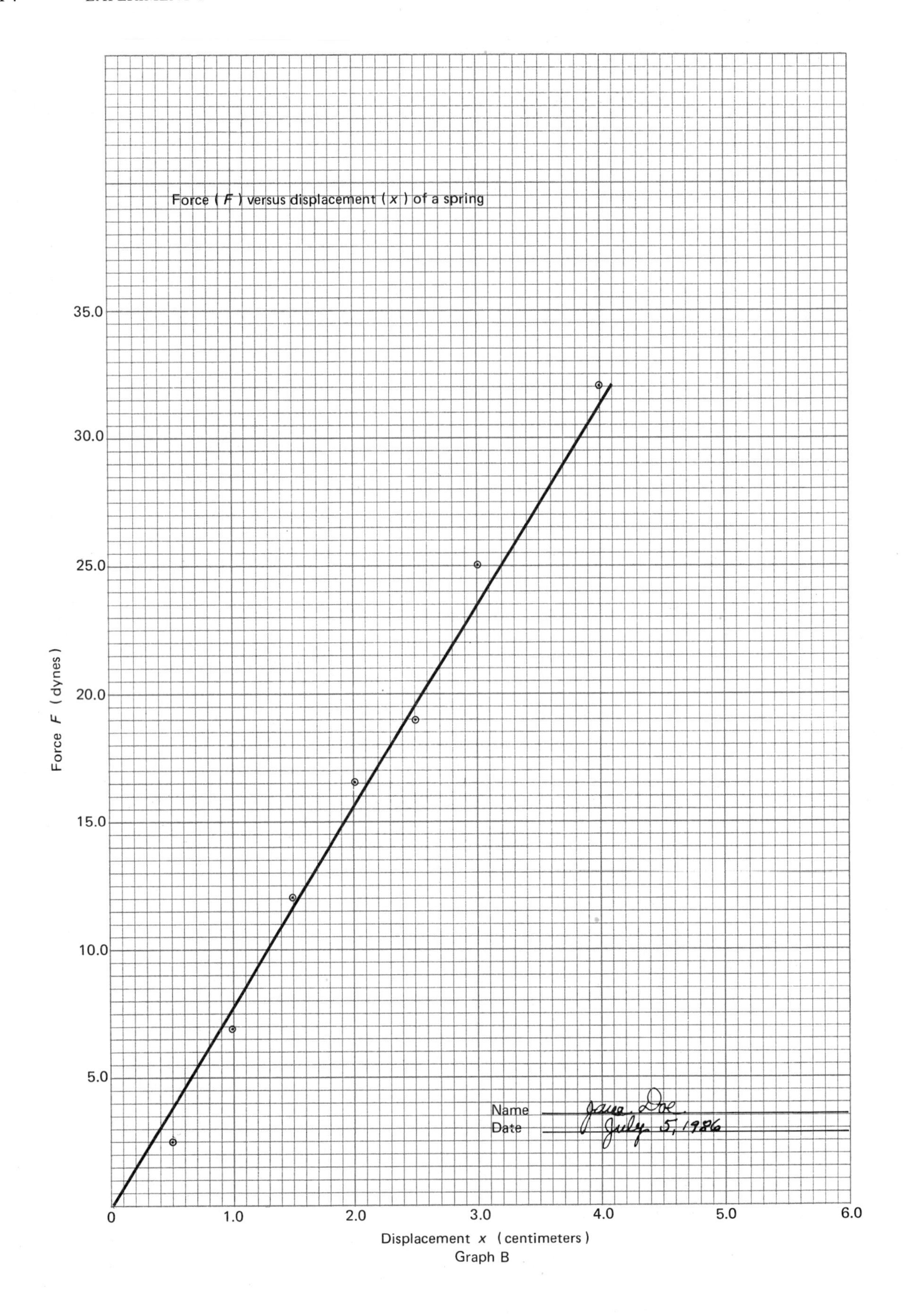
Force (F) versus displacement (x) of a spring
35.0
30.0
25.0
20.0
15.0
10.0
5.0
Force F (dynes)
Name
Date
0
1.0
2.0
3.0
4.0
5.0
6.0
Displacement x (centimeters)
Graph B

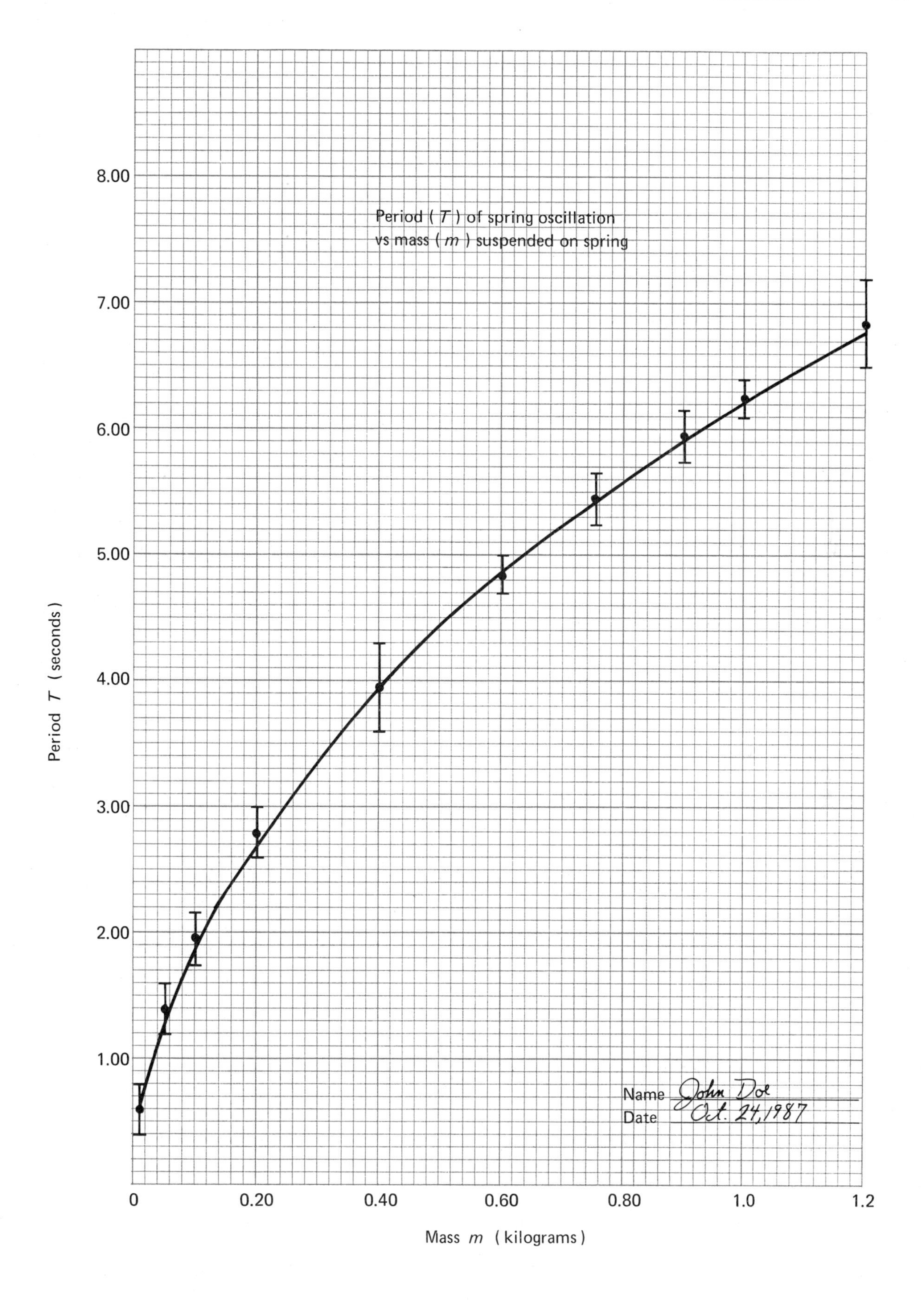

Fig. 1-7 An example of graphical data with error bars. The error bars indicate the precision of the measurement. In this case, the precision is indicated by the mean deviation.

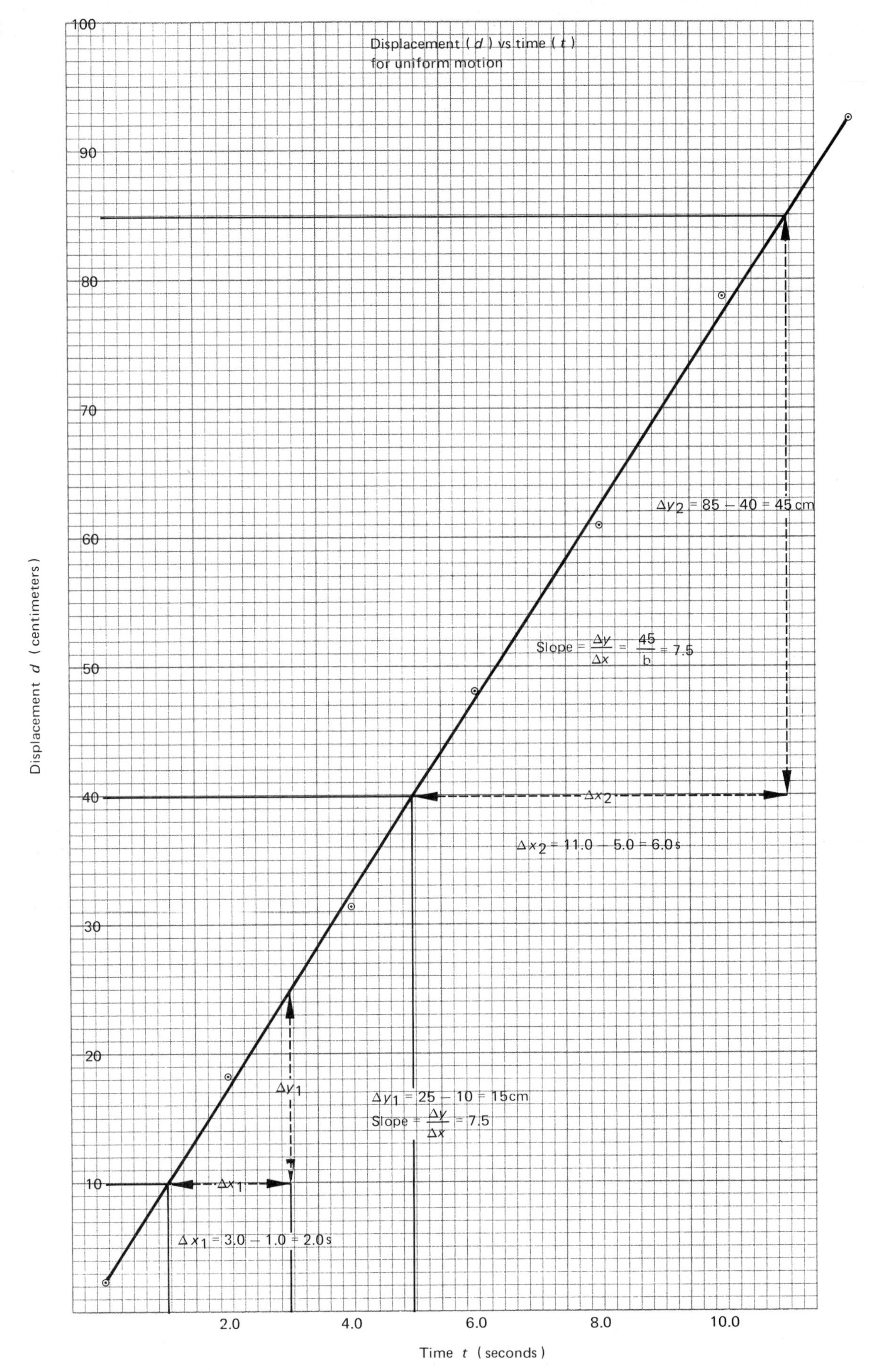

Fig. 1-8 Examples of intervals for determining the slope of a straight line. The slope is the ratio of $\Delta y/\Delta x$. Any set of intervals can be used, but the end points of the interval should be relatively far apart.

IV. EXPERIMENTAL PROCEDURE

Complete the exercises in the Laboratory Report, showing calculations and attaching graphs as required.

LABORATORY REPORT

1. Significant Figures
   (a) Express the numbers listed in Data Table 1 to three significant figures, writing the numbers for the first column in normal notation and the numbers for the second column in powers-of-10 (scientific) notation.

DATA TABLE 1

1.065	________	67,000	________
2347	________	0.3140	________
10.07	________	29.35	________
0.1133	________	0.007865	________
26,302	________	970,000,000	________

   (b) A rectangular block of wood is measured to have the dimensions 11.2 cm $\times$ 3.8 cm $\times$ 4.10 cm. Compute the volume of the block, showing explicitly how doubtful figures are carried through the calculation (by underlining) and report the final answer with the correct number of significant figures.

Calculations (show work)

Computed volume (in powers of 10) ....
(units)

(continued)

(c) In an experiment to determine the value of π, a cylinder is measured to have an average value of 4.25 cm for its diameter and an average value of 13.39 cm for its circumference. What is the experimental value of π to the correct number of significant figures?

Calculations (show work)

Experimental value of π ....

2. Expressing Experimental Error
   (a) If the accepted value of π is 3.14, what is the fractional error and the percent error of the experimental value found in 1(c)?

Calculations (show work)

Fractional error ....

Percent error ....

(b) In an experiment to measure the acceleration due to gravity g, two values, 972 cm/s^2 and 967 cm/s^2, are determined. Find (1) the percent difference of the measurements, (2) the percent error of each measurement, and (3) the percent error of their mean. (Accepted value: $g = 980\ cm/s^2$.)

Calculations (show work)

Percent difference ....

Percent error of E_1 ....

Percent error of E_2 ....

Percent error of mean ....

(c) Data Table 2 shows data taken in a free-fall experiment. Measurements were made of the distance of fall (y) at each of four precisely measured times. Complete the table. Use only the proper number of significant figures in your table entries, even if you carry extra digits during your intermediate calculations.

DATA TABLE 2

Time, t (s)	Distance (cm)								
	y_1	y_2	y_3	y_4	y_5	$\bar{y}$	$\bar{d}$	σ	t^2
0	0	0	0	0	0				
0.50	1.0	1.4	1.1	1.4	1.5				
0.75	2.6	3.2	2.8	2.5	3.1				
1.00	4.8	4.9	5.1	4.7	4.8				
1.25	8.2	7.9	7.5	8.1	7.4				

(d) Plot a graph of $\bar{y}$ versus t with $\bar{d}$ error bars for the free-fall data in part (c). Remember that $t = 0$ is a known point.

(e) The equation of motion for an object in free fall starting from rest is $y = \frac{1}{2}gt^2$, where g is the acceleration due to gravity. The graphical form of this equation and of the data plotted in part (d) is called a **parabola**, which has the general form $y = ax^2$.

Plot a graph of $\bar{y}$ versus t^2 for the data in Data Table 2; that is, plot the square of the time on the abscissa. (Use regional circles around the data points rather than error bars.) This will then be a plot of $\bar{y}$ versus t', where $t' = t^2$. [$y = \frac{1}{2}gt^2 = \frac{1}{2}gt'$, which is the general form $y = ax$ or that of a straight line with a slope $a = \frac{1}{2}g$ (and intercept $b = 0$).] Determine the slope of your graph and compute the experimental value of g.

Calculations (show work)

Experimental value of g from graph ....
(units)

(f) Compute the percent error of the experimental value of g determined from the graph in part (e). (Accepted value: $g = 9.8\ \text{m/s}^2$.)

Calculations (show work)

Percent error ....

(continued)

(g) The relationship of the applied force F and the displacement x of a spring has the general form $F = kx$, where the constant k is called the **spring constant** and is a measure of the "stiffness" of the spring. Notice that this equation has the form of a straight line. Find the value of the spring constant k of the spring used in determining the experimental data plotted in Graph B of Fig. 1-6. (*Note:* Since $k = F/x$, the units of k in the graph are dynes/cm.)

Calculations (show work)

Value of spring constant of spring in Graph B of Fig. 1-6 ....
(units)

(h) The general relationship of the period of oscillation T of a mass suspended on a spring is $T = 2\pi\sqrt{m/k}$, where k is the spring constant. Replot the data in Fig. 1-7 so as to obtain a straight-line graph and determine the value of the spring used in the experiment. [*Hint:* Square both sides of the equation and put in the form $y = ax + b$, as was done in part (e).] Show the final form of the equation and calculations.

Value of spring constant of spring in Fig. 1-7 ....
(units)

Name .... Section .... Date ....

Lab Partner(s) ....

EXPERIMENT 2 *Mass, Volume, and Density*

ADVANCE STUDY ASSIGNMENT

Read the experiment and answer the following questions.

1. What is the least count of a measurement instrument?

2. Does a laboratory balance measure weight or mass? Explain.

3. What is the function of the vernier scale on the vernier caliper?

4. What is meant by a negative zero error and how is the zero correction made in this case? What kind of error does a zero correction correct for?

5. What is the purpose of the ratchet mechanism on a micrometer caliper?

6. Explain how readings from 0.00 through 1.00 mm are obtained from the micrometer thimble scale when it is only calibrated from 0.00 through 0.50 mm?

(continued)

7. What is the density of a material, and what does it indicate?

8. How is the volume of a heavy irregularly shaped object determined experimentally?

EXPERIMENT **2**

Mass, Volume, and Density

I. INTRODUCTION

Common laboratory measurements involve the determination of the fundamental properties of mass and length. Most people are familiar with the use of scales and rulers or meter sticks. However, for more accurate and precise measurements, laboratory balances and vernier calipers or micrometer calipers are often used, particularly in measurements involving small objects. In this initial experiment on measurement you will learn how to use these instruments and their advantages. Also, the relationship of mass to size (volume), or density, will be considered, and the densities of several materials will be determined experimentally.

II. EQUIPMENT NEEDED

- Laboratory balance
- Vernier caliper
- Micrometer caliper (metric)
- Meter stick
- Graduated cylinder
- Cylindrical metal rod (e.g., aluminum, brass, or copper)
- Sphere (metal or glass, e.g., ball bearing or marble)
- Short piece of solid copper wire
- Rectangular piece of metal sheet (e.g., aluminum)
- Irregularly shaped metal object

III. THEORY

A. *Least Count of an Instrument Scale*

The **least count** is the smallest subdivision marked on an instrument scale. This is the unit of the smallest reading that can be made without estimating. For example, the least count or smallest subdivision of a meter stick is usually the millimeter (mm). See Fig. 2-1. We commonly say "the instrument is calibrated in centimeters (numbered major divisions) with a millimeter least count."

A measurement reading usually has one more significant figure than the least count reading of the instrument scale. This is the estimated fractional part (doubtful figure) of the smallest subdivision. In the case of a meter stick with the least count of 1 mm, a reading can be made to 0.1 mm (or 0.01 cm or 0.0001 m).

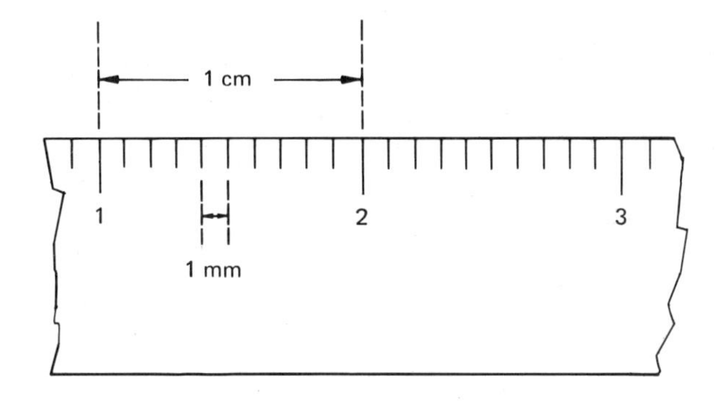

Fig. 2-1 Meter sticks are commonly calibrated in centimeters (numbered major divisions) with a least count of millimeters.

B. *Laboratory Balances*

The common types of laboratory balances are shown in Fig. 2-2. Balances are used to balance the weight of a unknown mass m against that of a known mass m_1 (i.e., $mg = m_1 g$ or $m = m_1$), and the mass of the unknown is read directly in mass units, usually grams. [The weight w of an object is its mass m times a constant g, the acceleration due to gravity, $g = 32\ \text{ft/s}^2 = 9.8\ \text{m/s}^2 = 980\ \text{cm/s}^2$ (i.e., $w = mg$ or $m = w/g$). Common scales, such as bathroom scales, are calibrated in weight (force) units (pounds) rather than in mass units.]

A set of known masses is used to balance an unknown mass on a pan balance [Fig. 2-2(a)]. On a beam balance the riders on the beams are used to balance the unknown mass on the platform [Fig. 2-2(b)]. The common laboratory beam balance is calibrated in grams. In this case, the least count is 0.1 gm and a reading can be made to 0.01 gm.*

Before making a mass determination, check to see if the balance is zeroed. Adjustments can be made by means of knurled balance nuts [as seen in Fig. 2.2(a) at the right of the balance beam].

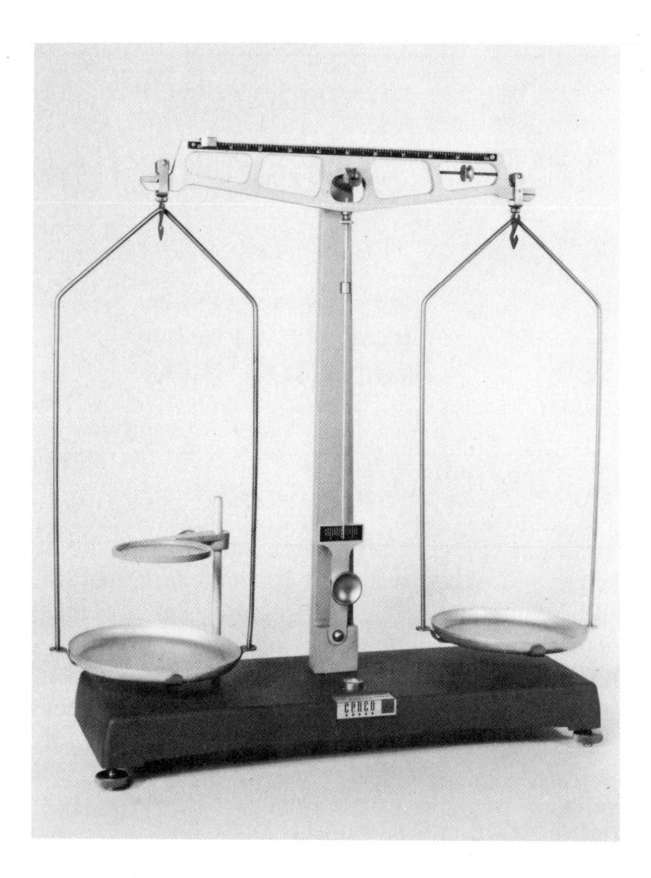

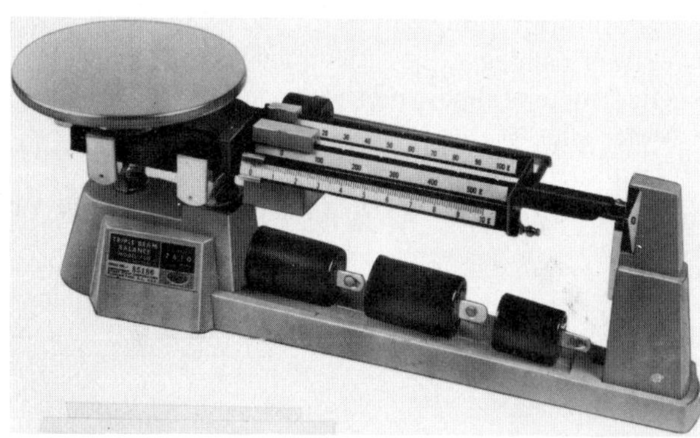

Fig. 2-2 Laboratory balances. (a) A double-hanging-pan balance. (b) A triple-beam balance, platform model (also available in a hanging-pan model). (Photos courtesy of Central Scientific Co.)

C. *The Vernier Caliper*

The **vernier caliper** (Fig. 2-3), commonly called a **vernier,** consists of a rule with a main engraved scale and a movable jaw with an engraved vernier scale. The span of the lower jaw is used to measure length and is particularly convenient for measuring the diameter of a cylindrical object. The span of the upper jaw is used to measure distances between two surfaces (e.g., the inside diameter of a hollow cylindrical object).

The main scale is calibrated in centimeters with a millimeter least count, and the movable vernier scale has 10 divisions that cover 9 divisions on the main scale. When making a measurement with a meter stick, it is necessary to estimate or "eyeball" the fractional part of the smallest scale division (the tenth of a millimeter). The function of the vernier scale is to assist in the accurate reading of the fractional part of the scale division.

The leftmost mark on the vernier scale is the zero mark (lower scale for metric reading and upper scale for inches). The zero mark is often unlabeled. A

*The official abbreviation of mass (m) in grams is g. The standard symbol for acceleration due to gravity is g, where weight is given by mg. To avoid confusion with these symbols, since the acceleration due to gravity is often left in symbol form, gram will be abbreviated gm in some cases in this book for distinction. However, it should be kept in mind that the official abbreviation is g.

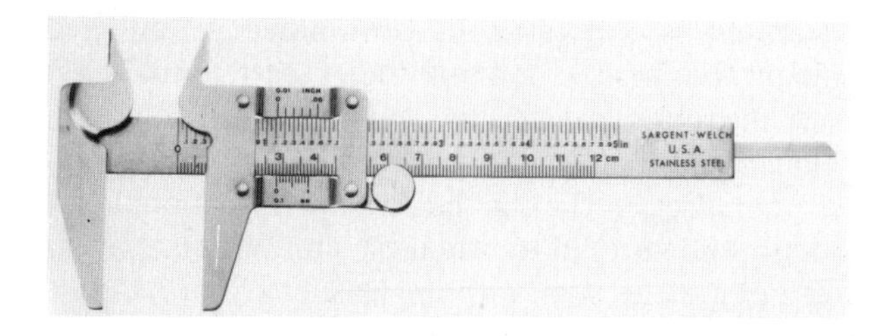

Fig. 2-3 The vernier caliper. See the text for a description. (Photo courtesy of Sargent-Welch Scientific Co.)

measurement is made by closing the jaws on the object to be measured and reading where the zero mark on the vernier scale falls on the main scale (see Fig. 2-4).

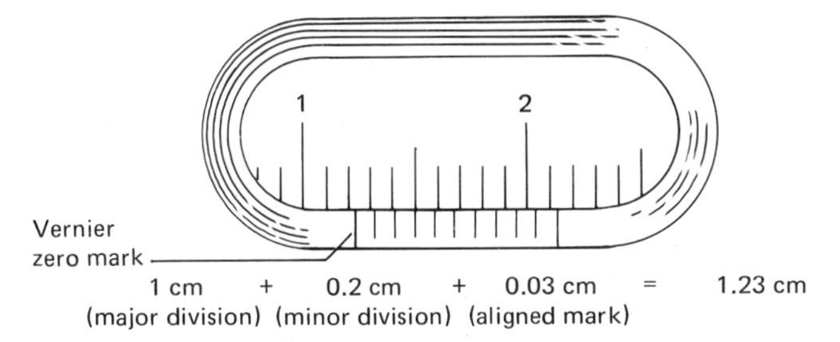

Fig. 2-4 An example of a vernier caliper reading. (See text for description.)

The major and minor division marks (significant figures) are read directly from the main scale (1.2 cm in Fig. 2-4). The estimated significant figure or the fractional part of the smallest subdivision (least count) is obtained by noting which line on the vernier scale is closest to or coincides with a mark on the main scale. (In Fig. 2-4 this is the third mark to the right of the vernier zero.) The mark number is the fractional part of the smallest subdivision (0.03 cm in Fig. 2-4 for a measurement of 1.23 cm).

Before making a measurement, the zero of the vernier caliper should be checked with the jaws completely closed. It is possible that through misuse the caliper is no longer zeroed and thus gives erroneous readings (systematic error). If this is the case, a zero correction must be made for each reading. If the vernier zero lies to the right of the main scale zero, measurements will be too large and the error is said to be *positive*. In this case, the zero correction is made by subtracting the zero reading from the measurement reading. For example, the "zero" reading in Fig. 2-5 is +0.05 cm, and this amount must be subtracted from each measurement reading for more accurate results.

Similarly, if the error is *negative*, or the vernier zero lies to the left of the main scale zero, measurements will be too small and the zero correction must be added to the measurement readings.

Summarizing these corrections in equation form

$$\text{Corrected reading} = \text{actual reading} - \text{zero reading}.$$

For example, for a positive error of +0.05 cm as in Fig. 2-5,

$$\text{Corrected reading} = \text{actual reading} - 0.05 \text{ cm}.$$

If there were a negative correction of −0.05 cm, then

$$\begin{aligned}\text{Corrected reading} &= \text{actual reading} - (-0.05) \text{ cm} \\ &= \text{actual reading} + 0.05 \text{ cm}.\end{aligned}$$

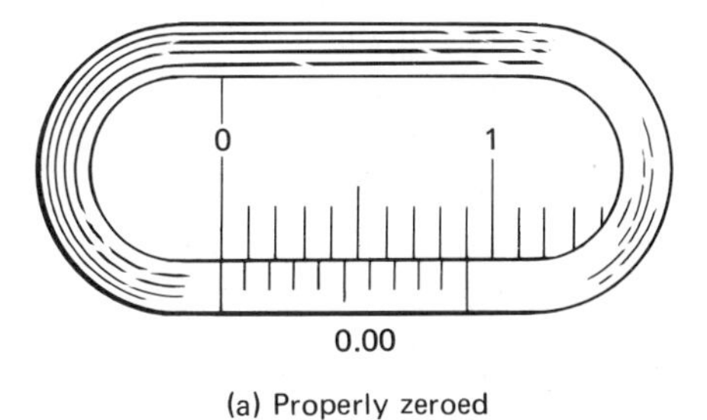

(a) Properly zeroed

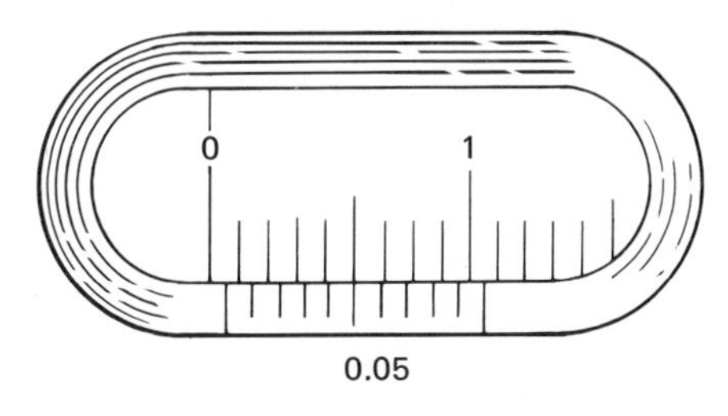

(b) Positive error +0.05 cm
(subtracted from measurement reading)

Fig. 2-5 Zeroing vernier caliper with jaws closed. (a) Zero error. (b) Positive error, +0.05 cm.

D. *Micrometer Caliper*

The **micrometer caliper** (Fig. 2-6), commonly called a **mike,** consists of a movable spindle (jaw) that is advanced toward another parallel-faced jaw (called an anvil) by rotating the thimble. The thimble rotates over an engraved sleeve (or "barrel") that is mounted on a solid frame.

Most micrometers are equipped with a ratchet (ratchet handle to far right) which allows slippage of the screw mechanism when a small and constant force is exerted on the jaw. This permits the jaw to be tightened on an object with the same amount of force each time. Care should be taken not to force the screw (particularly if the micrometer does not have a ratchet mechanism), so as not to damage the measurement object and/or the micrometer. The micrometer caliper provides for accurate measurements of small lengths and is particularly convenient in measuring the diameters of thin wires and the thickness of thin sheets.

The axial main scale on the sleeve is calibrated in millimeters and the thimble scale is calibrated in 0.01 mm (hundredths of a millimeter). The movement mechanism of the micrometer is a carefully machined screw with a pitch of 0.5 mm. The pitch of a screw, or the distance between screw threads, is the lateral linear distance the screw moves when turned through one rotation.

The axial line on the sleeve main scale serves as a reading line. Since the pitch of the screw is 0.5 mm

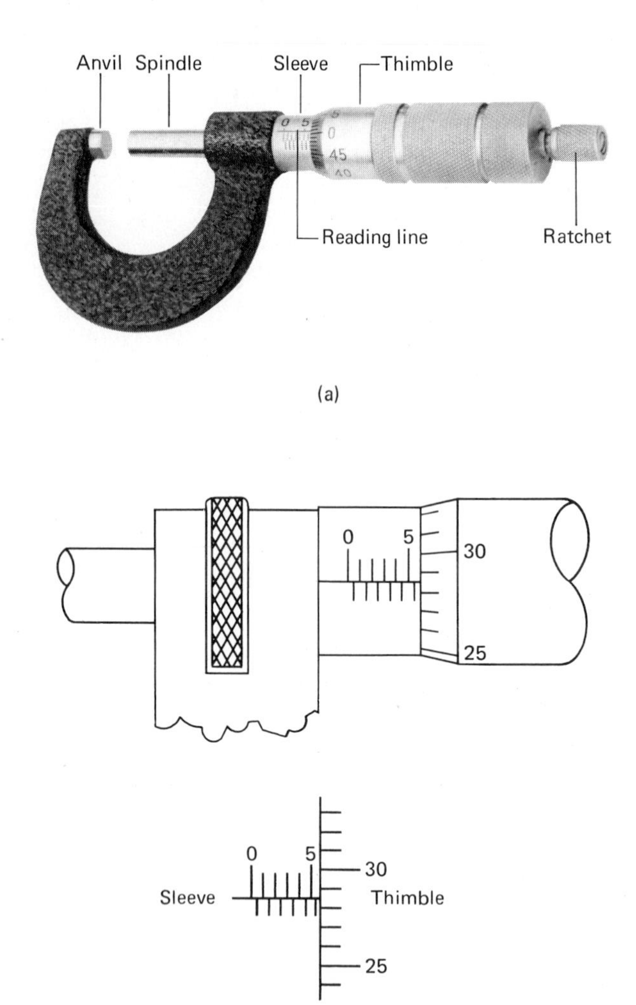

Fig. 2-6 A micrometer caliper and an example of a micrometer reading. This particular mike has the 1.0-mm and 0.5-mm scale divisions below the reading line. In the diagram, as on some mikes, the 1.0-mm divisions are above the line and the 0.5-mm divisions below it. The thimble in the diagram is in its second rotation, as indicated by being past the 0.5-mm mark, so the reading is 5.500 + 0.285 mm, or 5.785 mm, where the last 5 is the estimated figure. (Photo courtesy of Sargent-Welch Scientific Co.)

and there are 50 divisions on the thimble, when the thimble is turned through one of its divisions, the thimble moves (and the jaws open or close) $\frac{1}{50}$ of 0.5 mm or 0.01 mm ($\frac{1}{50} \times 0.5$ mm $= 0.01$ mm). One complete rotation of the thimble (50 divisions) moves it through 0.5 mm, and a second rotation through another 0.5 mm, for a total of 1.0 mm or one scale division along the main scale. That is, the first rotation moves the thimble from 0.00 through 0.50 mm and the second rotation moves the thimble from 0.50 through 1.00 mm.

It is sometimes instructive to think of the 1-mm main-scale divisions as being analogous to dollar ($) divisions and the thimble scale divisions as being cents ($0.01). The first rotation of the thimble corresponds to going from $0.00 to $0.50 (50 cents) and the second rotation corresponds to going from $0.50 to $1.00, so that two complete rotations go through 100 cents or $1.00 of the main scale.

Some micrometers have a scale that indicates the 0.5-mm marks of the main-scale divisions, and hence tells which rotation the thimble is in (Fig. 2-6). Cheaper mikes do not have this extra graduation and the main scale must be closely examined to determine which rotation the thimble is in. If a mike does not have the 0.5-mm scale, you must determine whether the thimble is in its first rotation, in which case the thimble reading is between 0.00 and 0.50 mm (corresponding to the actual engraved numbers on the thimble), or in the second rotation, in which case the reading is between 0.50 and 1.00 mm (the actual thimble scale reading plus 0.50). This can be determined by judging whether the edge of the thimble is in the first or the second half of the main-scale division. Notice that the zero mark on the thimble is used to indicate both 0.00 mm (beginning of the first rotation) and 0.50 mm (beginning of the second rotation).

Measurements are taken by noting the position of the edge of the thimble on the main scale and the position of the reading line on the thimble scale. For example, for the drawing in Fig. 2-6, the mike has a reading of 5.785 mm. On the main scale is read 5.000 mm plus one 0.500-mm division (scale below reading line), giving 5.500 mm. That is, in the figure, the thimble is in the second rotation of a main-scale division.

The reading on the thimble scale is 0.285 mm, where the 5 is the estimated or doubtful figure (i.e., the reading line is estimated to be midway between the 28 and the 29 marks). Some mikes have vernier scales on the sleeves to help read this last significant figure.

A zero check should be made and a zero correction applied to each reading if necessary as described in Section III.C. A zero reading is made by rotating the screw until the jaw is closed or the spindle comes into contact with the anvil. The contacting surfaces of the spindle and anvil should be clean and free of dust. (Micrometers can be adjusted to zero readings by means of a spanner wrench. Do not attempt to do this without your instructor's permission or supervision.)

E. *Density*

The **density** ρ of a substance is defined as the mass m per unit volume V (i.e., $\rho = m/V$). This may be determined experimentally by measuring the mass and volume of a sample of a substance and forming the ratio m/V. The volume of regularly shaped objects may be calculated from length measurements; for example:

Rectangle: $V = \ell \times w \times h$ (length × width × height)

Cylinder: $V = A\ell = (\pi r^2)\ell$ (circular cross-sectional area $A = \pi r^2$, where r is the radius, times the length ℓ of the cylinder)

Sphere: $V = \frac{4}{3}\pi r^3$ (where r is the radius of the sphere)

Density is commonly expressed in grams per cubic centimeter (g/cm^3) or kilogram per cubic meter (kg/m^3).

Density provides a measure of the compactness of matter in a substance. For example, the marble and Styrofoam ball in Fig. 2-7 have the same mass (5.0 g), but the marble has greater density. With measured radii of $r_m = 0.75$ cm and $r_b = 6.0$ cm for the marble and ball, respectively, the calculated densities are

$$\rho_m = \frac{m_m}{V_m} = \frac{m_m}{\frac{4}{3}\pi r_m^3} = \frac{5.0}{\frac{4}{3}\pi(0.75)^3} = 2.8 \text{ g/cm}^3$$

$$\rho_b = \frac{m_b}{V_b} = \frac{m_b}{\frac{4}{3}\pi r_b^3} = \frac{5.0}{\frac{4}{3}\pi(6.0)^3} = 0.0055 \text{ g/cm}^3$$

(Notice that the calculated results have only two significant figures. Why?)

To find the density of a heavy irregularly shaped object, determine its volume by immersing it in water (or some other liquid) in a graduated cylinder. Since the object will displace a volume of water equal to its own volume, the difference in the cylinder readings before and after immersion is the volume of the object. Cylinders commonly have scale divisions of milliliters (ml) and 1 ml = 1 cc (cubic centimeter) = 1 cm^3.

Fig. 2-7 The marble and the styrofoam ball have equal masses, but different densities ($\rho = m/V$). Since the volume of the ball is larger than that of the marble, its density is less. (Photo courtesy of Gerald Taylor.)

The physical property of density can be used to identify substances in some cases. If a substance is not pure or it is inhomogeneous (mass not evenly distributed), an average density is obtained, which is generally different from that of a pure or homogeneous substance.

IV. EXPERIMENTAL PROCEDURE

A. *Least Count of an Instrument Scale*

1. List the least count and the estimated fraction of the least count of each of the measuring instruments in Data Table 1. For example, for a meter stick, 1 mm and 0.1 ($\frac{1}{10}$) mm, respectively.

B. *Thickness Measurements*

2. Using the micrometer caliper, take a zero reading and record in Data Table 3. Then, take several measurements of a single page of this manual to determine the average thickness per page incorporating the zero correction. Record the data and result in Data Table 2.

3. With the micrometer, take thickness measurements of a group of several pages together [e.g., 10 pages (sheets of paper)] and record the data in Data Table 2. Calculate the average thickness per page.

4. With the vernier caliper, take several measurements of the total thickness of the manual (*ex-*

cluding covers).* Record the data in Data Table 2 and compute the average overall thickness of the manual.

5. Using the values of the average thickness per page determined in procedures 1 and 2 and the overall average thickness of the manual from procedure 3, compute the number of pages (sheets of paper) in your manual. For example, if the average thickness per page is 0.150 mm and the average overall thickness is 35.5 mm (3.55 cm), the calculated number of papers is

$$\frac{35.5 \text{ mm}}{0.150 \text{ mm/page}} = 236.6666 = 237 \text{ pages}$$

6. Determine the actual number of pages (sheets of paper) in the manual. (Remember to subtract any pages handed in from Experiment 1.) Compute the percent error for each of the two experimentally determined values.

C. *Density Determinations*

7. The densities of the materials of the various objects are to be determined from mass and volume (length) measurements. The mass and length measurements will give you experience in using the laboratory balance and the vernier and micrometer calipers.

8. Using the appropriate measuring instrument(s), take several measurements to determine the average dimensions of the regularly shaped objects so that their volumes can be calculated. Record the data in Data Table 3. Read the dimensions in centimeters. Remember to make a zero correction for each reading if necessary.

9. Calculate the volumes of each of the objects and record in Data Table 4.

10. Determine the volume of the irregularly shaped metal object by the method described in Section III.E. Record the volume in Data Table 4.

11. Using a laboratory balance, determine the mass of each object and record the results in Data Table 4.

12. Calculate the density of the material of each object and find the percent error of each experimental result. (Accepted density values are given in Appendix A, Table A1.)

*Be sure the pages are compacted as much as possible before the measurements are taken.

Name .... Section .... Date ....

Lab Partner(s) ....

EXPERIMENT 2

LABORATORY REPORT

A. *Least Count of an Instrument Scale*

DATA TABLE 1

Instrument	Least count	Estimated fraction of least count
Meter stick		
Vernier caliper		
Micrometer caliper		
Balance		
Graduated cylinder		

B. *Thickness Measurements*

DATA TABLE 2

Reading	Thickness of single page (mm)	Thickness of ______ pages (mm)	Thickness of manual, excluding covers (mm)
1			
2			
3			
4			
Average			

Calculations (show work)

		Percent error
Actual number of pages (sheets) in manual	....	
Computed number of pages (from single-page measurement)	....	....
(from multiple-page measurement)	....	....

C. *Density Determination*

DATA TABLE 3

Zero reading: Vernier caliper .... Micrometer caliper ....

	Rod		Wire		Sphere	Rectangular sheet		
Instrument used								
Reading	Diameter (cm)	Length (cm)	Diameter (cm)	Length (cm)	Diameter (cm)	Length (cm)	Width (cm)	Thickness (cm)
1								
2								
3								
4								
Average								

DATA TABLE 4

Object	Mass (g)	Volume (cm^3)	Experimental density (g/cm^3)	Accepted density (g/cm^3) from Table A1	Percent error
Rod Type of material:					
Wire Type of material:					
Sphere Type of material:					
Rectangular sheet Type of material:					
Irregularly shaped object Type of material:					

Volume and Density Calculations (show work)

Name .... Section .... Date ....

Lab Partner(s) ....

EXPERIMENT 2 *Mass, Volume, and Density*

QUESTIONS

1. Explain the probable source of error in the experimental determination of the number of manual pages.

2. Based on your general experience in determining the length and width of the lengths in decimal fractions of both centimeters and inches, do you see any advantage in use of the metric system? Explain.

3. Suppose that you were given an irregularly shaped object that floated. Describe how you would experimentally determine its volume.

4. A thin circular sheet of aluminum has a radius of 20 cm and a thickness of 0.50 mm. Find the mass of the sheet in grams.

5. Archimedes, a famous Greek scientist, was given a problem by King Hieron II of Syracuse (Sicily). The king suspected that his crown, which was supposed to be made of pure gold, contained some silver alloy, and he asked Archimedes to prove or disprove his suspicion. (The crown did contain silver.) How would you have experimentally determined whether or not the crown was pure gold?

EXPERIMENT **3**

Measuring the Height of a Building

I. INTRODUCTION

Experimental information is sometimes required that is not convenient or possible to obtain by direct measurement. For example, it would be not only inconvenient, but also hazardous, to measure the height of a building or some other tall object with a meter stick by lowering an observer from the top. Of course, an obvious suggestion would be simply to lower a rope to the ground, then measure the length of the rope. However, let us add the additional restrictions that access cannot be gained because the building is locked and no one is available with a key to open it.

In this experiment you are asked to do just that — measure the height of a building or tower on campus without entering it. Also, the height of a flagpole or some other object such as a tree is to be measured. How is this to be done? Obviously, the heights must be determined by some indirect means through the measurement of some related parameters. This experiment challenges your ingenuity and demonstrates an experimental situation often encountered in physics — the measurement of physically inaccessible quantities. Measurements must then be done on accessible related parameters. Such is almost always the case in atomic and nuclear physics.

Consider a practical example of an indirect height measurement. Suppose a tall tree near a building or power line is to be cut down. By experimentally measuring the height of the tree, it can be determined whether the tree can be cut down safely without hitting anything and perhaps save the inconvenient and dangerous procedure of topping the tree.

II. EQUIPMENT NEEDED (Instructor's option)

- Two meter sticks
  and/or
- A meter stick and protractor
  (A piece of string with an attached weight might prove helpful in taking readings in this case.)

III. EXPERIMENTAL PROCEDURE

The instructor will designate (a) a building or tower, and (b) some other tall object such as a flagpole or tree. Using your ingenuity, knowledge, and the equipment provided, experimentally measure the heights of the designated objects. You are to devise the experimental method (using only the equipment provided) with the best (most accurate) technique. *Note:* Would it be good scientific procedure to make only one measurement of each object?

IV. LABORATORY REPORT

1. Prepare a written report that clearly presents and describes the experimental procedure used, the data obtained, and the calculations, results, and analysis. Your instructor may wish to suggest a particular format.

   Remember, a laboratory report should be comprehensible to someone who is not familiar with the experiment or the experimental method. A laboratory report reflects the quality of your work and the effectiveness of presentation.

2. (Optional) The instructor may appoint someone to list the results of each team of lab partners and compute the class averages of the heights of the objects. These class averages may be used in the error analysis of individual team measurements.

Name .... Section .... Date ....

Lab Partner(s) ....

EXPERIMENT 4 *The Scientific Method: The Simple Pendulum*

ADVANCE STUDY ASSIGNMENT

Read the experiment and answer the following questions.

1. What is the scientific method and how is it applied?

2. What are the physical parameters of a simple pendulum?

3. What is meant by the period of a pendulum?

4. How does the period of a pendulum vary theoretically with (a) length; (b) mass of bob; (c) angular displacement?

5. What is meant by a "small-angle approximation"?

6. Would the period of a particular pendulum be the same on the moon as on the earth? Explain.

EXPERIMENT **4**

The Scientific Method: The Simple Pendulum

I. INTRODUCTION

The laboratory is a place for the investigation of physical phenomena and principles. Originally, scientists studied physical phenomena in the laboratory in the hope that they might discover relationships and principles involved in the phenomena. This might be called a **trial-and-error approach.** Today, the physics laboratory is used in general to apply what is called the **scientific method.** This principle states that no theory or model of nature is tenable unless the results it predicts are in accord with the experiment. That is, rather than applying the somewhat haphazard trial-and-error approach, scientists now try to theoretically predict physical phenomena, then test their theories against planned experiments in the laboratory. If enough experimental results agree with the theoretical predictions, we consider the theory to be valid and believe that we have an accurate description of certain physical phenomena.

To illustrate the scientific method, in this experiment a theoretical expression or equation that describes the behavior of a simple pendulum will be given. You will test the validity of this relationship experimentally. In the process you will learn the variable that influences the period of a simple pendulum and how the physical relationship and experimental data can be used to determine other useful information (e.g., the value of the acceleration due to gravity).

II. EQUIPMENT NEEDED

- Meter stick
- Laboratory timer or stopwatch
- Protractor
- String
- Three or more pendulum bobs of different masses
- Pendulum clamp (if available)
- 1 sheet of Cartesian graph paper

III. THEORY

A **pendulum** consists of a "bob" (a mass) attached to a string that is fastened so that the pendulum assembly can swing or oscillate in a plane (Fig. 4-1). For a simple or ideal pendulum, all the mass is considered to be concentrated at a point at the center (of mass) of the bob.

The physical quantities of a simple pendulum are (1) the length L of the pendulum, (2) the mass m of the pendulum bob, (3) the angular displacement θ through which the pendulum swings, and (4) the period T of the pendulum, which is the time it takes for the pendulum to swing through one complete oscillation (e.g., from A to B and back to A in Fig. 4-1).

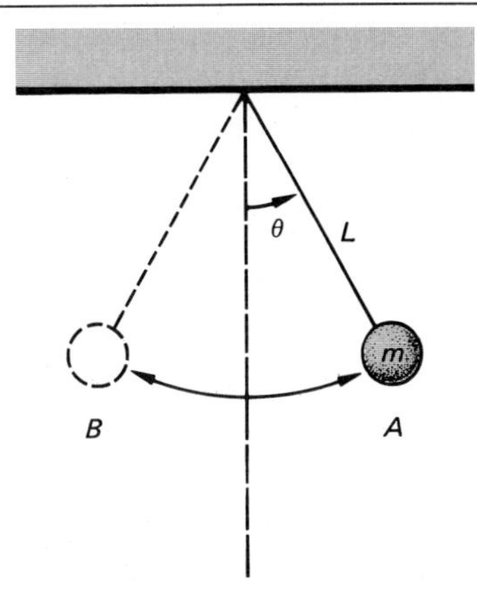

Fig. 4-1 A simple pendulum. The period T of the pendulum is the time it takes to swing from A to B and back to A.

From experience or preliminary investigation, it should be evident that the period of a pendulum is proportional to its length (i.e., the longer the pendulum, the greater its period). How do you think the other parameters (m and θ) affect the period?

From physical principles and advanced mathematics, the theoretical expression for the period of a simple pendulum oscillating in a plane is

$$T = 2\pi \sqrt{\frac{L}{g}}\left(1 + \frac{1}{4}\sin^2\frac{\theta}{2} + \frac{9}{64}\sin^4\frac{\theta}{2} + \cdots\right) \quad \textbf{(4-1)}$$

where g is the acceleration due to gravity and the terms in parentheses are part of an infinite series. In calculating T for a given angular displacement θ, the more terms of the series that are evaluated, the greater the accuracy of the theoretical result.

For small angles ($\theta \lesssim 10°$), the θ terms in the series are small compared to unity (i.e., $<<1$) and in this case to a good approximation

$$T = 2\pi \sqrt{\frac{L}{g}} \quad \textbf{(4-2)}$$

(This is called the **first-order approximation.** If the second term in the series is retained, the approximation is to second order, and so on.)

Notice that even without an approximation (Eq. 4-1), the period is theoretically independent of the mass of the pendulum bob. Also, in the limits of the small-angle approximation (Eq. 4-2), the period is independent of the displacement angle.

It is sometimes helpful to visualize a physical system as a "black box" with inputs and outputs.* The black box is the relationship connecting the input and output parameters. The term parameter refers to anything in the physical system that can be measured. The input parameters are the physical variables that may *control or influence* the behavior of the output parameters (the physical quantities that are measured and *describe* the resulting behavior of the system). The input parameters are often called **independent variables** because they can (and should) be varied independently of each other. The output parameters, on the other hand, may be called **dependent variables** because their values depend on the inputs. In any given system, some of the inputs may have little or no effect on the outputs.

In the case of a simple pendulum, a black box diagram with a little insight might look like this:

Input

$m \quad \theta \quad L$

$\downarrow \quad \downarrow \quad \downarrow$

Simple Pendulum

$$T = 2\pi\sqrt{L/g}\left(1 + \frac{1}{4}\sin^2\frac{\theta}{2} + \cdots\right)$$

$\downarrow$

T

Output

You may find drawing black box diagrams helpful in understanding the physical systems to be investigated in later experiments.

*Suggested by I. L. Fischer, Bergen Community College.

IV. EXPERIMENTAL PROCEDURE

1. Set up a simple pendulum arrangement. If a pendulum clamp is not available and the string must be tied around something (e.g., a lab stand), make sure that the string is secure and does not slip on the support.

2. Experimentally investigate the small-angle approximation (Eq. 4-2) and the theoretical prediction (Eq. 4-1) that the period increases with larger angles. Do this by determining the pendulum period for the several angles listed in Data Table 1, keeping the length and mass of the pendulum constant. [Rather than timing only one oscillation, time several (e.g., four or five) and determine the average period.]

   Measure and record the pendulum length. The length should be measured to the center of the pendulum bob. Compute the percent error for each angle θ using Eq. 4-2 as the accepted value. (In this case, do not use the absolute difference so that each percent error will have signs, + or −. There is a question relating to this at the end of the experiment.) Draw a conclusion from the data.

3. Experimentally investigate if the period is independent of the mass of the pendulum bob. Using the three masses provided, determine the periods of a pendulum with each mass as the bob (constant length L and small angle of oscilla-

tion). Record your results in Data Table 2 and draw a conclusion from the data.

4. Experimentally investigate the relationship between the period and length of a pendulum. Using five different lengths (e.g., 20, 40, 60, 80, and 100 cm), determine the average period of a pendulum of each length (constant mass and small angle of oscillation). Record the data in Data Table 3.

5. Compute the theoretical period for each pendulum length (Eq. 4-2) and enter the results in Data Table 3 ($g = 9.8\ \text{m/s}^2 = 980\ \text{cm/s}^2$).

6. Compute the percent error between the experimental and the theoretical values of the period for each pendulum length and record in Data Table 3. Draw conclusions about the validity or applicability of Eq. 4-2.

7. The object of the preceding experimental procedures was to determine the validity or applicability of Eq. 4-2, that is, whether the experimental results agreed with the theoretical predictions as required by the scientific method. Once found acceptable, a theoretical expression can then be used to experimentally determine other quantities occurring in the expression. For example, Eq. 4-2 provides a means for experimentally determining g, the acceleration due to gravity, by measuring the pendulum parameters of length and period as was done previously.

   Squaring both sides of Eq. 4-2, we have

$$T^2 = \frac{4\pi^2}{g} L \qquad \textbf{(4-3)}$$

or

$$L = \frac{g}{4\pi^2} T^2$$

Hence, the equation has the form $y = ax^2$, that of a parabola. As was learned in Experiment 1, this can be plotted as a straight line with the general form $y = ax'$ by letting $x' = T^2$, and the line will have a slope of $a = g/4\pi^2$.

8. Plot L versus T^2 from the experimental data in Data Table 3, determine the slope of the graph, and compute the experimental value of g. Record this in the Laboratory Report and compute the percent error of the result.

Name .... Section .... Date ....

Lab Partner(s) ....

EXPERIMENT 4 *The Scientific Method: The Simple Pendulum*

LABORATORY REPORT

DATA TABLE 1 Mass, m .... Pendulum length, L ....

Angle θ	Period, T (s)		Percent error
	Experimental	Theoretical	
5°			
10°			
20°			
30°			
45°			
60°			

Conclusion:

DATA TABLE 2 θ .... L ....

m	T (s)		Percent error
	Experimental	Theoretical	

Conclusion:

DATA TABLE 3 θ .... m ....

L	T (s)		Percent error	T^2 (s^2)
	Experimental	Theoretical		

Conclusion:

Name .... Section .... Date ....

Lab Partner(s) ....

Value of g from experimental data (slope of graph) .... (units)

Percent error ....

QUESTIONS

1. It was suggested that the time for several periods be measured and the average period determined, rather than timing only one period.
   (a) What are the advantages of this method?

   (b) How would the result be affected if a large number of periods were timed?

2. Thomas Jefferson once suggested that the period of a simple pendulum be used to define the standard unit of length. What would be the length of a "1-second" pendulum (a pendulum with a period of 1 s)?

3. Review the discussion of the *order* approximations for Eq. 4-1 in the Theory section. Calling the first-order approximation T_1 and the second-order approximation T_2,

   (a) for a 1-meter pendulum with an angular displacement of 10°, compute T_1 and T_2, and their percent difference. Use five-digit accuracy when computing these T's.

   (b) Repeat the calculations for an angle of 60° and comment on the results.

(continued)

4. Examine your percent errors listed in Data Table 1. (Remember that these numbers should have signs, + or −.)

   (a) Calculate the mean of these numbers. This mean or average represents the systematic error. Why?

   (b) Calculate the average deviation from the mean. (See Experiment 1.) This represents the random error. Why?

Name .... Section .... Date ....

Lab Partner(s) ....

EXPERIMENT 5 *Uniformly Accelerated Motion*

ADVANCE STUDY ASSIGNMENT

Read the experiment and answer the following questions.

A. *Object in Free Fall*

1. From what height should you drop the object to obtain good experimental results?

2. Should the person dropping the object or a lab partner do the timing? Explain.

3. Suppose that the initial height of the object was measured from the top of the object at the release point to the floor. How would this affect your experimental result? Is this a random or a systematic error?

B. *Free-Fall Spark-Timer Apparatus*

4. How are data recorded on the tape strip, and what information does the data tape give?

5. What precautions should be taken in using the apparatus?

6. What is the equation describing the instantaneous velocity of an object in free fall, and what is the shape of the graph of the instantaneous velocity versus time?

(continued)

7. What is the difference between average velocity and instantaneous velocity?

8. Describe how the instantaneous velocity of an object in free fall can be calculated from displacement and time data.

C. *Linear Air Track*

9. What is the function of the air in the air-track operation?

10. How is the air track positioned to study uniformly accelerated motion? Explain.

11. How is the acceleration of a glider traveling on an elevated air track (a) related to the angle of elevation; (b) the height of elevation?

12. (Omit if question 6 answered.) What is the equation describing the instantaneous velocity of a glider on an elevated air track, and what is the shape of the graph of the instantaneous velocity versus time?

13. (Omit if question 7 answered.) What is the difference between the average velocity and instantaneous velocity?

14. (Omit if question 8 answered.) Describe how the instantaneous velocity of a glider traveling on an elevated air track can be calculated from displacement and time data.

EXPERIMENT 5

Uniformly Accelerated Motion

I. INTRODUCTION

An important case in kinematics is that of an object in uniformly accelerated motion. Probably the most common example of an object moving with a constant or uniform acceleration is **free fall,** in which an object falls with the acceleration due to gravity g. Near the surface of the earth the acceleration due to gravity is essentially constant, with a value of

$$g = 980 \text{ cm/s}^2 = 9.8 \text{ m/s}^2 = 32.2 \text{ ft/s}^2$$

In this experiment we make use of the acceleration due to gravity to investigate an object undergoing uniformly accelerated motion to see how its velocity and displacement change with time. Conversely, with displacement and time measures, the value of g can be determined. The experimental data and its analysis will provide a better understanding of the kinetic equations describing the motion. Investigations will be made using one or more of the following: (1) an object in free fall, (2) a free-fall spark-timer apparatus, and/or (3) a linear air track.

II. EQUIPMENT NEEDED

A. *Object in Free Fall*

- Lead weight or small wooden block
- Meter stick
- Laboratory timer or stopwatch

B. *Free-Fall Spark-Timer Apparatus*

- Free-fall and spark-timer apparatus
- Meter stick
- 1 sheet of Cartesian graph paper

C. *Linear Air Track*

- Linear air track
- Several laboratory timers or stopwatches
- Wooden blocks of two different heights
- 1 sheet of Cartesian graph paper

III. THEORY

A. *Object in Free Fall*

An object in free fall (air resistance neglected) falls under the influence of gravity, or with an acceleration of 980 cm/s^2. The distance y an object falls in a time t is given by

$$y = v_0 t + \tfrac{1}{2} g t^2 \qquad \textbf{(5-1)}$$

where v_0 is the initial velocity.

If an object is dropped from rest ($v_0 = 0$), then

$$y = \tfrac{1}{2}gt^2 \tag{5-2}$$

(downward taken as positive to avoid minus sign). Hence, by measuring the time t it takes for an object to fall a distance y, the acceleration due to gravity g can be easily calculated.

B. *Free-Fall Spark-Timer Apparatus*

Several types of free-fall spark-timer apparatus are shown in Fig. 5-1. The free-fall assembly consists of a metal object that falls freely between two wires with a tape strip of specially treated paper between the object and one of the wires. The spark timer is a fast timing device that supplies a high voltage across the wires periodically for preset time intervals (e.g., a frequency of 60 Hz or time interval of $\frac{1}{60}$ s, since $t = \frac{1}{f}$). The free-fall apparatus is equipped with an electromagnet that releases the metal object when the spark timer is activated.

A high voltage will cause a spark to jump between two electrical conductors in close proximity. The wires are too far apart for a spark to jump directly from one wire to the other; however, as the metal object falls between the wires, the spark (electrical current) jumps from one wire to the metal object, travels through the object, and jumps to the other wire. In so doing, the spark burns a spot on the paper tape strip. The spots on the tape are then a certain time interval apart, as selected and set on the spark timer. An appropriate time interval is set on the spark timer so as to give a sufficient number of spots spaced at convenient intervals. The series of spots on the tape gives the vertical distance of fall as a function of time, from which can be measured the distance y_i the object falls in a time t_i.

The instantaneous velocity v of a free-falling object (neglecting air resistance) at a time t is given theoretically by

$$v = v_0 + gt \tag{5-3}$$

(where downward is taken as the positive direction). Hence, a graph of v versus t is a straight line ($y = ax + b$) with a slope $a = \Delta v/\Delta t = g$ and an intercept $b = v_0$, the initial velocity. Recall that t in Eq. 5-3 is really a time *interval* measured from an arbitrary starting time, $t_0 = 0$. At this time, the velocity of the object is v_0, which may or may not be zero ($v_0 = 0$ if dropped from rest).

The motion of the falling object as recorded on the experimental data tape is analyzed as follows. The average velocity $\bar{v}$ of an object traveling a distance y_i in a time t_i is defined as

$$\bar{v} = \frac{y_i}{t_i} \tag{5-4}$$

Keep in mind that y_i and t_i are really length and time *intervals* or the differences between corresponding instantaneous lengths and times. Referenced to an initial position and time (y_0 and t_0),

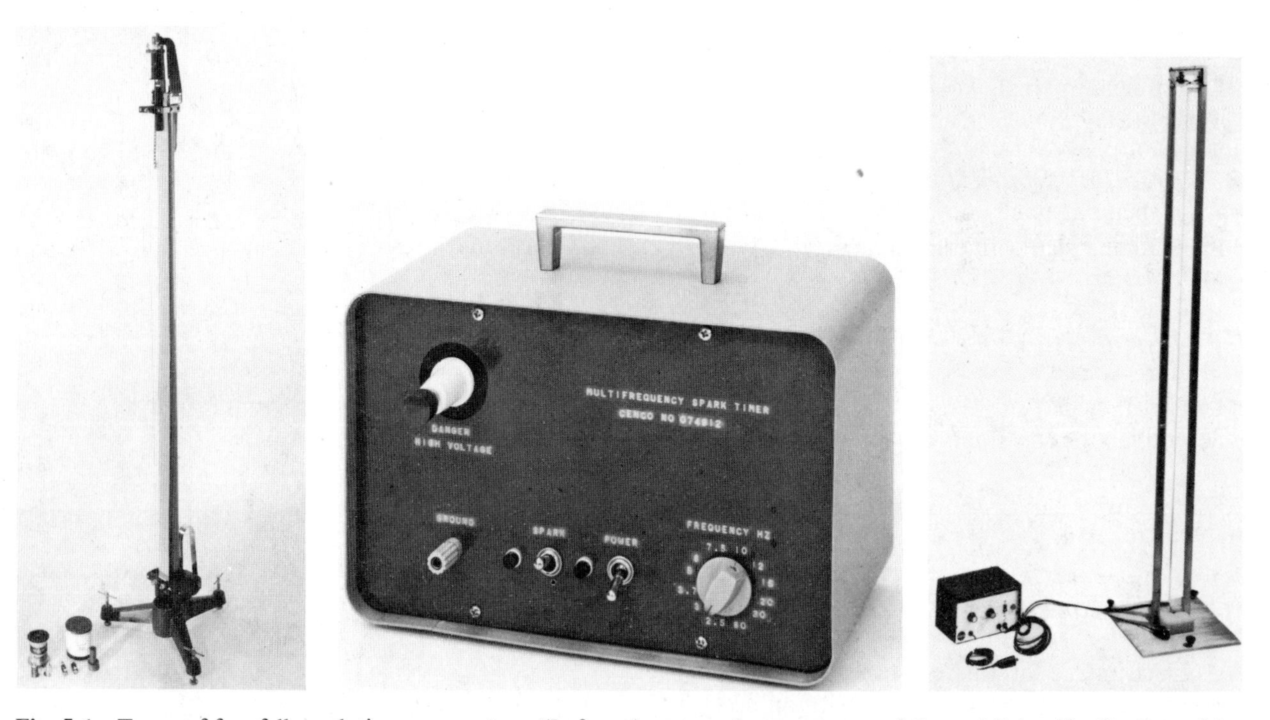

Fig. 5-1 Types of free-fall spark-timer apparatus. (Left and center: photo courtesy of Central Scientific Co. Inc.; right: photo courtesy of the Ealing Corporation, S. Natick, MA 01760)

$\Delta y_i = y_i - y_0$ and $\Delta t_i = t_i - t_0$. Arbitrarily taking $y_0 = 0$ and $t_0 = 0$, we have $\Delta y_i = y_i$ and $\Delta t_i = t_i$. (It is these intervals that will be measured from the data tape.)

For a uniformly accelerated object (moving with a constant acceleration) as in the case of free fall, the average velocity is given by

$$\bar{v} = \frac{v_i + v_0}{2} \tag{5-5}$$

where v_i and v_0 are the instantaneous velocities at times t_i and t_0, respectively. (Why is this? Consult your textbook.) Then, equating the expressions for $\bar{v}$ given by Eq. 5-4 and Eq. 5-5 and solving for v_i, we have

$$\frac{v_i + v_0}{2} = \frac{y_i}{t_i}$$

and

$$v_i = \frac{2y_i}{t_i} - v_0 \tag{5-6}$$

If $v_0 = 0$ (i.e., the object falls from rest), then

$$v_i = \frac{2y_i}{t_i} \tag{5-7}$$

Fig. 5-2 A linear air track. (Photo courtesy of The Ealing Corporation, S. Natick, MA 01760)

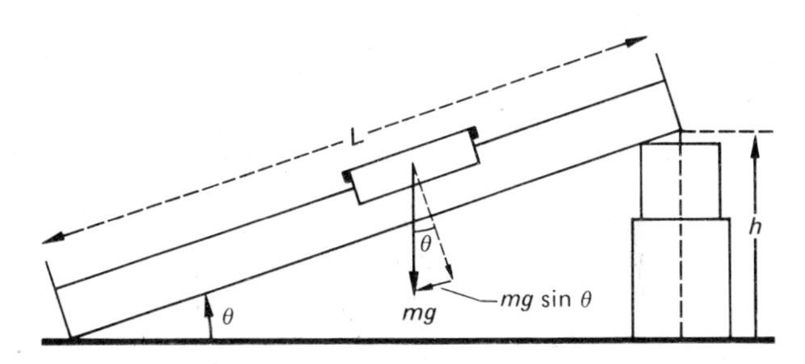

Fig. 5-3 By elevating one end of the air track, the acceleration of the glider is due to a component of the force due to gravity, $a = g \sin \theta$.

C. *Linear Air Track*

A type of linear air track is shown in Fig. 5-2. Air is supplied to the interior of the hollow track and emerges through a series of small holes in the track. This provides a cushion of air on which a glider travels along the track with very little friction (an example of the use of a gaseous lubricant).

To have the glider move under the influence of gravity, one end of the air track is elevated on a block. The acceleration of the glider along the air track is then due to a component of the force due to gravity, $F = ma = mg \sin \theta$ (Fig. 5-3). The acceleration a of the glider along the air track is

$$a = g \sin \theta \tag{5-8}$$

and from the geometry, $\sin \theta = h/L$ (side opposite the angle over the hypotenuse). Hence,

$$a = \frac{gh}{L} \tag{5-9}$$

The instantaneous velocity v at a time t of the uniformly accelerating glider is given theoretically by

$$v = v_0 + at \tag{5-10}$$

Hence, a graph of v versus t is a straight line ($y = ax + b$) with a slope $a = \Delta v/\Delta t$ and an intercept $b = v_0$. If the glider starts from rest, the initial velocity v_0 is zero, and

$$v = at \tag{5-11}$$

The instantaneous velocities of the glider can be found from the experimental data of the measured displacements y_i of the glider along the air track at times t_i by the method given in Section III.B.

IV. EXPERIMENTAL PROCEDURE

A. *Object in Free Fall*

1. Drop the lead weight or small wooden block from a fixed height y above the floor and measure its time of fall. Record the data for five trials in Data Table 1.
2. See Question 3 at the end of experiment for time correction determination.
3. Compute the acceleration due to gravity g from Eq. 5-2 using the corrected time of fall. Find the average or mean value and the mean deviation of your results (see Experiment 1).

B. *Free-Fall Spark-Timer Apparatus*

4. Your laboratory instructor will make a data tape for you or assist you in obtaining one. Care must be taken in aligning the apparatus, and when working with high voltages, one must be careful not to receive an electrical shock.

5. Record the time interval of the spark timer used on the data tape and draw small circles around the burn spots so that their locations can be easily seen. Occasionally, a spot of the sequence may be missing (e.g., due to local misalignment of the wires); however, it is usually easy to tell that a spot is missing by observation of the tape. Do not try to guess where the spot should be.

6. Draw a straight line through each spot perpendicular to the length of the tape. Using the line through the beginning spot as a reference line ($y_0 = 0$), measure the distance of each spot line from the reference line (y_1, y_2, y_3, etc.). Write the measured value of the distance on the tape by each respective spot line.

   Making use of the known spark-timer interval, write the time taken for the object to fall a given distance on the tape by each spot line taking $t_0 = 0$ at $y_0 = 0$. For example, if the timer interval is $\frac{1}{60}$ s, the time interval between the reference line ($y_0 = 0$) and the first spot line (y_1) is $t_1 = \frac{1}{60}$ s, and the time taken to fall to the second spot line (y_2) is $t_2 = \frac{1}{60} + \frac{1}{60} = \frac{2}{60} = \frac{1}{30}$ s. (Do not forget to account for the time intervals associated with missing spots should this occur.)

7. Record the data on the tape in Data Table 2. Using Eq. 5-7, compute the instantaneous velocity of the falling object at each spot line from the experimental data and record.

8. At this point, you should realize that the instantaneous velocities given by Eq. 5-7 ($v_i = 2y_i/t_i$) are *not* the actual instantaneous velocities of the falling object, since it had a non-zero initial velocity or was in motion at the first spot line (y_0). Eq. 5-6 really applies to the situation, and $2y_i/t_i = v_i + v_0$. Note that the instantaneous velocities you computed ($2y_i/t_i$) included v_0.

   Even so, plot the computed v_i's on a v versus t graph and determine the slope. This will still be an experimental value of g. Compute the percent error of your experimental result. (Accepted value, $g = 980$ cm/s^2.)

9. You will notice on your graph that the line does not intercept the y axis at the origin ($t = 0$). This is because $t = 0$ was not measured at the actual time of release, but at some time later. From Eq. 5-6 and 5-7 we see that at $t = 0$ in the measurement time frame

$$v_{i_{(t=0)}} = \frac{2y_0}{t_0} = 2v_0$$

   where y_0 and t_0 are the distance and time measured by the zero values *from* the point of release. The initial velocity at the first line spot is then $v_0 = y_0/t_0$. This gives you the extra bonus of being able to determine v_0 from your graph, since

$$v_0 = \frac{v_{i_{(t=0)}}}{2}$$

   where $v_{i_{(t=0)}}$ is the intercept value. Compute the initial velocity the falling object had at your first line spot and record in the Data Table.

C. *Linear Air Track*

10. The air track should be set up and leveled by the instructor or laboratory assistant. *Do not* attempt to make any adjustments of the air track. Ask your instructor for assistance if you need it.

11. Turn on the air supply and place the glider in motion by applying a small force on the glider in a direction parallel to the air track. *Do not* attempt to move the glider on the air track if the air supply is not turned on. Use the same force for each trial, e.g., by compressing a spring attached to the glider.

12. Using laboratory timers or stopwatches, determine and record the times required for the glider to travel several of the distances listed in Data Table 3 for three different trials. Several students should work together, each with a timer and each taking a time reading as it passes his or her assigned distance mark. Make several practice trials before taking the actual data. (Remember that the distances are length intervals and need not be measured from the end of the air track. If you have a short air track, choose a shorter set of distances. Make use of as much of the air-track length as conveniently possible.)

13. After completing procedure 12, ask the instructor to elevate one end of the air track on a block or obtain permission to do so.

14. Start the glider from rest near the elevated end of the air track. If you have a shorter air track, measure and record the times required for the glider to travel several of the distances listed in Data Table 3. Use the experimental method described in procedure 12.

15. Have the end of the air track elevated to a different height and repeat the time measurements for this height.

16. Using Eq. 5-7, compute the instantaneous velocity of the glider for each of the times in the three experimental sets of data in Data Table 3.

17. Plot v versus t for each case on the same graph and determine the slope of each line.

18. Using Eq. 5-9, compute an experimental value of the acceleration due to gravity g for each of the elevated air-track cases. Compute the percent error for each experimental result. (Accepted value of $g = 9.8$ m/s^2 = 980 cm/s^2.)

LABORATORY REPORT

A. *Object in Free Fall*

DATA TABLE 1

y ....

Trial	Time of fall, t (s)	Corrected t (s)	Calculated g	Deviation
1				
2				
3				
4				
5				
	Mean (average) value			

Result (average value ± mean deviation) ....

Calculations (show work)

(continued)

B. *Free-Fall Spark-Timer Apparatus*

DATA TABLE 2

Spark-timer interval ....

Distance, y_i (cm)		Time, t_i (s)		Computed velocity, v_i (cm/s)	
y_1		t_1		v_1	
y_2		t_2		v_2	
y_3		t_3		v_3	
y_4		t_4		v_4	
y_5		t_5		v_5	
y_6		t_6		v_6	
y_7		t_7		v_7	
y_8		t_8		v_8	
y_9		t_9		v_9	
y_{10}		t_{10}		v_{10}	
y_{11}		t_{11}		v_{11}	
y_{12}		t_{12}		v_{12}	
y_{13}		t_{13}		v_{13}	
y_{14}		t_{14}		v_{14}	
y_{15}		t_{15}		v_{15}	

Value of g from graph (attach graph to lab report) ....
(units)

Percent error ....

Initial velocity at y_0 ....

Lab Partner(s) ....

EXPERIMENT 5 *Uniformly Accelerated Motion*

C. *Linear Air Track*

DATA TABLE 3

Distances		100 cm	150 cm	200 cm	250 cm	300 cm
1. Time, t_i (s)	1					
Level air track	2					
	3					
	Av.					
Computed v_i (cm/s)						
2. Time, t_i (s)	1					
Elevated air track	2					
h_1 ________	3					
d ________	Av.					
Computed v_i (cm/s)						
3. Time, t_i (s)	1					
Elevated air track	2					
h_2 ________	3					
d ________	Av.					
Computed v_i (cm/s)						

Calculations (show work)

Slopes of graphs 1. ....

2. ....

3. ....

Experimental values of g
(computed from data) 2. ....

3. ....

Percent error 2. ....

3. ....

(continued)

QUESTIONS

A. *Object in Free Fall*

1. In the experiment, the air resistance is considered to be negligible. Suppose that it were not. How would this affect the experimentally determined value of g?

2. What is probably the greatest source of error in the experimental procedure? Is it random or systematic error?

3. (a) Have someone hold a meter stick vertically by one end with the person who did the timing close by and his or her open hand cupped around the meter stick ready to grasp the stick when it is dropped. The person holding the meter stick does not tell the other person when the stick will be released and the stick is to be grasped "as soon as he or she can." (Several trials should be made so as to obtain an average reaction time.) From the distance of fall and the accepted value of g, obtain the estimate of the reaction time of the person grasping the meter stick.

   (b) Subtract this time from the times of fall in Data Table 1 to account for the reaction time in stopping the timer or clock.

B. *Free-Fall Spark-Timer Apparatus*

1. Suppose that a different spark-timer interval were used. How would this affect the slope of the graph of v versus t?

2. What would be the shape of the curve of a graph y versus t of the experimental data? How would you determine the value of g from a graph using only the y and t values (i.e., not computing v_i)?

3. If $t = 0$ were taken to be associated with some line spot other than y_0 (e.g., y_3 instead), how would this affect the v versus t graph?

4. Calculate v_0 directly from the first two measurement entries in Data Table 2, using the equation $v_0 = 2y_1/t_1 - y_2/t_2$. (Your instructor can derive this for you.) How does this compare with the value determined from your graph?

C. *Linear Air Track*

1. What would be the shapes of the curves of a graph of y versus t of the data in each experimental case? How would you determine the value of the glider acceleration from a graph using only y and t values (i.e., not computing v_i)?

2. What is the physical significance of the slope of the graph for the case of the level air track?

3. What is the maximum possible value of the slope of a v versus t curve for a glider released from rest on an air track elevated at one end? Describe the experimental setup in this case.

Name .... Section .... Date ....

Lab Partner(s) ....

EXPERIMENT 6 *The Addition and Resolution of Vectors: The Force Table*

ADVANCE STUDY ASSIGNMENT

Read the experiment and answer the following questions.

1. What is the difference between a scalar and a vector?

2. How are vectors represented graphically?

3. What is meant by drawing a vector to scale?

4. Describe the parallelogram method of vector addition.

5. How may the resultant of two vectors be computed analytically from vector parallelogram?

6. What is the cosine of 225° (cos 225°)?

7. What is the polygon method of vector addition?

(continued)

8. What is meant by resolving a vector into components?

9. Describe the component method of vector addition.

10. On the force table, what is the difference between the equilibrant and resultant, and why is only one actually determined experimentally?

EXPERIMENT **6**

The Addition and Resolution of Vectors: The Force Table

I. INTRODUCTION

A **vector** is a mathematical concept to represent quantities that have magnitude *and* direction. Such quantities include displacement, velocity, acceleration, and force. Because vectors have the property of direction, the common method of scalar addition is not applicable to vector quantities. (A scalar is a quantity with magnitude only.)

To find the resultant or vector sum of two or more vectors, special methods of vector addition must be used. Two basic methods of vector addition are the graphical parallelogram method and the analytical component method. In this experiment we shall investigate these methods in the addition of force vectors. The results of the graphical and analytical methods will be compared with experimental results obtained from a force table. The experimental arrangements of forces (vectors) will physically illustrate the principles of the methods of vector addition.

II. EQUIPMENT NEEDED

- Force table with four pulleys
- Four weight hangers
- Set of slotted weights (masses), including three 50 g and three 100 g
- String
- Protractor
- Ruler
- Level
- 3 sheets of Cartesian graph paper

III. THEORY

A. *Methods of Vector Addition: Graphical*

PARALLELOGRAM METHOD

Vectors are represented graphically by arrows. The length of a vector arrow (drawn to scale on graph paper) is proportional to the magnitude of the vector and the arrow points in the direction of the vector.

To add two vectors $\mathbf{A} + \mathbf{B}$ [Fig. 6-1(a)], the parallelogram of which **A** and **B** are adjacent sides is formed. The arrow diagonal of the parallelogram **R** is the resultant or vector sum of $\mathbf{A} + \mathbf{B}$ or by vector addition, $\mathbf{R} = \mathbf{A} + \mathbf{B}$. The magnitude R of the resultant vector is proportional to the length of the diagonal arrow, and the direction of the resultant vector is that of the diagonal arrow **R**. The direction of **R** may be specified as being at an angle θ relative to **A**.

TRIANGLE METHOD

An equivalent method of finding **R** is to place the vectors to be added "head to tail" [head of **A** to tail of **B**, Fig. 6-1(b)]. Vector arrows may be moved as long as they remain pointed in the same direction. The head-to-tail method gives the same resultant as forming a parallelogram.

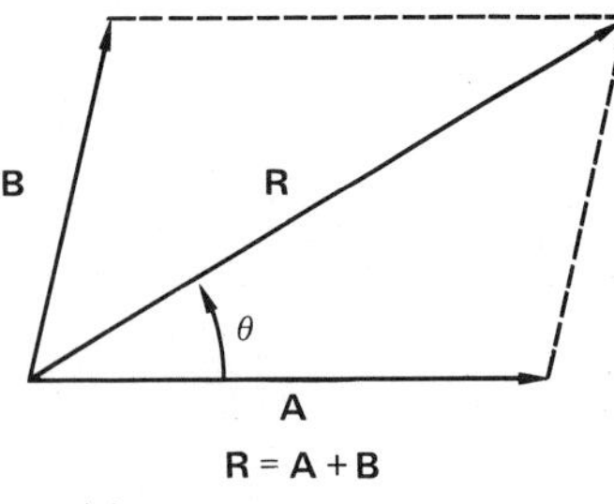

(a) Parallelogram Method

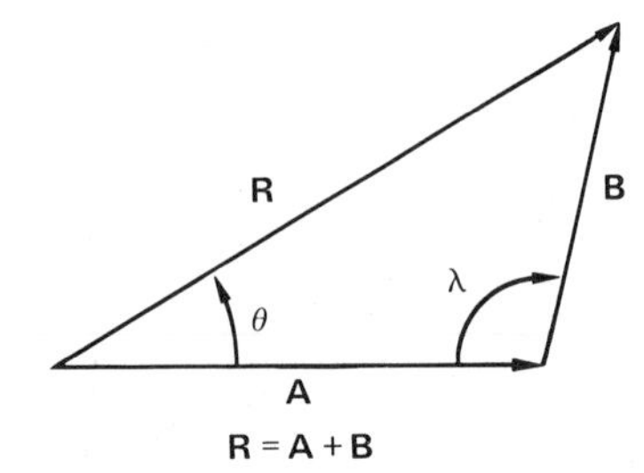

(b) Triangle "Head-To-Tail" Method

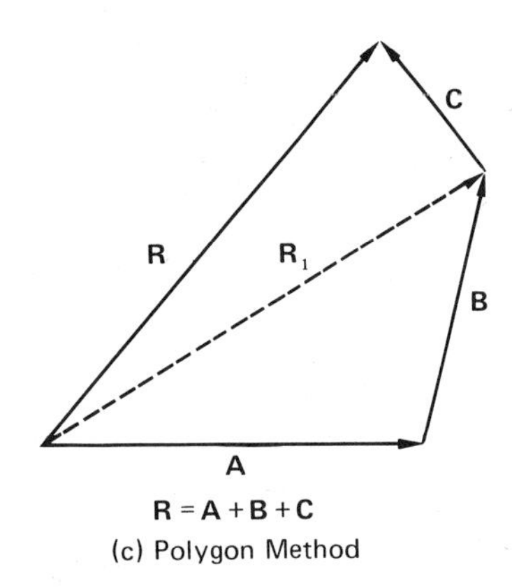

(c) Polygon Method

Fig. 6-1 Methods of vector addition. See the text for a description.

POLYGON METHOD

If more than two vectors are added, the head-to-tail method forms a polygon [Fig. 6-1(c) for three vectors]. The resultant $\mathbf{R} = \mathbf{A} + \mathbf{B} + \mathbf{C}$ is the vector arrow from the tail of the **A** arrow to the head of the **C** vector. The length (magnitude) and the angle of orientation of **R** can be measured from the vector diagram. Note that this is equivalent to applying the head-to-tail method (two vectors) twice (i.e., **A** and **B** are added to give $\mathbf{R}_1$ and then **C** is added to $\mathbf{R}_1$ to give **R**).

The magnitude (length) R and the orientation angle θ of the resultant vector **R** in a graphical method can be measured directly from the vector diagram using a ruler and a protractor. For example, suppose that the vectors **A** and **B** represent forces, in particular, weight forces ($\mathbf{F} = m\mathbf{g}$) and a length (L) of 0.10 m (or 10 cm) on the vector diagram represents the weight of a mass of 0.50 kg. Then, in terms of the force magnitude ($F = mg = 0.50g$ newtons), the length scale or scaling factor is $F/L = 0.50g/0.10 = 5.0g$ newtons per meter (N/m) in SI units. *Note:* it is convenient to leave g, the acceleration due to gravity (9.8 m/s^2) in symbol form so as to avoid numerical calculations until necessary. Also, the masses of laboratory "weights" are usually stamped in grams. Should your instructor allow you the option of using these non-SI units and if you so choose, then the scaling factor for the weight force of a 500 gm mass would be $F/L = 500g/10$ cm $= 50g$ dyn/cm.* Using g in symbol form here avoids large numbers since $g = 980$ cm/s^2.

Then, having computed the scaling factor, if the length of **R** is measured to be 0.060 m, the magnitude of R is

$$\begin{aligned} R &= \text{(scaling factor)(measured length)} \\ &= (5.0g \text{ N/m})(0.060 \text{ m}) = 0.30g \text{ N } (= 2.9 \text{ N}) \end{aligned}$$

Analytical

TRIANGLE METHOD

The magnitude of **R** (Fig. 6-1b) can be computed from the law of cosines if the angle λ (*angle opposite* **R**) is known. Knowing λ and with the given magnitudes of **A** and **B**,

$$R^2 = A^2 + B^2 - 2AB\cos\lambda \qquad \textbf{(6-1)}$$

The angle θ (the angle between **R** and **A**) can then be computed using the law of sines because the magnitudes of sides **B** and **R** are known, and

$$\frac{B}{\sin\theta} = \frac{R}{\sin\lambda}$$

For example, suppose that the magnitudes of **A** and **B** are $3.0g$ N and $2.0g$ N, respectively, and λ was known (perhaps measured) to be 125°. Then,

$$\begin{aligned} R^2 &= (3.0g)^2 + (2.0g)^2 - 2(3.0g)(2.0g)\cos 125^\circ \\ &= 9.0g^2 + 4.0g^2 + 12g^2\cos 55^{\circ\dagger} \\ &= 20g^2 \end{aligned}$$

and

$$R = 4.5g = (4.5)(9.8) = 44 \text{ N}$$

Then the angle θ can be found using the law of sines:

$$\theta = \sin^{-1}\left(\frac{B\sin\lambda}{R}\right) = \sin^{-1}\left(\frac{2.0g\sin 125^\circ}{4.5g}\right) = 21^\circ$$

*It is unfortunate that the commonly used symbol for the acceleration due to gravity (g) and the official abbreviation for gram (g) are the same. To avoid confusion, g will be used for the acceleration due to gravity and gm for gram.

†$\cos 125^\circ = \cos(180^\circ - 55^\circ) = -\cos 55^\circ$, from the trigonometric identity $\cos(A - B) = \cos A\cos B + \sin A\sin B$.

Remember that this is the angle between vectors **R** and **A**.

COMPONENT METHOD

If two vectors **A** and **B** are at right (90°) angles [Fig. 6-2(a)], then the magnitude of their resultant is given by the **Pythagorean theorem**, $R = \sqrt{A^2 + B^2}$ (the hypotenuse of a right triangle is equal to the square root of the sum of the squares of the legs of the triangle). Notice that the law of cosines reduces to this formula with $\lambda = 90°$ (because $\cos 90° = 0$). The angle of orientation is given by $\tan \theta = B/A$ or $\theta = \tan^{-1} B/A$.

By a somewhat inverse process, we may resolve a vector into x and y components [Fig. 6-2(b)]. That is, the vector **R** is the resultant of $\mathbf{R}_x$ and $\mathbf{R}_y$, and $\mathbf{R} = \mathbf{R}_x + \mathbf{R}_y$, where $R_x = R \cos \theta$ and $R_y = R \sin \theta$. The magnitude of R is given by $R = \sqrt{R_x^2 + R_y^2}$ and $\tan \theta = R_y/R_x$ or $\theta = \tan^{-1} R_y/R_x$.

The vector sum of any number of vectors can be obtained by adding the x and y components of the vectors. This is illustrated in Fig. 6-3. The magnitude of the resultant is given by $R = \sqrt{R_x^2 + R_y^2}$ where $\mathbf{R}_x = \mathbf{A}_x + \mathbf{B}_x$ and $\mathbf{R}_y = \mathbf{A}_y + \mathbf{B}_y$, and $\theta = \tan^{-1} R_y/R_x$.

However, rather than placing the vectors head to tail as in Fig. 6-3, it is usually more convenient to have all the vectors to be added to originate from the origin when resolving into components (Fig. 6-4). Applying this method analytically to $\mathbf{R} = \mathbf{A} + \mathbf{B} + \mathbf{C}$ in Fig. 6-4, we have

$$\begin{aligned} R_x &= A_x + B_x + C_x \\ &= 6 \cos 60° + 0 - 10 \cos 30° \\ &= -5.7 \\ R_y &= A_y + B_y + C_y \\ &= 6 \sin 60° + 5 - 10 \sin 30° \\ &= 5.2 \end{aligned}$$

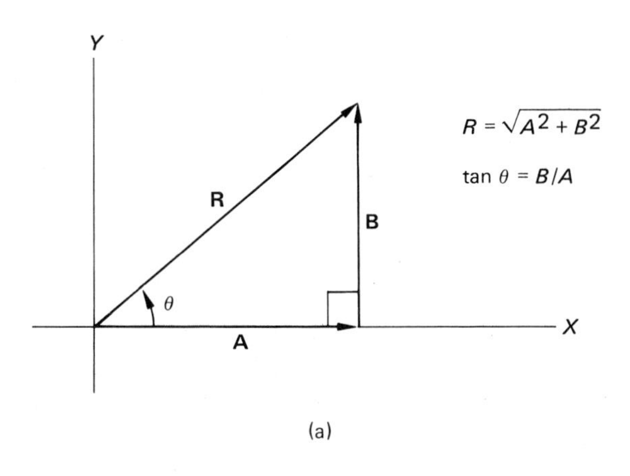

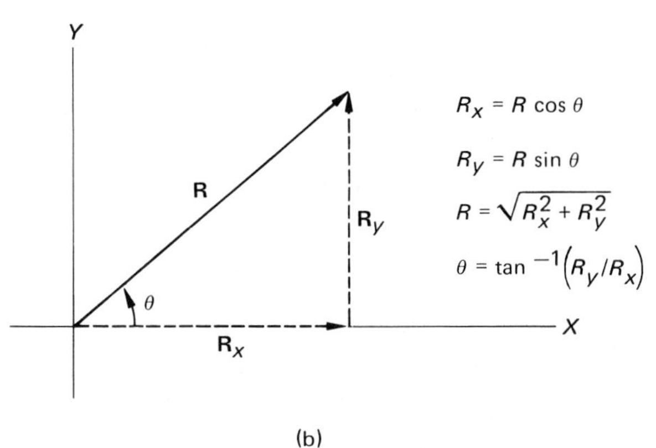

Fig. 6-2 (a) The vector addition of **A** and **B** gives the resultant **R**. (b) A vector **R** can be resolved into x and y components, $\mathbf{R}_x$ and $\mathbf{R}_y$, respectively.

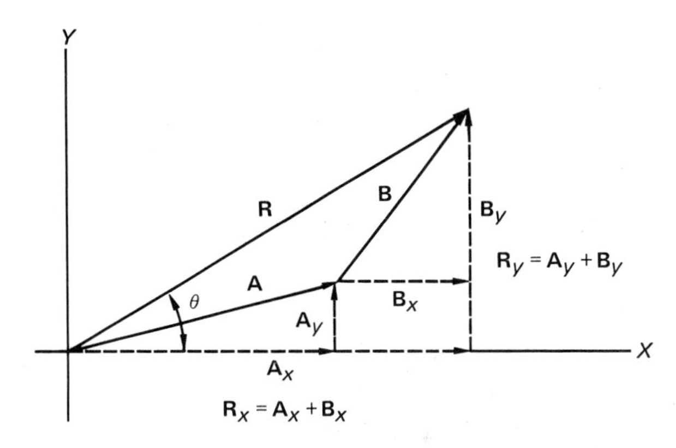

Fig. 6-3 The resultant or vector sum of any number of vectors can be obtained by adding the x and y components of the vectors, as illustrated here for two vectors.

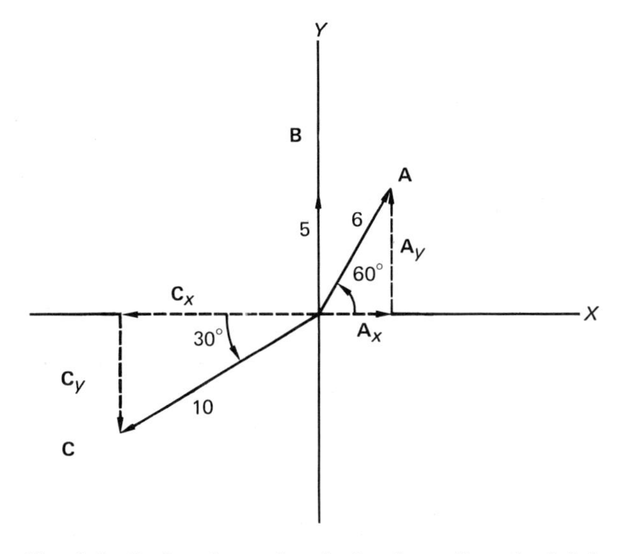

Fig. 6-4 Rather than using the head-to-tail method, it is usually more convenient to have all vectors originating from the origin when resolving into components. See the text for a description.

where the directions are indicated by positive and negative signs. Note B has no x component and C_x and C_y are in the negative x and y directions, as indicated by the minus signs. Then the magnitude of R is

$$R = \sqrt{R_x^2 + R_y^2} = \sqrt{(-5.7)^2 + (5.2)^2} = 7.7$$

and

$$\theta = \tan^{-1}\left|\frac{R_y}{R_x}\right| = \tan^{-1}\frac{5.2}{5.7} = 42°$$

relative to the negative X axis (or $180° - 42° = 138°$ relative to the $+X$ axis). The minus R_x and positive R_y indicate that the resultant is in the second quadrant.

Experimental

THE FORCE TABLE

The **force table** is an apparatus that allows the experimental determination of the resultant of force vectors (Fig. 6-5). The rim of the circular table is calibrated in degrees. Weight forces are applied to a central ring by means of strings running over pulleys and attached to weight hangers. The magnitude of a force (vector) is varied by adding or removing slotted weights and the direction is varied by moving the pulley.

The resultant of two or more forces (vectors) is found by balancing the forces with another force (weights on a hanger) so that the ring is centered around the central pin. The balancing force is *not* the resultant **R**, but rather the equilibrant **E**, or the force that balances the other forces and holds the ring in equilibrium.

The equilibrant is the vector force of equal magnitude, but in the opposite direction to that of the resultant (i.e., **R** = −**E**), (Fig. 6-6). For example, if an equilibrant has a magnitude of $0.30g$ N in a direction of 225° on the circular scale, the resultant of the forces has a magnitude of $0.30g$ N in the opposite direction, $225° - 180° = 45°$. It should be evident that the resultant cannot be determined directly from the force table.

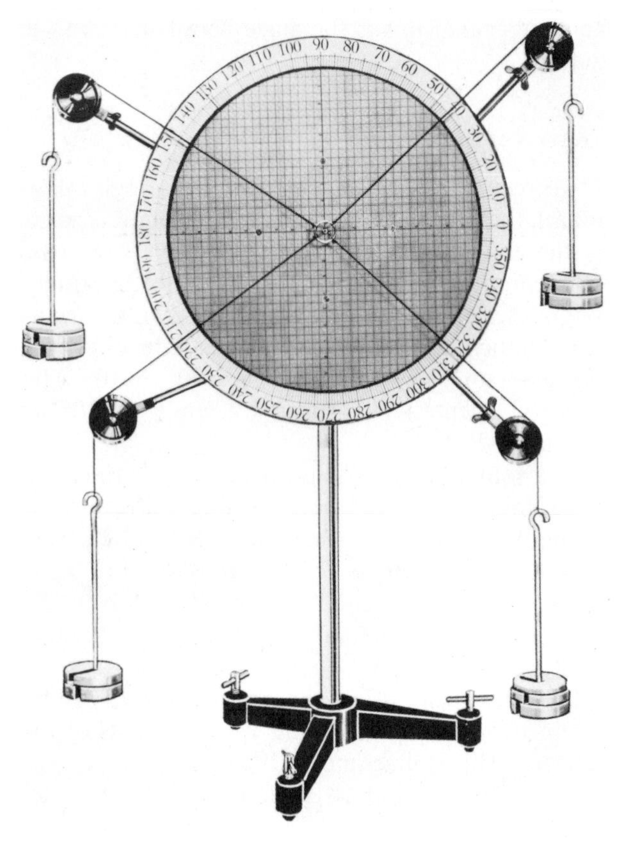

Fig. 6-5 A force table. The circular table is shown here vertically for demonstration. In the laboratory, the table is horizontal. (Photo courtesy of Sargent-Welch Scientific Company)

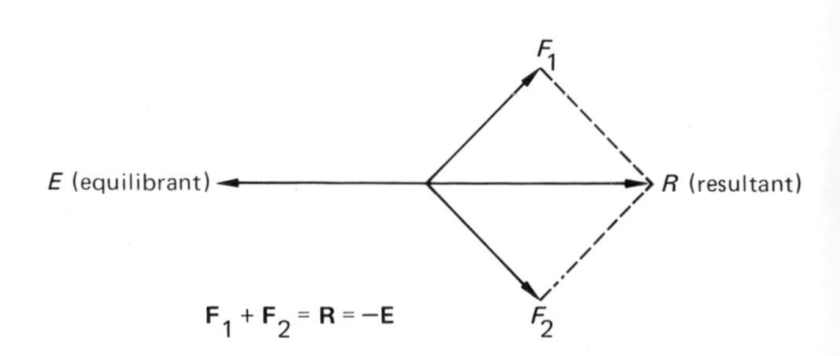

Fig. 6-6 On the force table, the magnitude and direction of the equilibrant **E** rather than the resultant **R** are measured, and **R** = −**E**.

IV. EXPERIMENTAL PROCEDURE

1. Check to see if the force table is level. (Ask your instructor for a level if this has not been done.) Make any necessary adjustments by means of the leveling screws in the tripod base of the table.

2. **Vector addition 1.** Given two vectors $F_1 = 0.20g$ N ($200g$ dyn) at 30° and $F_2 = 0.20g$ N ($200g$ dyn) at 120°, find their vector sum or resultant $\mathbf{F} = \mathbf{F}_1 + \mathbf{F}_2$ by the following procedures. (*Note:* Orientation angles of vectors are measured from the 0° reference line or X axis.)

   (a) *Graphical.* Using the parallelogram method of vector addition, draw a vector diagram to scale. Use a scale such that the finished vector diagram fills about half a sheet of graph paper. Measure the magnitude and direction of the resultant (with ruler and protractor) and record the results in the data table.

(b) *Analytical.* Using the law of cosines, compute the magnitude of the resultant force. Compute the angle of orientation from the relationship $\tan\theta = F_2/F_1$. (Why can you use $\tan\theta$? Remember that θ is the angle between **F** and $\mathbf{F}_1$.) Record the results in the data table.

(c) *Experimental.* On the force table, clamp pulleys at 30° and 120° and add enough weights to each weight hanger to total 0.20 kg so as to give weight forces of $F_1 = F_2 = 0.20g$ N in these directions. (The weight hangers usually have masses of 50 gm.)

Using a third pulley and weights, determine the magnitude and direction of the equilibrant force that maintains the central ring centered in equilibrium around the center pin. Record the magnitude and direction of the *resultant* of the two forces in the data table. Remember, the resultant has the same magnitude as the equilibrant but is in the opposite direction. (*Note:* The string knots on the central ring should be of a non-tightening variety so that the strings will slip freely on the ring and allow the strings to pull directly away from the center. When the forces are balanced, the pin may be carefully removed to see if the ring is centered around the central hole.)

3. **Vector addition 2.** Repeat procedure 2 for $F_1 = 0.20g$ N at 20° and $F_2 = 0.15g$ N at 80°. Use the other half sheet of graph paper from Case 1 for the graphical analysis. Be careful in the analytical analysis. Can you use $\tan\theta = F_2/F_1$ in this case?

4. **Vector addition 3.** Repeat procedure 2 with $F_1 = F_x = 0.20g$ N (at 0°) and $F_2 = F_y$ (at 90°). In this case, $\mathbf{F} = \mathbf{F}_x + \mathbf{F}_y$ where $\mathbf{F}_x$ and $\mathbf{F}_y$ are the x and y components of **F**, respectively. That is, the resultant can be resolved into these components. Use one-half of another sheet of graph paper for the graphical method.

5. **Vector resolution.** Given a force vector of $F = 0.30g$ N at 60°, resolve the vector into its x and y components and find the magnitudes of $\mathbf{F}_x$ and $\mathbf{F}_y$ by the following procedures:

(a) *Graphical.* Draw a vector diagram to scale (on the other half sheet of graph paper used in Case 3) with the component vectors [see Fig. 6-2(b)] and measure the magnitudes of $\mathbf{F}_x$ and $\mathbf{F}_y$. Record the results in the data table.

(b) *Analytical.* Compute the magnitudes of $\mathbf{F}_x$ and $\mathbf{F}_y$ (see Section III). Record the results in the data table.

(c) *Experimental.* Clamp pulleys at 240°, 90°, and 0° on the force table. Place a *total* of 0.30 kg on the 240° pulley string using a weight hanger. This force is then the equilibrant of $F = 0.30g$ N at 60°(60° + 180° = 240°), which must be used on the force table rather than the force itself. Add weights to the 0° and 90° hangers until the system is in equilibrium. The 0° and 90° forces are then the $\mathbf{F}_x$ and $\mathbf{F}_y$ components, respectively, of **F**. Record their magnitudes in the data table.

6. **Vector addition 4.** Given the force vectors $F_1 = 0.10g$ N at 30°, $F_2 = 0.20g$ N at 90°, and $F_3 = 0.30g$ N at 225°, find the magnitude and direction of their resultant $\mathbf{F} = \mathbf{F}_1 + \mathbf{F}_2 + \mathbf{F}_3$ by the following procedures:

(a) *Graphical.* Use the polygon method.

(b) *Analytical.* Use the component method.

(c) *Experimental.* Use the force table.

Record the results in the data table.

LABORATORY REPORT

DATA TABLE

	Forces (N)	Resultant (magnitude and direction)		
		Graphical	Analytical*	Experimental
Vector addition 1	$F_1 = 0.20g, \theta_1 = 30°$ $F_2 = 0.20g, \theta_2 = 120°$			
Vector addition 2	$F_1 = 0.20g, \theta_1 = 20°$ $F_2 = 0.15g, \theta_2 = 80°$			
Vector addition 3	$F_1 = F_x = 0.20g, \theta_1 = 0°$ $F_2 = F_y = 0.15g, \theta_2 = 90°$			
Vector resolution	$F = 0.30g, \theta = 60°$	F_x F_y	F_x F_y	F_x F_y
Vector addition 4	$F_1 = 0.10g, \theta_1 = 30°$ $F_2 = 0.20g, \theta_2 = 90°$ $F_3 = 0.30g, \theta_3 = 225°$			

*Show analytical calculations below.

Calculations (show work)

QUESTIONS

1. Considering the graphical and analytical methods to give the true vector resultant in each case, which method is more accurate? Give the probable sources of error for each method.

2. Vector subtraction (e.g., **A** − **B**) is a special case of vector addition [i.e., **A** − **B** = **A** + (−**B**)]. Suppose that the cases of Vector Addition 1, 2, and 3 in this experiment were vector subtraction (i.e., $\mathbf{F}_1 - \mathbf{F}_2$).
   (a) What effect would this have on the directions of the resultants? (Do not calculate explicitly. Simply state in which quadrant the resultant would be in each case.)

   (b) Would the magnitude of the resultant be different for vector subtraction than for vector addition in each case? If so, state whether the subtractive resultant would be greater or less than the additive resultant.

3. A picture hangs on a nail as shown in Fig. 6-Q3. If the tension T in each string segment is 3.5 N (newtons), (a) what is the equilibrant or the upward reaction force of the nail? (b) What is the weight of the picture?

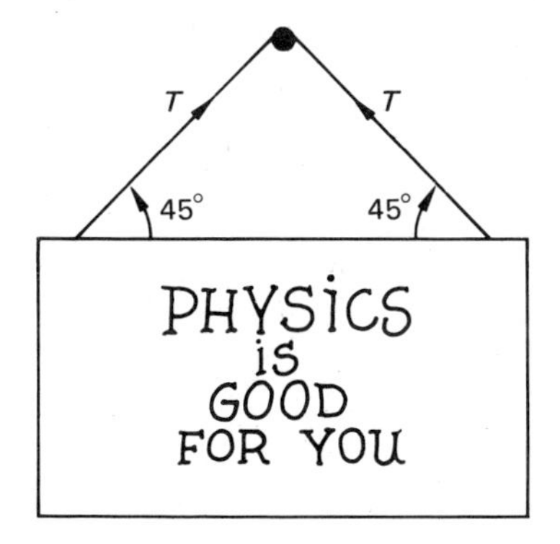

Fig. 6-Q3

4. A hospital traction device is shown in Fig. 6-Q4. What is (a) the horizontal force; (b) the vertical force exerted on the foot by the arrangement? (The cord transmits an undiminished tension T of 10 lb.)

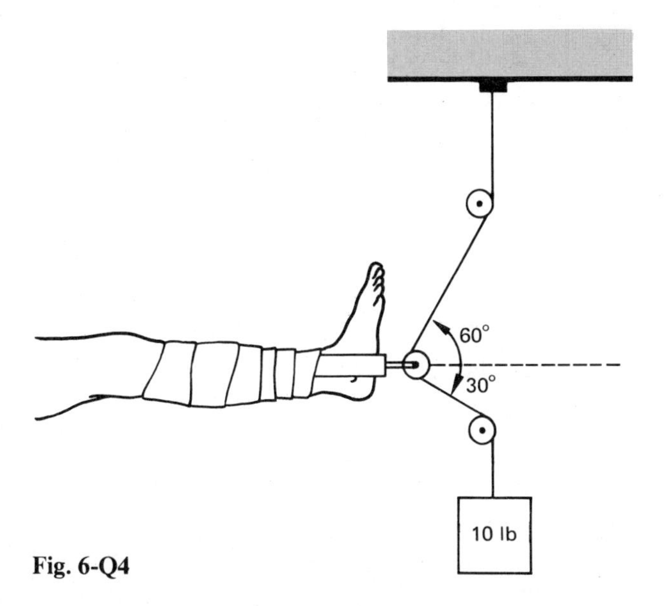

Fig. 6-Q4

Name .... Section .... Date ....

Lab Partner(s) ....

EXPERIMENT 7 *Newton's Second Law: The Atwood Machine*

ADVANCE STUDY ASSIGNMENT

Read the experiment and answer the following questions.

1. Give Newton's second law in mathematical form, and describe how the acceleration of an object or system varies with the unbalanced force and mass of the system.

2. What is the advantage of using an Atwood machine in experimentally measuring the acceleration of a system?

3. How does the frictional force of the pulley affect the acceleration of the system?

4. How can the frictional force be experimentally determined and how is this used in calculations?

5. What is F and m in Newton's second law in terms of the Atwood machine?

6. How is the acceleration of the Atwood system determined experimentally?

EXPERIMENT 7

Newton's Second Law: The Atwood Machine

I. INTRODUCTION

Newton's second law of motion states that the acceleration a of an object or system is directly proportional to the unbalanced or net force F_{net} acting on the system and inversely proportional to the total mass m of the system (i.e., $a \propto F_{net}/m$), or in equation form with standard units, $a = F_{net}/m$ (or $F_{net} = ma$). This relationship will be investigated using an *Atwood machine,* which consists of two masses connected by a string looped over a pulley (see Fig. 7-1). The Atwood machine was originally used, circa 1827, before the development of

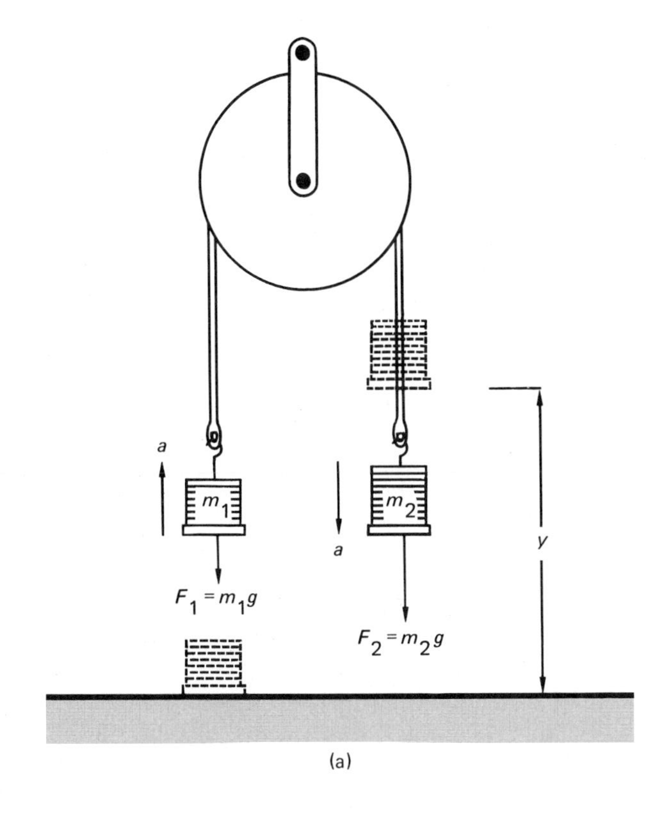

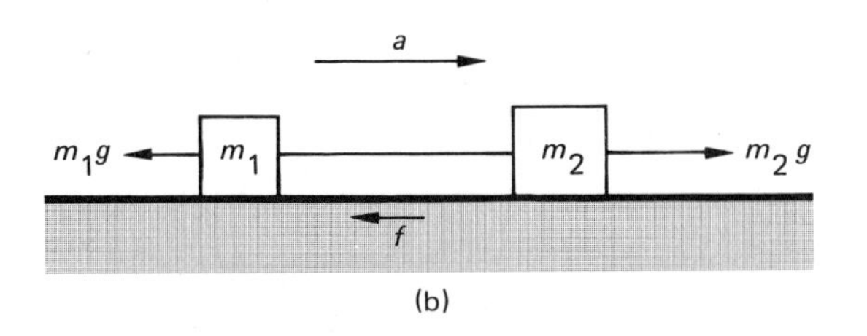

(c)

Fig. 7-1 The Atwood machine. The dynamics of the Atwood machine (a) are shown in diagram (b) in a horizontal position for convenience of analysis. (The fixed pulley is simply a "direction changer.") The force f represents the friction of the pulley.

fast-timing devices to determine the acceleration due to gravity g by experimentally measuring the relatively slow uniform acceleration of the system. By varying the unbalanced weight force and the total mass of the system, the resulting accelerations will be determined experimentally from the distance-time measurements and compared with the predictions of Newton's law.

II. EQUIPMENT NEEDED

- Pulley (preferably precision ball-bearing type with large diameter)
- Clamps and support rods
- Two weight hangers
- Set of slotted weights, including small 5-, 2-, and 1-gram weights
- Paper clips
- String
- Laboratory timer or stopwatch
- Meter stick
- 2 sheets of Cartesian graph paper

III. THEORY

The light string is considered to be of negligible mass. With masses m_1 and m_2 as the ascending and descending sides of the system, respectively (Fig.7-1), the unbalanced (net) force is

$$F_{\text{net}} = m_2 g - m_1 g = (m_2 - m_1)g \quad \textbf{(7-1)}$$

where the friction and inertia of the pulley are neglected.

By Newton's second law,

$$F_{\text{net}} = ma = (m_1 + m_2)a \quad \textbf{(7-2)}$$

where $m = m_1 + m_2$ is the total mass of the moving system. Then, equating Eqs. 7-1 and 7-2, we have

$$(m_2 - m_1)g = (m_1 + m_2)a$$

or

$$a = \frac{(m_2 - m_1)g}{m_1 + m_2} \quad \textbf{(7-3)}$$

In the experimental arrangement, there will be an appreciable frictional force f associated with the pulley that opposes the motion. Also, the pulley has inertia. This inertia can be taken into account by adding an equivalent mass m_{eq} to the total mass being accelerated. Hence, for better accuracy, the equation for the acceleration of the system should be modified as follows:

$$F_{\text{net}} = F - f = (m_2 - m_1)g - f$$
$$= ma = (m_1 + m_2 + m_{\text{eq}})a$$

or

$$a = \frac{(m_2 - m_1)g - f}{m_1 + m_2 + m_{\text{eq}}} \quad \textbf{(7-4)}$$

If a mass system of the Atwood machine moves with a constant velocity, the acceleration a of the system is zero, and

$$a = 0 = \frac{(m_2 - m_1)g - f}{m_1 + m_2 + m_{\text{eq}}}$$

or

$$f = (m_2 - m_1)g = m_f g \quad \text{(uniform velocity)} \quad \textbf{(7-5)}$$

which provides a method of determining the magnitude of the frictional force of the pulley, or the mass m_f needed to balance the frictional force.

Hence, the expression for the theoretical acceleration of the system (Eq. 7-4) may be written

$$a_t = \frac{(m_2 - m_1 - m_f)g}{m_1 + m_2 + m_{\text{eq}}} \quad \textbf{(7-4a)}$$

where a_t is used to distinguish from the experimentally measured acceleration a_m. Thus, part of the weight of m_2 goes into balancing or canceling the frictional force of the pulley. In the experimental acceleration trials, the m_f determined in each case is left on the descending hanger as part of m_2.

To experimentally determine the acceleration of the system so that it may be compared to that predicted by Eq. 7-4a, the time t for the descending mass to fall through a given distance y is measured. Then using the kinematic equation

$$y = v_0 t + \tfrac{1}{2}at^2$$

with the mass starting from rest ($v_0 = 0$), we have

$$y = \tfrac{1}{2}at^2$$

or

$$a_m = \frac{2y}{t^2} \quad \text{(7-6)}$$

where a_m is the experimentally measured acceleration.

When a_m is determined experimentally using distance-time measurements, friction and pulley inertia are involved. These are taken into account in the theoretical expression (Eq. 7-4a), so the experimental and theoretical values of a will be more comparable.

IV. EXPERIMENTAL PROCEDURE*

1. Set up the Atwood machine as shown in Fig. 7-1. Use enough string so that the distance of travel (y) is slightly less than one meter for convenient measuring. (To measure y, hold one hanger against the floor and measure from the floor to the bottom of the other hanger.)

   As noted in the Theory section, the pulley contributes to the inertia of the system as though an "equivalent mass," m_{eq}, were part of the total mass being accelerated. The instructor will provide the value of m_{eq} or tell you how to measure it. Record the value of m_{eq} in the Data Table.

A. *Varying the Total Mass (unbalanced force constant)*

2. Begin with the descending mass (m_2) and the ascending mass (m_1) equal to 50 grams, i.e., the masses of the hangers alone. With $m_1 = m_2$, the system is in equilibrium — equal forces, $m_1 g = m_2 g$. In the absence of friction, a slight tap or momentary force applied to m_2 should set the system in uniform motion (constant speed). Why? However, because of the opposing frictional force, the motion will not persist.

3. Add small masses to m_2 until a downward push causes m_2 to descend with a uniform (constant) velocity. (See "Comments on Experimental Technique" at the end of the Procedure section.) Apply a sufficient push so the masses move at a reasonable speed, not too slowly. You may find it easier to recognize uniform motion by observing the rotating pulley rather than the masses. Record m_1 and m_2 in Data Table 1 in the (*) column. These values are used to calculate the frictional mass, $m_f = m_2 - m_1$, needed in the theoretical calculation of the acceleration of the system (Eq. 7-4a).

4. (a) Add 10 grams to m_2, leaving m_f in place. This creates an unbalanced force that should cause the system to accelerate from rest. Measure the distance y. Record y, m_1, and the new value of m_2 in the Data Table (Trial 1).†
   (b) Make three independent measurements of the time it takes for m_2 to travel the distance y from rest. Record the data under Trial 1. (See "Comments on Experimental Technique" at the end of the Procedure section.)
   (c) Remove m_f and the 10-gram mass before proceeding to the next Trial.

5. (a) Add 100 grams to each hanger for a total of 150 gm each.
   (b) Repeat procedure 3 (measurement of frictional mass) and record data in the next (*) in Data Table 1.
   (c) Repeat procedure 4 (measurement of acceleration with a net 10-gm mass imbalance). Record the data in the Trial 2 column. The calculations for Trial 2 should utilize the value of m_f obtained for the immediately preceding (*) column. *Note:* The values of m_f and y should be remeasured for each of the trials in Data Table 1. As the total mass is changed, the friction will change likewise. Even the length of the string (y distance) may vary noticeably.

6. Repeat procedures 3 and 4 with another 100 grams added to each hanger for a total of 250 grams each. (The mass of m_1 should be 250 grams for this Trial.)

7. Repeat procedures 3 and 4 with another 100 grams added to each hanger for a total of 350 grams each. (The mass of m_1 should be 350 grams for this Trial.)

*The refinements in this Experimental Procedure were developed by Professor I. L. Fischer, Bergen Community College.

†The Data Tables are arranged to facilitate data-taking and analysis. The upper (seven) rows include all the experimental measurements, while the lower (six) rows are for calculations based on these measurements.

B. *Varying the Unbalanced Force (total mass constant)*

8. Begin with an ascending mass $m_1 = 260$ grams (50-gram hanger + 200 + 5 + 2 + 2 + 1-gram masses) and a similar descending mass $m_2 = 260$ grams (50-gram hanger + 200 + 10 gram masses).

9. Measure the frictional mass as in procedure 3. Record the data in the (*) column in Data Table 2. The value of m_f from these data may be used in the calculations for all trials in Data Table 2, since the total mass (and presumably the friction) will now be constant.

10. Leaving m_f in place, transfer 1 gram from m_1 to m_2 in order to create a net unbalanced force without affecting the total mass. Make three measurements of the travel time as in procedure 4. Record all pertinent data in the Trial 5 column.

11. Leaving m_f and the previously transferred 1-gm mass in place,
    (a) transfer an additional 2 grams for Trial 6;
    (b) transfer an additional 2 grams for Trial 7;
    (c) transfer an additional 5 grams for Trial 8. (The final net unbalanced mass, $m_2 - m_1 - m_f$, should be 20 grams.)

C. *Comments on Experimental Technique*

12. Measure the frictional mass to a precision of $\pm\frac{1}{2}$ gram. Fine adjustment of the descending mass may be made by using small "custom" masses (paper clips) as needed. These paper clips can be attached to the cord just above the m_2 hanger. Good precision is necessary for good results because the frictional force is comparable in magnitude to the accelerating force. Small errors in the frictional masses may create large experimental errors.

13. The masses must start from rest during the acceleration trials. A good technique is as follows:
    (a) Hold m_1 down against the floor;
    (b) Simultaneously release m_1 and start the timer;
    (c) Stop the timer at the instant m_2 strikes the floor.

    The best results are obtained when the same person releases m_1 and operates the timer.

14. One lab partner should stop the rotating pulley as soon as the timing is completed. Otherwise, masses may be jolted off the hangers by the impact. It may be helpful to place a shock-absorbing pad on the floor.

15. Take turns at each task.

Name .... Section .... Date ....

Lab Partner(s) ....

EXPERIMENT 7 *Newton's Second Law: The Atwood Machine*

LABORATORY REPORT

m_{eq} ________ (gm)	DATA TABLE 1 Varying the Total Mass (unbalanced force constant)								DATA TABLE 2 Varying the Unbalanced Force (total mass constant)				
	Trial								Trial				
	*	1	*	2	*	3	*	4	*	5	6	7	8
Descending mass m_2 (gm)													
Ascending mass m_1 (gm)													
Distance of travel y (cm)													
Time of travel t (s) — Run 1													
Run 2													
Run 3													
Average													
Measured acceleration $a_m = 2y/t^2$ (cm/s^2)													
Total mass $= m_1 + m_2 + m_{eq}$													
Measured frictional mass $m_f = m_2 - m_1$													
Net force (dynes) $= (m_2 - m_1 - m_f)g$													
Theoretical acceleration $a_t = \frac{\text{Net force}}{\text{total mass}}$													
Percent difference between a_m and a_t													

*Measurement of frictional mass, m_f. Masses move with constant velocities when given an initial push.

(continued)

QUESTIONS

1. In the experiment, should the mass of the string be added to the total mass moved by the unbalanced force for better accuracy? Explain.

2. Complete the following sentences:
   (a) When the unbalanced force increases (total mass constant), the acceleration of the system

   ....

   (b) When the total mass that is accelerating increases (unbalanced force constant), the acceleration of the system ....

3. Using the data in Data Table 2 (constant total mass), plot a_m versus $(m_2 - m_1)$ for each trial and draw a straight line that best fits the data. Find the slope and intercept of the line and enter the values below.
   Rewrite Eq. 7-4a in slope-intercept form ($y = ax + b$), and using the data in Trial 6, compute the slope and intercept. (Show calculations.) Compare and comment on your results.

	(From graph)	(From Eq. 7-4a)
Slope	.... (units)	.... (units)
Intercept	.... (units)	.... (units)

4. Using the data in Data Table 1 (constant unbalanced force), plot $1/a_m$ versus $(m_1 + m_2)$ for each trial and draw a straight line that best fits the data. Perform the same kind of analysis as in Question 3 using the data in Trial 2 for computing the slope and intercept.

	(From graph)	(From Eq. 7-4a)
Slope	.... (units)	.... (units)
Intercept	.... (units)	.... (units)

Name .... Section .... Date ....

Lab Partner(s) ....

EXPERIMENT 8 *Conservation of Linear Momentum: The Air Track*

ADVANCE STUDY ASSIGNMENT

Read the experiment and answer the following questions.

1. What is the condition for the conservation of linear momentum of a system?

2. Is linear momentum conserved in common applications? Explain.

3. What is meant when it is said that a quantity, such as linear momentum, is conserved?

4. Is the conservation of linear momentum consistent with Newton's first law of motion? Explain.

5. In a system of particles for which the total linear momentum is conserved, is the linear momentum of the individual particles constant? Explain.

6. Suppose that a particle of mass m_1 approaches a stationary mass m_2 and $m_2 >> m_1$.

   (a) Describe the velocity of m_2 after an elastic collision; that is, is there a transfer of momentum? Justify your answer mathematically.

   (b) What is the approximate momentum of m_1 after collision?

EXPERIMENT **8**

Conservation of Linear Momentum: The Air Track

I. INTRODUCTION

The conservation of linear momentum ($\mathbf{p} = m\mathbf{v}$) is an important physical concept. However, the experimental investigation of this concept in an introductory physics laboratory is hampered by ever-present frictional forces.

An air track provides one of the best methods to investigate linear momentum (see Fig. 5-2). Aluminum gliders riding on a cushion of air on the track approaches frictionless motion — a necessary condition for the conservation of linear momentum.

In the absence of friction, the total linear momentum of a system of two gliders should be conserved during a collision of the gliders. That is, the total linear momentum of the system should be the same after collision as before collision. By measuring the velocities of gliders of the same and different masses before and after collision, the total momentum of a system can be determined and the conservation of linear momentum investigated.

II. EQUIPMENT NEEDED

- Air track
- Three gliders (two of similar mass)
- Four laboratory timers
- Laboratory balance
- Masking tape
- Meter stick (if length scale not on air track)

III. THEORY

The linear momentum $\mathbf{p}$ of a particle or object is defined as

$$\mathbf{p} = m\mathbf{v} \tag{8-1}$$

where m is the mass of the object and $\mathbf{v}$ its velocity. Since velocity is a vector quantity, so is linear momentum.

Newton's second law of motion, commonly expressed in the form $\mathbf{F} = m\mathbf{a}$, can also be written in terms of momentum:

$$\mathbf{F} = \frac{\Delta \mathbf{p}}{\Delta t} \tag{8-2}$$

If there is no net or unbalanced external force acting on the object ($F = 0$), then

$$\mathbf{F} = \frac{\Delta \mathbf{p}}{\Delta t} = 0$$

or

$$\Delta \mathbf{p} = 0$$

That is, the change in the momentum is zero, or the momentum is conserved. By "conserved" it is meant that the momentum remains constant (in time).

$$\Delta \mathbf{p} = \mathbf{p}_f - \mathbf{p}_i = 0$$

or

$$\mathbf{p}_f = \mathbf{p}_i \tag{8-3}$$

and the "final" momentum $\mathbf{p}_f$ at any time t_f is the same as the initial momentum $\mathbf{p}_i$ at time t_i.

Notice that this is consistent with Newton's first law of motion, since

$$\mathbf{p}_f = \mathbf{p}_i \quad \text{or} \quad m\mathbf{v}_f = m\mathbf{v}_i$$

$$\mathbf{v}_f = \mathbf{v}_i$$

That is, an object remains at rest ($\mathbf{v}_i = 0$) or in uniform motion ($\mathbf{v}_i = \mathbf{v}_f$) unless acted on by some external force.

The previous development also applies to the total momentum of a system of particles or objects. For example, the total linear momentum of a system of two objects m_1 and m_2 is $\mathbf{P} = \mathbf{p}_1 + \mathbf{p}_2$, and if there is no net external force acting on the system, then

$$\Delta\mathbf{P} = 0$$

In the case of a collision between the two objects (with no net external force acting), the initial momentum before the collision is the same as the final momentum after the collision. That is,

$$\overset{\textit{Before}}{\mathbf{p}_{1_i} + \mathbf{p}_{2_i}} = \overset{\textit{After}}{\mathbf{p}_{1_f} + \mathbf{p}_{2_f}} \tag{8-4}$$

or

$$m_1\mathbf{v}_{1_i} + m_2\mathbf{v}_{2_i} = m_1\mathbf{v}_{1_f} + m_2\mathbf{v}_{2_f}$$

In one dimension, the directions of the velocity and momentum vectors are commonly indicated by plus and minus signs.

IV. EXPERIMENTAL PROCEDURES

1. Determine the mass of each glider and record in the Trial Data Table. Let the masses of the two gliders of nearly equal mass be m_1 and m_2 and the mass of the third glider be m_3.

2. With pieces of tape, mark off two equal and convenient lengths (e.g., $\frac{1}{2}$ or 1 m) on both sides of the center position of the air track. Make full use of the length of the track, but leave some space near the ends of the track. Place the four tape reference marks at the lower edges of the track so as not to interfere with the glider motion.

3. *Time trials.* By measuring the time interval Δt it takes a glider to move the reference mark length d, the velocity of the glider can be determined, $v = d/\Delta t$, where $\Delta t = t_2 - t_1$. The actual timing of the motion of a glider moving between the two sets of reference marks is done by either method (A) four observers, each with a timer and assigned to an individual reference mark, or method (B) two observers, each with a timer and assigned to one set of reference marks, as described below.

   In addition to giving timing practice and determining the better method of timing, the time trials check out the experimental setup for possible systematic errors. The time intervals for the individual gliders to travel the equal distances between the reference marks should be very similar for any one trial. If not, the air track may need leveling and/or there may be some frictional problem with part of the track. Should this be the case, call your instructor. *Do not* attempt to level the air track on your own.

   Experimentally carry out each of the following timing methods to determine which is better.

   Method A. Set one of the gliders in motion with a *slight* push so that it moves with moderate speed up and down the track. (A few practice starts help.) As the glider hits the bumper at one end of the track, all four observers should start their timers. As the leading edge of the glider passes the reference marks, each respective observer stops his or her timer. (A dry run or two to become familiar with the timing sequence is helpful.) Carry out this procedure two times for each of the three gliders and record the data in the Trial Data Table.

   Method B. Set the glider into motion. The two observers should start and stop their individual timers as the leading edge of the glider passes their respective reference marks. Carry out this procedure two times for each of the three gliders and record the data in the Trial Data Table.

4. Compute the Δt's for each trial and calculate the percent difference for each trial set. From the data, decide which timing method should be used based on consistency or precision.

CASE 1: COLLISION BETWEEN TWO GLIDERS OF (NEARLY) EQUAL MASS, WITH ONE INITIALLY AT REST

5. With one of the gliders (m_2) of nearly equal mass stationary at the center position of the air track, start the other glider (m_1) moving toward the stationary glider. See Fig. 8-1. (It may be more convenient to start m_1 moving away from m_2 and take measurements as m_1 returns from rebound-

ing from the end of the track.) A trial run should show that m_1 remains at rest after collision and that m_2 is in motion. Determine the time it takes for m_1 to travel between the reference marks as it approaches m_2 and the time it takes for m_2 to travel between the other set of reference marks after collision. Carry out this procedure three times and record the data in Data Table 1. Compute the total momentum before and after collision and the percent difference in these values for each trial.

CASE 2: COLLISION BETWEEN TWO GLIDERS OF UNEQUAL MASS, WITH THE MORE MASSIVE GLIDER INITIALLY AT REST

6. Repeat procedure 5 with m_2 replaced by m_3 (more massive than m_1 and m_2). See Fig. 8-1. In this case, m_1 will travel in the opposite direction after collision, as a trial run will show. Make appropriate adjustments in the timing procedure to measure the velocity of m_1 before *and* after collision. Record the data and the required calculations in Data Table 2.

CASE 3: COLLISION BETWEEN TWO GLIDERS OF (NEARLY) EQUAL MASS INITIALLY TRAVELING IN OPPOSITE DIRECTIONS

7. With m_1 and m_2 initially moving toward each other (Fig. 8-1), determine the total momentum before and after collision. Make appropriate adjustments in the timing procedure to measure the velocities of m_1 and m_2 before and after collision. Carry out the procedure three times and record the data in Data Table 3. Compute the percent difference for the total momentum before and after collision for each trial.

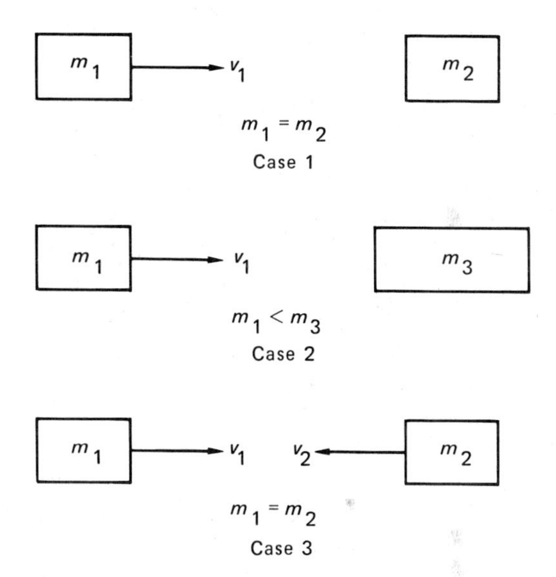

Fig. 8-1 Experimental collision cases.

LABORATORY REPORT

TRIAL DATA TABLE

Glider mass (gm)		Method A							Method B		
		t_1	t_2	Δt_{12}	t_3	t_4	Δt_{34}	Percent diff.	Δt_{12}	Δt_{34}	Percent diff.
m_1											
m_2											
m_3											

(continued)

DATA TABLE 1

Trial	Before collision			After collision			
	m_1			m_2			
	Δt_1 (s)	v_{1_i} (cm/s)	p_1	Δt_2 (s)	v_{2_i} (cm/s)	p_2	Percent diff.
1							
2							
3							

DATA TABLE 2

Trial	Before collision			After collision							Percent diff.
	m_1			m_1			m_3			Total momentum	
	t_{1_i} (s)	v_{1_i} (cm/s)	Total momentum	t_{1_f} (s)	v_{1_f} (cm/s)	p_{1_f}	t_{3_f} (s)	v_{3_f} (cm/s)	p_{3_f}		
1											
2											
3											

DATA TABLE 3

Trial	Before collision							After collision							Percent diff.
	m_1			m_2			Total momentum	m_1			m_2			Total momentum	
	Δt_{1_i}	v_{1_i}	p_{1_i}	Δt_{2_i}	v_{2_i}	p_{2_i}		Δt_{1_f}	v_{2_f}	p_{1_f}	Δt_{2_f}	v_{2_f}	p_{2_f}		
1															
2															
3															

Name .... Section .... Date ....

Lab Partner(s) ....

EXPERIMENT 8 *Conservation of Linear Momentum: The Air Track*

QUESTIONS

1. Do the results of the experiment support the conservation of linear momentum? Consider possible sources of error.

2. Was it necessary to have equal length intervals in the experiment to properly investigate the conservation of momentum? Explain.

3. In Cases 1 and 2 one of the gliders was initially at rest, so it must have received an acceleration. Is the glider accelerating as it passes between the reference marks? Explain.

4. Was the kinetic energy conserved in any of the three cases? Justify your answers with a sample calculation for a trial from each case. If the kinetic energy is not conserved, where did it go?

Name .... Section .... Date ....

Lab Partner(s) ....

EXPERIMENT 9 *Projectile Motion: The Ballistic Pendulum*

ADVANCE STUDY ASSIGNMENT

Read the experiment and answer the following questions.

A. *The Ballistic Pendulum*

1. What conservation laws are involved in the computation of the initial velocity of a projectile using the ballistic pendulum?

2. When removing the ball from the pendulum bob of the ballistic pendulum, what is it important to do to prevent damage to the apparatus?

3. Is the center of mass of the pendulum-ball system at the center of the ball? If not, how is the center of mass located?

B. *Determination of the Initial Velocity of a Projectile from Range-Fall Measurements*

4. What is the range of a projectile?

5. After the projectile leaves the gun, what are the accelerations in the x and y directions?

6. How is the location where the ball strikes the floor determined?

(continued)

C. *Projectile Range Dependence on the Angle of Projection*

7. How does the range of a projectile vary with the angle of projection θ?

8. Theoretically, at what angle of projection is the range a maximum?

EXPERIMENT 9

Projectile Motion: The Ballistic Pendulum

I. INTRODUCTION

Projectile motion refers to the motion of an object in a plane (two dimensions) under the influence of gravity. The kinematic equations of motion describe the components of such motion, and may be used to analyze projectile motion. In most textbook cases, the initial velocity of a projectile is given and the motion is described through the equations of motion. However, in this laboratory experiment, the unknown initial velocity will be determined from experimental measurements. This will be done (1) through the use of the ballistic pendulum and (2) from range-fall distance measurements. The dependence of the projectile range on the angle of projection will also be investigated so as to experimentally determine the angle of projection that gives the maximum range.

II. EQUIPMENT NEEDED

- Ballistic pendulum
- Sheets of plain paper and carbon paper
- Meter stick
- Protractor
- Laboratory balance
- Masking tape
- Wooden blocks
- 1 sheet of Cartesian graph paper

III. THEORY

A. *The Ballistic Pendulum*

Types of ballistic pendula apparatus are shown in Fig. 9-1. The ballistic pendulum is used to measure the initial velocity of a horizontally projected object (a metal ball) fired from a spring gun. The projectile is fired into a stationary, hollow pendulum bob suspended by a light rod, and the pendulum and the embedded projectile swing upward. A catch mechanism stops the pendulum at its highest position of swing. By measuring the vertical distance that the center of mass of the pendulum-ball system rises, one can compute the initial velocity of the projectile through the use of the conservation of linear momentum and the conservation of mechanical energy.

Consider the schematic diagram of a ballistic pendulum shown in Fig. 9-2. A projectile of mass m with an initial horizontal velocity v_{x_0} is fired into and becomes embedded in a stationary pendulum of mass M.

The conservation of linear momentum in the x direction applies just before and after collision. The velocity of the pendulum bob is initially zero and the combined system $(m + M)$ has a velocity V just after collision. Hence, we may write

$$mv_{x_0} = (m + M)V \qquad \textbf{(9-1)}$$

(before) (after)

After the collision, the pendulum with the embedded projectile swings upward (momentum of the system no longer conserved) and the center of mass of the system is raised a maximum vertical distance h. By the conservation of mechanical energy, the increase in potential energy is equal to the kinetic

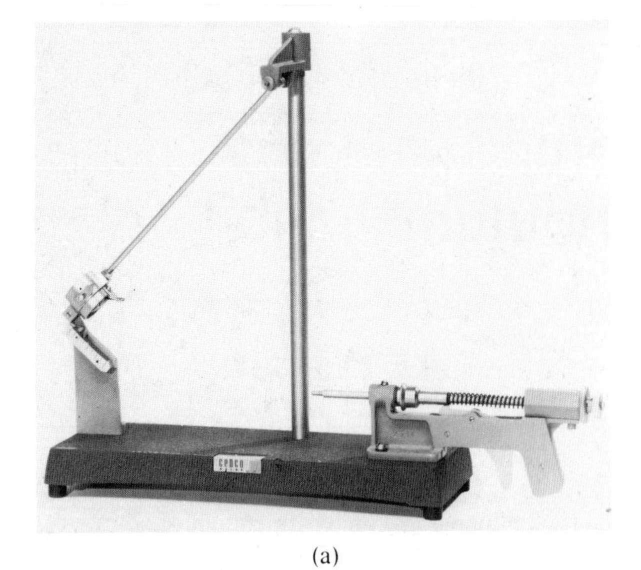

(a)

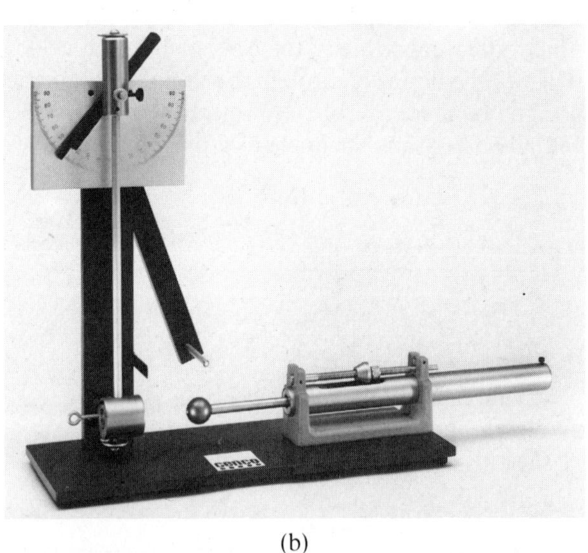

(b)

(c)

Fig. 9-1 Types of ballistic pendula. (a) and (b) Central Scientific Co. models. (Photos courtesy of Central Scientific Co., Inc.) (c) The Beck model.

energy of the system just after collision (the friction of the support is considered negligible). Hence,

$$\underset{\text{(kinetic energy)}}{\tfrac{1}{2}(m + M)V^2} = \underset{\text{(potential energy)}}{(m + M)gh} \tag{9-2}$$

Solving Eq. 9-2 for V, we find that

$$V = \sqrt{2gh} \tag{9-3}$$

Substituting this expression into Eq. 9-1 and solving for v_{x_0} yields

$$v_{x_0} = \frac{m + M}{m}\sqrt{2gh} \tag{9-4}$$

Hence, by measuring m, M, and h, the initial velocity of the projectile can be computed.

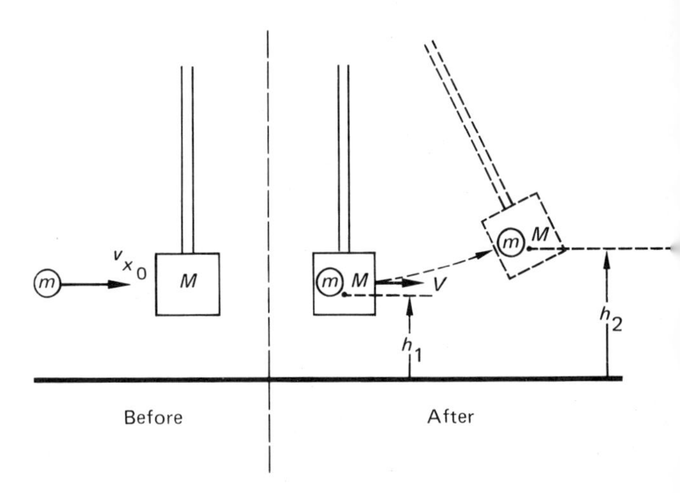

Fig. 9-2 Ballistic pendulum action.

B. *Determination of the Initial Velocity of a Projectile from Range-Fall Measurements*

If a projectile is projected horizontally with an initial velocity v_{x_0} from a height y, it will describe an arc as illustrated in Fig. 9-3. The projectile will travel a horizontal distance x (called the *range* of the projectile) while falling a vertical distance y.

The initial vertical velocity is zero, $v_{y_0} = 0$, and the acceleration in the y direction is $a_y = g$ (acceleration due to gravity — downward taken as positive in this case for convenience). There is no horizontal acceleration, $a_x = 0$; hence, the components of the motion are described by

$$x = v_{x_0}\,\mathrm{t} \tag{9-5}$$

and

$$y = \tfrac{1}{2}gt^2 \tag{9-6}$$

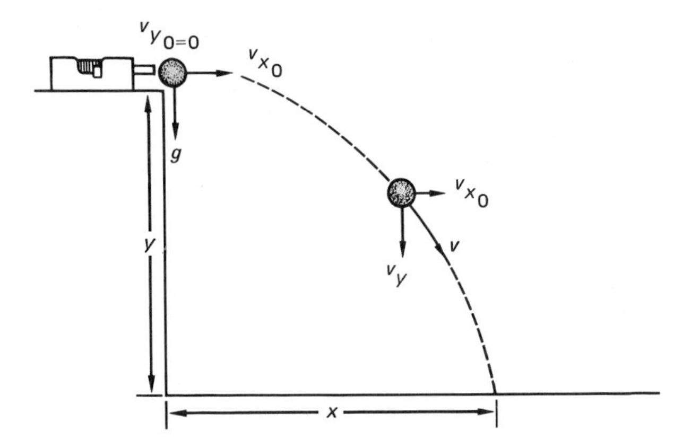

Fig. 9-3 The configuration for range-fall measurements.

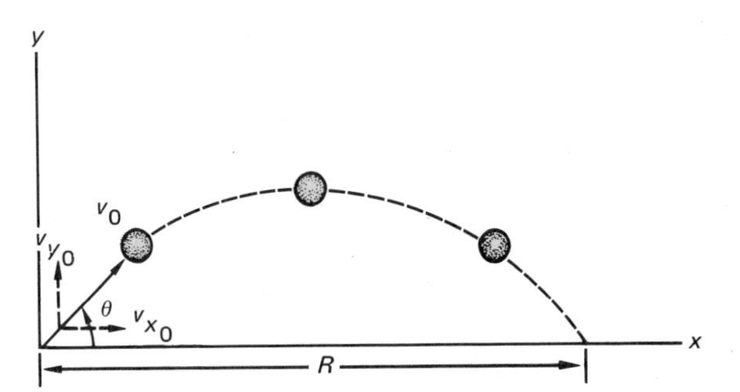

Fig. 9-4 Projectile motion at an arbitrary projection angle.

Eliminating t from these equations and solving for v_{x_0}, we have

$$v_{x_0} = \sqrt{\frac{gx^2}{2y}} = \left(\frac{g}{2y}\right)^{1/2} x \qquad \textbf{(9-7)}$$

Hence, by measuring the range x and the distance of fall y, the initial velocity of the projectile can be computed.

C. *Projectile Range Dependence on the Angle of Projection*

The projectile path for a general angle of projection θ is shown in Fig. 9-4. The components of the initial velocity are

$$v_{x_0} = v_0 \cos\theta$$
$$v_{y_0} = v_0 \sin\theta \qquad \textbf{(9-8)}$$

At the top of the arc path, $v_y = 0$, and since

$$v_y = v_{y_0} - gt$$
$$= v_0 \sin\theta - gt$$

(downward taken as negative) we have

$$v_0 \sin\theta - gt_m = 0$$

or

$$t_m = \frac{v_0 \sin\theta}{g} \qquad \textbf{(9-9)}$$

where t_m is the time for the projectile to reach the maximum height y_m. The total time of flight t is then

$$t = 2t_m = \frac{2v_0 \sin\theta}{g} \qquad \textbf{(9-10)}$$

During the time t, the projectile travels a distance R (range) in the x direction:

$$R = v_{x_0} t = \frac{2v_0^2 \sin\theta \cos\theta}{g}$$

But using the trigonometric identity $2 \sin\theta \cos\theta = \sin 2\theta$,

$$R = \frac{v_0^2 \sin 2\theta}{g} \qquad \textbf{(9-11)}$$

From Eq. 9-11, we see that the range of the projectile depends on the angle of projection θ. The maximum range R_{max} occurs when $\sin 2\theta = 1$. Since $\sin 90° = 1$, by comparison

$$2\theta = 90° \quad \text{or} \quad \theta = 45°$$

Hence, a projectile has a maximum range for $\theta = 45°$, and

$$R_{max} = \frac{v_0^2}{g} \qquad \textbf{(9-12)}$$

which provides another method to determine experimentally the initial velocity of a projectile.

IV. EXPERIMENTAL PROCEDURE

A. *The Ballistic Pendulum*

1. Obtain the projectile ball, which may be in the pendulum bob. (*Note:* When removing the ball from the pendulum bob of some types of ballistic pendula, be sure to push up on the spring catch that holds the ball in the pendulum so as not to damage it.) Place the projectile ball on the ball rod of the spring gun and cock the gun by pushing with the palm of the hand. (On the Beck model of the ballistic pendulum, it is necessary to hold down the trigger mechanism of the gun while pushing the ball rod back to one of three available firing positions.)

Keeping your fingers away from the projectile end of the gun, fire the projectile in the pendulum to see how the apparatus operates. If the catch mechanism does not catch on the notched track, you should adjust the thumb screw of the pendulum suspension to obtain the proper alignment.

2. The preset pointer or a dot on the side of the pendulum bob indicates the position of the center of mass of the pendulum-ball system. With the pendulum hanging freely, measure the height h_1 of the pointer above the base surface (Fig. 9-2) and record in Data Table 1.

3. Fire the ball into the freely hanging, stationary pendulum and note the notch at which the catch mechanism stops on the curved track. Counting upward on the curved track, record the notch number in Data Table 1. Repeat this procedure four times and record the notch number for each trial in the data table.

4. Determine the average of these observations, which is the average highest position of the pendulum. Place the catch mechanism in the notch corresponding most closely to the average and measure the height h_2 of the pointer or dot above the base surface (Fig. 9-2), or above another reference horizontal plane such as the lab table. For the model shown in Fig. 9-1(b), the angle θ is measured on the attached protractor scale and the height h is computed by trigonometry.*

5. Loosen the thumb screw of the pendulum support and carefully remove the pendulum. Weigh and record the masses of the pendulum and the ball. (*Note:* The mass of the pendulum is that of the bob and the support rod. Do not attempt to remove the support rod from the bob.) For the model shown in Fig. 9-1(b), a counterweight is employed so only the masses of the bob and the projectile need be taken into account. Consult your instructor for an explanation in this case.

6. From the data, compute the initial velocity using Eq. 9-4 ($g = 9.8\ \text{m/s}^2 = 980\ \text{cm/s}^2$).

*$h = L - L\cos\theta$, where L is the length of the pendulum.

B. *Determination of the Initial Velocity of a Projectile from Range-Fall Measurements*

7. With the pendulum removed, position the apparatus near one edge of the laboratory table as shown in Fig. 9-3. Fire the ball from the gun and note where the ball strikes the floor. (The range of the ball is appreciable, so you will probably have to shoot the ball down an aisle. Be careful not to hit anyone with the ball, particularly the instructor.)

8. Place a sheet of paper where the ball hit the floor and cover this with carbon paper (carbon side down). Tape the papers to the floor (or weight them down) so that they will not move. When the ball strikes the carbon paper, the mark made on the sheet of paper below will help you determine the range of the projectile.† Also, mark the position of the apparatus on the table (e.g., using a piece of tape as a reference). It is important that the gun be fired from the same position each time.

9. Fire the ball five times and measure the horizontal distance or range x the ball travels for each trial (see Fig. 9-3). Record the measurements in Data Table 2 and find the average range. Also, measure the height y of the ball rod from the floor and record in the data table. The height y is measured from the bottom of the ball as it rests on the gun to the floor.

10. Using Eq. 9-7, compute the initial velocity of the ball ($g = 9.8\ \text{m/s}^2 = 980\ \text{cm/s}^2$). Compare this to the velocity determined in part A and compute the percent difference.

C. *Projectile Range Dependence on the Angle of Projection*

11. Position the ballistic pendulum apparatus (with the pendulum removed) on the floor with several books under one end in such a way that it can be fired at an angle θ relative to the horizon-

†The range will be measured from the center of the ball just as it leaves the gun (put the ball on the gun without loading the spring) to the marks that are on the paper on the floor.

tal. Aim the projectile down an aisle or hallway.

12. Using a protractor to set the angles of projection, fire the projectile at angles of 20°, 30°, 40°, 45°, 50°, 60°, and 70° with two or three trials for each angle. The projectile should be aimed so that it lands as close as possible to the same horizontal reference line each time. Station one or more lab partners near where the projectile strikes the floor. They are to judge the average range of the two or three trials. Measure the average range for each angle of projection and record the data in Data Table 3.

13. Plot the range versus the angle of projection, drawing a smooth curve through the series of points. From the graph, determine the angle of projection for the maximum range and record in the data table.

LABORATORY REPORT

A. *The Ballistic Pendulum*

DATA TABLE 1

Trials	Notch number of pendulum catch	
		Height h_2 of pointer with pendulum catch in closest to average notch number ....
1		Height h_1 of pointer with pendulum freely suspended
2		....
3		$h = h_2 - h_1$ ....
4		Mass of ball m
5		....
		Mass of pendulum, M (bob and support)
Average		....

Computed initial velocity, v_{x_0} ....
(units)

Calculations (show work)

(continued)

B. *Determination of the Initial Velocity of a Projectile from Range-Fall Measurements*

DATA TABLE 2

Trial	Range
1	
2	
3	
4	
5	
Average	

Vertical distance of fall, y ....

Computed initial velocity, v_{x_0} ....
(units)

Percent difference between results of parts A and B ....

Calculations (show work)

C. *Projectile Range Dependence on the Angle of Projection*

DATA TABLE 3

Angle of projection	Average range
20°	
30°	
40°	
45°	
50°	
60°	
70°	

Angle of projection for maximum range from graph ....

Lab Partner(s) ....

EXPERIMENT 9 *Projectile Motion: The Ballistic Pendulum*

QUESTIONS

A. *The Ballistic Pendulum*

1. Is the collision between the ball and the pendulum elastic or inelastic? Justify your answer by calculating the kinetic energy of the system before collision using the value of v_{x_0} found in the experiment and the kinetic energy just after collision using the experimental value of h in Eq. 9-2.

2. Using the results of Question 1, find the fractional kinetic-energy loss during collision. Express the "loss" as a percent. What became of the "lost energy"?

3. Expressing the kinetic energy in terms of momentum (i.e., $K = \frac{1}{2}mv^2 = p^2/2m$), prove that the fractional loss during the collision is equal to $M/(m + M)$ using symbols, not numbers.

(continued)

4. Compute the fractional energy loss from the experimental mass values using the equation developed in Question 3, and compare this to the result in Question 2. Explain the difference, if any.

B. *Determination of the Initial Velocity of a Projectile from Range-Fall Measurements*

5. What effect does the force of gravity have on the horizontal velocity of the projectile? Explain.

C. *Projectile Range Dependence on the Angle of Projection*

6. Using experimental data, compute the magnitude of the initial velocity v_0 of the projectile from Eq. 9-12 and compare this to the results of parts A and B of the procedure.

Name .... Section .... Date ....

Lab Partner(s) ....

EXPERIMENT 10 *Centripetal Force*

ADVANCE STUDY ASSIGNMENT

Read the experiment and answer the following questions.

1. Define centripetal force.

2. What supplies the centripetal force for (a) a satellite in orbit about the earth; (b) the mass in uniform circular motion in this experiment?

3. An object moving in *uniform* circular motion is accelerated. How can this be, since uniform motion implies constant motion?

4. For an object in uniform circular motion, on what parameters does the measurement of the centripetal force depend?

5. If the centripetal force acting on an object in uniform motion suddenly ceased to act or went to zero, what would happen to the object? That is, what would be its subsequent motion?

6. Suppose that the centripetal force acting on an object in circular motion were increased and the object remained in the same circular path. How would its motion be affected?

7. Explain how the centripetal force is directly determined for each apparatus described in the experiment.

EXPERIMENT **10**

Centripetal Force

I. INTRODUCTION

The earth revolves about the sun and atomic electrons revolve around the nucleus. What keeps these objects in orbit. The answer is **centripetal (center-seeking) force.** The centripetal force is supplied by gravitational and electrical interactions, respectively, for each of these cases.

The study of centripetal force in the laboratory is simplified by considering objects in uniform circular motion. An object in uniform circular motion moves with a constant speed (a scalar) but has a changing velocity (a vector) because of the continuing change in direction. This change in velocity results from a centripetal acceleration that is due to a centripetal force.

In the experimental situation(s) of this experiment, the centripetal force will be supplied by a spring, which can be readily measured. However, the magnitude of the centripetal force can also be determined from other experimental parameters (e.g., the frequency of rotation of the object, its mass, and its radius of orbit). Centripetal force will be experimentally investigated by measuring these parameters and comparing the calculated results with the direct measurement of the spring force, which mechanically supplies the center-seeking centripetal force.

II. EQUIPMENT NEEDED

A. *Centripetal Force Apparatus with Variable-Speed Rotor and Counter*

- Laboratory timer or stopwatch
- Weight hanger and slotted weights
- Vernier caliper
- Support rod and clamp
- String
- Safety glasses

B. *Manual Centripetal Force Apparatus*

- Laboratory timer or stopwatch
- Meter stick
- Weight hanger and slotted weights
- String
- Laboratory balance

III. THEORY

An object in uniform circular motion requires a centripetal or center-seeking force to "hold" it in orbit. For example, when one swings a ball on a rope around one's head (Fig. 10-1), the **centripetal force** $F_c = ma_c$ is supplied by the person and transmitted to the *ball* through the rope. In the absence of the centripetal force (e.g., if the rope breaks or if the person releases the rope), the ball would no longer be held in orbit, and would fly off in the direction of its tangential velocity v.

An object in uniform circular motion moves with a constant speed. Even though the object's speed is constant, its velocity is continually changing because the direction of the motion is continually changing.

Fig. 10-1 An object in uniform circular motion must have a centripetal acceleration, $a_c = v^2/r$. In the case of swinging a ball on a rope around one's head, the centripetal force, $F_c = ma_c$, is supplied by the person and transmitted to the ball through the rope.

This change in velocity results from a centripetal acceleration a_c that is due to the applied centripetal force F_c. The direction of the acceleration (and force) is always toward the center of the object's circular path, and it can be shown (see your text) that the magnitude of the acceleration is given by

$$a_c = \frac{v^2}{r} \tag{10-1}$$

where v is the tangential or orbital speed of the object and r the radius of the circular orbit. By Newton's second law, $F = ma$, the magnitude of the centripetal force is

$$F_c = ma_c = \frac{mv^2}{r} \tag{10-2}$$

where m is the mass of the object. In terms of distance and time, the orbital speed v is given by $v = 2\pi r/T$, where T is the period of the orbit.

Notice that this general expression describes the centripetal force acting on an object in uniform circular motion in terms of the properties of the motion and orbit. It is equal to the expression of a physical force that actually supplies the centripetal action. For example, in the case of a satellite in uniform circular motion around the earth, the centripetal force is supplied by gravity, which is generally expressed $F_g = Gm_1m_2/r^2$, and $F_c = F_g$. Similarly, for an object being held in uniform circular motion by the tension force of a string, the spring force is equal to Eq. 10-2.

The centripetal force given by Eq. 10-2 can also be expressed in terms of the angular speed ω or frequency f of rotation using the expression $v = r\omega$ and $\omega = 2\pi f$:

$$F_c = \frac{mv^2}{r} = \frac{m(r\omega)^2}{r} = mr\omega^2$$

and

$$F_c = mr(2\pi f)^2 = 4\pi^2 mrf^2 \tag{10-3}$$

where ω is in radians per second and f is in hertz (cycles per second). In this experiment it is convenient to think of f as being in revolutions per second.

IV. EXPERIMENTAL PROCEDURE

A. *Centripetal Force Apparatus with Variable-Speed Rotor*

1. The centripetal force apparatus mounted on a variable-speed rotor is shown in Fig. 10-2.* **Before turning on the rotor,**
   (a) By means of the threaded collar on the centripetal force apparatus, adjust the spring to a minimum tension (0–5 on the scale above the threaded collar).
   (b) By means of the milled screw head near the base of the rotor, move the rubber friction disk to near the center of the driving disk. (The driving disk can be pushed back so that the friction disk can be moved freely.) This will give a low angular starting speed when the rotor is turned on (but don't turn it on yet!). The speed of the rotor is increased or decreased by moving the friction disk in (up) or out (down), respectively, along the radius of the driving disk. (*Note:* Excessive speeds can be dangerous. Do not go beyond the speeds needed.)
   (c) Make certain that the force apparatus is locked securely in the rotor mount by means of the locking screw. Have the instructor check your setup at this point.

*The following procedures apply particularly to the belt-driven rotor model.

(a) (b) (c) (d)

Fig. 10-2 Centripetal force apparatus. Older model. (a) The speed of the rotor is adjusted by moving the rubber friction disk by means of a milled screw head, as illustrated in the photo. (b) When the centripetal force is equal to the spring force, the pointer P will rise and point horizontally toward the tip of the index screw I. The pointer and screw may be seen more clearly in (a) on the lower horizontal frame of the rotor. See also Fig. 10-3. (c) Motor with belt guard and rotating arm in horizontal storage position. The belt guard has been removed in (a) and (b) to provide a clearer illustration. *When in operation, the motor should always be equipped with a belt guard for safety.* (d) A newer, self-contained centripetal force apparatus that eliminates this problem. (Courtesy of Central Scientific Co., Inc.)

2. Referring to Fig. 10-3, when the rotor is turned on and adjusted to the proper speed, the cylindrical mass m in the centripetal force apparatus in contact with pointer P will cause the pointer to rise and horizontally point toward the index screw I. In this condition, the mass will be in uniform circular motion around the axis of rotation through I.

   When taking measurements, be careful not to come in contact with the rotating force apparatus and motor belt drive. It is recommended that a belt guard be installed on the variable-speed rotor if this has not already been done.

3. Put on your safety glasses and turn on the rotor. Adjust the speed until the pointer rises and is opposite the head of the index screw I. Observe this with your eyes on a level with the index. (Why is it a good precaution to wear safety glasses while doing this?) The pointer will be slightly erratic, and as a particular speed is reached, it will "jump" and point horizontally toward the index screw I. Do not exceed this speed. The pointer would be aimed at the head of the index screw when the rotor is spinning at higher speeds too. Why?

   Do not lock the friction disk. Rather, ob-

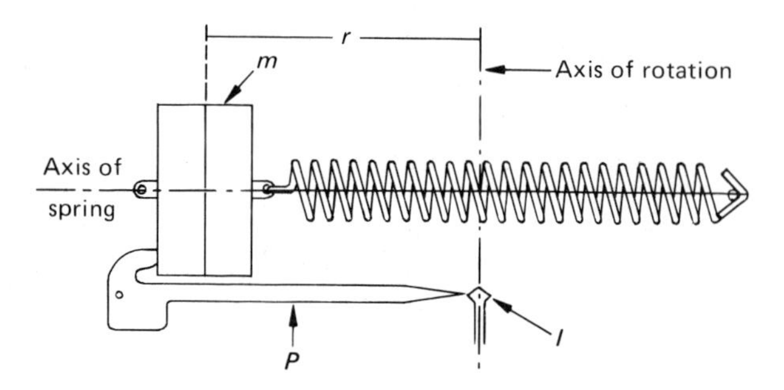

Fig. 10-3 When the apparatus is rotating, the mass acting against the pointer P will cause it to rise and point toward the index screw I.

serve and adjust the speed of the rotor *continuously* during each timed interval in order to keep the pointer as steady as possible. Continuous adjustment is necessary because the rotor speed varies when the counter is engaged. Because the pointer will point horizontally at excessive speeds and induce experimental error, an alternate technique is to continually adjust the rotor speed so that the pointer is not quite horizontal, e.g., aimed midway or just below the head of the index screw. Experiment with your apparatus and see which technique is better, trying to maintain the pointer horizontally at the critical "jump" speed or aiming the pointer at a lower position on the screw at a slightly slower speed.

4. Practice engaging the counter and adjusting the rotor speed. (*Do not* engage the counter too forcefully or you will overly slow down the rotor, yet don't engage the counter so lightly so that you will accidentally cause the rotor to lose contact with the rotor gear.) When satisfied with your technique, record the (initial) counter reading in the Laboratory Report. Using a laboratory timer or stopwatch, measure (count) the number of rotations for a one-minute time interval. One lab partner should engage the counter for the timing interval, while the other adjusts the rotor speed.

   Repeat this procedure for four more one-minute time intervals, but *do not* use the previous final counter reading for the next initial interval reading. Advance the counter to a new arbitrary initial reading for each trial. Share the action. One lab partner should be the "speed controller" who constantly watches and adjusts the rotor speed as described in procedure 3. Another partner should be the "timer" who engages the counter and times the interval. If there are three lab partners, the third partner may handle the counter engagement and disengagement in response to the timer's instructions. Rotate team responsibilities periodically. (Why is this a desirable practice with regard to experimental results?)

5. Subtract the counter readings to find the number of rotations for each timed interval. (They should be similar.) Then, compute the average number of rotations of the five one-minute time intervals (average rotations per minute).

   Divide the average value by 60 (1 min = 60 s) to obtain the average rotation frequency in rotations (cycles) per second or hertz (Hz).

6. Without altering the spring tension setting, remove the centripetal force apparatus from the rotor and suspend it from a support as shown in Fig. 10-4. Suspend enough mass on the hanger to produce the same extension of the spring as when on the rotor (pointer aimed at the index screw position). Record this mass M' (includes mass of hanger) in the Laboratory Report. Also, record the mass of the cylinder m in the force apparatus (stamped on the end of the cylinder). Add to find the total suspended mass, $M = M' + m$, and compute the direct measure of F_c = weight of total suspended mass = Mg.

   With the spring at the same tension setting and the apparatus still hanging from the support

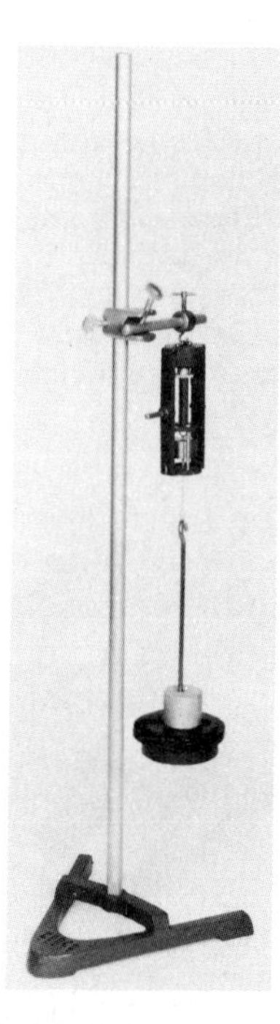

Fig. 10-4 Arrangement for application of gravitational force. (Photo courtesy of Central Scientific Co., Inc.)

with the same mass M' applied, use a vernier caliper measure and record the distance r, or the radius of the circular rotational path. This is the distance between the axis of rotation (line through the index screw) and the center of mass of the cylinder (see Fig. 10-3). This distance is conveniently measured between a line scribed on the upper part of the force apparatus frame above the index screw and a line scribed on the center of the cylinder.

7. Using Eq. 10-3, compute the magnitude of the centripetal force. Compare this with the directly measured value given by the weight force required to produce the same extension of the spring by computing the percent difference.

8. Change the spring tension to a maximum setting (about 20 on the scale above the threaded collar) and repeat procedures 3 to 7.

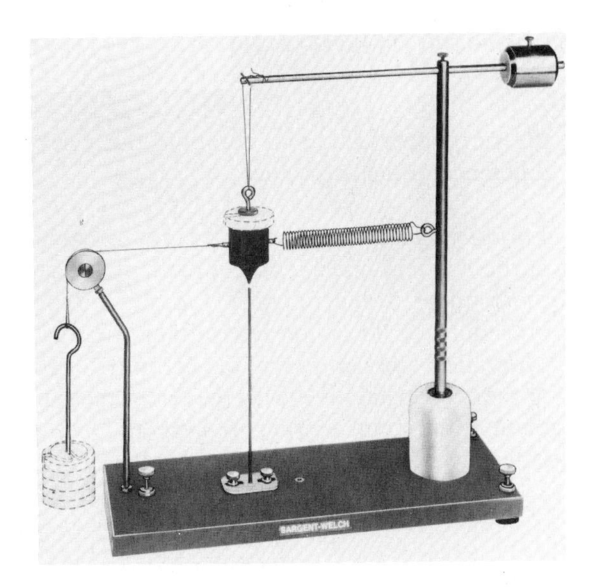

Fig. 10-5 A hand-operated centripetal force apparatus. The suspended weights used to determine the centripetal force supplied by the spring are not attached to the bob when the apparatus is rotated. (Photo courtesy of Sargent-Welch Scientific Company)

B. *Manual Centripetal Force Apparatus*

9. A type of hand-operated centripetal force apparatus is shown in Fig. 10-5. By rolling the rotor between the thumb and fingers, a suspended mass bob is set into circular motion with the centripetal force being supplied by a spring. The horizontal support arm is counterbalanced for ease of operation, with the position of the counterbalance not being critical.

   A pulley mounted to the base of the apparatus is used to make direct measurement of the spring tension supplying the centripetal force for uniform circular motion of a particular radius indicated by the distance between the vertical pointer rod P and the axis of rotation.

10. Remove the bob and determine its mass on a laboratory balance. Record the mass in the Laboratory Report. Adjust the position of the vertical pointer rod, if possible, to the smallest possible radius (distance between pointer tip and the center of vertical rotor shaft). Measure this distance and record in the data table.

11. Attach the bob to the string on the horizontal support arm, and with the bob hanging freely (spring unattached), adjust the support arm so that the bob is suspended directly over the pointer. Attach the spring to the bob and practice rolling the rotor between the thumb and fingers so that the bob revolves in a circular path and passes over the pointer on each revolution in uniform circular motion. (Adjust the position of the counterbalance on the support arm if necessary for ease of operation.)

    While one lab partner operates the rotor, another lab partner with a laboratory timer or stopwatch times the interval for the bob to make about 25 revolutions. The number of counted revolutions may have to be varied depending on the speed of the rotor. Count enough revolutions for an interval of at least 10 s. Record the data in the data table. Practice the procedure before making an actual measurement.

12. Repeat the counting-timing procedure two times. Compute the time per revolution of the bob for each trial and determine the average time per revolution of the three trials. From the data, calculate the average speed of the bob. Recall $v = c/t = 2\pi r/T$, where c is the circumference of the circular orbit, r is the radius of the orbit, and T is the average time per revolution or period. Then, using Eq. 10-2, calculate the centripetal force.

13. Attach a string to the bob opposite the spring and suspend a weight hanger over the pulley. Add weights to the hanger until the bob is directly over the pointer. Record the weight Mg in the data table. (Do not forget to add the mass of the weight hanger.) This weight is a direct measure of the centripetal force supplied by the spring during rotation. Compare this with

the calculated value and compute the percent difference of the two values.

14. *Variation of mass.* Unscrew the nut on the top of the bob and insert a slotted mass of 100 g or more under it and retighten the nut. Repeat procedures 11 to 13 for determining the period of rotation and comparing the computed value of the centripetal force with the direct measurement of the spring tension. (*Question:* Does the latter measurement need to be repeated?)

15. *Variation of radius.* Remove the slotted masses from the bob and if pointer P is adjustable, move it farther away from the axis of rotation to provide a larger path radius. Measure and record this distance in the data table. Repeat procedures 11 to 13 for this experimental condition.

16. *Variation of spring tension (optional).* Replace the spring with another spring of different stiffness. Repeat procedures 11 to 13.

LABORATORY REPORT

A. *Centripetal Force Apparatus with Variable-Speed Rotor*

1. Minimum spring tension: scale reading ....

Trial	Counter readings		Difference in readings (rotations/min)
	Final	Initial	
1			
2			
3			
4			
5			
		Average N	

Computation of centripetal force (show work)

Average rotational frequency $f = N/60$ ....

Suspended mass M' ....

Cylinder mass m ....

Total suspended mass, $M = M' + m$ ....

Direct measure of F_c = total suspended weight, $F_c = Mg$ ....

Lab Partner(s) ....

EXPERIMENT 10 *Centripetal Force*

Radius of circular path ....

Computed F_c ....

Percent difference ....

2. Maximum spring tension: scale reading ....

Trial	Counter readings		Difference in readings (rotations/min)
	Final	Initial	
1			
2			
3			
4			
5			
		Average N	

Computation of centripetal force (show work)

Average rotational frequency $f = N/60$ ....

Suspended mass M' ....

Cylinder mass m ....

Total suspended mass, $M = M' + m$ ....

Direct measure of F_c = total suspended weight, $F_c = Mg$ ....

Radius of circular path ....

Computed F_c ....

Percent difference ....

(continued)

B. *Manual Centripetal Force Apparatus*

1. Mass of bob ....
   Radius of circular path ....

	Trial 1	Trial 2	Trial 3
Number of revolutions			
Total time (s)			
Time/revolution (s)			

Average time per revolution ....

Average speed of bob (v) ....

Computed value of centripetal force ....

Direct measurement of centripetal force ....

Percent difference ....

Computation of centripetal force (show work)

2. Variation of mass

   Mass of bob plus slotted mass ....

   Radius of circular path ....

	Trial 1	Trial 2	Trial 3
Number of revolutions			
Total time (s)			
Time/revolution (s)			

Average time/revolution ....

Average speed of bob (v) ....

Computed value of centripetal force ....

Direct measurement of centripetal force ....

Percent difference ....

Computation of centripetal force (show work)

Lab Partner(s) ....

EXPERIMENT 10 *Centripetal Force*

3. Variation of radius

Mass of bob ....

Radius of circular path ....

	Trial 1	Trial 2	Trial 3
Number of revolutions			
Total time (s)			
Time/revolution (s)			

Average time/revolution ....

Average speed of bob (v) ....

Computed value of centripetal force ....

Direct measurement of centripetal force ....

Percent difference ....

Computation of centripetal force (show work)

4. Variation of spring tension

Mass of bob ....

Radius of circular path ....

	Trial 1	Trial 2	Trial 3
Number of revolutions			
Total time (s)			
Time/revolution (s)			

Average time/revolution ....

Average speed of bob (v) ....

Computed value of centripetal force ....

Direct measurement of centripetal force ....

Percent difference ....

Computation of centripetal force (show work)

(continued)

QUESTIONS

1. How does the centripetal force vary with the radius of the circular path? Consider constant frequency and constant speed. Was this substantiated by experimental results?

2. If the centripetal force is increased for circular motion, what is the effect on (a) the frequency of rotation f (with r constant); and (b) f and r when both are free to vary?

3. In part A of the experimental procedure, the counter was advanced to a new arbitrary initial reading for the time interval trials. This is because if consecutive counter readings had been used (i.e., final reading for one interval taken as the initial reading for the next interval), then in averaging the differences all readings except the first and last would have been "thrown out" by the averaging process. That is, the result is essentially a one 5-minute time interval. Prove this explicitly. (*Hint:* Let the counter readings be $n_0, n_1, n_2, \ldots,$ and then $\Delta n_1 = n_1 - n_0$, $\Delta n_2 = n_2 - n_1$, etc. Perform the averaging process on these values.)

Name .... Section .... Date ....

Lab Partner(s) ....

EXPERIMENT 11 *Friction*

ADVANCE STUDY ASSIGNMENT

Read the experiment and answer the following questions.

1. State the three general empirical rules used to describe friction.

2. What is the normal force?

3. What is the coefficient of friction, and in what units is it expressed?

4. Distinguish between μ_s and μ_k. Which is generally greater?

5. When is (a) $f_s < \mu_s N$; and (b) $f_s = \mu_s N$?

6. What is the relationship between the applied force and the force of kinetic friction when a block slides on a surface with a constant velocity?

EXPERIMENT **11**

Friction

I. INTRODUCTION

Friction refers to the resistance of motion occurring between contacting surfaces. The friction between unlubricated solids is a broad and complicated topic, since it depends on the contacting surfaces and the material properties of the solids. Three general empirical "rules" often used to describe friction between solids are that the frictional force is:

1. Independent of the surface area of contact.
2. Proportional to the load (or, the normal component of the force pressing the surfaces together).
3. Independent of the sliding speed.

It would seem logical that friction would depend on the *actual* area of contact of the irregularities of the surfaces. But this contact area depends on the load, or the normal component of the force with which the surfaces are pressed together. That is, increasing the load should simply increase the amount of contact surface area. Then, too, for an object sliding on a surface, is it consistent that the friction should be independent of the sliding speed? With these thoughts in mind, the validity of the foregoing empirical rules will be investigated in this experiment.

II. EQUIPMENT NEEDED

- Board with attached pulley
- Wooden block with hook (e.g., piece of 2 by 4 lumber or commercially available block)
- Weight hanger and set of weights
- String
- Protractor
- Laboratory balance
- Table clamp and support
- Meter stick
- Masking tape
- 2 sheets of Cartesian graph paper

(Optional)
- Phenolic block
- Aluminum block
- Wheel cart
- Lubricating powder

III. THEORY

Considering the magnitude of the force of friction f to be proportional to the magnitude of the load or normal force N, we may write

$$f \propto N$$

or

$$f = \mu N \qquad \textbf{(11-1)}$$

where the Greek letter μ (mu) is a dimensionless constant of proportionality called the **coefficient of friction.** Referring to the case in Fig. 11-1(a), the normal reaction force N acts on the block and is equal and opposite to the downward force (weight) of the block on the horizontal surface (i.e., $N = mg$).

When a force F is applied to the block parallel to the surface and no motion occurs, we say that the

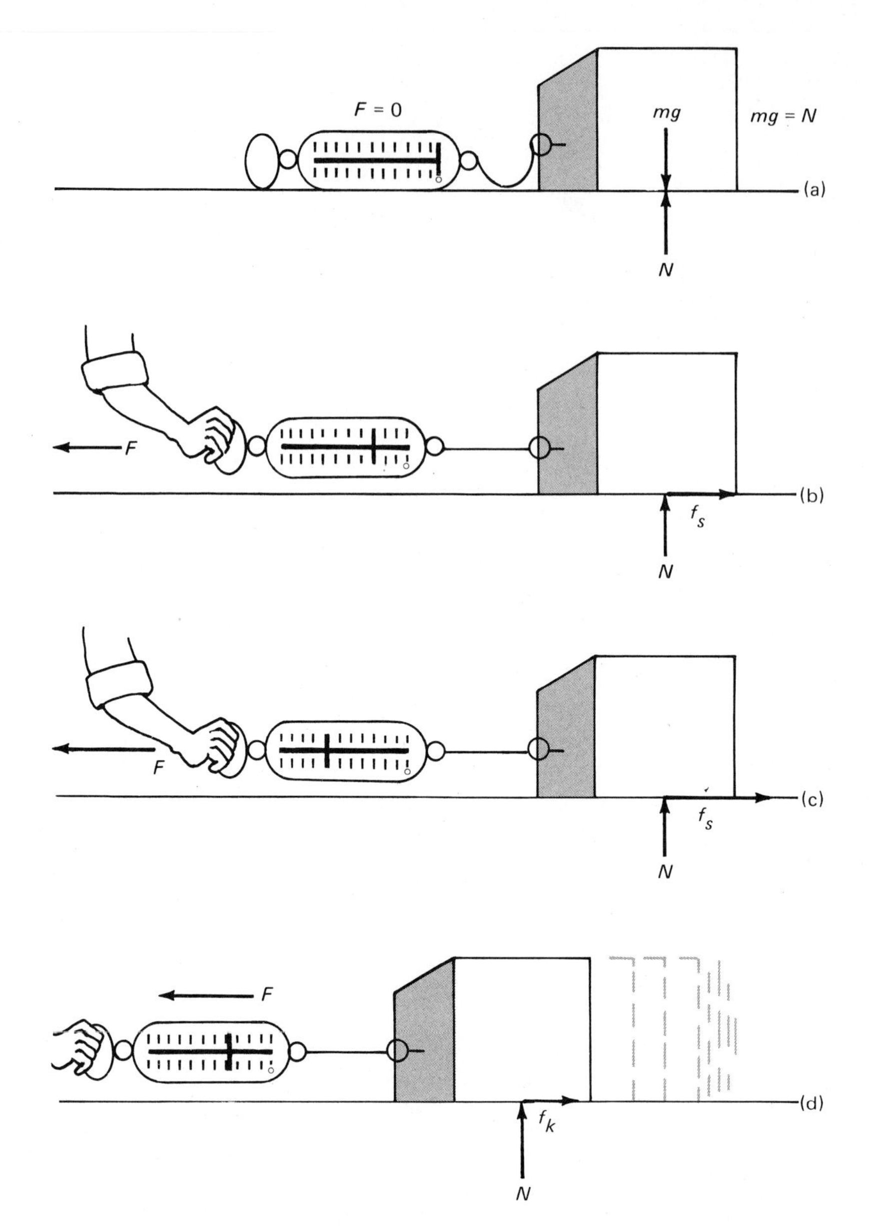

Fig. 11-1 Applied and frictional forces.

applied force is balanced by an opposite force of static friction, f_s, which is exerted on the block by the table along the surface of contact [Fig. 11-1(b)]. As the magnitude of the applied force is increased, f_s increases to a *maximum* value given by

$$f_s = \mu_s N \qquad \textbf{(11-2)}$$

where μ_s is the coefficient of static friction. The maximum force of static friction will be the same as the smallest force applied parallel to the surface necessary to set the block into motion.*

At the instant F becomes greater than $f_s = \mu_s N$, the block is set into motion, and the motion is opposed by the force of kinetic (sliding) friction f_k and

$$f_k = \mu_k N \qquad \textbf{(11-3)}$$

where μ_k is the coefficient of kinetic (sliding) friction.

In general, $\mu_k N < \mu_s N$, and the unbalanced force causes the block to accelerate. However, if the applied force is reduced so that the block moves with a uniform velocity, then $F = f_k = \mu_k N$.

Usually for a given pair of surfaces, $\mu_k < \mu_s$. That is, it takes more force to overcome static friction (get an object moving) than to overcome kinetic friction (keep it moving). Both coefficients may be greater than one, but they are commonly less than one. The actual values depend on the nature and roughness of the surfaces.

*These conditions of f_s are sometimes written $f_s \leq \mu_s N$; that is, f_s is less than or equal to $\mu_s N$.

IV. EXPERIMENTAL PROCEDURE

A. *Determination of* μ_s

1. Determine the mass of the wooden block on a laboratory balance and record its weight [expressed as mass times g (i.e., mg)] in the Laboratory Report. The reported mass and force may be expressed in kilograms or grams and newtons or dynes, respectively, depending on your instructor's preference and instructions.

2. Clean the surfaces of the board and block so they are free from dust and other contaminants. Place the board with the pulley near the edge of the table so that the pulley projects over the table's edge. Attach one end of a length of string to the wooden block and the other end to a weight hanger. Place the block flat on the board and run the string over the pulley so that the weight hanger is suspended over the end of the table (Fig. 11-2). Be sure that the string is parallel to the board.

3. With the block lying on one of its sides of larger area, add weights to the hanger until the block just begins to move. (If the 50-gm weight hanger causes the block to move, add some weights to the block.) Determine the required suspended mass within 1 gm. Record the weight force required to move the block in Data Table 1.

   Suggested experimental technique:
   (a) Keep the block in the middle of the plane.
   (b) Lift the block, gently lower it onto the plane and *restrain* from moving for a count of five (*do not* press it against the plane), then release the block. If the block moves, the suspended mass M is too large; if it doesn't move, M is too small; if the block moves about half the time, M is about right.

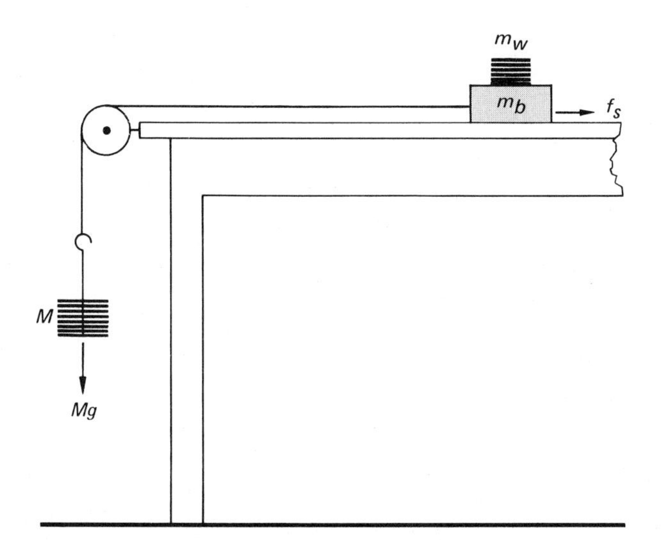

Fig. 11-2 Experimental setup to determine μ.

4. Repeat procedure 3 with 100-, 200-, 300-, 400-, and 500-gram masses, respectively, added to the block. Record the results in Data Table 1.

5. Plot the weight force just required to move the block ($F = f_s$) versus the total block weight moved (normal force N). Since $f_s = \mu_s N$, the curve should be a straight line with a slope of μ_s. Determine the slope and record in Data Table 1. (*Hint:* Plot in units of mass times g, where g is retained explicitly. Include the point (0, 0) as datum.)

B. *Determination of* μ_k

HORIZONTAL BOARD

6. When the block moves with a uniform speed, $F = f_k = \mu_k N$; that is, the weight force F is just balanced by the frictional force f_k and there is no acceleration.

7. Using the larger side (surface area) of the block and the series of added masses as in part A, add mass to the weight hanger until a slight push on the block will cause it to move with a uniform speed. It may be helpful to tape the weights to the block. The required weight force for this in each case should be less than that for the corresponding case in part A. Why? Record the data in Data Table 2.

   Suggested experimental technique:
   (a) Begin with the block at one end of the plane and give it a push so that it slides across the entire plane.
   (b) Observe the behavior of the block in the same region as before, namely in the middle of the plane. This is where the block should be observed for constant speed.

8. Plot the weight ($F = f_k$) versus the total block weight moved (normal force N) for these data on the same graph as before. Since $f_k = \mu_k N$, the curve should be a straight line with a slope of μ_k. Determine the slope of the curve and record in Data Table 2. Calculate the percent decrease of μ_k from the μ_s value.

ELEVATED BOARD (inclined plane)

9. Elevate the pulley end of the board on a support to form an inclined plane (Fig. 11-3). Note in the figure that the magnitude of the normal force is equal to a *component* of the weight force.

   With the block lying on a side of its larger surface area, determine the angle (θ) of incline that will allow the block to slide down the plane with a constant speed after being given a slight tap. (No suspended weight is used in this case.)

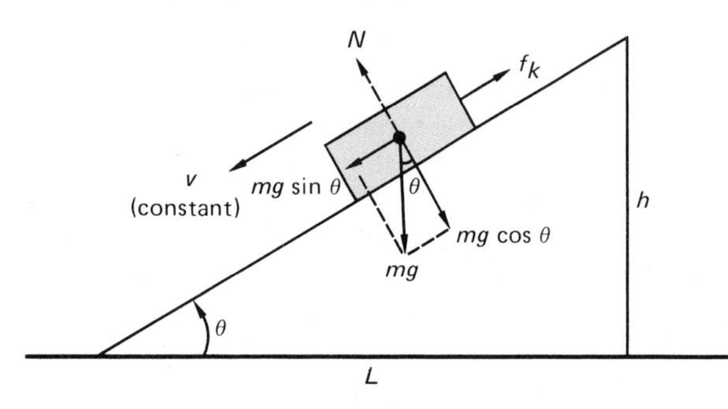

Fig. 11-3 Experimental setup to detemine μ from the angle of incline.

10. Using a protractor, measure the angle θ and record in Data Table 3. Also, with a meter stick, measure the length L of the base (along the table) and the height h of the inclined plane. Record the ratio h/L in Data Table 3.

11. Repeat this procedure for the block with the series of added masses as in the procedure for the horizontal board and record in Data Table 3. It may be helpful to tape the masses to the block.

12. Find the tangents of the θ's in Appendix B, Table B6 or using a calculator, and record. Compare the average of these values with the average of the ratios of h/L. These averages should be similar.

13. Compare the average value of tan θ with the value of μ_k found in the procedure for the horizontal board and calculate the percent difference.

C. *Dependences of μ**

14. Use the *inclined plane method* to investigate the dependence of μ on area, material, velocity, rolling, and lubrication. The experimental setups are described in Data Table 4. Answer the questions listed after the Data Table.

*This experimental procedure and modifications were suggested by Professor I. L. Fischer, Bergen Community College.

LABORATORY REPORT

Mass of block, m_b ....

A. *Determination of μ_s*

DATA TABLE 1

Measurement of μ_s on level plane. (See Fig. 11-2.) Purpose: To show $f_s = \mu_s N$, where N depends on $m_b + m_w$.

m_w	0					
$N = (m_b + m_w)g$*						
$f_s = F = Mg$						

*It is convenient to express the force in terms of mg, where g is left in symbol form. Don't forget to show mass and force units in Tables.

Calculations (show work)

μ_s ....
(from graph)

B. *Determination of μ_k*

DATA TABLE 2

Measurement of μ_k on level plane. Purpose: To show $f_k = \mu_k N$, where N depends on $m_b + m_w$.

m_w	0					
$N = (m_b + m_w)g$						
$f_k = F = Mg$						

Calculations (show work)

μ_k ....
(from graph)

Percent decrease of μ_k relative to μ_s ....

DATA TABLE 3

Measurement of μ_k by inclined plane method. (See Fig. 11-3.) Purpose: To show $\mu_k = \tan\theta$, where θ is independent of $m_b + m_w$.

m_w	0						
θ							Average
h/L							
$\tan\theta$							

Calculations (show work)

Percent difference between
$\tan\theta = \mu_k$ and μ_k from Data Table 2. ....

C. *Dependences of* μ

DATA TABLE 4

Various measurements using inclined plane method.

No.	Conditions	θ	$\mu = \tan\theta$
1	Wooden block on larger area, static (μ_s)		
2	Wooden block on smaller area, static (μ_s)		
3	Wooden block on smaller area, kinetic (μ_k)		
Optional: other measurements of kinetic friction (μ_k)			
4	Phenolic block		
5	Aluminum block, moving slowly		
6	Aluminum block, moving faster		
7	Wheeled cart		
8	Aluminum block with dry lubricating powder		
9	Phenolic block with dry lubricating powder		

(a) Compare No. 1 with Data Table 1: Is the inclined plane method valid for μ_s?

(b) Compare No. 2 with No. 1, and No. 3 with Data Table 3: Does μ depend on area?

(c) Compare Nos. 3, 4, and 5: Does μ_k depend on material?

(d) Compare No. 5 with No. 6: Does μ_k depend on velocity?

(e) Compare No. 7 with anything: Does rolling friction compare with other types of friction?

(f) Compare Nos. 8 and 9 with Nos. 5 and 4: What is the effect of adding the lubricant?

QUESTIONS

1. Explain why $f_s \leq \mu_s N$; that is, why is f_s less than or equal to $\mu_s N$?

2. Speculate in terms of the microscopic surface irregularities as to why $\mu_k < \mu_s$ and what effect a lubricant has on the coefficient of friction.

(continued)

3. Why was it experimentally convenient to have the block move along the board with a uniform speed when determining the coefficient of kinetic friction?

4. Prove that the tan θ is equal to μ_k when the block slides down the incline with a constant speed. (Use symbols, not numbers.)

5. Suppose that the block were made to move up the inclined plane with a uniform speed by suspending masses on a string over the pulley. Derive an equation for the coefficient of kinetic friction for this case in terms of the suspended masses, the mass of the block, and the angle of incline.

6. Draw and justify conclusions about the validity of the empirical rules for friction based on your experimental results.

Name .... Section .... Date ....

Lab Partner(s) ....

EXPERIMENT 12 *Work and Energy*

ADVANCE STUDY ASSIGNMENT

Read the experiment and answer the following questions.

1. What is the difference between the conservation of mechanical energy and the conservation of total energy?

2. Is mechanical energy conserved in any actual situation? Explain.

3. Discuss the difference in the work done or the energy expended in pushing a car up an incline (a) with a constant speed, and (b) with an acceleration, i.e., where does the energy go?

4. Discuss the difference in the energy lost when a car rolls down an incline (a) with a constant speed and (b) with an acceleration, i.e., where does the energy go?

5. Under what conditions would the frictional forces be expected to be equal in magnitude for a car moving up an incline and a car moving down an incline?

6. Is the force of friction the same for different angles of incline with all other parameters being equal? Explain by specifically considering the angles used in the experiment.

EXPERIMENT **12**

Work and Energy

I. INTRODUCTION

Work and **energy** are intimately related, as pointed out in a common definition of energy as being the ability to do work. When work is done by a system, energy is expended or the system loses energy. Conversely, when there is work input to a system, the system gains energy.

In an ideal conservative system, energy is transferred back and forth between kinetic energy and potential energy. In such a system, the sum of the kinetic and potential energies is constant, as expressed by the law of conservation of mechanical energy. However, in actual systems, friction is always present and these systems are nonconservative. That is, some energy is lost as a result of the work done against frictional forces. Even so, the *total* energy is conserved.

In this experiment you will make use of the conservation of energy to study the relationship between work and energy in the cases of a car rolling up and down an inclined plane. The ever-present frictional forces and the work done against friction will be investigated and taken into account so as to provide a better understanding of the concept of work-energy. To simplify matters, experimental conditions with constant speeds will be used so that only the relationship between work and changes in gravitational potential energy will have to be considered.

II. EQUIPMENT NEEDED

- Inclined plane with pulley and Hall's carriage (car)
- Weight hanger and slotted weights
- String
- Meter stick
- Protractor (if plane not so equipped)
- Laboratory balance

III. THEORY

A. *Work of Friction: Force–Distance Method*

(a) The situation for a car moving up an inclined plane with a constant speed is illustrated in Fig. 12-1(a). Since the car is not accelerating, the upward force parallel to the plane F must be equal in magnitude to that of the sum of the forces down (parallel to) the plane:

$$F = F_{\parallel} + f$$

where f is the force of friction and $F_{\parallel}$ is the component of the car's weight parallel to the plane.

Since the magnitude of F is equal to that of the descending weight w_1, we may write

$$w_1 = F_{\parallel} + f \qquad \textbf{(12-1)}$$

where the direction of f is down the plane.

The magnitude of the component of the car's weight perpendicular to the plane $F_{\perp}$ is equal to the magnitude of the normal force N exerted by the plane on the car (i.e., $F_{\perp} = N$), and $f = \mu N$, where μ is the

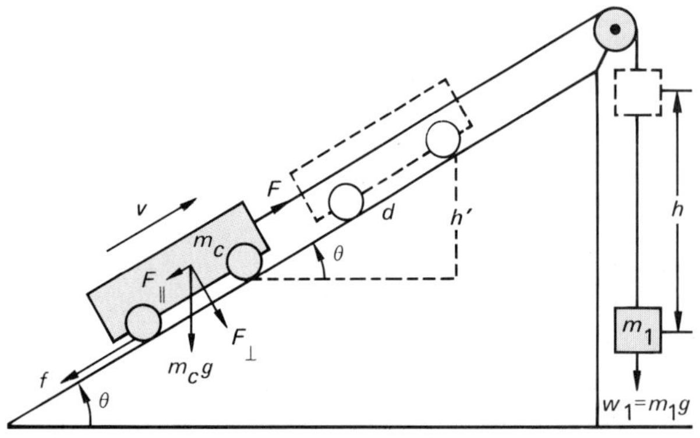

$F = w_1 = F_{\|} + f$ (not to scale)

(a) Car moving upward with constant velocity.

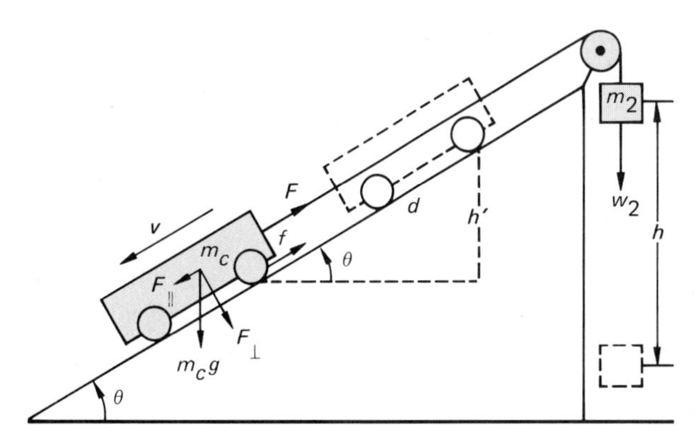

$F_{\|} = F + f = w_2 + f$ (not to scale)

(b) Car moving downward with constant velocity.

Fig. 12-1 Arrangement for determining the force of friction.

coefficient of (rolling) friction. (See Experiment 11.)

(b) The situation for a car moving down an inclined plane with the same constant speed is illustrated in Fig. 12-1(b). Again, since the car is not accelerating, the forces up the plane must be equal and opposite to the forces down the plane, and

$$F_{\|} = F + f$$

where, in this case, the direction of f is up the plane. Since $F = w_2$,

$$F_{\|} = w_2 + f$$

or

$$w_2 = F_{\|} - f \qquad \textbf{(12-2)}$$

If the car moves at approximately the same constant speed in each case, it is reasonable to assume that the magnitude of the frictional force f is the same in each case. Subtracting Eq. 12-2 from Eq. 12-1 to eliminate $F_{\|}$, we have

$$\begin{aligned} w_1 &= F_{\|} + f \\ w_2 &= F_{\|} - f \\ \hline w_1 - w_2 &= 2f \end{aligned}$$

or

$$f = \frac{w_1 - w_2}{2} \qquad \textbf{(12-3)}$$

Hence, the magnitude of the force of (rolling) friction can be determined from the experimental parameters of the two situations (the same angle of incline in each case). Then, knowing the distance d the car moves up the plane, the frictional work may be determined

$$W_f = fd \qquad \textbf{(12-4)}$$

B. *Work of Friction: Energy Method*

(a) For the case of the car moving up the plane, by the conservation of energy, the *decrease* in the potential energy of the descending weight on the weight hanger $\Delta V_w = m_1 gh$ is equal to the *increase* in the potential energy of the car $\Delta V_c = m_c gh'$ *plus* the energy lost to friction, which is equal to the work done against the force of friction W_f. That is,

$$\Delta V_w = \Delta V_c + W_f \qquad \text{(car moving up)}$$

or

$$\begin{aligned} W_f &= \Delta V_w - \Delta V_c \\ &= m_1 gh - m_c gh' \end{aligned} \qquad \textbf{(12-5)}$$

(b) Similarly, for the case of the car moving down the plane, by the conservation of energy, the decrease in the potential energy of the descending car is equal to the increase in the potential energy of the ascending weight plus the work done against the force of friction:

$$\Delta V_c = \Delta V_w + W_f \qquad \text{(car moving down)}$$

or

$$\begin{aligned} W_f &= \Delta V_c - \Delta V_w \\ &= m_c gh' - m_2 gh \end{aligned} \qquad \textbf{(12-6)}$$

Hence, the experimental setup provides two methods of determining the work of friction or the work done against the frictional force.

IV. EXPERIMENTAL PROCEDURE

1. Using a laboratory balance, determine the mass of the car m_c and record in the Laboratory Report.

2. Arrange the inclined plane and the car as shown in Fig. 12-2, with an angle of incline of $\theta = 30°$. Make certain that the pulley is adjusted so that the string attached to the car is parallel to the plane. (Should the car accelerate up the plane by the weight of the weight hanger alone, place some weights in the car so that the car is initially stationary. Add the additional mass to that of the car in Data Table 1.)

3. Add enough weights to the weight hanger so that the car moves up the incline with a slow uniform speed when the car is given a slight tap. Record the total suspended mass in Data Table 1.

4. With the car positioned near the bottom of the incline, mark the position of the car's front wheels and give the car a slight tap to set it into motion. Stop the car near the top of the plane after it moves up the plane (with a constant speed), and measure the distance d it moved up the plane as determined by the stopped position of the car's front wheels. Also, measure the height h the weight hanger descends. This corresponds to the situation in Fig. 12-1(a). Both d and h should be the same. Why? Record the data in Data Table 1.

5. With the car near the top of the plane, remove enough weights from the weight hanger so that the car rolls down the inclined plane with a slow uniform speed on being given a slight tap. This corresponds to the situation in Fig. 12-1(b). Record the total suspended mass in Data Table 1. For convenience, use the same d (and h) as in procedure 4.

6. Compute the frictional force f (Eq. 12-3) and then the work done against friction W_f (Eq. 12-4). Show your calculations and record the results in the Data Table.

7. Compute the work done against friction (W_f) from the changes in potential energies for both the car moving up the plane and moving down the plane (Eqs. 12-5 and 12-6, respectively).

8. Compare the frictional work computed by the force-distance method (procedure 6) with the average frictional work for the up and down cases computed by the energy method (procedure 7) by finding the percent difference between the two values.

9. Adjust the angle of the inclined plane to $\theta = 45°$ and repeat procedures 3 to 8 recording your measurements in Data Table 2.

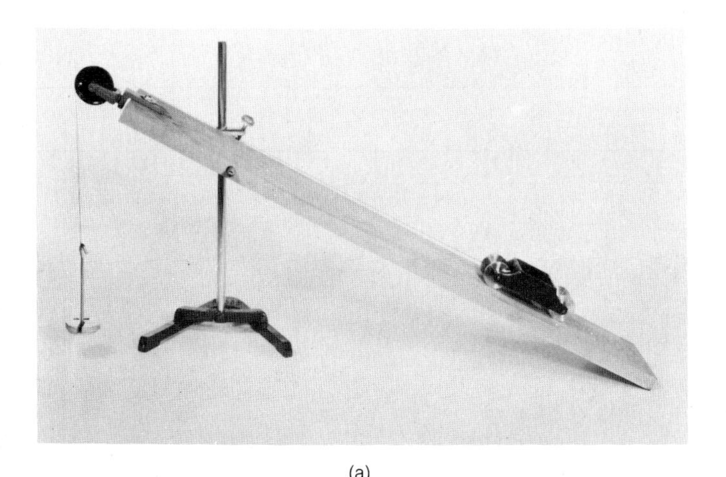

(a)

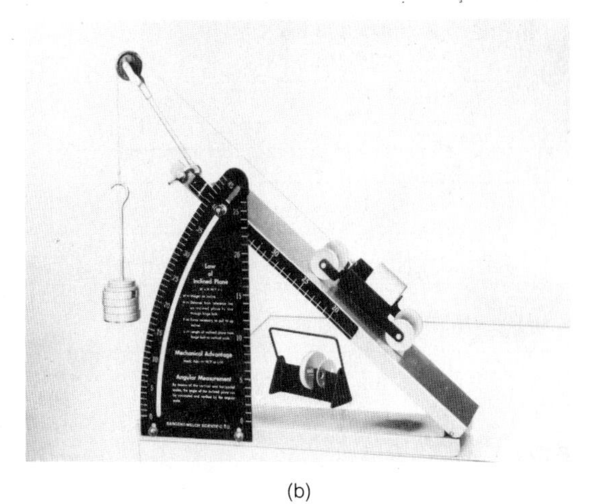

(b)

Fig. 12-2 Types of inclined plane apparatus. (a) Inclined plane board and stand. (Photo courtesy of Central Scientific Co., Inc.) (b) Calibrated inclined plane. (Photo courtesy of Sargent-Welch Scientific Company)

Name .... Section .... Date ....

Lab Partner(s) ....

EXPERIMENT 12 *Work and Energy*

LABORATORY REPORT

DATA TABLE 1

Mass of car, m_c ....

Angle of incline	Car moving up incline			Car moving down incline				
	Suspended mass m_1	d	h	Suspended mass m_2	d	h	f	W_f

h'	Up incline	Down incline	Average W_f
	$W_f = \Delta V_w - \Delta V_c$	$W_f = \Delta V_c - \Delta V_w$	

(h' Is calculated by trigonometry.)

Calculations (show work)

Percent difference in W_f by two methods ....

DATA TABLE 2

Mass of car, m_c....................................

Angle of incline	Car moving up incline			Car moving down incline				
	Suspended mass m_1	d	h	Suspended mass m_2	d	h	f	W_f

h'	Up incline	Down incline	Average W_f
	$W_f = \Delta V_w - \Delta V_c$	$W_f = \Delta V_c - \Delta V_w$	

(h' Is calculated by trigonometry.)

Calculations (show work)

Percent difference in W_f by two methods ....

Name .... Section .... Date ....

Lab Partner(s) ....

EXPERIMENT 12 *Work and Energy*

QUESTIONS

1. What is the percent energy lost to friction in each of the experimental cases? (Show your calculations.)

	$\theta = 30°$	$\theta = 45°$
Car moving up incline	....	....
Car moving down incline	....	....

2. In the experiment, what is the net force on the car when (a) moving up the incline; and (b) moving down the incline? Justify your answers with equations.

3. What is the net work done by the forces acting on the car for the car (a) moving up the incline; and (b) moving down the incline? Justify your answers. Consider positive and negative work.

4. Does a normal force $N = F_{\perp}$ act on each wheel of the car? Explain.

5. Prove that the coefficient of rolling friction for the car moving down the inclined plane with a constant speed is given by $\mu = \tan\theta - \dfrac{m_2}{m_c \cos\theta}$. (Use symbols, not numbers.)

6. Is the friction of the pulley taken into account in the experiment? Explain how this affects the experimental results.

Name .... Section .... Date ....

Lab Partner(s) ....

EXPERIMENT 13 *Torques, Equilibrium, and Center of Gravity*

ADVANCE STUDY ASSIGNMENT

Read the experiment and answer the following questions.

1. What conditions must be present for (a) translational equilibrium; (b) rotational equilibrium of a rigid body?

2. If these conditions for equilibrium are satisfied, is the rigid body necessarily in static equilibrium? Explain.

3. Write a definition and a mathematical expression for torque.

4. What is meant by clockwise and counterclockwise torques, and when are the sums of these torques on a rigid body equal?

5. What defines the center of gravity of rigid body, and how is it related to the center of mass?

6. Define linear mass density. Also, what is implied if it is assumed that the linear mass density of an object is constant?

EXPERIMENT **13**

Torques, Equilibrium, and Center of Gravity

I. INTRODUCTION

An important condition of rigid bodies in many practical applications is that of **static equilibrium.** Examples include beams in bridges and beam balances. When a rigid body such as a beam or a rod is "in balance," it is at rest or in static equilibrium. In particular, the beam is in rotational static equilibrium. That is, it does not rotate about some point or axis of rotation.

The criterion for rotational static equilibrium is that the sum of the torques, or moments of force acting on a rigid body, be equal to zero. To study torques and rotational equilibrium, a "beam" balance in the form of a meter stick and suspended weights will be used. By using the "moment-of-force" method, the mass of an object will be determined experimentally and the experimental value compared to the mass of the object as measured on a laboratory balance. Also, the concepts of center of gravity and center of mass will be investigated.

II. EQUIPMENT NEEDED

A. *Apparatus with One Support Point*

- Meter stick
- Support stand
- String and one knife-edge clamp *or* four knife-edge clamps (three with wire hangers)
- Four hooked weights (50-g, two 100-g, and 200-g)
- Unknown mass with hook
- Laboratory balance

B.* *Apparatus with Two Support Points*

- Two meter sticks
- Four knife-edge clamps with hangers
- Hangers and slotted masses (enough for 0.5 kg and 1.0 kg)
- Two spring balances (1.5 kg or 15 *N* capacity)
- Unknown mass with hook
- Laboratory balance
- Protractor

(*Note:* If both parts of the experiment are done, the equipment needs may be combined.)

III. THEORY

The conditions for the mechanical equilibrium of a rigid body are

$$\Sigma \mathbf{F} = 0 \qquad \textbf{(13-1a)}$$

$$\Sigma \tau = 0 \qquad \textbf{(13-1b)}$$

That is, the (vector) sums of the forces **F** and torques τ acting on the body are zero.

*The ranges of the equipment are given as examples. These may be varied to apply to available equipment.

The first condition, $\Sigma\ \mathbf{F} = 0$, is concerned with **translational equilibrium** and ensures that the object is at a particular location (not moving linearly) or that it is moving with a uniform linear velocity (Newton's first law of motion). In this experiment, the rigid body (the meter stick) is restricted from linear motion and $\Sigma\ \mathbf{F}$ is automatically satisfied.

To be in static equilibrium, a rigid body must also be in rotational static equilibrium. Although the sum of the forces on the body may be zero and it is not moving linearly, it is possible that it may be rotating about some fixed axis of rotation. However, if the sum of the torques is zero, $\Sigma\tau = 0$, the object is in **rotational equilibrium,** and either it does not rotate (static case) or it rotates with a uniform angular velocity. (Forces produce linear motion and torques produce rotational motion.)

A **torque,** or **moment of force,** results from the application of a force acting at a distance from an axis of rotation. The magnitude of the torque is equal to the product of the force's magnitude and the perpendicular distance from the axis of rotation to the force's line of action, or $\tau = Fd$ (Fig. 13-1). The distance d is called the **lever arm** or the **moment arm** of the force.

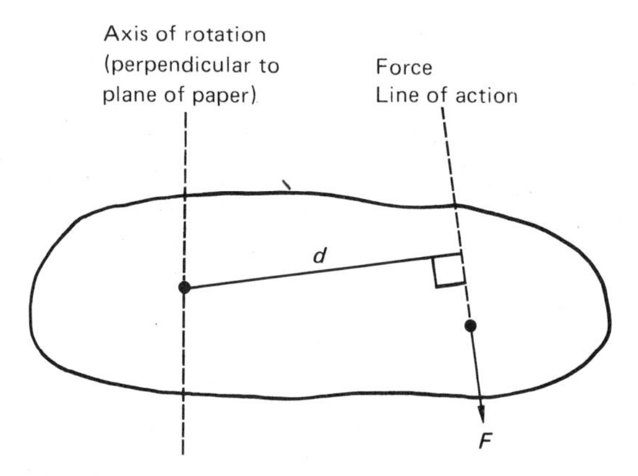

Fig. 13-1 The magnitude of the torque is equal to the product of F and the perpendicular distance d from the axis of rotation to the force's line of action $\tau = Fd$.

Relative to an axis of rotation, a rigid body can rotate in only two directions, clockwise and counterclockwise. It is therefore customary to refer to clockwise torques and counterclockwise torques, that is, torques that may produce clockwise rotations and torques that may produce counterclockwise rotations about an axis of rotation. For example, in Fig. 13-2, F_1 and F_2 produce counterclockwise torques and F_3 and F_4 produce clockwise torques, but no rotation would take place if the system were in rotational equilibrium.

The condition for rotational equilibrium is

$$\Sigma\,\tau = \Sigma\,\tau_{cc} + \Sigma\,\tau_{cw} = 0 \qquad \textbf{(13-2)}$$

where τ_{cc} and τ_{cw} are counterclockwise and clockwise torques, respectively. Designating the directions arbitrarily by plus and minus signs, Equation 13-2 can be written

$$\Sigma\tau_{cc} - \Sigma\tau_{cw} = 0$$

or

$$\Sigma\tau_{cc} = \Sigma\tau_{cw} \qquad \textbf{(13-3)}$$

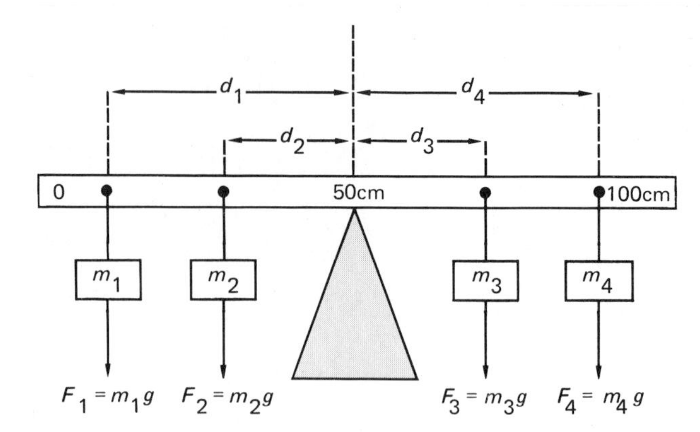

Fig. 13-2 An example of torque in different directions. F_1 and F_2 give rise to counterclockwise torques and F_3 and F_4 to clockwise torques.

sum of counterclockwise torques = sum of clockwise torques.

For example, for the rod in Fig. 13-2, we have

Counterclockwise   *Clockwise*

$$\tau_1 + \tau_2 = \tau_3 + \tau_4$$

or

$$F_1d_1 + F_2d_2 = F_3d_3 + F_4d_4$$

The forces are due to weights suspended from the rod. Then, with $F = mg$,

$$\begin{aligned} m_1gd_1 + m_2gd_2 &= m_3gd_3 + m_4gd_4 \\ m_1d_1 + m_2d_2 &= m_3d_3 + m_4d_4 \end{aligned} \qquad \textbf{(13-4)}$$

☐ **Example 13.1** Let $m_1 = m_3 = 50$ gm,* $m_2 = m_4 = 100$ gm in Fig. 13-2, where m_1, m_2, and m_3 are at the 10, 40, and 60 cm marks or positions, respectively, on the meter stick. Where would m_4 have to be suspended for the stick to be in static equilibrium?

Solution

$$\Sigma\,\tau_{cc} = \Sigma\,\tau_{cw}$$

or, using Eq. 13-4,

$$m_1g(50 - 10) + m_2g(50 - 40) = m_3g(60 - 50) + m_4gd_4$$

where the lever arms (d's) are measured from the pivot point (50 cm) of the meter stick. Then, putting in the mass values,

$$(50)g(40) + (100)g(10) = (50)g(10) + (100)gd_4$$

*The official abbreviation of mass (m) in grams is g. The standard symbol for acceleration due to gravity is g, where weight is given by mg. To avoid confusion with these symbols, since the acceleration due to gravity is often left in symbol form, gm will be used in this book as the abbreviation for gram in some cases for distinction. However, it should be kept in mind that the official abbreviation is g.

Dividing all the terms in the equation by the acceleration due to gravity g,

$$2000 + 1000 = 500 + 100d_4$$

or

$$d_4 = \frac{2500}{100} = 25 \text{ cm}$$

Hence, for rotational equilibrium m_4 is 25 cm from the support position (axis of rotation), or at the 75-cm position on the meter stick (measured from the zero end). □

A. *Center of Gravity and Center of Mass*

The gravitational torques due to "individual" mass particles of rigid body define what is known as the body's **center of gravity.** The center of gravity is the point of the body about which the sum of the gravitational torques about an axis through this point is zero. For example, consider the rod shown in Fig. 13-3. If the rod is visualized as being made up of individual mass particles and the point of support is selected such that $\Sigma\, \tau = 0$, then

$$\Sigma\, \tau_{cc} = \Sigma\, \tau_{cw}$$

or

$$\sum_{cc} (m_i g)d_i = \sum_{cw} (m_i g)d_i$$

and

$$(m_1 d_1 + m_2 d_2 + m_3 d_3 + \cdots)_{cc} = (m_1 d_1 + m_2 d_2 + m_3 d_3 + \cdots)_{cw}$$

With the rod in rotational equilibrium, it may be

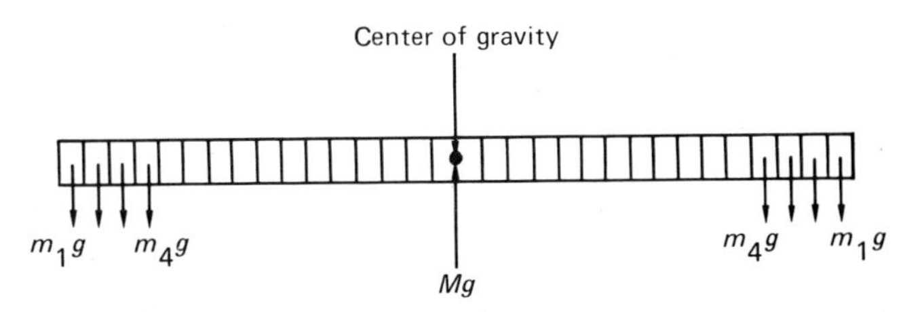

Fig. 13-3 A rod can be considered as being made up of individual masses in rotational equilibrium when the vertical support force is directed through the center of gravity.

supported by a force equal to its weight, where the support force is directed through the center of gravity. Hence, it is as though all of the object's weight (Mg) is concentrated at the center of gravity. That is, if you were blindfolded and supported an object at its center of gravity on your finger, weightwise you would not be able to tell if it were perhaps a rod or a block of equal mass.

If an object's weight were concentrated at its center of gravity, so would be its mass, and we often refer to an object's **center of mass** instead of center of gravity. These points are the same as long as the acceleration due to gravity g is constant (uniform gravitational field). Notice how g can be factored and divided out of the previous "weight" equations, leaving "mass" equations.

Also, it should be evident that for a symmetric object with a uniform mass distribution, the center of gravity is located at the center of symmetry. For example, if a rod has a uniform mass distribution, its center of gravity is located at the center of the rod's length. Why?

IV. EXPERIMENTAL PROCEDURE

A. *Apparatus with One Support Point*

1. A general experimental setup is illustrated in Fig. 13-4, where the masses or weights are suspended by clamp weight hangers. The hooked masses may also be suspended from small loops of string, which can be slid easily along the meter stick. The string allows the position of a mass to be easily read and may be held in place by a small piece of masking tape.
   (a) Determine the mass of the meter stick (without any clamps) and record in the Laboratory Report.
   (b) Weights may be suspended by loops of string or clamps with weight hangers. The string method is simpler; however, if you choose or are instructed to use weight hangers, weigh the three clamps together on a laboratory balance and compute the average mass of a clamp. Record in the data table.

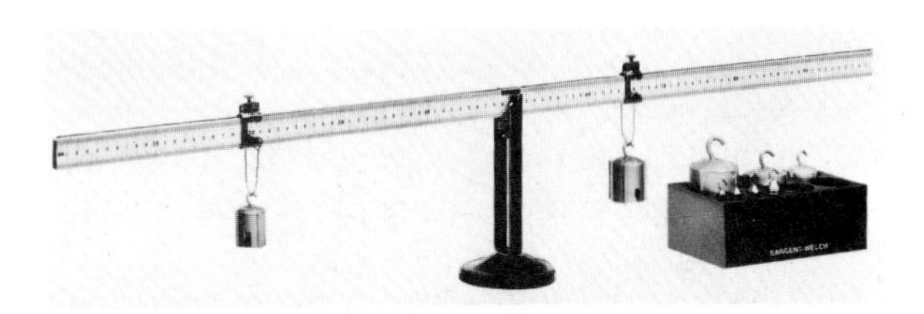

Fig. 13-4 Example of experimental setup for torque and equilibrium conditions. (Photo courtesy of Sargent-Welch Scientific Company)

2. With a knife-edge clamp on the meter stick near its center, place the meter stick (without any suspended weights) on the support stand. Make certain that the knife-edges are on the support stand. (The tightening screw head on the clamp will be down.) Adjust the meter stick through the clamp until the stick is balanced on the stand. Tighten the clamp screw and record the meter stick reading or the distance of the balancing point x_0 from the zero end of the meter stick in Data Table 1.

3. *Case 1: Two known masses.*
   (a) With the meter stick on the support stand at x_0, suspend a mass $m_1 = 100$ gm at the 15-cm position on the meter stick, that is, 15 cm from the zero end of the meter stick.
   (b) Set up the conditions for static equilibrium by adjusting the moment arm of a mass $m_2 = 200$ gm suspended on the side of the meter stick opposite m_1. Record the masses and moment arms in Data Table 1. Remember the moment arms are the distances from the pivot point to the masses (i.e., $d_i = |x_i - x_0|$).
   (c) Compute the torques and find the percent difference in the computed values (i.e., compare the clockwise torque with the counterclockwise torque). If clamps are used instead of string, do not forget to add the masses of the clamps.

4. *Case 2: Three known masses.*
   Case (a)
   (i) With the meter stick on the support stand at x_0, suspend $m_1 = 100$ gm at the 30-cm position and $m_2 = 200$ gm at the 70-cm position. Suspend $m_3 = 50$ gm and adjust to the appropriate moment arm of this mass so that the meter stick is in static equilibrium. Record the data in Data Table 1.
   (ii) Compute the torques and compare as in procedure 3.
   Case (b)
   (i) Calculate theoretically the lever arm (d_3) for the mass $m_3 = 50$ gm for the system to be in equilibrium if $m_1 = 100$ gm is at the 20-cm position and $m_2 = 200$ gm is at the 60-cm position. (Remember to add the masses of the hanger clamps if used.) Record in the data table.
   (ii) Check your results experimentally and compute the percent error of the experimental value of d_3, taking the previously calculated value as the accepted value.

5. *Case 3: Unknown mass — The balance principle.*
   (a) With the meter stick on the support stand at x_0, suspend the unknown mass (m_1) near one end of the meter stick (e.g., at the 10-cm position). Suspend from the other side of the meter stick an appropriate known countermass m_2 (e.g., 200 gm) and adjust its position until the meter stick is "in balance." Record the known mass and moment arms in Data Table 1.
   (b) Remove the unknown mass and determine its mass on a laboratory balance.
   (c) Compute the value of the unknown mass by the method of moments and compare with the measured value by calculating the percent error.

In the previous cases, the mass of the meter stick was not explicitly taken into account since the fulcrum or the position of the support was at the meter stick's center of gravity or center of mass. In effect, the torques due to the mass of the meter stick on either side of the support position canceled each other. For the following cases, the center of gravity of a "loaded" meter stick system will be used as the point of support. That is, the masses will be suspended at definite positions and the point of balance or the center of gravity will be determined experimentally. A loaded meter stick is analogous to a body whose mass is not uniformly distributed.

6. *Case 4: Meter stick with one mass.*
   (a) Suspend a mass $m_1 = 100$ gm at or near the zero end of the meter stick (Fig. 13-5). If a string loop is used, a piece of tape to hold the string in position helps. Move the meter stick in the support clamp until the meter stick is balanced. (This case is analogous to the solitary see-saw — sitting on one side of a balanced see-saw with no one on the other

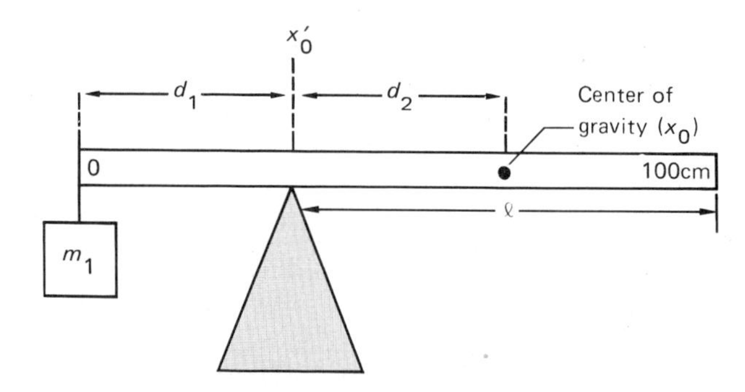

Fig. 13-5 A meter stick in equilibrium with one suspended mass.

side.) Record the position of the center of gravity of the loaded meter stick x_0' in Data Table 2.

Calculations: Since the meter stick is in balance (static equilibrium), the point of support must be at the center of gravity of the loaded stick [i.e., the torques (clockwise and counterclockwise) on either side of the meter stick must be equal]. But where is the mass or force on the side of the meter stick opposite the suspended mass? The balancing torque must be due to the mass of length ℓ of the meter stick (Fig. 13-5). To investigate this:

(i) Compute the average linear density μ of the meter stick, $\mu = M/L$ (gm/cm), where M is the total mass of the meter stick (measured previously) and L is the total length of the stick or 100 cm. Record in Data Table 2. The average linear mass density is the average mass (gm) per unit length (cm).

(ii) Assuming the mass of the meter stick to be uniformly distributed, the mass m_2 of the portion of the meter stick on the side of the support stand opposite m_1 can be considered to be concentrated at the center of mass (or center of gravity) of that length of the meter stick (Fig. 13-5).
Compute the mass m_2 of length ℓ of the meter stick. The length ℓ is given by $\ell = 100\text{ cm} - x_0'$, and $m_2 = \mu\ell$.

(iii) Compute the moment arm d_2 of the center of gravity of length ℓ of the meter stick and record this value in the data table ($d_2 = \ell/2$).

(iv) Compute the torque due to m_2 and compare it to the torque due to the suspended mass m_1 on the opposite side of the support stand by calculating the percent difference. (Is the percent difference comparable to those of the previous cases?)

(b) Take into account the mass m_3 of the length (d_1) of the meter stick on the same side of the support stand as the suspended mass m_1. Assume this mass to be concentrated at the center of this length (i.e., $d_1/2$).
The condition for equilibrium is then

$$m_1 d_1 + \frac{m_3 d_1}{2} = m_2 d_2$$

$$\tau_{cc} = \tau_{cw}$$

where m_3 is determined from the linear mass density. Compute the percent difference between the counterclockwise and clockwise torques with this consideration.

(c) Instead of considering the meter stick as consisting of two masses m_2 and m_3, now represent it by a single mass M located at x_0, where M is the total mass of the meter stick and x_0 is the location of its center of mass as determined in procedure 2.
Using the values of M and x_0, calculate the counterclockwise and clockwise torques and the percent difference between them.

7. *Case 5: Center of gravity.*
(a) With a mass $m_1 = 100$ gm positioned at or near one end of the meter stick as in Case 4, suspend a mass $m_2 = 100$ gm on the opposite side of the support stand at the 60-cm position. Adjust the meter stick in the support-stand clamp until the stick is in balance. This locates the center of gravity (x_0') of the configuration. Record d_1 and d_2 in Data Table 2. Repeat the procedure with m_2 positioned at
(b) 70 cm and
(c) 80 cm.
Notice how the position of the center of gravity moves as the mass distribution is varied.
(d) Based on the experimental data, what would you predict the position of the center of gravity (x_0') of the system to be if m_2 were moved to the 90-cm position? Record your prediction in the data table.
Using your prediction, compute the counterclockwise and clockwise torques, taking into account the mass of the meter stick as in procedure 6(c). Compare the torques by computing the percent difference.
Experimentally determine the position of the center of gravity of the system and compute the percent difference between the experimental and predicted values.

B. *Apparatus with Two Points of Support*

1. The experimental arrangement is illustrated in Fig. 13-6. Loops of string may be used instead of clamps for suspension as in part A, however, clamps are generally more convenient in this case.
(a) Determine the mass of the meter stick (without any clamps) using a laboratory balance and record in the Laboratory Report.
(b) Determine the mass of each of the four clamps with hangers using a laboratory balance and record (if not using string loops). Since the clamps usually have slightly differ-

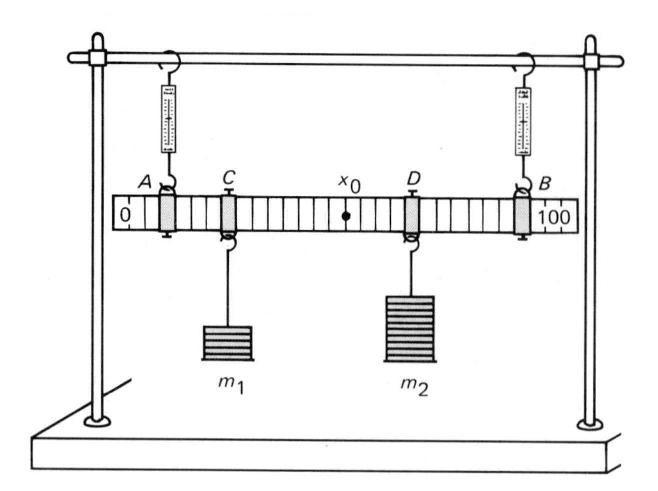

Fig. 13-6 Illustration of experimental setup for apparatus with two points of support (suspension). Equilibrium conditions require that $\Sigma \mathbf{F} = 0$ and $\Sigma \tau = 0$.

ent masses, it is suggested that the clamps be labeled with a pencil (A, B, C, and D) so they are not mixed up during the experiment.

2. Determine the location of the center of mass of the meter stick (if not done in procedure 2, part A). Suspend the meter stick by a single clamp positioned near the center of the stick. Adjust the clamp position until the meter stick is in balanced equilibrium. Record the clamp position or the distance of the balancing point x_0 from the zero end of the meter stick. (If the mass distribution of the meter stick is uniform, this should be at the 50-cm position. Why?)

3. Hang the two spring balances from a horizontal support(s) above the table. Check the "zero" readings of the balances with no mass suspended from them. If the readings are not zero, record the zero corrections (− or +). These corrections should be added or subtracted to the balance readings in the experimental procedures.

   Place the four clamps on the meter stick and suspend the meter stick from the balances as in Fig. 13-6, with the suspension clamps at the 10-cm and 90-cm positions on the meter stick. Placing clamps C and D together in contact on either side of the x_0 (center of mass) position, make sure that the meter stick is horizontal. Check this by using the second meter to see if the ends of the suspended meter stick are the same distance from the table top. If not, adjust the spring balance suspensions accordingly. (*Note:* Make certain that the spring balances are in vertical positions directly above the 10-cm and 90-cm marks.)

4. *Case 1.* Suspend masses $m_1 = 0.50$ kg and $m_2 = 1.0$ kg from the intermediate clamps. (Do not forget to include the mass of the hanger.) Place and secure m_2 at the 60-cm position on the meter stick. Adjust the position of m_1 along the meter stick until the stick is horizontal. (Check this as above.) Record the pertinent data in Data Table 3.

5. (In the following calculations, do not forget to include the masses of the clamps. *Hint:* For m_A and m_B in torque calculations, consider net forces, e.g., $F_A - m_A g$. Also, it is convenient to leave the acceleration due to gravity in symbol form unless the instructor requests otherwise.)
   (a) Compute the downward forces and compare their sum with the sum of the upward forces by finding the percent difference.
   (b) Compute the clockwise and counterclockwise torques *about an axis through the zero end of the meter stick.* Find the percent difference between the sums of these torques.

6. *Case 2.* Suspend the unknown mass m_u in place of m_1 and again make the stick horizontal by adjusting the position of m_u. Record the pertinent data in Data Table 4.

7. Using the conditions of equilibrium, compute the value of the unknown mass. Record the result from each condition in the data table. *(In this case, for the summation of the torques, take the axis of rotation to be through the center of mass of the meter stick.)*

8. Determine the value of the unknown mass using a laboratory balance. Compare this with the two values computed in procedure 7 by finding the percent differences.

9. *Case 3.*
   (a) Again using m_1 and m_2, suspend m_1 at the 20-cm position and m_2 at the 70-cm position on the meter stick. (The stick should not be horizontal. Why?) Move the spring balance clamps on the meter stick *or* the upper supports of the spring balances (if movable) so that the balances are at angles (θ_A and θ_B) other than 90° to the horizontal until the meter stick is horizontal. (If the balance clamps are moved, this may necessitate moving the m_1 clamp to a position

nearer the center of the meter stick. Adjust accordingly.)

(b) Using a protractor, measure the angles the balance clamp hangers make with the horizontal meter stick and record in Data Table 5.

10. Analyze this case of nonparallel suspension forces in terms of the conditions of equilibrium. Use the zero end of the meter stick as the axis of rotation. (Recall that the condition $\Sigma \mathbf{F} = 0$ requires that $\Sigma \mathbf{F}_x = 0$ and $\Sigma \mathbf{F}_y = 0$, or that the sums of the components be zero. Also recall that a torque is defined as the product of the force and the perpendicular distance from the axis of rotation to the force's line of action, *or equivalently,* the product of the distance from the axis of rotation to the force's point of application and the perpendicular component of the force.)

Lab Partner(s) ....

EXPERIMENT 13 *Torques, Equilibrium, and Center of Gravity*

LABORATORY REPORT

A. *Apparatus with One Point of Support*

Mass of meter stick, M ....

Total mass of clamps ....

Average mass of one clamp m_c ....

DATA TABLE 1

Balancing position (center of gravity) of meter stick x_0 ....

Diagram*	Values (add m_c to masses if clamps used)	Intermediate calculations	Final** results
Case 1 d_1 d_2 x_1 x_0 x_2 0 100 cm m_1 m_2	m_1 .... x_1 = 15 cm m_2 .... x_2	d_1 .... d_2	τ_{cc} .... τ_{cw} .... Percent diff. ....
Case 2(a)	m_1 .... x_1 = 30 cm m_2 .... x_2 = 70 cm m_3 .... x_3	d_1 .... d_2 .... d_3	τ_{cc} .... τ_{cw} .... Percent diff. ....
Case 2(b)	m_1 .... x_1 = 20 cm m_2 .... x_2 = 60 cm m_3 .... x_3	d_1 .... d_2	d_3 .... (calc.) d_3 .... (meas.) Percent diff. ....
Case 3	m_1 .... (unknown) x_1 .... (known) m_2 .... (known) x_2 .... (from expt.) m_1 .... (measured)	d_1 .... d_2	m_1 .... (calc.) m_1 .... (meas.) Percent diff. ....

*Draw a diagram to illustrate each case using Case 1 diagram as an example.
**Attach a sheet to the Laboratory Report showing calculations for each case.

Name .... Section .... Date ....

Lab Partner(s) ....

EXPERIMENT 13 *Torques, Equilibrium, and Center of Gravity*

DATA TABLE 2 Linear mass density of meter stick, $\mu = M/L$ ....

Draw a diagram for each case as in Case 4(a). In these diagrams, m_i represents a physical mass suspended at x_i, where $d_i = |x_i - x_0'|$. $<m_i>$ is an "equivalent" mass representing a part of the meter stick mass, for example in Case 4(a), $<m_2> = \rho\ell$ located at the position $<x_2> = \ell/2$, where ℓ is the length of the meter stick on that side of the support. M is the total mass of the meter stick, effective located at its center of gravity, x_0, and $d_M = |x_0 - x_0'|$.

Diagram*	Values (add m_c if appl.)	Calculations	Results**
Case 4(a) d_1 (d_2) x_1 x_0' (x_2) 0 100 cm m_1 (m_2)	m_1 .... $x_1 = 0$ cm x_0'	d_1 .... $<m_2>$ .... $<x_2>$ .... $<d_2>$	τ_{cc} .... τ_{cw} .... Percent diff. ....
Case 4(b)	m_1 .... $x_1 = 0$ cm x_0'	$<m_3>$ .... $<x_3>$ .... $<d_3>$	τ_{cc} .... τ_{cw} .... Percent diff. ....
Case 4(c)	m_1 .... $x_1 = 0$ cm x_0'	M .... x_0 .... d_M	τ_{cc} .... τ_{cw} .... Percent diff. ....
Case 5(a)	m_1 .... $x_1 = 0$ cm m_2 .... $x_2 = 60$ cm x_0'	d_1 .... d_2 .... d_M	
Case 5(b)	same except $x_2 = 70$ cm x_0'	d_1 .... d_2 .... d_M	
Case 5(c)	same except $x_2 = 80$ cm x_0'	d_1 .... d_2 .... d_M	
Case 5(d)	same except $x_2 = 90$ cm x_0' .... (predicted)	τ_{cc} .... τ_{cw} .... Percent diff.	x_0' .... (measured) Percent diff. ....

*Draw a diagram to illustrate each case using Case 1 diagram as an example.
**Attach a sheet to the Laboratory Report showing calculations for each case.

B. *Apparatus with Two Points of Suspension*

Mass of meter stick, M ....

Meter stick center of mass position x_0 ....

Hanger masses m_A .... m_C ....

m_B .... m_D ....

Spring balance zero corrections

m_1 .... A ....

m_2 .... B ....

DATA TABLE 3

	Forces up (N)	Forces down (N)	Position on meter stick	Lever arm (d)	τ^*_{cw} (N–m)	τ^*_{cc} (N–m)
$m_1 + m_C$						
$m_2 + m_D$						
M						
F_A						
F_B						
Sum						
Percent diff.						

*Axis of rotation through zero end of meter stick.

Calculations of percent differences (show work)

DATA TABLE 4

	Forces up (N)	Forces down (N)	Position on meter stick	Lever arm (d)	τ^*_{cw} (N–m)	τ^*_{cc} (N–m)
$m_u + m_C$						
$m_2 + m_D$						
M						
F_A						
F_B						

*Axis of rotation through center of mass of meter stick.

Name .... Section .... Date ....

Lab Partner(s) ....

EXPERIMENT 13 *Torques, Equilibrium, and Center of Gravity*

Calculations of m_u (show work)

	m_u	Percent difference (with measured value)
From $\Sigma \mathbf{F} = 0$	....	....
From $\Sigma \tau = 0$	....	....
Measured	....	

DATA TABLE 5

	Spring balance reading (N)	Forces down (N)	Position on meter stick	Lever arm (d)
$m_1 + m_C$	X			
$m_2 + m_D$	X			
M	X			
F_A		X		
F_B		X		

θ_A ....

θ_B ....

Analyze this case in terms of the conditions of equilibrium below, with the axis of rotation through the zero end of the meter stick. (Show work clearly with equations written in symbol form before doing calculations.)

(continued)

QUESTIONS (answer those applicable)

1. Explain how the condition $\Sigma \mathbf{F} = 0$ is satisfied for the meter stick in part A of the experiment.

2. In many instances, the balancing position x_0 of the meter stick by itself (procedure 2) is not at 50 cm. Explain why this is the case.

3. Explain the principle of the triple-beam laboratory balance (see Experiment 2).

4. Suppose in a situation as in Case 2, part A, in the experiment, $m_1 = 200$ gm were at the 20-cm position and $m_2 = 100$ gm at the 65-cm position. Would there be a problem in experimentally balancing the system with $m_3 = 50$ gm? Explain. If so, how might the problem be resolved?

5. In part B of the experiment, is it necessary that the spring balance positions be equidistant from the ends of the meter stick? Explain.

Lab Partner(s) ....

EXPERIMENT 14 *Simple Machines: Mechanical Advantage and Efficiency*

ADVANCE STUDY ASSIGNMENT

Read the experiment and answer the following questions.

1. Distinguish between AMA and TMA and explain what these quantities tell you about a machine. Support your explanation with an example.

2. What is the efficiency of a machine, and what does it tell you about the machine?

3. Machines make it easier to perform some tasks and we commonly say that machines do work for us. One might think from these statements that a machine increases or multiplies its work input. Is this the case? Explain.

4. On what physical conditions do the AMA and TMA of an inclined plane depend?

5. Does the movable pulley(s) of a block-and-tackle system have any effect on the weight of the load? Explain.

(continued)

6. Explain how the TMA of a block and tackle is obtained from the configuration of the system.

7. A machine is commonly used as a force "multiplier," but suppose for some reason that a wheel and axle were to be used as a force "reducer" (i.e., the output force was to be less than the applied input force). How could this be accomplished?

EXPERIMENT 14

Simple Machines: Mechanical Advantage and Efficiency

I. INTRODUCTION

Machines are used daily by each of us to "do work." Upon analysis, all mechanical machines, however complex, are combinations of simple machines of which there are six classes: (1) inclined planes, (2) wedges, (3) screws, (4) levers, (5) pulleys, and (6) wheels and axles.

Although used to perform work, a simple machine is basically a device that is used to change the magnitude (or direction) of a force. Essentially, a machine is a force multiplier. The magnitude of this multiplication is given by a machine's **mechanical advantage,** that is, the **actual mechanical advantage (AMA),** which takes into account frictional forces, or the **theoretical mechanical advantage (TMA),** which expresses the ideal, nonfrictional situation. The relative amount of useful work done by a machine [total work input = useful work + work (energy) lost to friction] is expressed by the ratio of the useful work output and the total work input, which is called the *efficiency.*

In this experiment, the AMA's, TMA's, and efficiencies of several simple machines will be experimentally determined in order to illustrate these concepts and to show the parameters on which the force multiplication of the machines depends.

II. EQUIPMENT NEEDED

- Inclined plane with pulley and Hall's carriage (car)
- Two single pulleys and two double- or triple-sheave pulleys*
- Wheel and axle
- Two weight hangers and slotted weights
- String
- Meter stick
- Vernier calipers
- Protractor

III. THEORY

The actual mechanical advantage (AMA) of a machine is defined as

$$\text{AMA} = \frac{F_o}{F_i} \qquad \textbf{(14-1)}$$

where F_o and F_i are the actual output and input forces, respectively. The AMA is the force multiplication factor of the machine. For example, if AMA $= 2 = F_o/F_i$, then $F_o = 2F_i$, or the output force is twice the input force.

In no case is the work multiplied by a machine. However, the total work (energy) is conserved:

$$\begin{aligned}\text{total work in} &= \text{total work done}\\ &= \text{useful work} + \\ &\quad\ \text{work done against friction}\end{aligned}$$

or

$$F_i d_i = F_o d_o + W_f \qquad \textbf{(14-2)}$$

where d_i and d_o are the parallel distances through which the input force F_i and the output force F_o act,

*The single pulleys are not really necessary, as one sheave of the multiple pulleys can be used as a single pulley. The single pulleys are convenient for instruction.

respectively. (Work = force × parallel distance, $W = Fd$.) W_f is the work done against friction. In actual situations, there is always some work (energy) input lost to friction.

If a machine were frictionless, $W_f = 0$, then $F_i d_i = F_o d_o$. For this theoretical situation, a **theoretical mechanical advantage (TMA)** can be expressed as

$$\text{TMA} = \left(\frac{F_o}{F_i}\right)_{\text{theoretical}} = \frac{d_i}{d_o} \qquad \textbf{(14-3)}$$

This is an ideal situation, and the theoretical mechanical advantage is sometimes called the **ideal mechanical advantage** (IMA).

Note that the TMA is the ratio of the distances through which the forces act, and thus depends on the geometrical configuration of the machine. That is, from the configuration of the machine, the distances through which the forces would act can be determined, and hence the TMA.

The **efficiency (*E*)** of a machine is defined as the ratio of its work output and its work input:

$$E = \frac{\text{work output } (W_o)}{\text{work input } (W_i)} = \frac{F_o d_o}{F_i d_i}$$

$$= \frac{F_o/F_i}{d_i/d_o} = \frac{\text{AMA}}{\text{TMA}} \qquad \textbf{(14-4)}$$

The efficiency is often expressed as a percent. Because of friction, AMA < TMA, and the efficiency is always less than 1 or 100 percent. The efficiency tells what part of the work input is useful work (work output):

$$\text{work output} = E\,(\text{work input})$$

For example, if $E = 0.7$ or 70 percent, then 0.7 or 70 percent of the work input is used by the machine to do useful work. The rest of the work input, 0.3 or 30 percent, is lost to friction.

A. *Inclined Plane*

Inclined planes in the form of ramps are commonly used to move loads to higher and lower elevations. A load is illustrated in Fig. 14-1 being pulled up an inclined plane (see also Fig. 12-1). The work output W_o is the moving of the car's weight w_1 through a vertical distance h (increase in potential energy),

$$W_o = F_o d_o = w_1 h \qquad \textbf{(14-5)}$$

The work input W_i in moving the car up the plane is equal to the applied force F times the parallel distance ℓ through which it acts,

$$W_i = F_i d_i = F\ell \qquad \textbf{(14-6)}$$

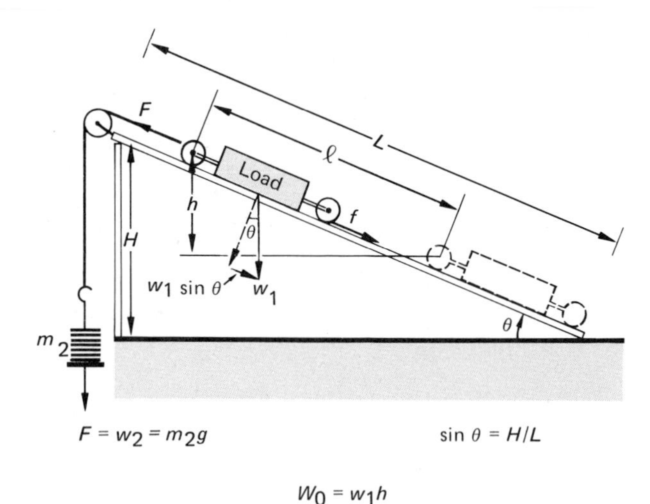

Fig. 14-1 Arrangement for determining the mechanical advantage of a inclined plane.

This work is done against the weight component of the car acting parallel to the plane, $w_1 \sin\theta$, and the force of friction f. When the car moves up the plane with a uniform speed, the vector sum of the forces acting on the car parallel to the plane is zero, and $F = w_1 \sin\theta + f$. If the car were accelerated up the plane, the input force (and input work) would be greater than actually needed to move the car up the plane. The mechanical advantage is based on the minimum input force.

The AMA of an inclined plane is then,

$$\underset{\textit{(inclined plane)}}{\text{AMA}} = \frac{F_o}{F_i} = \frac{w_1}{F} = \frac{w_1}{w_2} \qquad \textbf{(14-7)}$$

where $w_2 = F$ is the weight of the suspended mass that moves the car up the plane with a uniform speed.

If there were no frictional forces acting on the car (the theoretical mechanical situation), an input force $F = w_1 \sin\theta$ would move the block up the plane with a uniform speed. In this case, from $W_o = W_i$ we have $F_o d_o = F_i d_i$, and

$$\frac{F_o}{F_i} = \frac{w_1}{F} = \frac{w_1}{w_1 \sin\theta} = \frac{1}{\sin\theta} = \frac{d_i}{d_o} \qquad \textbf{(14-8)}$$

Hence, the TMA of an inclined plane is

$$\underset{\textit{(inclined plane)}}{\text{TMA}} = \left(\frac{F_o}{F_i}\right)_{\text{theoretical}} = \frac{d_i}{d_o} = \frac{1}{\sin\theta}$$

where θ is the angle of inclination of the plane.

Ideally, the inclined plane allows the load or car to be raised with a force of $w_1 \sin\theta$ instead of w_1, which would be required to raise it a distance h directly. However, the load must be moved a greater distance by the smaller force (through length ℓ instead of h). Notice from Fig. 14-1 that $\sin\theta = H/L$ and $\text{TMA} = 1/\sin\theta = L/H$, where L is the total length of the plane and H is the maximum height. Thus, for a given incline, the TMA can be determined from the geometry of the plane.

B. *Pulleys*

A *pulley* is actually a continuous lever with equal lever arms [Fig. 14-2(a)]. When a pulley or a system of pulleys is used to lift a load of weight w by an applied force F, the AMA is

$$\underset{[\text{pulley(s)}]}{\text{AMA}} = \frac{F_o}{F_i} = \frac{w}{F} \quad \textbf{(14-9)}$$

The TMA is the ratio of the distance through which the forces act, since ideally $W_o = W_i$ as in the previous case.

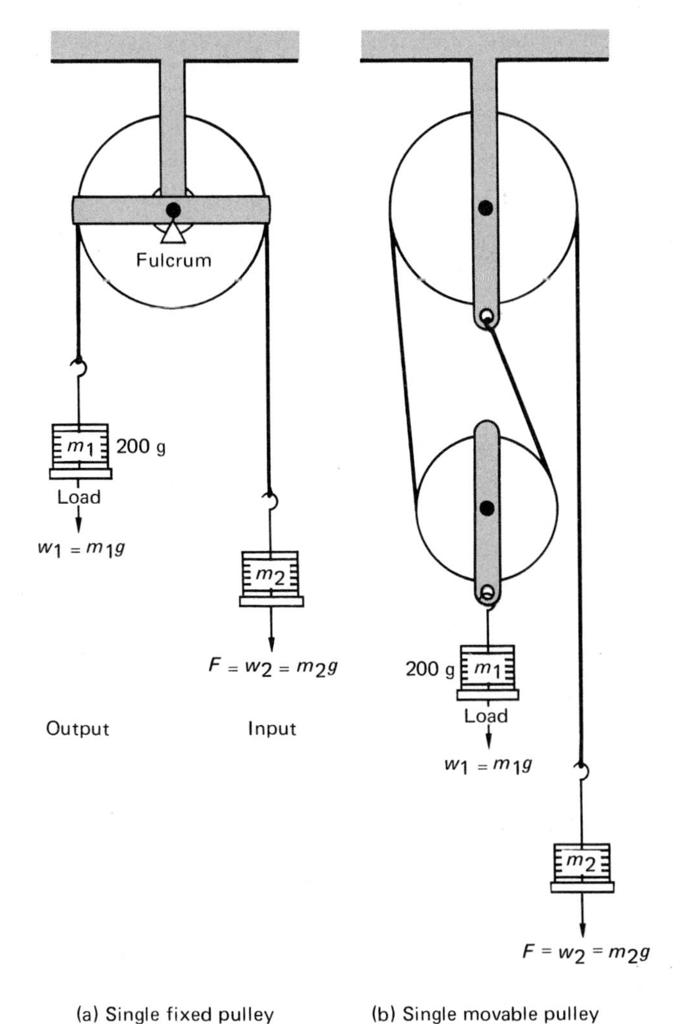

Fig. 14-2 Pulley arrangements. (a) Single fixed pulley. (b) Single movable pulley.

$$\underset{[\text{pulley(s)}]}{\text{TMA}} = \frac{d_i}{d_o} = \frac{h_i}{h_o} \quad \textbf{(14-10)}$$

where h_i and h_o are the vertical heights through which the input and output forces act, respectively. For example, for a single movable pulley [Fig. 14-2(b)], suppose that the load is moved a distance h. The applied force must move through a distance $2h$, because to move the load a distance h, a length of string $2h$ must be "pulled up" by the applied force (i.e., each support string of the pulley must be shortened by a length h).

The AMA is measured when the load is lifted with a uniform speed so that acceleration is not a consideration. When the load is not accelerated, there is no net force on the load. The load could be accelerated by applying a greater input force, but in this case, the work input would be greater than actually needed to just lift the load. As stated previously, the mechanical advantage is based on the minimum input force. An equivalent measurement of the minimum input force is obtained when the system is in static equilibrium, in which case the net vertical force of the system is zero. (Of course, to lift the load a slight tap or force would have to be given to put the system in motion.)

Note that a suspended movable pulley adds to the weight of the load, since the weight of the pulley is also lifted by the applied force. A set of fixed and movable pulleys (Fig. 14-3) is commonly called a **block and tackle.** The pulleys, commonly called sheaves (pronounced "shivs"), may be arranged in tandem, which is the configuration commonly used in the laboratory to make it easier to thread the pulleys, or they may have a common axis of rotation for compactness in practical application.

C. *Wheel and Axle*

The combined wheel and axle has many applications. For example, when you open a door by turning a door knob, you are using a wheel and axle. This simple machine consists of a wheel fixed to a shaft or axle with the same axis of rotation (Fig. 14-4). Essentially, it is equivalent to a lever with unequal lever arms.

As shown in Fig. 14-4, a force F applied tangentially to the wheel with a radius R can lift a load w by means of a string or rope wrapped around the axle (radius r). The AMA of the wheel and axle is

$$\underset{(\text{wheel and axle})}{\text{AMA}} = \frac{F_o}{F_i} = \frac{w}{F} \quad \textbf{(14-11)}$$

In one revolution, the input force acts through a dis-

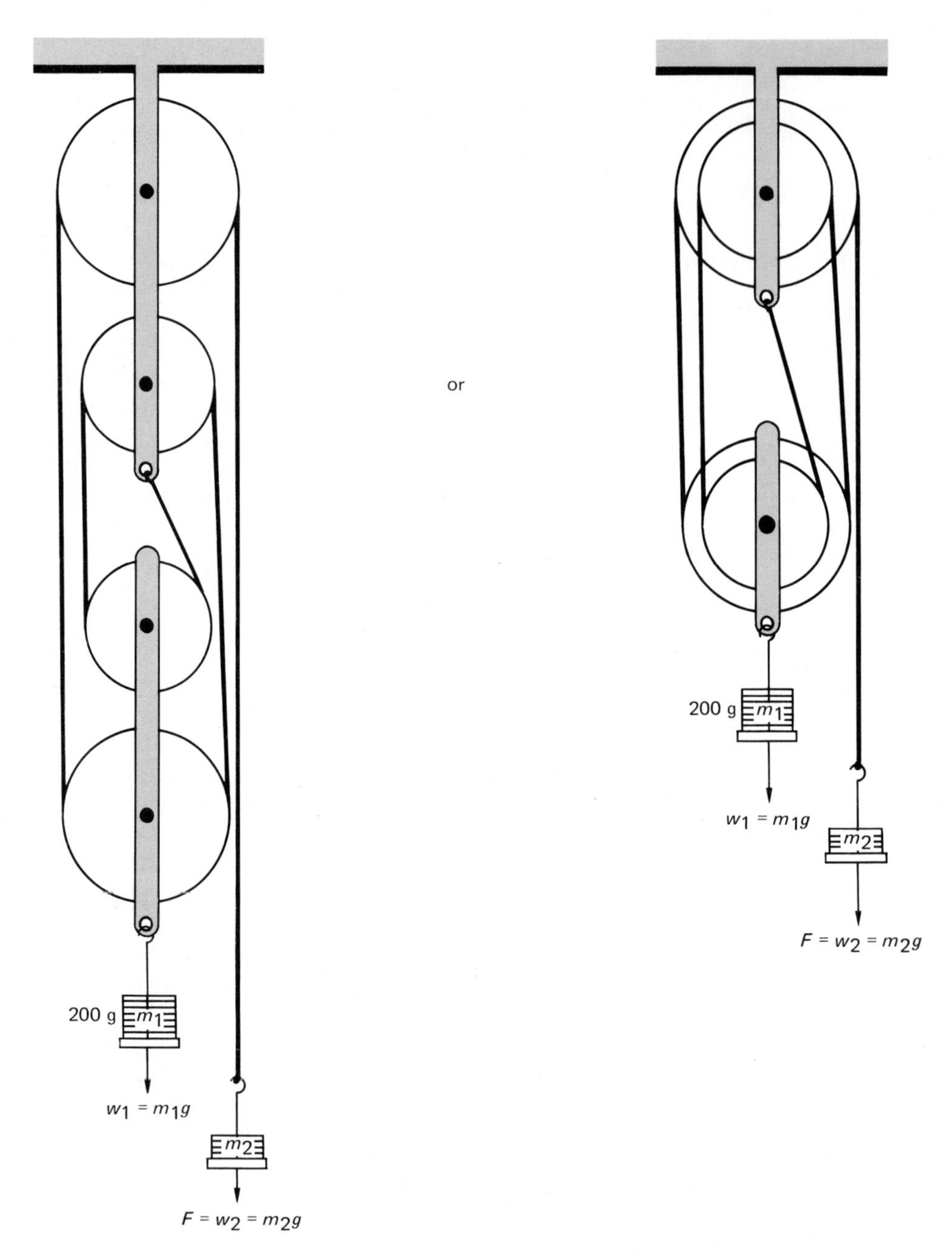

Fig. 14-3 Pulley arrangements with double movable pulleys. The pulleys may be arranged in tandem or have a common axis for compactness.

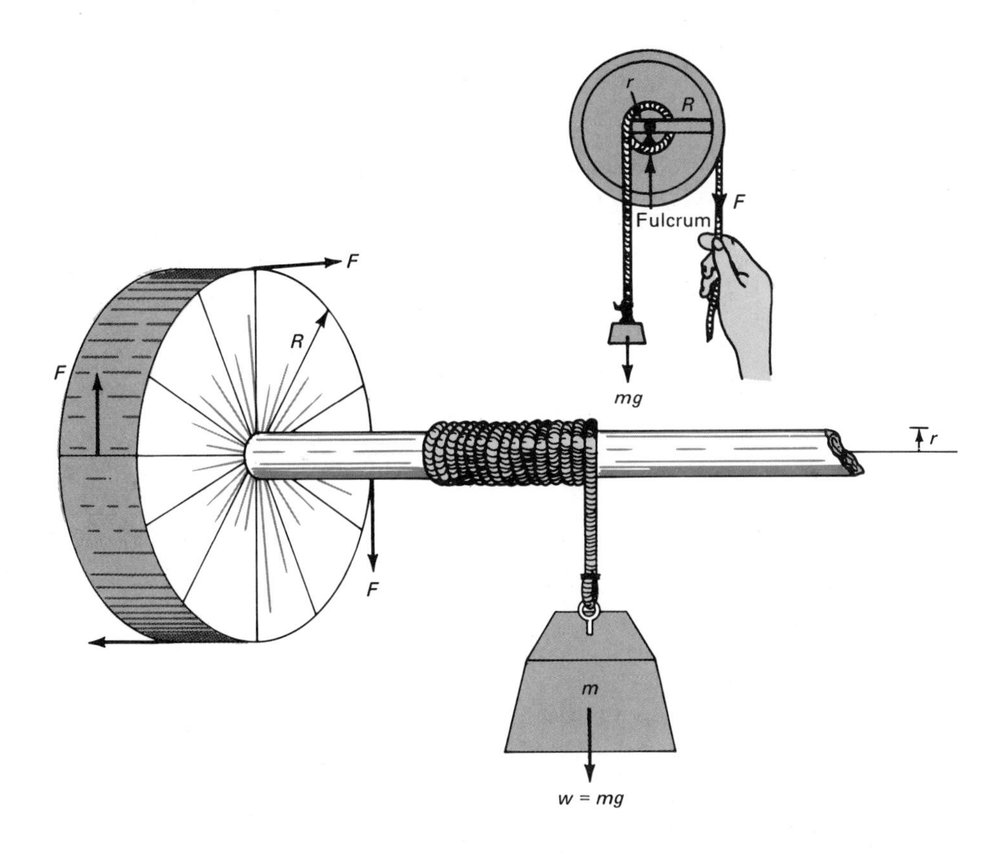

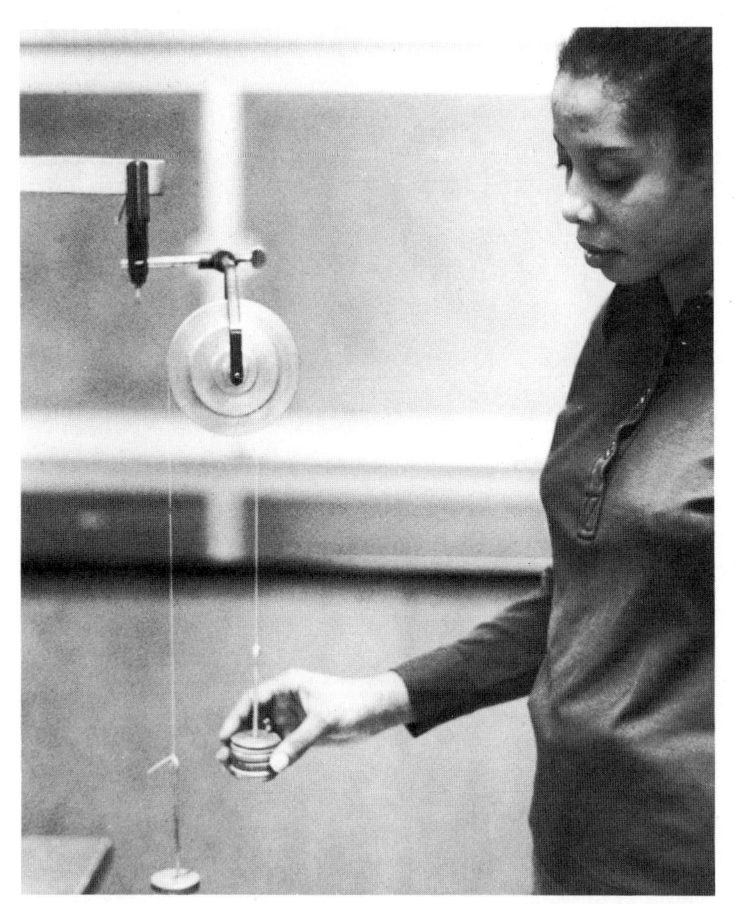

Fig. 14-4 A wheel and axle arrangement. The torques may be equal, but less force is supplied to the wheel than to the axle because of the wheel's larger radius.

tance $2\pi R$ and the output force through a distance $2\pi r$. For the ideal, nonfrictional case, $F_i d_i = F(2\pi R) = F_o d_o = w(2\pi r)$, and the TMA is

$$\underset{(wheel\ and\ axle)}{\text{TMA}} = \frac{d_i}{d_o} = \frac{R}{r} \quad \textbf{(14-12)}$$

In measuring the force to determine the AMA of a wheel and axle, it is convenient to use the static equilibrium case for a pulley system (see Section III.B).

IV. EXPERIMENTAL PROCEDURE

A. *Inclined Plane*

1. Determine the mass of the car on a laboratory balance and record in Data Table 1.

2. Adjust the inclined plane to an angle of $\theta = 10°$ (Fig. 14-1). With the car near the bottom of the plane, attach the car to a weight hanger by means of string suspended over a pulley, as shown in the figure. Adjust the pulley so that the string is parallel to the inclined plane.

3. Add weights to the weight hanger until the car moves slowly up the plane with a uniform speed after being given a slight tap. If not enough mass is suspended, the car will stop moving after being given a slight push. Record the suspended mass required for this in Data Table 1.

4. Calculate the AMA, TMA, and efficiency of the inclined plane for this case.

5. Adjust the plane to an incline angle of 20° and repeat procedures 2 to 4.

6. Adjust the plane to an incline angle of 30° and repeat procedures 2 to 4.

B. *Pulleys*

7. Determine the mass of one of the single pulleys and one of the multiple-sheave pulleys on a laboratory balance and record in Data Table 2. These pulleys will be used as the movable pulleys in the following situations and their weights must be included in the loads.

8. Assemble a single fixed pulley, as illustrated in Fig. 14-2(a), with enough weights on the force input weight hanger so that it moves downward with a slow uniform speed when given a slight tap. (A single pulley or one pulley of a multiple-sheave pulley may be used.) Record the masses in Data Table 2.

9. The next step is to measure the distances d_i and d_o with a meter stick. Pull down the weight hanger supplying the input force F a distance (d_i or h_i), say 10 or 20 cm, and note the distance the load moves upward (d_o or h_o). Record these distances in Data Table 2.

10. Calculate the AMA, TMA, and efficiency for this case.

11. Assemble a pulley system as illustrated in Fig. 14-2(b). Repeat procedures 8 to 10 for this case. (Don't forget to include the mass of the movable pulley as part of the load since it, too, is being raised.)

12. Assemble a pulley system as illustrated in Fig. 14-3 and repeat procedures 8 to 10.

C. *Wheel and Axle*

13. Using the vernier calipers, determine the radii of the wheel (pulley with the largest diameter) and of the largest and smallest axles. The wheel and axle apparatus commonly used has three sizes of axles.

14. Tying and wrapping strings around the wheel and axle, set up a wheel and axle as shown in Fig. 14-4, with enough weights on the input force weight hanger so that it descends with a slow uniform velocity. Start with the largest-diameter axle if your wheel and axle has multiple axles. Record the masses of the applied force and the load in Data Table 3.

15. Calculate the AMA, TMA, and efficiency of the wheel and axle for this case.

16. Repeat procedures 14 and 15 with the load suspended from the axle with the smallest diameter (if your wheel and axle has multiple axles).

Name .... Section .... Date ....

Lab Partner(s) ....

EXPERIMENT 14 *Simple Machines: Mechanical Advantage and Efficiency*

LABORATORY REPORT

A. *Inclined Plane*

DATA TABLE 1 Mass of car, m_1 ....

Angle of incline	Output force or load,* $w_1 = m_1 g$	Input force,* $w_2 = m_2 g$	AMA	TMA	E

*It is convenient to express the acceleration due to gravity in symbol form (e.g., if $m_1 = 100$ gm, $w_1 = m_1 g = 100g$ dynes, or $m_1 = 0.1$ kg and $w_1 = m_1 g = 0.1g$ N).

Calculations (show work)

(continued)

B. *Pulleys*

DATA TABLE 2

Mass(es) of movable pulley(s) ....

....

Pulleys	Output force or load, $w_1 = m_1 g$	Input force, F, $w_2 = m_2 g$	Output distance, d_o or h_o	Input distance, d_i or h_i	AMA	TMA	E
Single fixed							
Single movable							
Double movable							

Calculations (show work)

C. *Wheel and Axle*

DATA TABLE 3

Axle radius	Wheel radius	Output force or load, $w_1 = m_1 g$	Input force, F, $w_2 = m_2 g$	AMA	TMA	E

Calculations (show work)

Name .... Section .... Date ....

Lab Partner(s) ....

EXPERIMENT 14 *Simple Machines: Mechanical Advantage and Efficiency*

QUESTIONS

1. Simple machines are often divided into two basic classes, inclined planes and levers, where the wedge and screw are included in the first class, and the pulley and the wheel and axle are included in the latter. Explain why these four simple machines can be included in the basic classes of inclined planes and levers.

2. A machine multiplies the force, but what is "sacrificed" for this force multiplication? Give a specific example.

3. (a) State how the AMA and TMA of an inclined plane vary with the inclination of the plane.

   (b) State how the efficiency of an inclined plane varies with the inclination of the plane and explain the reason for this variation.

4. A single fixed pulley is often called a "direction changer." Explain why this is an appropriate name.

(continued)

5. The TMA of a pulley system with movable pulley(s) or a block and tackle is equal to the number of supporting strands of the movable pulley or block.

   (a) Do your experimental results support this statement?

   (b) Explain the physical basis of the statement.

6. Suppose that a block-and-tackle system similar to the one in Fig. 14-3 were used to lift a 500-lb engine from an automobile a distance of 3 ft. If the block and tackle had the same AMA and efficiency as that you determined experimentally for the similar pulley system, (a) what is the applied force needed to lift the engine; (b) how much of the input work is lost to friction?

7. Estimating the radii of a common door knob on its shaft, how much force is applied to the shaft mechanism when the knob is turned with an applied tangential force of 2.0 N?

Name .... Section .... Date ....

Lab Partner(s) ....

EXPERIMENT 15 *Rotational Motion and Moment of Inertia*

ADVANCE STUDY ASSIGNMENT

Read the experiment and answer the following questions.

1. What are the rotational analogs of force, mass, and Newton's second law?

2. Define torque.

3. Describe two methods by which the moment of inertia of a body may be experimentally determined.

4. How is the frictional torque of a rotating system experimentally determined or compensated for?

5. When a body or system is a composite of regularly shaped bodies, what is the moment of inertia of the system about an axis of rotation along the component bodies' axes of symmetry?

6. A cylindrical object has a mass of 2 kg and a diameter and length of 0.10 and 0.18 m, respectively. What is the moment of inertia of the body about an axis along the axis of the cylinder? Show your work.

EXPERIMENT **15**

Rotational Motion and Moment of Inertia

I. INTRODUCTION

In the case of linear motion, an unbalanced force F acting on an object gives it an acceleration a, and by Newton's second law,

$$F = ma$$

where m is the mass of the object. For **rotational motion,** an unbalanced torque τ acting on a body causes it to rotate (i.e., gives it an angular acceleration α). The rotational analog of Newton's second law is

$$\tau = I\alpha$$

where I is the moment of inertia of the body.

The **moment of inertia** of a body depends on its mass distribution and shape. For a symmetrical, homogeneous object, the moment of inertia can be derived theoretically using calculus methods, and is expressed in terms of the object's total mass and dimensions. By determining these quantities, the moment of inertia can be calculated. However, the moment of inertia of an object can also be determined experimentally from the dynamical equation $\tau = I\alpha$ by measuring τ and α, or from energy considerations where the rotational kinetic energy is $\frac{1}{2}I\omega^2$.

In this experiment the theoretical and experimental values of some common regularly shaped objects will be determined and compared. The experimental procedures will provide a good insight into the concepts of rotational motion and the moment of inertia.

II. EQUIPMENT NEEDED

General

- Meter stick
- String
- Weight hanger and weights
- Laboratory timer or stopwatch
- Laboratory balance (kilogram range)
- Vernier calipers
- Large calipers

A. *Rotational inertia apparatus (Fig. 15-6) or similar apparatus*

B. *Rotating support with accessories (Fig. 15-4):*

- Cylindrical ring
- Disk
- Bar attachment (optional)
- Two cylinders
- Table clamp
- Support rod
- Two right-angle clamps
- One or two pulleys (depending on setup)

III. THEORY

The basic kinematic equations for linear and rotational motions are as follows:

Linear $s = \bar{v}t$ **(15-1)**

$$\bar{v} = \frac{v + v_o}{2} \quad \text{(constant } a\text{)} \qquad \textbf{(15-2)}$$

$$v = v_o + at \qquad \textbf{(15-3)}$$

Rotational $\theta = \bar{\omega}t$ **(15-1)**

$$\bar{\omega} = \frac{\omega + \omega_o}{2} \quad \text{(constant } \alpha\text{)} \qquad \textbf{(15-2)}$$

$$\omega = \omega_o + \alpha t \qquad \textbf{(15-3)}$$

where s and θ are the linear and angular distances, v and ω the magnitudes of the linear and angular velocities, $\bar{v}$ and $\bar{\omega}$ the magnitudes of the average linear and angular velocities for constant accelerations, and a and α the magnitudes of the linear and angular accelerations, respectively. (The angular distance θ is in radian measure.) The relationships between the linear and angular quantities are

$$s = r\theta \qquad \textbf{(15-4)}$$

$$v = r\omega \qquad \textbf{(15-5)}$$

$$a = r\alpha \qquad \textbf{(15-6)}$$

An unbalanced force F applied any distance from a body's axis of rotation produces an unbalanced torque τ, the magnitude of which is given by

$$\tau = rF \qquad \textbf{(15-7)}$$

where r is the perpendicular distance from the axis of rotation to the force's line of action (Fig. 15-1). For a particle of mass m,

$$\tau = rF = r(ma) = mra$$

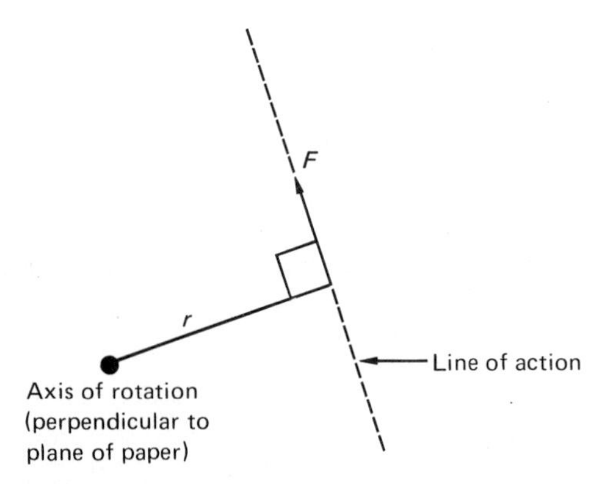

Fig. 15-1 The magnitude of the torque is the product of the force F and the perpendicular distance from the axis of rotation to the force's line of action.

where a is the tangential acceleration. Using the relationship $a = r\alpha$, we obtain

$$\tau = mr(r\alpha) = (mr^2)\alpha = I\alpha$$

where the moment of inertia $I = mr^2$. A rigid body is a system of particles, and summing this product for all the individual particles yields

$$\tau = (\textstyle\sum_i m_i r_i^2)\alpha$$
$$= I\alpha \qquad \textbf{(15-8)}$$

where the **moment of inertia** for a rigid body is $I = \sum_i m_i r_i^2$. Using calculus methods to evaluate this expression, moments of inertia of several regularly shaped bodies are given in Fig. 15.2.

When the torque is supplied by a constant weight force as in Fig. 15-3, the magnitudes of the angular velocity and angular acceleration of the rotating body can be found from measurements of the dis-

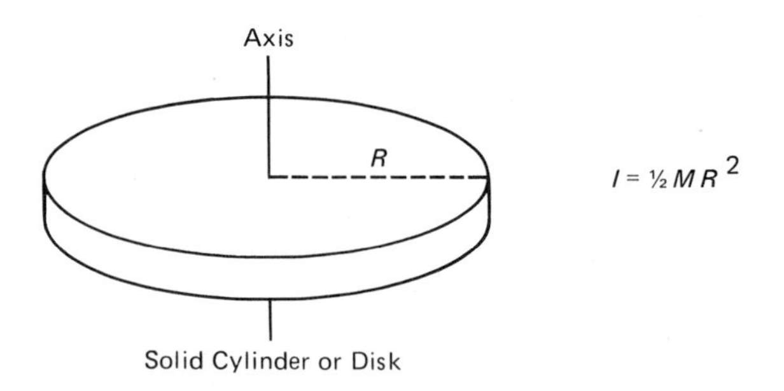

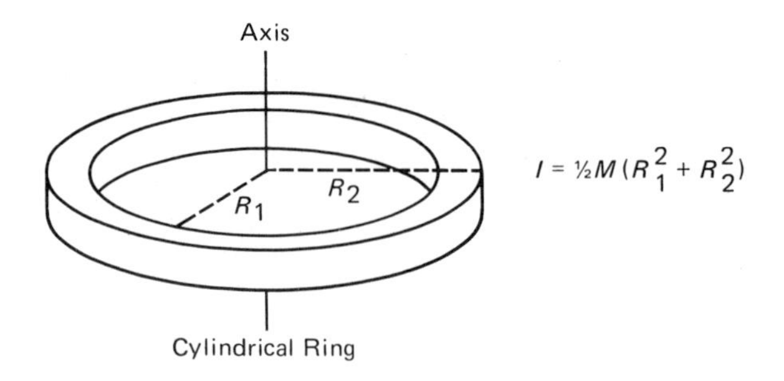

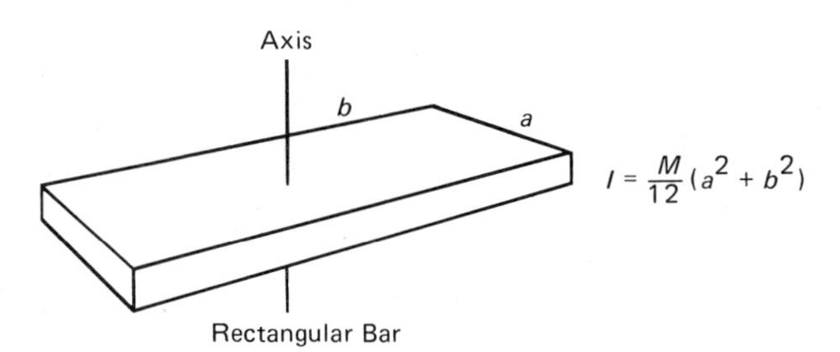

Fig. 15-2 Moments of inertia of some regularly shaped objects.

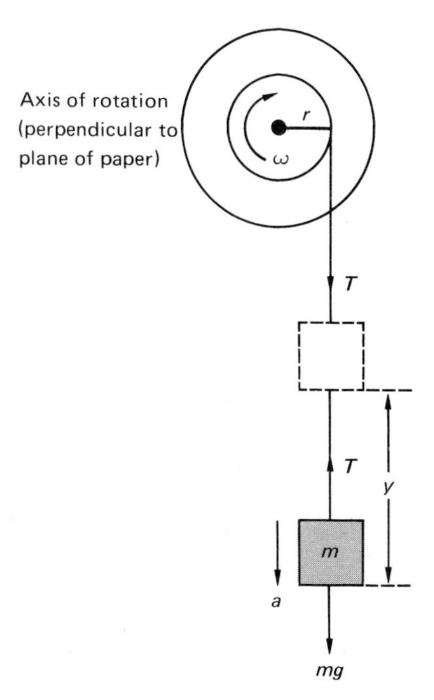

Fig. 15-3 A torque being applied by a constant-weight force.

tance y the weight descends from rest and the time t it takes the weight to travel this distance. From the preceding equations, the magnitude of the average velocity $\bar{v}$ is given by

$$\bar{v} = \frac{y}{t} \tag{15-9}$$

and the magnitude of the final velocity v of the descending weight is given by

$$\bar{v} = \frac{v}{2} \quad \text{or} \quad v = 2\bar{v} \tag{15-10}$$

where $v_0 = 0$ and the acceleration is constant (Eq. 15.2).

The angular velocity is then

$$\omega = \frac{v}{r} = \frac{2\bar{v}}{r} \tag{15-11}$$

where r is the radius of the axle. Also,

$$v = at \quad \text{or} \quad a = \frac{v}{t} \tag{15-12}$$

and

$$\alpha = \frac{a}{r}$$

From the forces acting on the descending mass (Fig. 15-3), the acceleration is given by Newton's second law:

$$mg - T = ma$$

or

$$T = mg - ma = m(g - a) \tag{15-13}$$

where T is the tension force in the string. The applied torque τ_a is then

$$\tau_a = rT \tag{15-14}$$

There are two torques acting on the wheel and axle, the applied torque τ_a and an opposing frictional torque τ_f due to the friction on the bearings of the rotating axle. The frictional torque is assumed to be constant. The net torque $\tau = I\alpha$ may then be written

$$\tau = \tau_a - \tau_f = I\alpha$$

or

$$\tau_a = I\alpha + \tau_f \tag{15-15}$$

Determination of *I*. The moment of inertia I of a system can be found from Eq. 15-15 using experimentally determined values of τ_a, α, and τ_f. The frictional torque can be determined or taken into account as follows:

(a) Determine the suspended mass m_0 needed to cause the body to rotate with a constant angular speed ω ($\alpha = 0$); then $\tau_f = \tau_a = rT = r(m_0 g)$. In this case, the frictional torque is balanced or vectorially canceled by the applied torque due to m_0.

(b) Since I and τ_f are constants, by determining the α's for two or more τ_a's, I and τ_f can be determined from the set of equations (Eq. 15-15) for the different conditions or solved for graphically.

In some instances, the rotating body may be symmetric, but a moment arm cannot be measured for the complete body (e.g., the rotating support in Fig. 15-4). In this case, the moment of inertia can be found from energy considerations, which can also be used to find the moment of inertia of a body with a well-defined moment arm. The potential energy lost by the descending mass in the setup in Fig. 15-4 goes into the linear kinetic energy of the mass and the rotational kinetic energy of the rotating body (neglecting frictional losses). By the conservation of energy,

$$\text{energy lost} = \text{energy gained}$$

$$mgy = \tfrac{1}{2}mv^2 + \tfrac{1}{2}I\omega^2 \tag{15-16}$$

where all the quantities except I are known or can be calculated from experimental data.

The energy lost to friction can be accounted for by adding the term $m_0 gy$ to the right side of Eq. 15-16, where m_0 is the mass determined in (a) above.

When a body is a composite of two or more regular bodies (e.g., a wheel and axle), the moment of in-

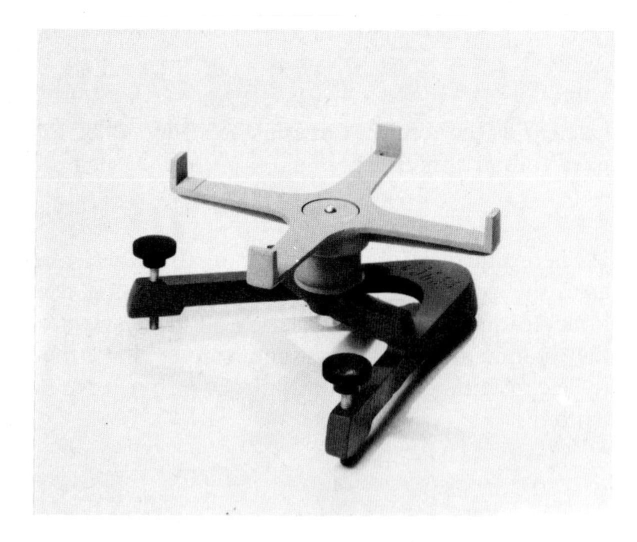

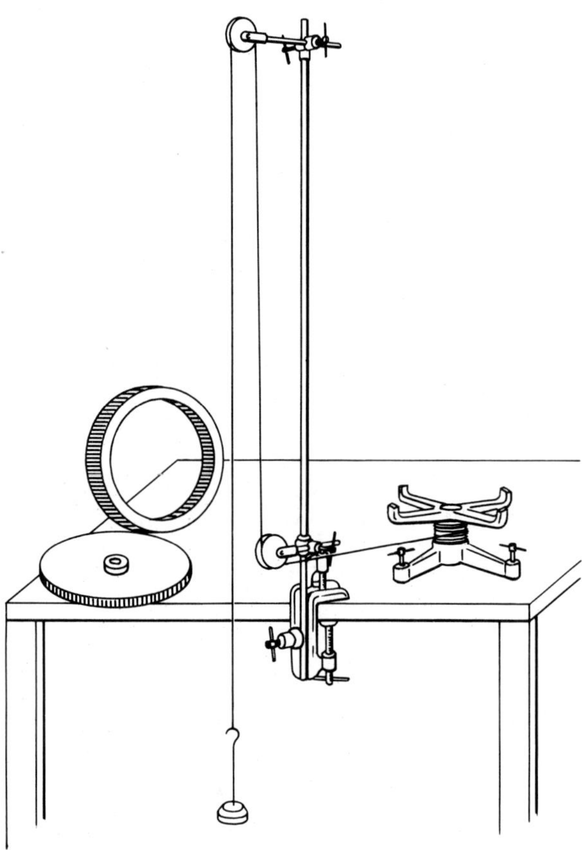

Fig. 15-4 A rotating support and illustration of experimental setup and accessory bodies. (Courtesy of Central Scientific Co., Inc.)

ertia of the system about an axis along the component bodies' axes of symmetry is the sum of the moments of inertia of each of the component bodies (i.e., $I = I_1 + I_2 + \cdots$).

When the system consists of a regularly shaped accessory body mounted on the rotating support with the axis of rotation through the center of mass (symmetry) of the accessory body, the total moment of inertia I of the system is $I = I_s + I_a$, where I_s and I_a are the moments of inertia of the support and accessory body, respectively. The moment of inertia of the accessory body is then $I_a = I - I_s$, where I and I_s are measured independently by one of the previously described methods.

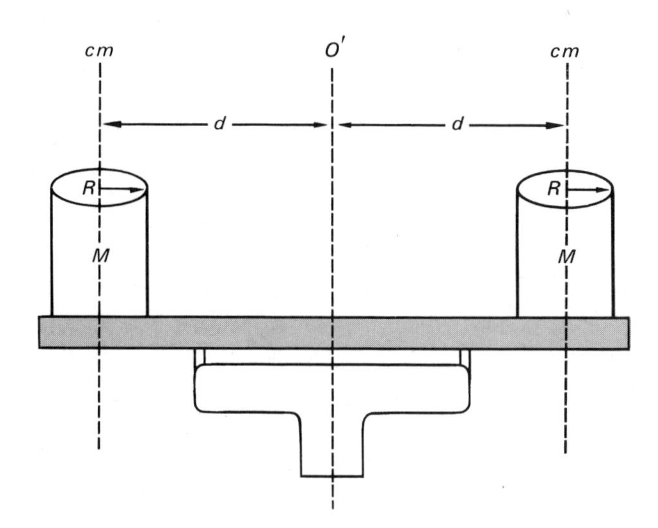

Fig. 15-5 The moment of inertia of a system of regular objects can be calculated using the parallel axis theorem. See the text for a description.

Parallel axis theorem. For a system of regular bodies, some of which do not have the axis of rotation through their centers of mass, the moment of inertia of the system can be calculated using the parallel axis theorem. Such a system is that of the cylindrical masses placed on opposite arms of the metal cross of the rotating support or a bar attachment, as shown in Fig. 15.5.

The *parallel axis theorem* states that the moment of inertia of a body about any axis O' parallel to an axis through the body's center of mass is equal to the moment of inertia about the axis through the center of mass plus the mass of the body times the square of the distance between the axes:

$$I_{o'} = I_{\text{cm}} + Md^2 \qquad \textbf{(15-17)}$$

Then, the total moment of inertia I for the system shown in Fig. 15-5 (identical cylinders) about an axis through O' is

$$\begin{aligned} I &= I_s + I_b + I_{c'} + I_{c'} \\ &= I_s + I_b + 2I_{c'} \end{aligned}$$

where I_s is the moment of inertia of the rotating support, I_b the moment of inertia of the bar, and $I_{c'}$ the moment of inertia of one of the cylinders.

The axis of rotation is through the centers of mass of the support and the bar, so I_s and I_b can be determined by previously described methods. By the parallel axis theorem, $I_{c'} = I_{cm} + Md^2$, where I_{cm} for a cylinder is $I_{cm} = \frac{1}{2}MR^2$ (Fig. 15-2). Although convenient, it is not necessary that the cylinders be placed equal distances from the axis of rotation (different d's are used if they are not equal distances).

IV. EXPERIMENTAL PROCEDURE

A. *Rotational Inertia Apparatus (Wheel and Axle)*

1. A rotational inertia apparatus is shown in Fig. 15-6. It is essentially a wheel and axle(s), a large heavy wheel or disk with two concentric axles of different radii that rotate freely on bearing mounts.

2. Measure the diameter of the disk with the large calipers and the diameters of each axle with the vernier calipers and record in Data Table 1. Also record the total mass of the wheel and axle, which is usually stamped on the wheel.

3. Fix one end of a length of string to the set pin in the axle and wrap the string around the smaller axle in a single layer. The string should be sufficiently long to allow weights attached to the free end of the string to descend to the floor. Making a loop in the free end of the string, attach just enough weight to the string so that the disk, when given a slight start, rotates with a uniform angular speed as the weight descends. (The ordinary 50-gm weight hanger will probably be too large. Put some smaller weights in the string loop.)

   The weight also descends with a uniform speed, which may be easier to judge than the disk's rotational speed. Record the mass of the descending weight in Data Table 1. (This will be used in a direct calculation of τ_f.)

Fig. 15-6 Rotational inertia apparatus (wheel and axle). (Photo courtesy of Sargent-Welch Scientific Company.)

4. Remove the weights from the string loop and attach a 0.050-kg (50-gm) weight hanger. Measure the distance of the bottom of the weight hanger from the floor with the string wrapped around the smaller axle, and record in Data Table 1. Releasing the weight hanger, measure the time it takes to reach the floor and record. (Some rotational inertia apparatus may be equipped with spark timers. Consult your instructor for operating instructions.) Repeat the timing procedure two more times for the same distance of descent.

5. Repeat the timing procedures (three trials) for a 0.10-kg (100-gm) mass (total) with the string wrapped around the smaller axle.

6. Repeat the timing procedures for the same masses (three trials each) with the string wrapped around the large axle.

7. Compute the average time of descent for each case, and using Eqs. 15-9 through 15-14, calculate the quantities called for in the Laboratory Report.

8. Notice that Eq. 15-15, $\tau_a = I\alpha + \tau_f$, is of the form of a straight line,

$$y = bx + c$$

   where b is the slope of the line and c its y intercept. Plot τ_a versus α for the data in Data Table 1 and draw the straight line that best fits the data. Determine the slope I and the y intercept τ_f. Compare these values with the computed theoretical values by calculating the percent differences.

9. *Moment of inertia from conservation of energy.* Using Eq. 15-16, find the moment of inertia of the wheel and axle using the data from one of the descending weight cases. (*Note:* The mass used on the left side of this equation should be the mass of the descending weight *minus* the mass of the descending weight in the constant angular speed case of procedure 3. Why?) Compare the moment of inertia determined from Eq. 15-16 with the computed theoretical value by calculating the percent difference.

B. *Rotating Support and Accessory Bodies*

10. A rotating support and some accessories are shown in Fig. 15-4. The rotating support consists of a light metal cross mounted on ball bearings so as to rotate in a horizontal plane about a vertical axis.

11. With the vernier calipers, measure the diameter of the drum of the rotating support and record in Data Table 2. (The string unwinds from the drum.) With the appropriate instruments, measure the linear dimensions of the accessory bodies needed to calculate the theoretical moments of inertia. (Refer to Fig. 15-2.) Also, determine the masses of the accessory bodies on a laboratory balance and record.*

12. Use a pulley arrangement similar to the one in Fig. 15-4 so that the descent distance of the driving weight is at least 1 meter. Attach one end of a length of string to the rotating support drum and wrap the string around the drum in a single layer. The length of the string should be sufficient to allow weights attached to the free end of the string to descend to the floor.

    Making a loop in the free end of the string, attach just enough weight (mass m_0) to the string so that when given a slight start, the cross of the rotating support rotates with a uniform speed as the weight descends. (The ordinary 50-gm weight hanger will probably be too large. If so, put some smaller weights in the string loop.) The weight should also descend with a uniform speed, which may be easier to judge. Make certain the length of string unwinding from the drum is parallel to the table top. Record the mass of the descending weight m_0 in Data Table 2.

13. Remove the weights from the string loop, attach a 50-gm weight hanger to the loop, and place the removed weights on the weight hanger. Record the total mass (m_1) in the data table. Measure the distance y from the bottom of the weight hanger to the floor with the string wrapped around the drum and record in the data table. After releasing the weight hanger, measure the time it takes to reach the floor and record. Repeat the timing procedure two more times for the same distance of descent and record.

14. Using the distance of descent and the average time of descent of the three trials, calculate the final velocity v of the descending weight and the final angular speed ω of the rotating support from the measured data from Eqs. 15-10 and 15-11. Using Eq. 15-16, calculate the moment of inertia I_s of the rotating support. [*Note:* The mass m in this equation does *not* include the mass on the weight hanger used to determine the constant speed in procedure 12 (i.e., $m = m_1 - m_0$). Why?]

15. Place one of the regularly shaped accessory bodies (e.g., the cylindrical ring) on the metal cross of the rotating support so that the axis of rotation is through the body's center of mass. Add just enough weight to the weight hanger so that the system rotates (or the weight hanger descends) with a constant speed when given a slight start. Record the total mass m_0 of the descending weight.

16. Add additional mass (0.10 to 0.20 kg) to the weight hanger so that it descends slowly to the floor when released from rest. Record the total mass m_1 of the descending weight. Measure the time (t) it takes for the weight hanger to descend a distance y to the floor for three trials and record these data.

17. Repeat procedures 15 and 16 for the other accessory bodies (disk and bar) if available.

18. Compute the average time of descent, and using Eqs. 15-9 to 15-11, find the final velocity of the descending weight and the final angular speed ω of the rotating system for each case. Calculate the moments of inertia I for the combined accessory-support systems from the experimental data using Eq. 15-16. Determine the moments of inertia of the individual accessory objects

*The values of the masses may be given to you by your instructor if the masses of the accessories are too large for regular laboratory balances.

($I_a = I - I_s$), using the previously determined value of I_s.

19. Using the measured masses and appropriate length measurements of the accessory objects, compute the theoretical values of their moments of inertia (see Fig. 15-2), and compare with the experimentally determined values by calculating the percent differences.

20. *Parallel axis theorem.* Set up the system as shown in Fig. 15-5, using the bar attachment if available. Measure the distance d from the axis of rotation to the parallel axes through the centers of masses of the cylinders and record. Repeat procedures 15, 16, and 18 for this system, where, in procedure 18, $I_a = I_{c'}$ and $I = I_s + I_b + 2I_{c'}$.

21. Applying the parallel axis theorem (Eq. 15-17), calculate the theoretical moment of inertia of the cylinders about the axis of rotation. Compare this result with the experimentally determined value of I by calculating the percent difference.

Name .... Section .... Date ....

Lab Partner(s) ....

LABORATORY REPORT

A. *Rotational Inertia Apparatus (Wheel and Axle)*

DATA TABLE 1

Diameter of smaller axle ....

Diameter of larger axle ....

Total mass of wheel and axle ....

Mass for constant v, m_0 ....

Radius r ....

Radius R ....

(Needed for some apparatus)
Thickness of wheel ....

Length of axles (beyond wheel) ....

Diameter of wheel ....

		Smaller axle		Larger axle	
Mass					
Distance of descent, y					
Time of descent, t (s)	1.				
	2.				
	3.				
	Average				
Final velocity, v					
Final angular velocity, ω (rad/s)					
Acceleration a					
Angular acceleration, α (rad/s^2)					
Tension, T					
Applied torque, τ_a					

Name .... Section .... Date ....

Lab Partner(s) ....

EXPERIMENT 15 *Rotational Motion and Moment of Inertia*

Calculations (show calculations for all cases)

1. Final velocity, v, and angular velocity, ω

2. Acceleration, a, and angular acceleration, α

3. Tension in string, T, and applied torque, τ_a

4. Theoretical moment of inertia, I (Fig. 15-2), and frictional torque, τ_f

5. Moment of inertia, I (energy method)

(continued)

Results (don't forget units)

Using the torque method:

	Experimental value from plotted data	Computed value	Percent difference
Moment of inertia, I	....	....	....
Frictional torque, τ_f	....	....	....

Using the energy method:

Moment of inertia, I ....

Percent of difference (between experimental and theoretical values) ....

B. *Rotational Support and Accessory Bodies*

DATA TABLE 2

Diameter of support drum		....	Radius, r	....		
Accessory objects:						
1. Cylindrical ring	Mass	....	Inner radius, R_1	....		
			Outer radius, R_2	....		
2. Disk	Mass	....	Radius, R	....		
3. Bar	Mass	....	Width, a	....		
			Length, b	....		
4. Solid cylinder	Mass	....	Diameter	....	Radius, R	....

(Both cylinders should be approximately the same)

Procedure 20: Distance from axis of rotation to parallel axis through center of mass of cylinders d ....

Name .... Section .... Date ....

Lab Partner(s) ....

EXPERIMENT 15 *Rotational Motion and Moment of Inertia*

	(a) Rotating support	(b) Support with cylindrical ring	(c) Support with disk	(d) Support with bar	(e) Support with (bar and) cylinders
m_0					
m_1					
$m = m_1 - m_0$					
Distance of descent, y					
Time of descent, t (s) 1.					
2.					
3.					
Average					
Final velocity, v					
Final angular velocity, ω (rad/s)					
Calculated I of system	(I_s)				
Moment of inertia of accessory, I_a $(I_a = I - I_s)$	X				$(I_a = I_{c'})$
Theoretical I_a	X				
Percent difference	X				

Space for calculations follows.

(continued)

Calculations (show calculations for all cases)

1. Final velocity, v, and final angular velocity, ω

2. I of the system

3. Theoretical I_a

QUESTIONS

A. *Rotational Inertia Apparatus (Wheel and Axle)*

1. The energy lost by the descending mass m_0 is m_0gy. Express this energy loss mathematically in terms of the fictional torque τ_f.

2. In a graph of τ_a versus α, what would be the implications if (a) the line went through the origin; and (b) the y intercept were negative?

B. *Rotating Support*

3. Answer Question 1.

4. What is the moment of inertia of the system shown in Fig. 15-Q4 (a) about the axis O'; and (b) about the axis O?

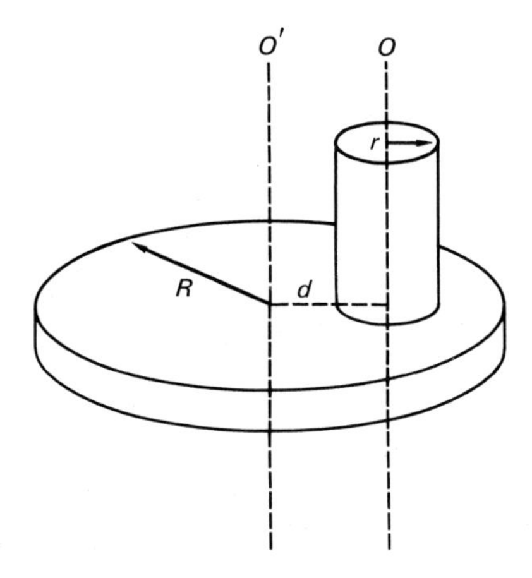

Fig. 15-Q4

Name .... Section .... Date ....

Lab Partner(s) ....

EXPERIMENT 16 *Elasticity: Young's Modulus*

ADVANCE STUDY ASSIGNMENT

Read the experiment and answer the following questions.

1. What is meant by "elasticity"?

2. Define the terms stress, strain, and Young's modulus.

3. Show that Hooke's law has the form of the general equation for a straight line. What would be the slope and y-intercept of this straight-line relationship?

4. What is the elastic limit of a material? Does Hooke's law hold for a material stressed beyond its elastic limit?

5. In the experiment, why is it permissible not to include the initial 1-kg weight and weight hanger in the loads for the corresponding elongations?

EXPERIMENT **16**

Elasticity: Young's Modulus

I. INTRODUCTION

All bodies are deformed to some extent when acted on by forces. If a body is deformed and it tends to return to its original dimensions, it is said to be **elastic.** Elasticity is a material property and is characterized by an **elastic modulus,** which is defined as the ratio of the stress to the strain. Stress is related to the deforming force and strain is the effect of relative changes in the dimensions of the body when subjected to a stress.

It is not commonly thought that materials such as steel are elastic. However, this is the case, as will be demonstrated. The deformation and recovery of a steel wire will be observed, and by experimentally determining the longitudinal strain for an applied tensile stress, the longitudinal elastic modulus, commonly called **Young's modulus,** will be computed.

II. EQUIPMENT NEEDED

- Young's modulus apparatus
- Optical lever attachment (mirror on tripod base and telescope and scale; a desk lamp may be needed to illuminate the scale if laboratory lighting is poor)

or

Micrometer screw attachment

- Slotted weights (ten 1-kg weights)
- Meter stick and 2-meter stick (if available)
- Micrometer calipers

III. THEORY

The elasticity of a material is a parameter that is not dependent on the shape or dimensions of a body. The elasticity is characterized by elastic moduli: Young's modulus (linear), shear modulus (two-dimensional), and bulk modulus (three-dimensional). Each modulus describes a different type of elastic deformation. Only Young's modulus will be considered in this experiment.

In general, an elastic modulus is defined as

$$\text{elastic modulus} = \frac{\text{stress}}{\text{strain}} \qquad \textbf{(16-1)}$$

For the linear case, the stress is equal to the applied longitudinal force F divided by the cross-sectional area A of the object normal to the force (Fig. 16-1). This is commonly called a **tensile** or **normal stress:**

$$\text{tensile stress} = \frac{F}{A} \qquad \textbf{(16-2)}$$

A **strain** is the effect or relative changes in the dimensions or shape of a body subjected to a stress.

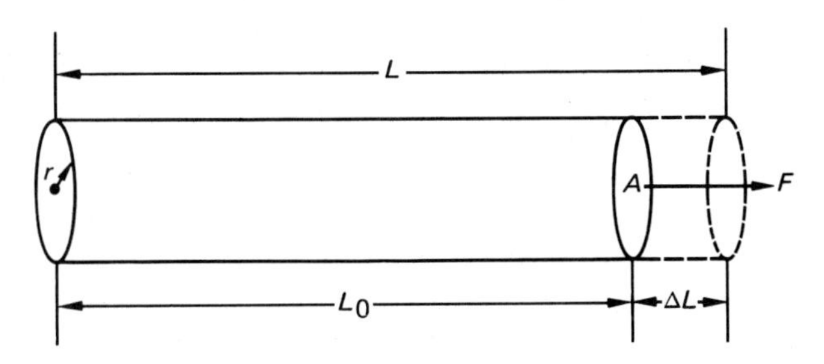

Fig. 16-1 Tensile stress.

For the linear case, this involves a change in length, and the tensile strain is defined as the ratio of the change in length ΔL to the original length L_0 (Fig. 16-1):

$$\text{tensile strain} = \frac{\Delta L}{L_0} = \frac{L - L_0}{L_0} \quad \textbf{(16-3)}$$

where L_0 is the original length. Longitudinal stresses and strains may be either tensile (elongating) or compressive.

The elastic modulus for the linear case is called *Young's modulus* Y and

$$Y = \frac{\text{tensile stress}}{\text{tensile strain}} = \frac{F/A}{\Delta L/L_0} \quad \textbf{(16-4)}$$

[It should be noted that it is the increment or change in stress $\Delta(F/A)$ that produces the strain ΔL.]

Since the strain is a ratio of lengths, it is dimensionless and the dimensions of Young's modulus are N/m^2, $dyne/cm^2$, and $lb/in.^2$ in the various systems of units. Provided that the elastic limit of the material is not exceeded, the ratio of the stress to the strain—Young's modulus—is found to be constant and characteristic of a given material. A material for which the stress is directly proportional to the strain is said to obey **Hooke's law,** which is commonly expressed as $F = -k\Delta x$, where k is a constant and the minus sign indicates that the spring force is opposite to the displacement. (See Experiment 17.) Equation 16-4 is of the form

$$F = \left(\frac{YA}{L_0}\right)\Delta L$$

A typical plot of stress versus strain is shown in Fig. 16-2. In the region of Hooke's law, the graph is a straight line with a slope equal to Young's modulus. If the stress is increased beyond the elastic limit (point A), the sample does not return to its original length when the stress is removed, but retains a permanent strain or is permanently deformed. If enough stress is applied, the sample will break (point B).

For a wire of radius r with an applied longitudinal weight force mg, Young's modulus has the form

$$Y = \frac{F/A}{\Delta L/L_0} = \frac{mg/\pi r^2}{\Delta L/L_0} \quad \textbf{(16-5)}$$

where ΔL is the elongation of the wire from its original length L_0.

OPTICAL LEVER

(For experimental apparatus with elongation measurements made by the optical lever method)

In the optical lever method, small elongations are measured by means of the deflection of a mirror. If the plane of a mirror is originally vertical (MM), as illustrated in Fig. 16-3, then when you look through a telescope, the cross-hairs of the telescope will be on a mark T on the scale. However, if the mirror becomes tilted through an angle θ (due to a change in the length of the wire), a reflected beam is turned through an angle of 2θ. The cross-hairs of the telescope will then be on a mark on the scale at point P. That is, looking through the telescope, you would see point P. From Fig. 16-3, it can be seen that the double angle of deflection 2θ can be written in terms of the deflection distance Δy on the scale and the distance D from the mirror to the scale:

$$\tan 2\theta = \frac{\Delta y}{D} \quad \textbf{(16-6)}$$

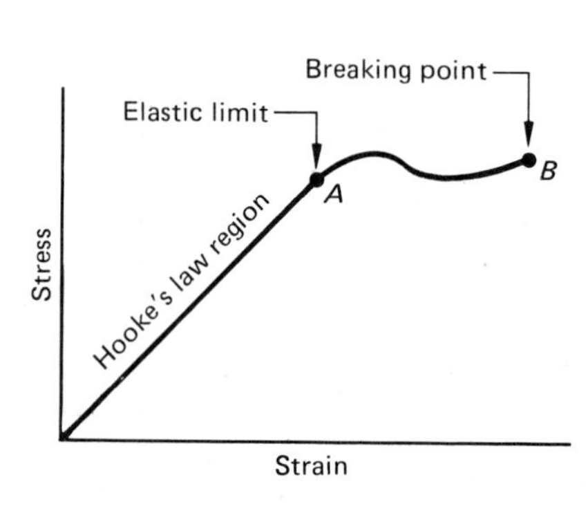

Fig. 16-2 A typical plot of stress versus strain. In the linear region, where Hooke's law is applicable, the slope of the line is equal to Young's modulus.

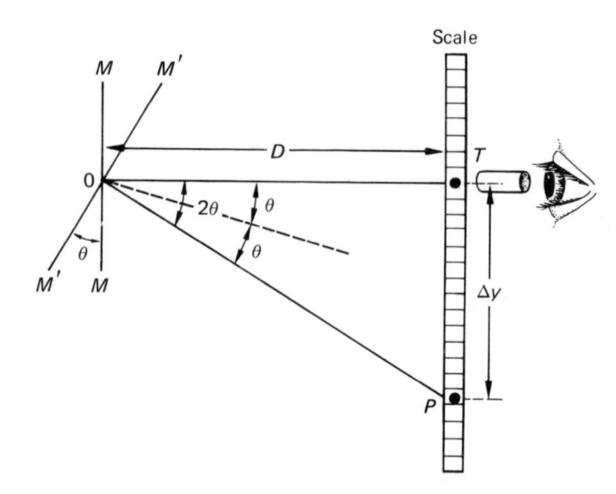

Fig. 16-3 The optical lever method of determining elongation.

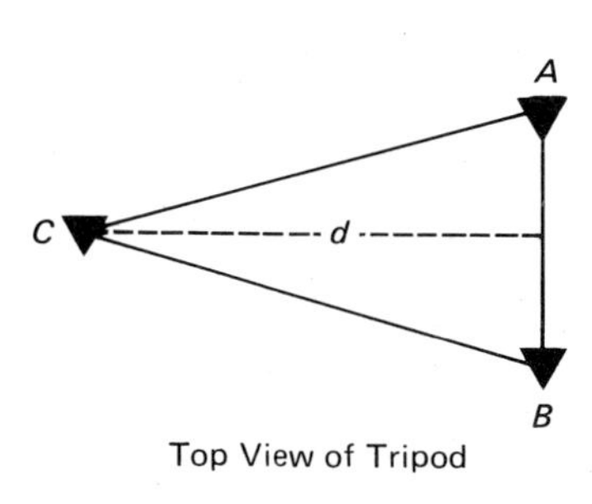

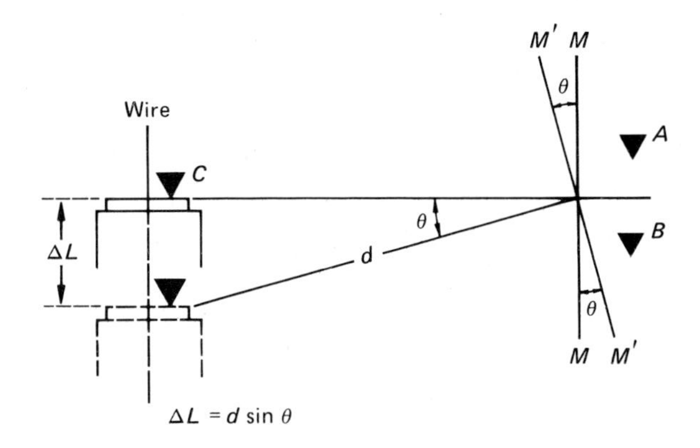

Fig. 16-4 The tilting of the optical lever mirror results from the lowering of one side of the tripod base of the mirror mount due to the elongation of the wire. (See Fig. 16-6.)

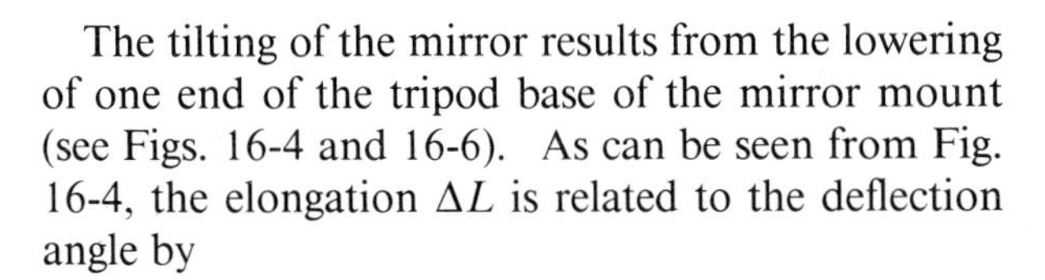

The tilting of the mirror results from the lowering of one end of the tripod base of the mirror mount (see Figs. 16-4 and 16-6). As can be seen from Fig. 16-4, the elongation ΔL is related to the deflection angle by

$$\Delta L = d \sin \theta \tag{16-7}$$

where d is the apex length of the tripod triangle. For a small angle, which is the case here, $\frac{1}{2} \tan 2\theta \simeq \tan \theta \simeq \sin \theta$. Hence, with Eq. 16-6, the elongation becomes

$$\Delta L = \frac{d\,\Delta y}{2D} \tag{16-8}$$

and the small elongation can be determined from easily measured experimental parameters.

IV. EXPERIMENTAL PROCEDURE

1. A common type of Young's modulus apparatus is shown in Fig. 16-5. The wire is supported by a chuck in a clamped yoke at the upper end of the apparatus support rods. A lower chuck that tightly grips the wire is positioned in the hole of an adjustable platform that is clamped to the support rods. If an optical lever attachment is used, the mount rests in the groove on the platform and on the face of the chuck [Fig. 16-6(a)].

   If a micrometer screw attachment is used [Fig. 16-6(b)], a lever with a spirit level on one end and a micrometer screw on the other is pivoted on the chuck. The micrometer screw is referenced to the adjustable platform. A weight hanger is suspended from the lower end of the wire, with the bottom of the hanger about 2 or 3 cm above the heavy tripod base of the apparatus.

2. Hang a 1-kg weight on the weight hanger and level the tripod base of the apparatus by means of the leveling screws. The weight should be centered between the support rods when the base is level. Measure the length L_0 of the wire from the bottom of the upper chuck to the top of the lower chuck and record in the appropriate data table. [*Note:* The length of wire between the gripping

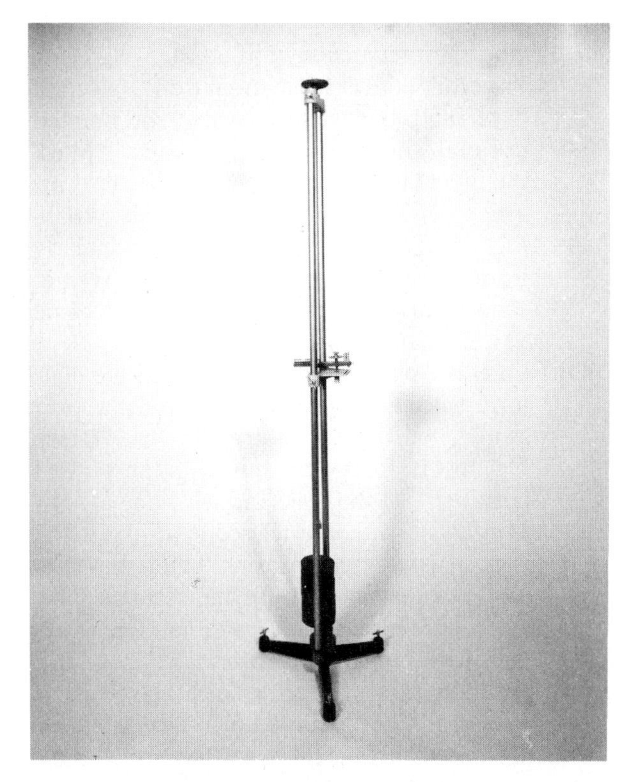

Fig. 16-5 A Young's modulus apparatus. (Photo courtesy of Central Scientific Co., Inc.)

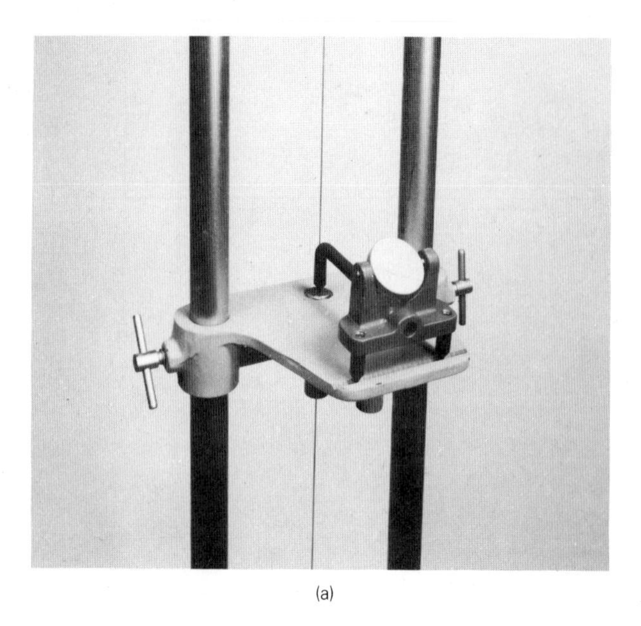

(a)

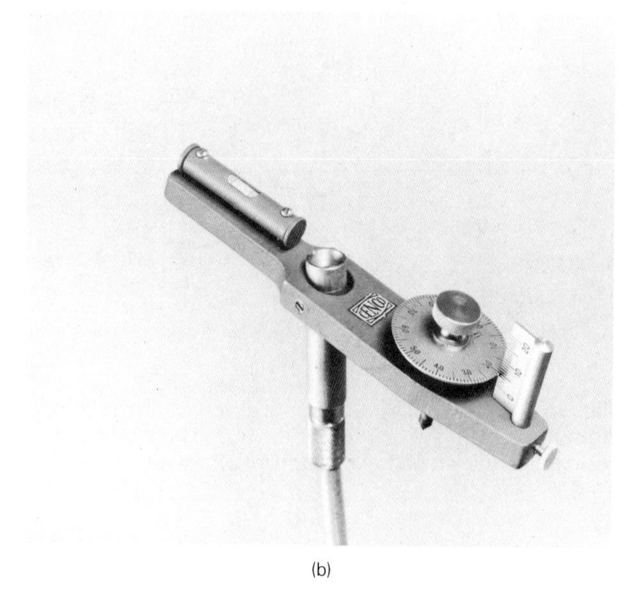

(b)

Fig. 16-6 (a) Optional lever attachment. (b) Micrometer screw attachment. (Photos courtesy of Central Scientific Co., Inc.)

chuck holders is taken as L_0 instead of the total length of the wire. Also, the measurement attachment zero reading, (y_0 below) is taken with an initial 1-kg mass on the weight hanger. These weights are *not* included as part of the individual load increments since the individual load increments (1 kg each) produce the corresponding ΔL's.]

3. (a) *Optical lever attachment* (if used).
   (i) Place the optical lever mount on a piece of paper and press lightly so that the pointed feet make indentations on the paper. Mark the indentations with a pencil. Then, draw a line between the two front points (AB in Fig. 16-4), and draw a line perpendicularly from this line to point C. With a vernier caliper, measure the length of the latter line (d in Fig. 16-4), and record.
   (ii) Carefully place the optical lever on the platform with its two front feet in the platform groove and its rear foot on the chuck (see Fig. 16-6). Rotate the mirror to the vertical position. Set the scale and telescope at least 1 m away, with the telescope at about the same height as the mirror. Adjust the telescope so that a mark on the scale as indicated by the telescope cross-hairs is easily seen. Record this zero reading (y_0).
   (iii) Carefully measure the distance between the mirror and the scale with a 2-meter stick (distance D in Fig. 16-3), and record. It may be necessary to illuminate the scale with a lamp.

   (b) *Micrometer screw attachment* (if used). Pivot the micrometer attachment lever on the lower chuck, and adjust the platform and the micrometer screw so that it has a mid-scale reading when the screw is balanced against the platform with the spirit level in a level position. Record this zero reading (L_0') of the micrometer. (*Note:* Always move the screw in the same direction just before each reading, e.g., toward the higher numbers, so as to avoid screw thread "backlash" or hysteresis.)

4. Carefully add another 1-kg weight to the load and record the scale reading seen through the telescope or the micrometer reading when again brought into level balance. The reading should not be taken immediately after increasing the load. A short time is required for the wire to stretch to equilibrium; this is called **elastic lag.**

5. Add 1-kg weights, one at a time, until there is a total of 10 kg on the hanger, recording the scale reading or the balanced micrometer reading after adding each 1-kg weight. (*Note:* These loads are appropriate for a steel wire. If another kind of wire is used, ask your instructor for the proper maximum load.)

   At some point in this procedure after a 5-kg load is on the wire, measure the diameter of the wire with the micrometer calipers and record in the appropriate data table. With a substantial load, error resulting from kinks in the wire is avoided.

6. Remove the weights one at a time and record the scale reading or balanced micrometer reading after each weight is removed (allowing for the elastic lag). Stop when only one 1-kg weight is left on the weight hanger. At this point, measure L_0 again.

7. Compute the elongations ΔL_i for each load from:

   (*Optical lever method*) the average of the scale readings for increasing and decreasing loads using Eq. 16-8, where $\Delta y_i = \overline{y}_i - \overline{y}_0$, where $\overline{y}_i$; and $\overline{y}_0$ are the averages of two scale readings (increasing and decreasing). That is, $\overline{y}_0$ is subtracted from each of the other average scale readings.

   (*Micrometer screw method*) the average micrometer readings for increasing and decreasing loads, $\Delta L_i = \overline{L}_i - \overline{L}_0'$ where $\overline{L}_i$ and $\overline{L}_0'$ are the averages of two micrometer readings (increasing and decreasing). That is, $\overline{L}_0'$ is subtracted from each of the other average lengths.

   Notice in the data table that the initial 1-kg weight and the weight of the weight hanger are not to be included as part of the individual loads.

8. Compute the stress F/A and the strain $\Delta L_i/L_0$ for each load.

   Use an average L_0 if the initial and final measurements are different. Plot a graph of stress versus strain. Draw a straight line that best fits the data points and determine the slope of the line. Compare this result with the value of Young's modulus of the wire material given in Appendix A, Table A2, by calculating the percent error.

Name .... Section .... Date ....

Lab Partner(s) ....

EXPERIMENT 16 *Elasticity: Young's Modulus*

LABORATORY REPORT

A. *Optical Lever Method*

DATA TABLE 1

d ....

D ....

Wire diameter ....

L_0 (initial) ....

L_0 (final) ....

Load (kg)	Increased load scale reading		Decreased load scale reading		Average scale reading	Computed elongation ΔL_i (mm)		Strain $\Delta L_i/L_0$	Stress
*	y_0		y_0						
1	y_1		y_1			ΔL_1			
2	y_2		y_2			ΔL_2			
3	y_3		y_3			ΔL_3			
4	y_4		y_4			ΔL_4			
5	y_5		y_5			ΔL_5			
6	y_6		y_6			ΔL_6			
7	y_7		y_7			ΔL_7			
8	y_8		y_8			ΔL_8			
9	y_9		y_9			ΔL_9			

*Initial 1-kg and weight hanger not included as part of loads.

Calculations (show work)

Slope of graph ....

Accepted Young's modulus ....

Percent error ....

Name .... Section .... Date ....

Lab Partner(s) ....

EXPERIMENT 16 *Elasticity: Young's Modulus*

B. *Micrometer Method*

DATA TABLE 2

Wire diameter ....

L_0 (initial) ....

L_0 (final) ....

Load (kg)	Increased load mike reading (mm)		Decreased load mike reading (mm)		Average reading	Elongation $\Delta L_i = \overline{L}_i - \overline{L}_0'$		Strain $\Delta L_i/L_0$	Stress $F/A = mg/\pi r^2$
*	L_0'		L_0'						
1	L_1		L_1			ΔL_1			
2	L_2		L_2			ΔL_2			
3	L_3		L_3			ΔL_3			
4	L_4		L_4			ΔL_4			
5	L_5		L_5			ΔL_5			
6	L_6		L_6			ΔL_6			
7	L_7		L_7			ΔL_7			
8	L_8		L_8			ΔL_8			
9	L_9		L_9			ΔL_9			

*Initial 1-kg and weight hanger are not included as part of loads.

(continued)

Calculations (show work)

Slope of graph ....

Accepted Young's modulus ....

Percent error ....

QUESTIONS

1. (a) Why is it proper to take L_0 as the length of wire between the chucks rather than the total length of the wire? (b) Why aren't the initial 1 kg and weight hanger included as part of the load?

2. Stress is defined as force/area (F/A), yet in graphically determining Young's modulus the load in kilograms was plotted versus strain. Does this give an erroneous value of Young's modulus? Explain.

3. Suppose that the stress versus strain graph for another wire had a steeper slope than that of the wire used in this experiment. What would this mean physically?

4. If the elastic limit of the wire had been exceeded, would the curve of the graph of stress versus strain go above or below an extension of the straight line in the Hooke's law region? Explain what each case would mean physically.

Lab Partner(s) ....

EXPERIMENT 16 *Elasticity: Young's Modulus*

5. (a) Suppose that the length of the wire between the chucks L_0 were twice the length used in the experiment (same wire). Would a particular load produce the same corresponding elongation as found in the experiment? If not, what would it be?

   (b) Suppose that the experiment were repeated (same L_0) with a wire of the same material, but with a diameter twice that of the wire used in the experiment. Would a particular load produce the same corresponding elongation as found in the experiment? If not, what would it be?

6. *Optional (optical lever method)* Prove that when a mirror is rotated (tilted) through an angle θ, a reflected beam is deflected by an angle 2θ (Fig. 16-3). *Hint:* Draw reflected rays for the same incident ray for each orientation.

7. *(Optional)* In the interest of speed and simplicity, "raw data" are often recorded directly on a graph. In this experiment the "increasing load" data might have been recorded on a graph where x = load = total mass (2 kg to 11 kg) and y = actual micrometer reading (in mm). Draw such a graph and use it to compute Y. Comment on the speed and validity of this method.

Name .... Section .... Date ....

Lab Partner(s) ....

EXPERIMENT 17 *Hooke's Law and Simple Harmonic Motion*

ADVANCE STUDY ASSIGNMENT

Read the experiment and answer the following questions.

1. What are Hooke's law and simple harmonic motion, and how are they related?

2. How is the "internal" restoring force of a spring measured experimentally?

3. What is the physical significance of the spring constant?

4. In the equation of motion for SHM, what physically determines A and T?

5. How is the period of a mass oscillating on a spring related to the spring constant? (Express mathematically and verbally.)

EXPERIMENT 17

Hooke's Law and Simple Harmonic Motion

I. INTRODUCTION

An elastic material tends to return to its original form or shape after being deformed. Hence, elasticity implies a restoring force that can give rise to vibrations or oscillations. For many elastic materials, the restoring force is proportional to the amount of deformation, if the deformation is not too great. This relationship for elastic behavior is known as Hooke's law. In one dimension, the restoring force F is proportional to the deformation displacement x, or $F \propto x$. In equation form, we have

$$F = -kx$$

where k is a constant of proportionality and the minus sign indicates that the displacement and the force are in opposite directions. In the case of the linear restoring force of a coil spring, k is called the **spring constant** and is a relative indication of the "stiffness" of the spring.

A particle or object in motion under the influence of a linear restoring force described by Hooke's law undergoes what is known as **simple harmonic motion (SHM).** This periodic oscillatory motion is one of the most common types found in nature. The period of oscillation of an object in simple harmonic motion is related to the constant of proportionality in Hooke's law.

In this experiment, the relationship of Hooke's law will be investigated along with the parameters and description of simple harmonic motion.

II. EQUIPMENT NEEDED

- Coil spring
- Wide rubber band
- Slotted weights and weight hanger
- Laboratory timer or stopwatch
- Meter stick
- Laboratory balance
- 2 sheets of Cartesian graph paper

III. THEORY

A. *Hooke's Law*

The fact that for many elastic substances the deformation is directly proportional to a restoring force that resists the deformation was first demonstrated by Robert Hooke (1635–1703), an English physicist. For one dimension, this relationship—known as **Hooke's law**—is expressed mathematically as

$$F = -k\,\Delta x = -k(x - x_0) \qquad \textbf{(17-1)}$$

$$F = -kx \qquad (\text{with } x_0 = 0)$$

where Δx is the linear deformation or displacement of the spring and x_0 is the initial position. The minus sign indicates that the force and displacement are in opposite directions. (For experimental convenience, the minus sign may be neglected.)

In terms of Young's modulus and stress and strain (see Experiment 16), Hooke's law has the form

$$F = \left(\frac{YA}{L_0}\right)\Delta L \qquad \textbf{(16-4)}$$

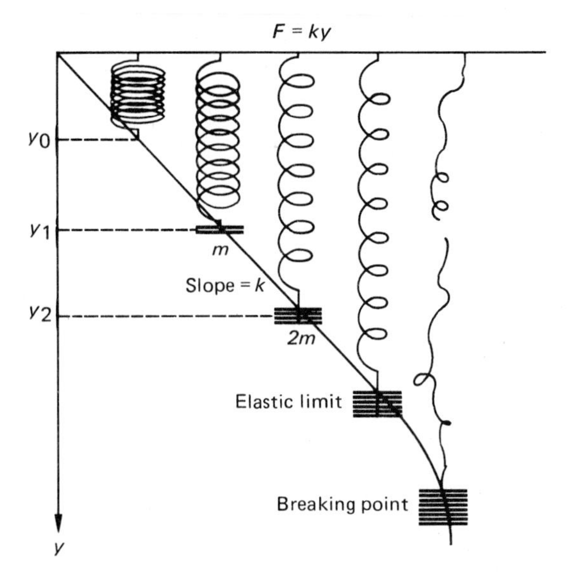

Fig. 17-1 The elongation of a spring versus force. Hooke's law holds up to thc clastic limit.

so $k = YA/L_0$. However, for a coil spring, the stress is practically a pure shear, and the constant k, called the spring constant, depends on the shear modulus of the wire, the radius of the wire, the radius of the coils, and the number of coils. The spring constant is sometimes called the "stiffness constant," since it gives an indication of the relative stiffness of a spring—the greater k, the greater the stiffness. As can be seen from Eq. 17-1, k may have units of N/m, dyne/cm, or lb/in.

According to Hooke's law, the elongation of a spring as a whole is directly proportional to the stretching force.* For example, as illustrated in Fig. 17-1, if an initially unstretched spring has a length y_0 and a suspended weight of mass m stretches the spring so that its length is y_1, then in equilibrium the weight force is balanced by the spring force and

$$F_1 = mg = k(y_1 - y_0)$$

Here we use y to indicate the vertical direction instead of x as in Eq. 17-1 which is usually used to mean the horizontal direction. Similarly, if another mass m is added and the spring is stretched to a length y_2,

$$F_2 = 2mg = k(y_2 - y_0)$$

and so on for more added weights. The linear relationship of Hooke's law holds, provided that the deformation or elongation is not too great. Beyond an elastic limit, a spring is permanently deformed and eventually breaks with increasing force (see Fig. 16-2, Expt. 16).

Notice that the general form of Hooke's law has the form of an equation for a straight line:

$$F = k(y - y_0)$$

or

$$F = ky - ky_0$$

which is of the general form $y = ax + b$

*The restoring spring force and the stretching force are equal in magnitude and opposite in direction (Newton's third law).

B. *Simple Harmonic Motion*

When the motion of an object is repeated in regular time intervals, the motion is referred to as being **periodic.** Such motion can be described in terms of trigonometric sine and cosine functions. These are referred to as **harmonic functions;** hence periodic motion is often called **harmonic motion.** A simple form of such motion is that of an object being acted upon by a force that obeys Hooke's law, and hence is referred to as **simple harmonic motion (SHM).** Examples of simple harmonic motion include the oscillations of a pendulum in small arcs and the oscillations of a mass on a spring.

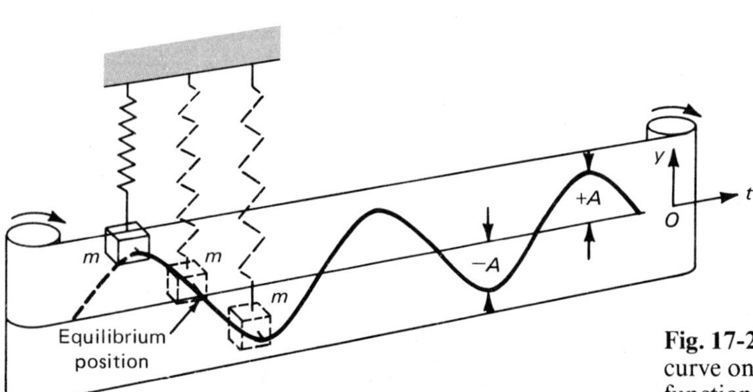

Fig. 17-2 A mass oscillating on a spring describes a wavy curve on a moving paper that can be described by a cosine function when $y = +A$ at $t = 0$.

As illustrated in Fig. 17-2, a mass oscillating on a spring would trace out a wavy, time-varying curve on a moving paper. The equation for this curve, which describes the oscillatory motion of the mass, can be written

$$y = A \cos \frac{2\pi t}{T} \qquad \textbf{(17-2)}$$

where T is the period of oscillation and A the amplitude or maximum displacement of the mass. The amplitude depends on the initial conditions of the system (i.e., how far the mass was initially displaced from its equilibrium position). If the mass were initially pulled below its equilibrium position (to $y = -A$) and released, the equation of motion would be $y = -A \cos 2\pi t/T$, which satisfies the initial condition, $t = 0$, $\cos 0 = 1$, and $y = -A$. The argument of the cosine (i.e., $2\pi t/T$) is in radians rather than degrees.

In actual practice, the amplitude decreases slowly as energy is lost to friction and the oscillatory motion is slowly "damped." In some applications, the simple harmonic motion of an object is intentionally damped (e.g., the spring-loaded needle indicator of an electrical measurement instrument of the scale on a common bathroom scale). Otherwise, the needle or scale would oscillate for some time about the equilibrium position, making it difficult to obtain a quick and accurate reading.

The period of oscillation depends on the parameters of the system. In general, the period of an object in SHM is given by

$$T = 2\pi \sqrt{-\frac{\text{displacement}}{\text{acceleration}}} \qquad \textbf{(17-3)}$$

In the case of mass on a spring, by Newton's second law and Hooke's law,

$$F = ma = -ky \quad \text{or} \quad \frac{-y}{a} = \frac{m}{k}$$

Hence, the period of oscillation of a mass on a spring is

$$T = 2\pi \sqrt{\frac{m}{k}} \qquad \textbf{(17-4)}$$

IV. EXPERIMENTAL PROCEDURE

A. *Rubber-Band Elongation*

1. Hang a rubber band on a support and suspend a weight hanger from the rubber band. Add an appropriate weight to the weight hanger (e.g., 100–300 gm) and record the total suspended weight ($m_1 g$) in Data Table 1. [It is convenient to leave the acceleration due to gravity in symbol form (e.g., if $m_1 = 100$ gm, then weight $= m_1 g = 100g$ dynes).] Fix a meter stick vertically alongside the weight hanger and note the position of the bottom of the weight hanger on the meter stick. Record this (y_1) in the data table.

2. Add appropriate weights to the weight hanger one at a time (e.g., 100 gm), and record the total suspended weight and the position of the bottom of the weight hanger on the meter stick after each elongation (y_2, y_3, etc.). The weights should be small enough so that seven or eight weights can be added without overstretching the rubber band.

3. Plot the total suspended weight force versus elongation position (mg versus y), and draw a smooth curve that best fits the data points.

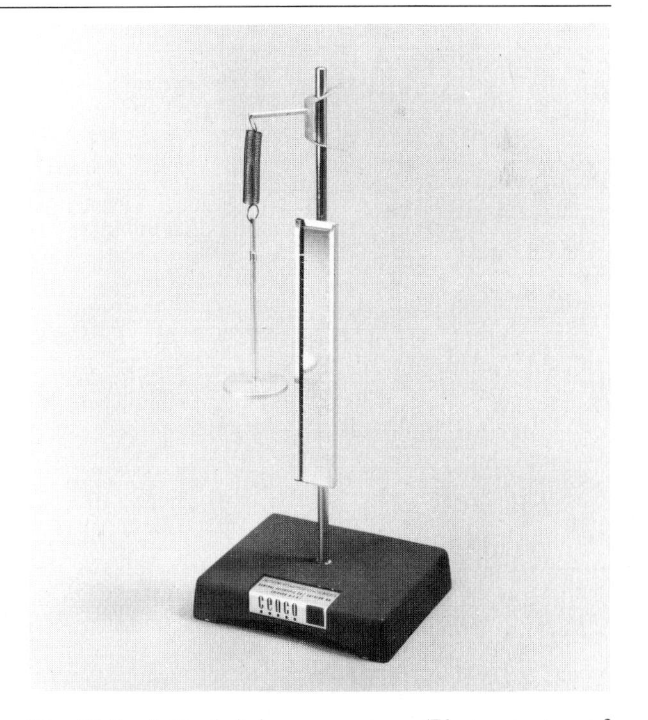

Fig. 17-3 A Hooke's law apparatus. (Photo courtesy of Central Scientific Co., Inc.)

B. *Spring Elongation*

4. Repeat procedures 1 and 2 for a coil spring. (A commercially available Hooke's law apparatus is shown in Fig. 17-3.)

5. Plot mg versus y on the same sheet of graph paper used in procedure 3 (double label axes if necessary) and draw a straight line that best fits the data. Determine the slope of the line (the spring constant k) and record in the data table. Answer the questions below the data tables.

C. *Period of Oscillation*

6. (a) With an appropriate small amount of mass on the weight hanger suspended from the spring (the mass should not oscillate too fast), pull the weight hanger down a certain distance (e.g., 5 to 10 cm) and release. Record the total mass in the data table.
(b) Using a laboratory timer or stopwatch, release the spring weight hanger from the predetermined initial displacement and determine the time it takes for the mass to make a number (5 to 10) of complete oscillations or cycles. The number of cycles timed will depend on how quickly the system loses energy or is damped. Make an effort to time enough cycles to get a good average period of oscillation. Record the total time and the number of oscillations in the data table.
Divide the total time by the number of oscillations to determine the average period.

7. Repeat procedure 7 for four more different mass values and record the results in Data Table 3. The initial displacement may be varied if necessary. (This should have no effect on the period. Why?)

8. Plot a graph of the period squared (T^2) versus the mass (m) and draw a straight line that best fits the data points. Determine the slope of the line and compute the spring constant k (Eq. 17-4). Compare this value k with that determined from the slope of the spring elongation graph in part B by computing the percent difference.

Lab Partner(s) ....

EXPERIMENT 17 *Hooke's Law and Simple Harmonic Motion*

LABORATORY REPORT

A. *Rubber-Band Elongation*

DATA TABLE 1

Total suspended weight*		Scale reading	
m_1g		y_1	
m_2g		y_2	
m_3g		y_3	
m_4g		y_4	
m_5g		y_5	
m_6g		y_6	
m_7g		y_7	
m_8g		y_8	

*It is convenient to leave g in symbol form, even when graphing.

B. *Spring Elongation*

DATA TABLE 2

Total suspended weight*		Scale reading	
m_1g		y_1	
m_2g		y_2	
m_3g		y_3	
m_4g		y_4	
m_5g		y_5	
m_6g		y_6	
m_7g		y_7	
m_8g		y_8	

k (slope of graph) ....
(units)

Calculations (show work)

(continued)

C. *Period of Oscillation*

DATA TABLE 3

Total suspended mass		Total time (s)	Number of oscillations	Average period T	T^2
m_1					
m_2					
m_3					
m_4					
m_5					

Calculations (show work)

Slope of graph ....

Computed spring constant k ....

Percent difference (of k's in B and C) ....

QUESTIONS

1. (a) Interpret the intercepts of the straight line for the spring elongation in the mg versus y graph of part B. Is the elastic property of the rubber band an example of Hooke's law? Explain. (b) Interpret the x-intercept of the straight line of the T^2 versus m graph of part C.

2. Draw a horizontal line through the y-intercept of the straight line graph of part B, and form a triangle by drawing a vertical line through the last data point. (a) Prove that the area of the triangle is the work done in stretching the spring. (*Hint:* $W = \frac{1}{2}kx^2$ and area of triangle $A = \frac{1}{2}ab$ or $\frac{1}{2}$ altitude times base.)

(b) Compute the work done in stretching the spring from the graph.

3. For a mass oscillating on a spring, at what positions do the (a) velocity and (b) acceleration of the mass have maximum values?

4. What is the form of the equation of motion for the SHM of a mass suspended on a spring when the mass is initially (a) released 10 cm above the equilibrium position; (b) given an upward push and it goes a maximum displacement of 8 cm; (c) given a downward push and it goes a maximum displacement of 12 cm? (*Hint:* Sketch the curve for the motion as in Fig. 17-2 and fit the appropriate trigonometric function to the curve.)

5. For only case (a) in Question 4, what is the displacement y of the mass at times (a) $t = T/2$; (b) $t = 3T/2$; (c) $t = 3T$?

6. *(Optional)* Using Eq. 17-3, show that the period of a simple pendulum oscillating in small arcs is given by $T = 2\pi \sqrt{L/g}$.

Lab Partner(s) ....

EXPERIMENT 18 *Standing Waves in a String*

ADVANCE STUDY ASSIGNMENT

Read the experiment and answer the following questions.

1. How is the wave speed related to the frequency and wavelength?

2. What is a standing wave, and what are nodes and antinodes?

3. What are normal nodes?

4. How does the wavelength of a standing wave in a vibrating string vary as the tension force in the string or the linear mass density of the string varies?

5. Standing waves in a string can be produced by oscillating the string at the various natural frequencies. However, in this experiment the string vibrator has only one frequency. How, then, are standing waves with different wavelengths produced?

EXPERIMENT **18**

Standing Waves in a String

I. INTRODUCTION

A wave is the propagation of a disturbance in a medium. When a stretched cord or string is disturbed, the wave travels along the string with a speed that depends on the tension in the string and its linear mass density. Upon reaching a fixed end of the string, the wave is reflected back along the string.

For a continuous disturbance, the propagating waves interfere with the oppositely moving reflected waves and a standing- or stationary-wave pattern is formed under certain conditions. These standing-wave patterns can be visually observed and the number of loops in a pattern depends on the length of the string and the wave velocity, which is dependent on the tension in the string.

The visual observation and measurement of standing waves serve to provide a better understanding of wave properties and characteristics. In this experiment, the relation between the tension force and the wavelength in a vibrating string will be studied as applied to the natural frequencies or normal modes of oscillation of the string.

II. EQUIPMENT NEEDED

- Electric string vibrator
- Clamps and support rod
- Pulley with rod support
- String
- Weight hanger and slotted weights
- Meter stick
- Laboratory balance
- 1 sheet of Cartesian graph paper

III. THEORY

A wave is characterized by its **wavelength** λ (cm), the **frequency of oscillation** f (Hz or 1/s = s^{-1}), and the **wave speed** v (cm/s). (See Fig. 18-1.) These quantities are related by the expression

$$\lambda f = v \qquad \textbf{(18-1)}$$

(Check to see if the equation is dimensionally correct.)

Waves in a stretched string are transverse waves; that is, the "particle" displacement is perpendicular to the direction of propagation. In longitudinal waves, the particle displacement is in the direction of wave propagation (e.g., sound waves). The maximum displacement of a wave is called its **amplitude,**

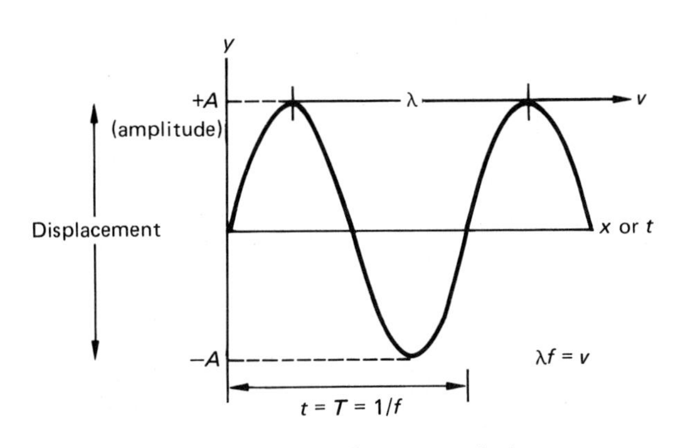

Fig. 18-1 The parameters of wave description.

which is related to the energy of the wave. The **period of oscillation T** is related to the frequency of oscillation, $T = 1/f$.

When two waves meet, they interfere and the combined wave form is a superposition of the two interfering waves. The superposition of two waves of equal amplitude and frequency traveling in opposite directions gives rise to what is known as a **standing** or **stationary wave.** The periodic constructive and destructive interference causes the formation of a standing-wave pattern as illustrated in Fig. 18-2. Notice that some of the "particles" on the axis are stationary. These positions are called *nodal points* or **nodes,** and the points of maximum displacements are called **antinodes.** We might say that the energy is "standing in the wave envelope as it alternates between the kinetic and potential energies of the particles."

In a stretched string being oscillated or shaken at one end, waves traveling outward from the oscillator interfere with waves that have been reflected at the other fixed end. However, standing waves in a given length of string occur only for certain definite wave frequencies. That is, for a given stretching tension or force, the string must be driven or oscillated with certain vibrational frequencies to produce standing waves. Since the string is fixed at each end, a standing wave must have a node at each end. As a result, only an integral number of half wavelengths may "fit" into the length L of the string, $L = \lambda/2, \lambda, 3\lambda/2, 2\lambda$, and so on, such that in general

$$L = \frac{n\lambda}{2} \qquad n = 1, 2, 3, 4, \ldots \qquad \textbf{(18-2)}$$

Figure 18-2 illustrates the case for $L = 3\lambda/2$.

The wave speed in a stretched string is given by

$$v = \sqrt{\frac{F}{\mu}} \qquad \textbf{(18-3)}$$

where F is the tension force in the string and μ is the linear mass density (mass per unit length, $\mu = m/L_0$) of the string. Using Eqs. 18-2 and 18-3 in $\lambda f = v$ (Eq. 18-1),

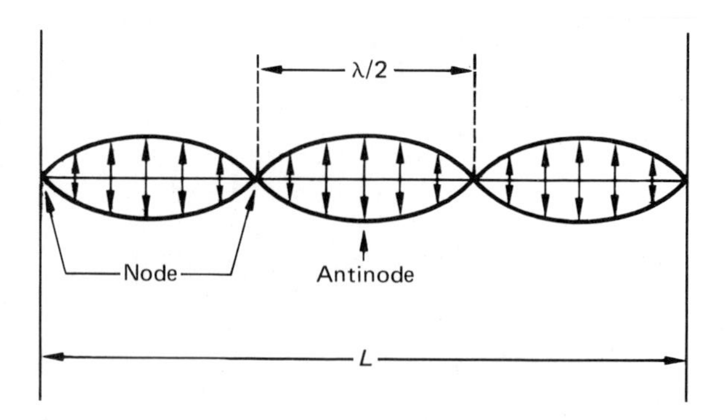

Fig. 18-2 A standing wave. Periodic constructive and destructive interference gives rise to a standing-wave pattern.

$$f_n = \frac{v}{\lambda_n} = \frac{n}{2L}\sqrt{\frac{F}{\mu}} \qquad n = 1, 2, 3, \ldots \qquad \textbf{(18-4)}$$

where f_n and λ_n are the frequency and wavelength, respectively, for a given integer n. Hence, only certain frequencies produce standing waves for a given string tension, density, and length. These frequencies are called the **natural frequencies** or **normal modes of oscillation** of the string. The lowest frequency f_1 is called the **fundamental frequency** or **first harmonic.** The higher frequencies, f_2, f_3, and so on, are higher harmonics (e.g., f_2 is the second harmonic, etc.).

In this experiment, the electrically driven string vibrator has a fixed frequency, so that the driving frequency cannot be varied to produce different normal-mode standing-wave patterns. However, by varying the string tension, the wave velocity can be varied so as to produce different standing-wave patterns. Since $v = \sqrt{F/\mu}$ (Eq.18-3),

$$\lambda = \frac{v}{f} = \frac{1}{f}\sqrt{\frac{F}{\mu}} \qquad \textbf{(18-5)}$$

where f and μ are constant. Hence, by varying F, the appropriate wavelengths can be selected that will "fit" into a given string length L to produce standing waves.

IV. EXPERIMENTAL PROCEDURE

1. Cut a piece of string long enough to be used in the experimental setup—long enough to be looped at each end so as to be attached to the vibrator and a weight hanger suspended from the end running over the pulley (Fig. 18-3). The vibrator and pulley should be clamped to opposite ends of a laboratory table, or about 150 cm apart. (This length may vary for a given setup.)

   Measure the total length of the string and determine its mass on a laboratory balance. Record these values in the data table and compute the linear mass density $\mu = m/L_0$. (*Note:* L_0 is the total length of the string.)

2. Attach the string to the vibrator and suspend a weight hanger from the other end as shown in

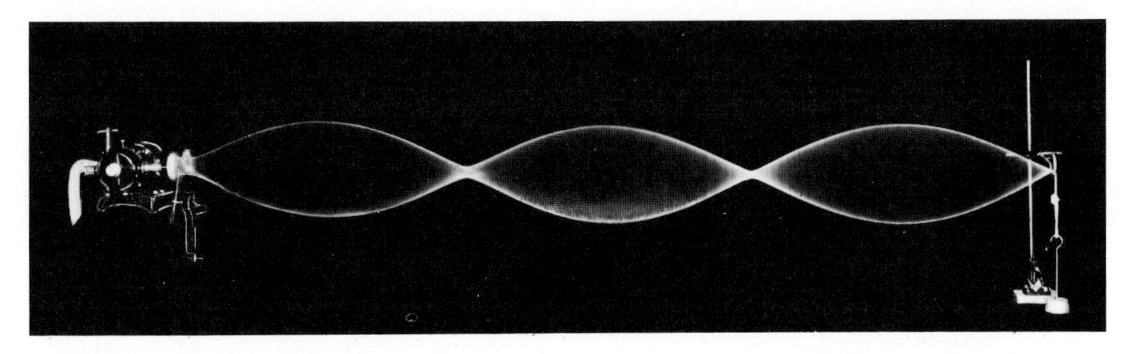

Fig. 18-3 A standing-wave apparatus. (Photo courtesy of Sargent-Welch Scientific Company)

Fig. 18-3. Make certain that the string is aligned directly between the vibrator and pulley and that it is parallel to the table surface. Measure the distance between the vibrator arm and the point of contact on the string on the pulley. Record this length L in Data Table 1.

Turn on the vibrator. Try to produce different standing-wave patterns in the string by alternatively lifting and pushing down on the weight hanger. It is helpful to place a folded thin strip of paper on the string to observe vibrating action. The number of loops increases with less tension. Why? Also, try grasping the string at a node and antinode of a given pattern to see what happens.

3. When you are familiar with the operation of the apparatus, add enough weights to the weight hanger so that a standing-wave pattern of two loops is formed in the string (nodal point at the center). Adjust the tension by adding or removing some small weights until the loops are of maximum amplitude. If sufficiently small weights are not available, a fine adjustment can be made by loosening the clamp holding the vibrator rod and sliding it slightly back and forth so as to find the optimum string length between the ends that gives the maximum loop width or amplitude for a given tension.

   When this is accomplished, measure with a meter stick the distance from the point where the string contacts the pulley to the center nodal point. The meter stick can be held alongside the vibrating string, or you may find it more convenient to grasp the string at the nodal point with your fingers, shut off the vibrator, and measure the distance from the pulley contact to the nodal point along the nonvibrating string. Make certain not to pull the string toward the vibrator and increase the length by raising the weight hanger. Apply a slight tension in the string *away* from the vibrator if necessary.

   Record this length ℓ_1 and the total suspended mass in Data Table 1. Since the length of one loop is one-half of a wavelength, $\ell_1 = \lambda/2$.

4. Remove enough weights from the weight hanger and adjust so that a standing-wave pattern of maximum amplitude with three loops (two nodal points in the string) is formed. Measure the distance from the pulley contact to the nodal point nearest the vibrator. (The fixed-end nodal point *at* the vibrator is not used because in vibrating up and down, it is not a "true" nodal point.) Record this length ℓ_2 and the total suspended mass in the data table. Since the length of two loops is equal to one wavelength, $\ell_2 = \lambda$.

5. Repeat procedure 4 for consecutive standing-wave patterns up to eight measured loops if possible. [The weight hanger by itself may supply too much tension for higher-order patterns, so it may have to be removed and smaller weight(s) suspended.] Compute the wavelength for each case.

   It should become evident that in general $\lambda = 2\ell_N/N$ or $\ell_N = N\lambda/2$, where N is the number of loops in a given ℓ. Notice the similarity of the latter form of this equation with Eq. 18-2, wherein the length L is the total vibrating length of the string.

6. Notice that Eq. 18-5 written as

$$\lambda = \frac{1}{f}\sqrt{\frac{F}{\mu}} = \left(\frac{1}{f\sqrt{\mu}}\right)\sqrt{F} \qquad \textbf{(18-5)}$$

   where f and μ are constants has the form of an equation of a straight line, $y = ax + b$, with $x = \sqrt{F}$.

   Plot the experimental data in a graph of λ versus $\sqrt{F}$. Draw the straight line that best fits the data and determine the slope of the line. From this value and the previously determined value of μ, compute the average frequency f of the oscillations.[†] [*Hint:* With the tension forces written with the symbol form of g as suggested in Data Table 1, it is convenient to plot $\sqrt{F}$ in the form $\sqrt{F} = \sqrt{mg} = \sqrt{m}\,\sqrt{g}$ (e.g., if $F = 100g$, $\sqrt{F} = \sqrt{100g} = 10\sqrt{g}$). The abscissa scale may be labeled in this form—a numerical prefix times the constant $\sqrt{g}$.]

[†]Should you have some scattered data points far from the straight line, see Question 2.

The string vibrator operates on 60-cycle ac current. The vibrating action is accomplished by means of an electromagnet operated by the input current. The vibrator arm is attracted toward an electromagnet during each half-cycle, or twice each cycle, so the vibrating frequency is $2 \times 60 = 120$ Hz (cycles per second). Using this as the accepted value of the vibrational frequency, compute the percent error of the experimentally determined value.

LABORATORY REPORT

DATA TABLE 1

Mass of string ....

Total length of string, L_0 ....

Linear mass density μ ....

Length of string between vibrator and pulley L ....

Number of loops measured N	Suspended mass	Tension force F*		Measured length		Wavelength (λ)	$\sqrt{F}$
1		F_1		ℓ_1			
2		F_2		ℓ_2			
3		F_3		ℓ_3			
4		F_4		ℓ_4			
5		F_5		ℓ_5			
6		F_6		ℓ_6			
7		F_7		ℓ_7			
8		F_8		ℓ_8			

*For convenience, express the tension weight force in terms of g (e.g., $m = 0.10$ kg, $F = mg = 0.10\,g$ N).

Lab Partner(s) ....

EXPERIMENT 18 *Standing Waves in a String*

Calculations (show work)

Slope of graph ....

Computed frequency f ....

Accepted frequency ....

Percent error ....

QUESTIONS

1. The wavelength associated with F_1 and ℓ_1 in the experiment does not correspond to the wavelength of the fundamental frequency of the string. (a) To which natural frequency does the wavelength associated with F_1 and ℓ_1 correspond?

   (b) What would be the tension in the string required to produce a standing wave with a wavelength corresponding to the fundamental frequency of the string? (*Hint:* Use Eq. 18-5.)

(continued)

2. Theoretically, the vibrator frequency is 120 Hz. However, sometimes the vibrator resonates with the string at a "subharmonic" of 60 Hz. (a) If this were the case in all instances, how would it affect the slope of the graph? (b) If you have some scattered data points far from the straight line on your graph, analyze the data for these points using Eq. 18-5 to determine the frequency.

3. How many normal modes of oscillation or natural frequencies does each of the following have? (a) A simple pendulum; and (b) a mass oscillating on a spring.

4. Stringed musical instruments, such as violins and guitars, use stretched strings. Explain (a) how tightening and loosening the strings tune them to their designated tone pitch or frequency; (b) why the strings of lower tones are thicker or heavier; (c) why notes of higher pitch or frequency are produced when the fingers are placed on the strings.

Name .... Section .... Date ....

Lab Partner(s) ....

EXPERIMENT 19 *Air Column Resonance: The Speed of Sound in Air*

ADVANCE STUDY ASSIGNMENT

Read the experiment and answer the following questions.

1. What is resonance?

2. What is the variable used in studying resonance in an air column in (a) a closed organ pipe of fixed length; (b) the resonance tube apparatus used in the experiment?

3. How does the speed of sound vary with temperature?

4. Should a tuning fork be set into oscillation by striking it with or on any object? Explain.

5. For a resonance tube apparatus with a total tube length of 1 m, how many resonance positions would be observed when the water level is lowered through the total length of the tube for a tuning fork with a frequency of (a) 500 Hz; (b) 1000 Hz? (Show your calculations.)

EXPERIMENT 19

Air Column Resonance: The Speed of Sound in Air

I. INTRODUCTION

All systems have one or more natural vibrating frequencies. When a system is driven at a natural frequency, there is a maximum energy transfer and the vibrational amplitude is a maximum. For example, when pushing a person on a swing, if the pushes are applied at the proper frequency, there will be maximum energy transfer and the person will swing higher (greater amplitude). When a system is driven at a natural frequency, we say that the system is in resonance (with the driving source) and refer to the particular frequency at which this occurs as a resonance frequency.

A swing or pendulum has only one natural frequency. However, a stretched string (see Experiment 18) and an air column of given lengths have several natural frequencies, depending on the number of wavelength segments that can be "fitted" into the system length. From the relationship between the frequency f, the wavelength λ, and the wave speed v, which is $\lambda f = v$, it can be seen that if the frequency and wavelength are known, the wave speed can be determined; and if the wavelength and wave speed are known, the frequency can be determined.

As an application of resonance, in this experiment the speed of sound in air will be determined by driving an air column in resonance. Also, knowing the speed of sound, the unknown frequency of a tuning fork will be found from resonance conditions.

II EQUIPMENT NEEDED

- Resonance tube apparatus
- Three tuning forks (500- to 1000-Hz range), with stamped frequency of one fork covered or masked so as to serve as an unknown
- Rubber mallet or block
- Meter stick (if measurement scale not on resonance tube)
- Thermometer (to determine air temperature—one per class is sufficient)
- Vernier calipers

III. THEORY

Air columns in pipes or tubes of fixed lengths have particular resonant frequencies. For example, in a closed organ pipe (closed at one end*) of length L, when the air column is driven at particular frequencies, the air column vibrates in resonance. The interference of the waves traveling down the tube and the reflected waves traveling up the tube produces (longitudinal) standing waves (see Experiment 18), which must have a node at the closed end of the tube and an antinode at the open end (Fig. 19-1).† This corresponds to transverse standing waves in a stiff string or rod fixed at one end. (Keep in mind that the waves in the air column are longitudinal—particle displacement is in the direction of wave propaga-

*A closed tube or organ pipe is open at one end. An open pipe is open at both ends.

†The antinode does not occur exactly at the open end of the tube but at a slight distance above it, which depends on the diameter of the tube.

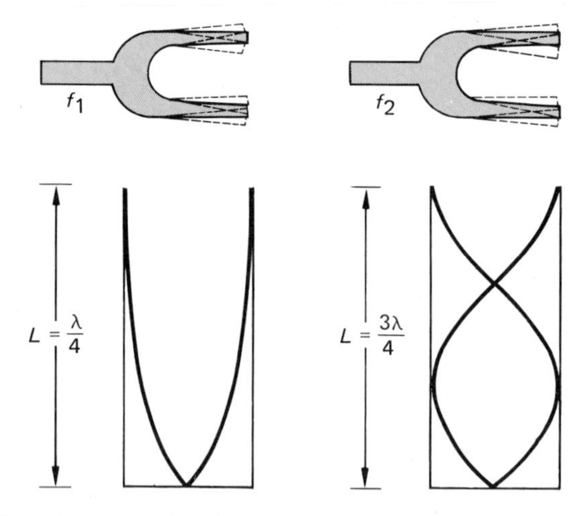

Fig. 19-1 Longitudinal standing waves of different frequencies in a tube. There must be a node at the closed end and an antinode near the open end.

tion. The transverse nature of the drawings is for illustration only.)

The resonance frequencies of a pipe or tube (air column) depend on its length L. As can be seen from Fig. 19-1, only a certain number of wavelengths or "loops" can be "fitted" into the tube length with the node–antinode requirements. Since each loop corresponds to one-half wavelength, resonance occurs when the length of the tube is nearly equal to an odd number of quarter wavelengths (i.e., $L = \lambda/4$, $L = 3\lambda/4$, $L = 5\lambda/4$, etc.) or in general $L = n\lambda/4$, $n = 1, 3, 5, \cdots$, and $\lambda = 4L/n$. Then with $\lambda f = v$, or $f = v/\lambda$,

$$f_n = \frac{nv}{4L} \qquad n = 1, 3, 5, \ldots \qquad \textbf{(19-1)}$$

Hence, an air column (tube) of length L has particular resonance frequencies and will be in resonance with the corresponding driving frequencies.

As in the case of standing waves in a string (Experiment 18), the lowest natural or resonant frequency f_1 is called the fundamental frequency or first harmonic. Successively higher frequencies are higher harmonics or overtones. For example, f_2 is the second harmonic or first overtone.

As can be seen from Eq. 19-1, the three experimental parameters involved in the resonance condition of an air column are f, v, and L. To study resonance in this experiment, the length L of an air column will be varied for a given driving frequency, instead of varying f for a fixed L as in the case of the closed organ pipe described above. (The wave speed in air is relatively constant in either case.) The length of an air column will be varied by raising and lowering the water level in a tube (see Fig. 19-3).

As the length of the air column is increased, more wavelength segments will fit into the tube, consistent with the node–antinode requirements at the ends (Fig. 19-2). The difference in the tube (air column) lengths when successive antinodes are at the open end of the tube and resonance occurs is equal to a half wavelength; for example,

$$\Delta L = L_2 - L_1 = \frac{3\lambda}{4} - \frac{\lambda}{4} = \frac{\lambda}{2}$$

(for the case illustrated in Fig. 19-2), and

$$\Delta L = L_3 - L_2 = \frac{5\lambda}{4} - \frac{3\lambda}{4} = \frac{\lambda}{2}$$

Since the antinodes are the positions of maximum amplitude (particle displacement), the antinodes correspond to the maximum sound intensity. As a result, when an antinode is at the open end of the tube, a loud resonance tone is heard. (If a node is at the open end—zero amplitude—there is no resonance sound coming from the tube.) Hence, the tube lengths for antinodes to be at the open end of the tube can be determined by lowering the water level in the tube and "listening" for successive antinodes. No end correction is needed for the antinode occurring slightly above the end of the tube in this case, since the difference in tube lengths for successive antinodes is equal to $\lambda/2$.

If the frequency f of the driving tuning fork is known and the wavelength is determined by measuring the difference in tube length between successive antinodes, $\Delta L = \lambda/2$ or $\lambda = 2\,\Delta L$, the speed of sound in air v_s can be determined from

$$v_s = \lambda f \qquad \textbf{(19-2)}$$

Once v_s is determined (for a given ambient temperature), the unknown frequency of a tuning fork

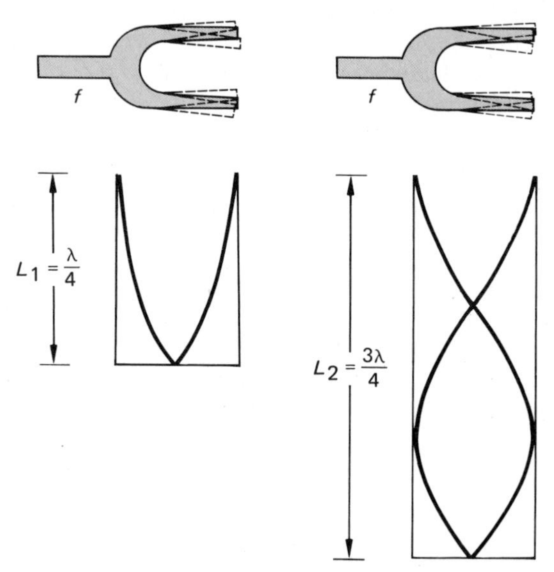

Fig. 19-2 Longitudinal standing waves of the same frequency in tubes of different lengths.

can be computed from Eq. 19-1 using the experimentally determined resonance wavelength in an air column for the unknown tuning fork frequency.

The speed of sound in air is temperature-dependent and is given to a good approximation by the relationship

$$v_s = 331.4 + 0.6T_c \text{ m/s} \tag{19-3}$$

where T_c is the air temperature in degrees Celsius. The equation shows that the speed of sound at 0°C is 331.4 m/s and increases by 0.6 m/s for each degree of temperature increase. For example, if the air temperature is $T_c = 20°C$ (room temperature, 68°F), then

$$v_s = 331.4 + (0.6)(20) = 343.4 \text{ m/s} = 34{,}340 \text{ cm/s}$$

IV. EXPERIMENTAL PROCEDURE

1. The resonance tube apparatus is shown in Fig. 19-3. Begin by raising the water level in the tube to near the top of the tube by raising the reservoir can on the support rod. This is done by depressing the can clamp.

   With the water level in the tube at this position, there should be little water in the reservoir can. If this is not the case, remove some water from the can to prevent overflow and spilling when the can is filled on lowering. Practice lowering and raising the water level in the tube to get the "feel" of the apparatus. (Practice also helps in reducing the squeaking noise usually associated with sliding the reservoir can on the support rod.)

   Measure and record the room temperature and the inside diameter (I.D.) of the tube.

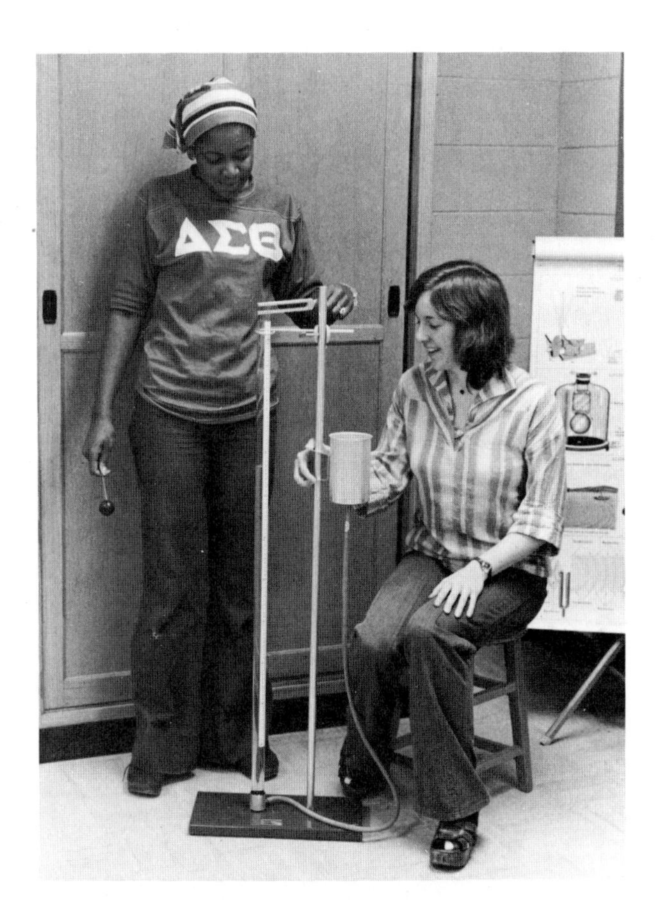

Fig. 19-3 A resonance tube apparatus. The "length" of the tube is varied by adjusting the water level in the tube.

2. With the water level in the tube near the top, take one of the tuning forks of known frequency (usually stamped on the fork) and set it into oscillation by striking it with a rubber mallet or on a rubber block, whichever is available. *Never strike the tuning fork on a hard object* (e.g., a table). This may damage the fork and cause a change in its characteristic frequency.

   Hold the vibrating fork slightly above the top of the tube so that the sound is directed down the tube. (A fork has directional sound-propagation characteristics. Experiment with a vibrating fork and your ear to determine these directional characteristics.) *If the tube is glass, do not let the vibrating fork strike the top of the tube.* This may chip the tube. Some resonance tubes are made of plastic.

3. With the vibrating fork above the tube, *quickly* lower the reservoir can to a low position on the support rod. The water level in the tube will fall slowly and the successive antinodes will be heard as the water level in the tube passes through the resonance length positions. It will probably be necessary to quickly and continuously strike the tuning fork to keep it vibrating sufficiently.

   As the water level passes through the successive resonance positions, note the approximate position of the water level for each on the length scale on the tube or on a meter stick held beside the tube if there is no scale on the tube. It is convenient to have one lab partner lower the can and observe the approximate resonance positions while another lab partner handles the tuning fork.

4. Raise the water level to the position of the first observed resonance. Move the fork back and forth (horizontally) past the tube opening to bet-

ter hear the sound boost at resonance. By slightly adjusting the water level up and down, determine the exact position of the water level on the scale (the length from the top of the tube) for the resonance condition and record this length in the data table. Repeat this procedure for the other observed resonance positions. (The number of observed lengths may be less than provided for in the data table.)

5. Repeat procedures 3 and 4 for the other tuning fork of known frequency.

6. (a) Compute the average wavelength for each fork from the average of the differences in the tube lengths between successive antinodes.
   (b) Using the known frequency of each fork, compute the speed of sound for each case.
   (c) Compare the average of these two values with the value of the speed of sound given by Eq. 19-3 by computing the percent error.

7. Repeat procedures 3 and 4 for the tuning fork of unknown frequency.

8. Compute the frequency of the tuning fork using the average experimental value of the wavelength and the velocity of sound as given by Eq. 19-3. Your instructor may supply you with the known frequency of the tuning fork. If so, compare your experimentally determined value with the known frequency by computing the percent error.

LABORATORY REPORT

Temperature ....

DATA TABLE

Inside diameter of tube ....

	Tuning fork 1 Frequency, f		Tuning fork 2 Frequency, f		Tuning fork 3 Unknown frequency	
	Position of resonance (cm)	ΔL	Position of resonance	ΔL	Position of resonance	ΔL
L_1		X		X		X
L_2						
L_3						
L_4						
Average ΔL						
Average λ						

Lab Partner(s) ....

EXPERIMENT 19 *Air Column Resonance: The Speed of Sound in Air*

Calculations (show work)

1. Speed of sound v_s

v_s (from fork 1 data) ....

v_s (from fork 2 data) ....

Average v_s ....

v_s (from Eq. 19-3) ....

Percent error ....

2. Frequency of tuning fork

Experimental value of f ....

Known value ....

Percent error ....

QUESTIONS

1. Suppose that the laboratory temperature were 5°C higher than the temperature at which you performed the experiment. Explain what effect(s) this would have on the experimental results.

2. With the water in the tube at the level for the tube length L_1 of the first resonance of the first tuning fork,

   (a) Would another tuning fork with a frequency lower than that of the first tuning fork produce a resonance?

   (b) Would another tuning fork of some higher frequency produce a resonance?
   If your answer is yes in either case, what would be the frequency of the other fork?

(continued)

3. What are the resonant frequencies for an open organ pipe of fixed length L? (Show your sketches and calculations.)

4. The uppermost antinode (Fig. 19-1) is slightly above the open end of the tube by an amount E (called the end correction). (a) Compute E from the data for each tuning fork. (b) Compare the computed E's with the theoretical values given by the equation $E = 0.4 \times$ I. D., where I. D. is the inner diameter of the tube.

Name .... Section .... Date ....

Lab Partner(s) ....

EXPERIMENT 20 *The Gas Laws*

ADVANCE STUDY ASSIGNMENT

Read the experiment and answer the following questions.

1. Which of the gas laws applies to a quantity of gas undergoing (a) an isometric (isovolumetric) process or change; (b) an isothermal change; (c) an isobaric change?

2. Show that the gas laws can be combined algebraically into the form of the ideal gas law.

3. Is millimeters of mercury (mm Hg) a technically correct unit of pressure? Explain.

4. Show that Gay-Lussac's law is equivalent to Charles' law.

5. Distinguish between a real and a perfect gas.

6. What does Gay-Lussac's law imply about the volume of a gas if its temperature could be lowered to absolute temperature and it still remained a gas?

EXPERIMENT **20**

The Gas Laws

I. INTRODUCTION

The **gas laws** are thermodynamic relationships that express the behavior of a quantity of gas in terms of the pressure p, volume V, and temperature T. The kinetic theory of gases expresses the behavior of a "perfect" or ideal gas to be $pV = NkT$, which is commonly referred to as the *perfect* or *ideal gas law*. However, the general relationships among p, V, and T contained in this equation had been expressed earlier in the classical gas laws for real gases. These gas laws were based on the empirical observations of early investigators, in particular the English scientist Robert Boyle (1627–1691) and the French scientists Jacques Charles (1747–1823) and Joseph Gay-Lussac (1778–1850).

In this experiment, the empirical gas laws and the relationship in defining absolute zero temperature by a constant-volume gas thermometer will be investigated.

II. EQUIPMENT NEEDED

- Boyle's law apparatus
- Glass bulb attachment
- Thermometer (0 to 110°C)
- Beaker and ring stand with wire screen
- Bunsen burner and striker
- Ice
- Barometer, wall-mounted
- 2 sheets of Cartesian graph paper

III. THEORY

A. *Boyle's Law*

Around 1660, Robert Boyle had established an empirical relationship between the pressure and the volume of a gas, which is known as **Boyle's law:**

At constant temperature, the volume occupied by a given mass of gas is inversely proportional to its pressure.

In mathematical notation, we write

$$V \propto \frac{1}{p}$$

or

$$V = \frac{k}{p}$$

and

$$pV = k \qquad \textbf{(20-1)}$$

where k is the constant of proportionality for a given temperature (see Fig. 20-1).

The pressure of a confined gas is commonly measured by means of a manometer (see Fig. 20-2). The gas pressure for such an open-tube manometer is the sum of the gauge pressure p_g and atmospheric pressure p_a,

$$p = p_g + p_a$$

where in terms of the density ρ of the manometer liquid (usually mercury), the acceleration due to gravity g, and Δh the height difference of the manometer columns,

$$p_g = \rho g \, \Delta h$$

The gauge pressure is often expressed in terms of the height difference $\Delta h = h_2 - h_1$ or in millimeters of mercury (mm Hg or torr) rather than pressure units.

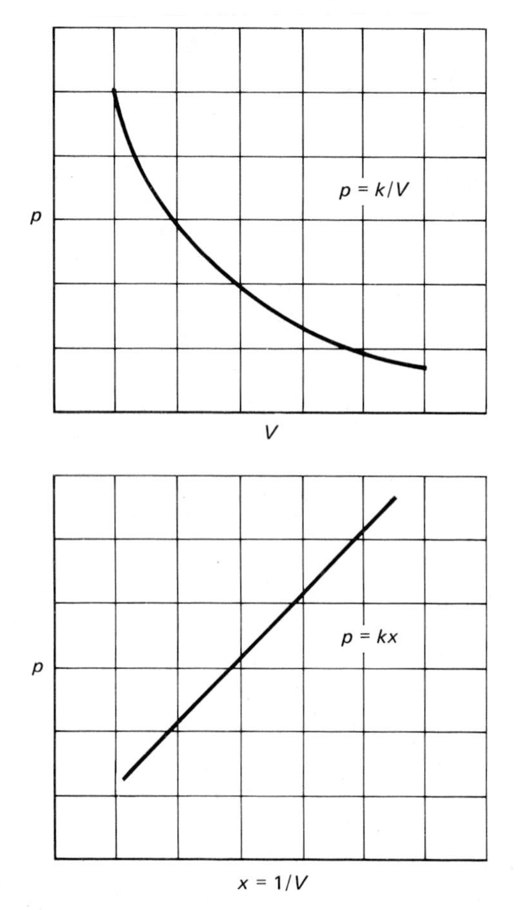

Fig. 20-1 Graphs of p versus V and p versus $1/V$ illustrating the relationship of Boyle's law.

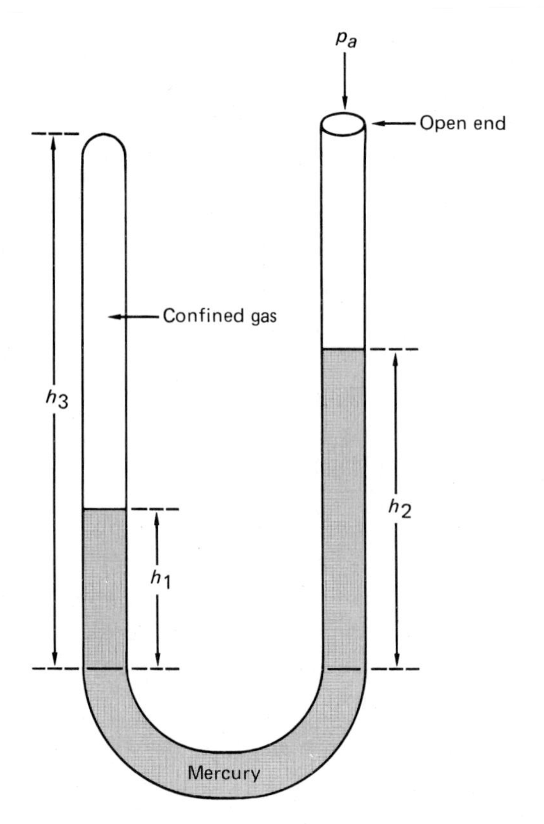

Fig. 20-2 An open-tube mercury manometer. The pressure is $p = p_g + p_a$, where $p_g = \rho_g\ \Delta h = \rho_g(h_2 - h_1)$ and p_a is atmospheric pressure.

In this case, Boyle's law becomes

$$pV = (p_g + p_a)V = k$$

or

$$p_g + p_a = \frac{k}{V} \qquad \textbf{(20-2)}$$

Letting $x = 1/V$, Eq. 20-2 becomes

$$p_g = kx - p_a \qquad \textbf{(20-3)}$$

which has the slope-intercept form of a straight line, $y = mx - b$. Note that as V approaches infinity, $x = 1/V$ approaches zero, and in the limit ($x = 0$) on a plot of p_g versus x, we have $p_g = -p_a$ or the y intercept is equal to the negative value of the atmospheric pressure. Of course, this is not possible experimentally, but the graph of experimental data may be extrapolated to the limit.

In an experiment using a manometer, the heights of the mercury columns (h_2 and h_1) are read and the pressure may be calculated from the height difference, $p_g = \rho g(h_2 - h_1)$. However, in analyzing the data it is more convenient to plot height differences rather than pressure versus (reciprocal) volume. Note that assuming the manometer tube to have a uniform cross-sectional area A, the volume of the confined gas in the closed tube end is $V = A(h_3 - h_1)$. (See Fig. 20-2.) Hence, an equivalent form of Eq. 20-3 in terms of the height difference of the columns (in mm Hg) is

$$h_2 - h_1 = k'x - p_a \qquad \textbf{(20-4)}$$

where $x = 1/(h_3 - h_1)$ and P_a is in mm Hg.

B. *Charles' and Gay-Lussac's Laws*

In 1787, the French physicist Jacques Charles reported the result of a series of experiments which is now known as **Charles' law:**

At constant volume, the pressure exerted by a given mass of gas is proportional to its absolute temperature.

In mathematical notation,

$$p \propto T$$

or

$$p = kT \qquad \textbf{(20-5)}$$

where k is a constant of proportionality.* In 1802, Gay-Lussac, using a somewhat different experimental approach, essentially restated Charles' law in a form more commonly used today (also known as **Gay-Lussac's law**):

At constant pressure, the volume of a given mass of gas is proportional to its absolute temperature.

That is,

$$V \propto T$$

or

$$V = k'T \qquad \textbf{(20-6)}$$

The temperature in the gas laws is expressed in terms of absolute temperature (kelvin, K). The relationship between the Kelvin (T_k) and Celsius (T_c) temperatures is $T_k = T_c + 273$, and Charles' law can be written in terms of the Celsius temperature:

$$p = kT_k = k(T_c + 273)$$

or

$$p = k(273) + kT_c \qquad \textbf{(20-7)}$$

At $T_c = 0°$,

$$p = k(273) = p_0$$

or

$$k = \frac{p_0}{273}$$

where p_0 is the pressure of the gas at $T_c = 0°\text{C}$. Hence Eq. 20-7 becomes

$$p = p_0 + \left(\frac{p_0}{273}\right) T_c$$

or

$$p = p_0(1 + \beta T_c) \qquad \textbf{(20-8)}$$

where $\beta = 1/273°\text{C}^{-1}$ is the temperature coefficient of pressure. For most common gases at low pressures, the experimental value of β is found to approximate $\frac{1}{273}$ per °C.

The preceding development is historically reversed, as Charles' law was originally used to define absolute zero temperature. The experimental value of $\beta = \frac{1}{273}$ per °C means that for every degree Celsius change in temperature, the pressure changes by $\frac{1}{273}$ of the pressure which the gas exerts at 0°C. Hence, if the temperature of a gas could be lowered 273 degrees below 0°C or to −273°C, the pressure would be zero!

$$p = p_0\left(1 + \frac{-273}{273}\right) = p_0(1 - 1) = 0$$

This irreducible minimum of temperature corresponding to $p = 0$ is called **absolute zero.**

Of course, a real gas would liquefy before such a temperature is reached, but in terms of an ideal or perfect gas which always remains a gas at any temperature, absolute zero is the temperature at which molecular activity ceases and the pressure is consequently zero.

Experimentally, absolute zero is defined by measuring p versus T_c of a real gas (at constant volume) over an appropriate temperature range and extrapolating the curve until it intersects the temperature axis that corresponds to the temperature of zero pressure (Fig. 20-3). This temperature is −273°C = 0 K.

Essentially, the experimental setup comprises a constant-volume gas thermometer (for normal temperature ranges), since the measured pressure p is calibrated in terms of temperature. That is, a p reading gives the temperature from the calibration curve. In the case of a constant-volume gas thermometer that uses an open-tube manometer to measure pressure, $p = p_g + p_a$, as in the discussion of Boyle's law.

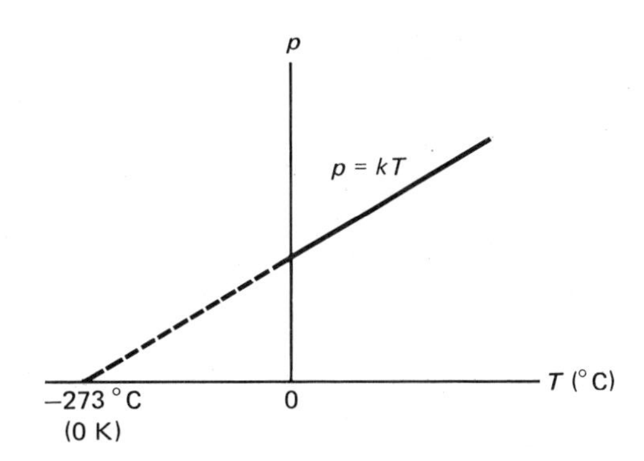

Fig. 20-3 A graph of p versus T illustrating the relationship of Charles' law. Extrapolation to the T axis, which corresponds to the temperature of zero pressure, defines absolute zero (0 K = −273°C).

*Note: k is a common symbol for a constant of proportionality, however, the k's are different in each case, usually in magnitude and definitely in units.

IV. EXPERIMENTAL PROCEDURE

1. Read the atmospheric pressure p_a (in mm Hg) on the wall-mounted barometer and record in the Laboratory Report.

A. *Boyle's Law*

Note: Mercury is an extremely dangerous substance. Should any be spilled, notify the instructor immediately.

2. Types of Boyle's law apparatus are shown in Fig. 20-4. One consists of a flexible-tube mercury manometer with a meter stick for column-height readings. The closed end of the manometer is equipped with a stopcock. The open end of the flexible-tube manometer may be raised or lowered to adjust the mercury levels in both glass tubes.

   If an older type nonflexible manometer is used in the experiment, only pressures above atmospheric pressure are attainable by adding mercury to the open end. (*Caution:* Mercury is a very poisonous substance.)

   New models include an adjustment knob connected to a mercury reservoir inside the base of the apparatus. Turning this knob allows more mercury to be added to both arms of the manometer, without the danger of having to handle mercury. Operating the knob is equivalent to (but easier and safer than) raising or lowering the manometer's glass tubes.

   Caution: Whichever model you use, always adjust (if possible) so that the mercury in both glass tubes is at about the same level ($h_2 = h_1$)

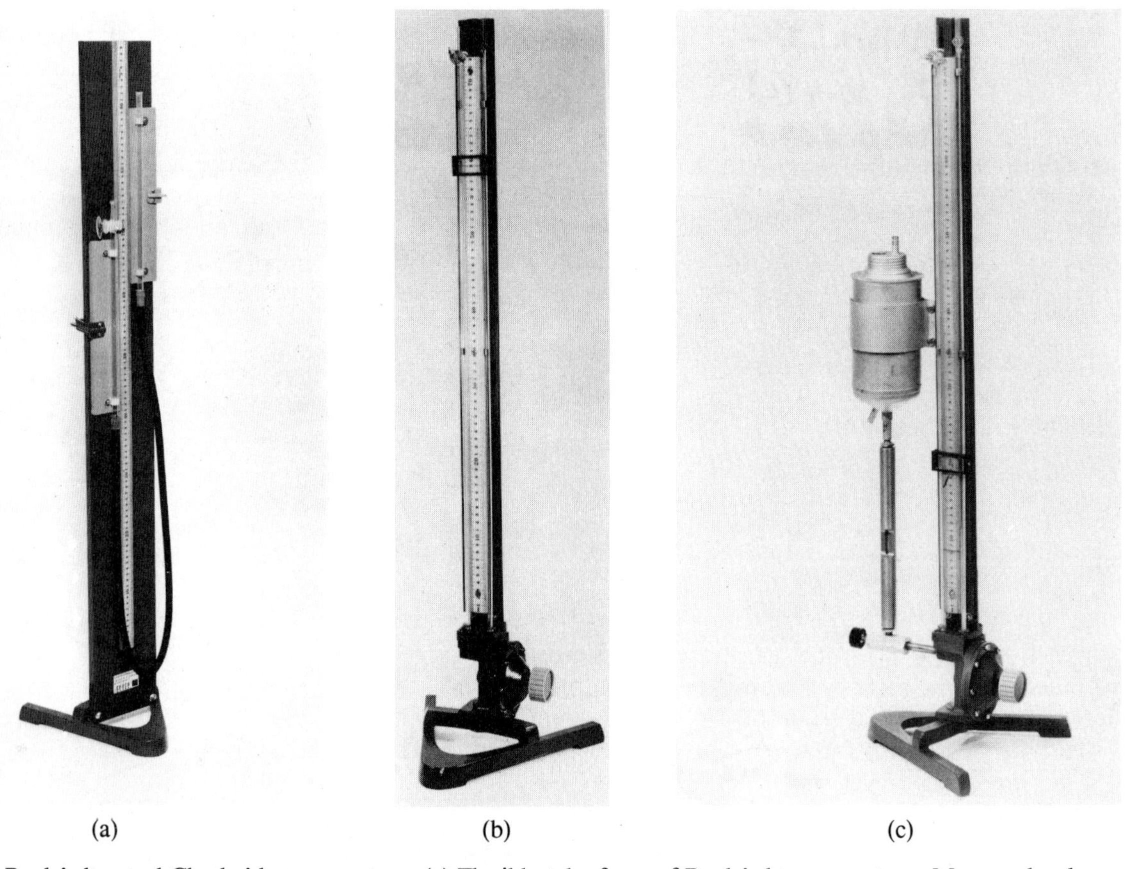

(a) (b) (c)

Fig. 20-4 Boyle's law and Charles' law apparatus. (a) Flexible tube form of Boyle's law apparatus. Mercury levels are adjusted by moving the glass tubes on slide mounts. (b) Diaphragm type of Boyle's law apparatus with enclosed mercury reservoir. The mercury levels in the tubes are adjusted by means of a diaphragm knob. (c) Apparatus for both laws, with metal bulb in copper jacket. Heated water may be added or removed from the jacket to control the temperature of the gas for Charles' law investigations. (Courtesy of Central Scientific Co., Inc.)

before opening the stopcock. Otherwise, mercury may overflow one of the tubes.

Open the stopcock and adjust the mercury levels so that the air space just below the stopcock ($h_3 - h_1$) is about 20 cm in length (or as close as the adjustment will allow).

3. Close the stopcock and keep it closed throughout the Boyle's law portion of the experiment. Read and record the level of the top of the closed tube h_3 (just below the stopcock) and the levels of the mercury in the closed tube h_1 and in the open tube h_2 ($h_2 = h_1$ in this instance). The latter data should be recorded in the next to bottom line in Data Table 1.

   It is helpful to fasten the stopcock with a stopcock clamp or rubber band. The fixed quantity of gas for the experiment is the air trapped in the closed end of the manometer. The stopcock should be well lubricated with stopcock grease so that it is airtight. Check the system for leaks by lowering the mercury to zero in the open-end arm and noting whether there is any change in the mercury levels after an appreciable time.

4. If there is no change in the mercury column heights (indicating that there are no leaks in the system), read and record the heights of the mercury columns h_1 and h_2.

   Then, raise the mercury in the open-tube arm of the manometer through a series of steps toward the initial equal-level condition ($h_1 = h_2$) and read the heights of the columns in each instance. Continue this procedure beyond the equal-level condition if possible, with the last reading being that with the mercury in the open-end arm at its highest attainable position short of overflowing.

5. (a) Compute the gauge pressures p_g in mm Hg ($h_2 - h_1$) and the "volumes" of the trapped air in terms of the closed tube lengths ($h_3 - h_1$).
   (b) Compute the reciprocal of the volume $x = 1/h_3 - h_1$).
   (c) Plot p_g (in mm Hg) versus x with the horizontal axis above the middle of the graph paper so as to allow for negative gauge pressures down to −800 torr (mm Hg). Draw the straight line that best fits the data and determine the value of its y intercept.

   According to theory, the negative value of the y intercept should be equal to the atmospheric pressure. Compare the experimental value to the barometer reading by computing the percent error.

B. *Charles' Law*

6. Return the mercury levels of the Boyle's law apparatus to the equal pressure condition ($h_2 = h_1$) and open the stopcock. Connect the glass bulb attachment to the stopcock nipple by means of a piece of rubber tube.* The apparatus now becomes a Charles' law apparatus that allows the volume of a gas to be kept constant while the temperature and pressure of the gas in the bulb is varied.

   A constant volume is maintained by adjusting for each measurement so that the mercury level of the closed tube side is even with a constant-volume mark. Choose such a mark a few centimeters below the stopcock and indicate it with a wax crayon or pencil if there are no scale marks on the glass. The stopcock remains open throughout the Charles' law procedure.

7. Adjust until the mercury level in the closed tube arm is even with the constant-volume mark. Read and record the heights h_1 and h_2 of the mercury columns in the closed and open arms, respectively, and the (room) temperature as indicated on the thermometer.

8. Place a beaker on a Bunsen burner stand with the glass bulb in the beaker. Fasten the bulb securely with clamps to prevent the bulb from floating in the beaker, and to prevent the bulb stopper (if your apparatus has one) from popping loose under pressure. Take care not to break the bulb.

   Carefully fill the beaker with ice and water (about half and half). Depending on the size of your beaker, it may be convenient to have some ice and water in the beaker before raising it over the bulb.

   According to Charles' law, you should observe the mercury column in the closed tube rise. Why? When the mercury column stops rising, adjust so that the mercury level in the closed tube is even with the constant-volume mark. Read and record the heights of the mercury columns. By definition, the temperature of the gas and the bulb is 0°C.

9. Carefully separate the beaker and the bulb. Remove any excessive ice remaining in the beaker, replace the beaker around the bulb, and add cold tap water until the bulb is completely im-

*A newer model has a metal bulb in a copper jacket, which is used for controlling temperature [see Fig. 20-4(c)]. The instructor will explain the procedure in this case.

mersed again. Place the thermometer in the beaker.

10. Light the Bunsen burner and heat the water slowly until the temperature is about 10°C. As the water is heated, the mercury level in the closed tube will begin to fall. Maintain the mercury level in the closed tube at the constant-volume mark.

    Remove the burner and stir the water with the thermometer. When equilibrium is reached, record the heights of the mercury columns (closed tube level at the constant volume mark) and the temperature.

11. Repeat procedure 10, raising the temperature of the water in 10°C steps until the boiling point is reached. The final reading should be taken when the water is boiling vigorously.

12. (a) Compute the gauge pressure p_g in mm Hg $(h_2 - h_1)$ and add the barometer reading to obtain the total pressure, $p = p_g + p_a$ for each measurement.
    (b) Plot p versus T. Use the long side of the graph paper as the horizontal axis with the temperature scale covering the range from −300 to +110°C. The vertical reference axis should intersect the horizontal axis at $T = 0°C$. Draw the straight line that best fits the data and extrapolate the line to the horizontal axis with a dashed line.
    (c) Determine the slope $p_0\beta$, vertical axis intercept p_0, and the temperature for $p = 0$ (horizontal axis intercept).

13. Compute the temperature coefficient of pressure β from the experimental results and compare it with $\beta = \frac{1}{273}\,°C^{-1}$ by computing the percent difference. Also, compare the temperature-axis-intercept value with that of absolute zero, −273°C.

Name .... Section .... Date ....

Lab Partner(s) ....

EXPERIMENT 20 *The Gas Laws*

LABORATORY REPORT

Laboratory barometer reading, p_a ....

A. *Boyle's Law*

DATA TABLE 1

Height of top of closed tube, h_3 ....

Mercury level		Gauge pressure, p_g $h_2 - h_1$ (mm Hg)	Column of gas $h_3 - h_1$	$x = 1/(h_3 - h_1)$
Open tube h_2 (mm)	Closed tube h_1 (mm)			
$h_2 = h_1$				

Calculations (show work and units)

Barometric pressure from graph intercept ....

Percent error ....

(continued)

B. *Charles' Law*

DATA TABLE 2

Mercury level		Gauge pressure, p_g $h_2 - h_1$ (mm Hg)	Total pressure $p = p_g + p_a$ (mm Hg)	Temperature of gas $T(°C)$
Open tube h_2 (mm)	Closed tube h_1 (mm)			

Calculations (show work and units)

Slope of graph ....

Vertical axis intercept ....

β ....

Percent difference with $\beta = \frac{1}{273}\,°C^{-1}$ ....

Experimental absolute zero ....

Percent error ....

Lab Partner(s) ....

EXPERIMENT 20 *The Gas Laws*

QUESTIONS

1. (a) The constant k in Boyle's law, $pV = k$, is not dimensionless. Show that the units of k are those of work.

   (b) What are the units of k' in Eq. 20-4?

2. In the experimental procedures for Boyle's law and Charles' law, it was assumed that the laboratory temperature and barometric pressure were respectively constant. (a) What error would be induced in each case if these assumptions were not correct?

   (b) What are other sources of error in the experimental procedures?

3. Devise a way to determine absolute zero using Gay-Lussac's law (i.e., describe and explain a constant-pressure thermometer).

(continued)

4. An automobile tire on a cold morning appears slightly flat and has a lower gauge pressure than the pressure to which it was previously inflated. Yet, after traveling several miles, the tire appears normal and has its recommended gauge pressure. Explain this behavior in terms of the gas laws. Assume that no air was lost from the tire.

Name .... Section .... Date ....

Lab Partner(s) ....

EXPERIMENT 21 *The Coefficient of Linear Expansion*

ADVANCE STUDY ASSIGNMENT

Read the experiment and answer the following questions.

1. What is meant by a negative expansion? Is the coefficient of linear expansion negative in this experiment?

2. What is isotropic expansion?

3. Explain how the coefficient of linear expansion is determined experimentally.

4. What are the units of the coefficient of linear expansion?

5. Would taking L_0 to be the length of the rod at 100°C (at the final temperature instead of the initial temperature) change the experimental value of α? Explain.

EXPERIMENT **21**

The Coefficient of Linear Expansion

I. INTRODUCTION

With few exceptions, solids increase in size or dimensions as the temperature is increased. Although this effect is relatively small, it is very important in material applications that undergo heating and cooling. Unless taken into account, material and structural damage can result: for example, a piston may become too tight in its cylinder, a rivet could loosen, or a bridge girder could produce damaging stress.

The expansion properties of a material depends on internal makeup and structure. Macroscopically, we express the thermal expansion in terms of coefficients of expansion, which are experimental quantities that represent the change in the dimensions of a material per degree temperature change. In this experiment, the thermal expansion of some metals will be investigated and their coefficients of linear expansion determined.

II. EQUIPMENT NEEDED*

- Linear expansion apparatus and accessories
- Steam generator and stand
- Bunsen burner and striker
- Rubber tubing
- Beaker
- Meter stick
- Thermometer (0 to 100°C)
- Two or three kinds of metal rods (e.g., iron and aluminum)

III. THEORY

For solids, a temperature change leads to the expansion of a body as a whole. (A contraction resulting from a temperature decrease is a negative expansion.) The change in one dimension of the solid, length, width, or thickness, is called **linear expansion.** This may be different for different directions; however, if the expansion is the same in all directions, it is referred to as **isotropic expansion.**

The fractional change in length $\Delta L/L_0$, where L_0 is the length of the object at the initial temperature T_0 (Fig. 21-1), is related to the change in temperature ΔT by

$$\frac{\Delta L}{L_0} = \alpha \Delta T \quad \text{or} \quad \Delta L = \alpha\, L_0\, \Delta T \qquad \textbf{(21-1)}$$

where $\Delta L = L - L_0$ and $\Delta T = T - T_0$ and α is the coefficient of linear expansion with units of inverse temperature (1/°C). The coefficient of expansion may vary slightly for different temperature ranges, but this variation is usually negligible for common applications and α is considered to be constant.

*To minimize the number of metal rods needed and for convenience, lab groups can exchange jacketed rods after making measurements to obtain different metal rods.

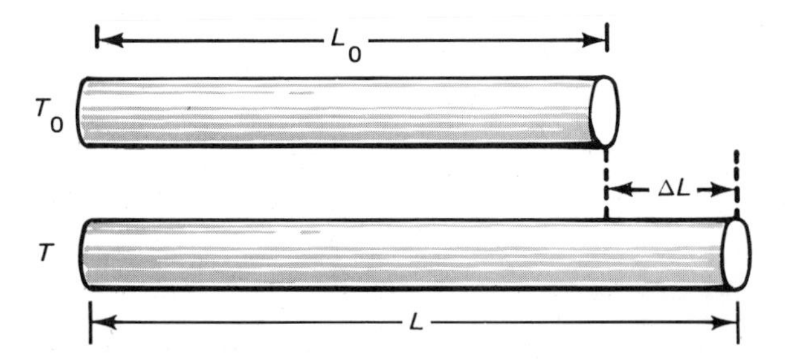

Fig. 21-1 Linear thermal expansion. At the initial temperature T_0 the length of the rod is L_0. At some higher temperature T, the rod has expanded to length L and $\Delta L = L - L_0$.

By Eq. 21-1, α is defined in terms of experimentally measurable quantities

$$\alpha = \frac{\Delta L}{L_0 \Delta T} \tag{21-2}$$

Hence, by measuring the initial length L_0 of an object (e.g., a metal rod) at an initial temperature T_0 and the change in its length ΔL for a corresponding temperature change ΔT, α can be computed.

This development may be extended to two dimensions. The linear expansion expression (Eq. 21-1) may be written

$$L = L_0(1 + \alpha \Delta T) \tag{21-3}$$

and for an isotropic material, its area is $A = L \times L$ or

$$\begin{aligned} A &= L^2 \\ &= L_0^2(1 + \alpha \Delta T)^2 \\ &= A_0(1 + 2\alpha \Delta T + \alpha^2 \Delta T^2) \end{aligned}$$

where $A_0 = L_0^2$. Since typical α's are of the order of $10^{-6}/°C$, the α^2 term may be dropped with negligible error, and to a good approximation

$$A = A_0(1 + 2\alpha \Delta T) \tag{21-4}$$

Comparing this expression with Eq. 21-3, the coefficient of area expansion is seen to be approximately twice the coefficient of linear expansion (i.e., 2α). A similar development can be carried out for the coefficient of volume expansion, which is approximately equal to 3α.

IV. EXPERIMENTAL PROCEDURE

1. A typical arrangement for determining the coefficients of linear expansion is shown in Fig. 21-2. The apparatus consists of a steam jacket with a micrometer attachment for measuring ΔL of a metal rod. A thermometer in the steam jacket measures the temperature of the rod. Steam is supplied to the jacket by a steam generator and a beaker is used to catch the condensate.

Fig. 21-2 Linear expansion apparatus. Steam admitted to the steam jacket causes the metal rod to expand and the expansion is measured by the micrometer attachment at the end of the rod.

2. Before assembling the apparatus, measure the lengths (L_0) of the metal rods with a meter stick to the nearest 0.1 mm and record these lengths. Avoid handling the rods with your bare hands so as not to raise their temperature. Use a paper towel or cloth.

3. Assemble the apparatus, placing one of the rods in the steam jacket. Initially, have one end of the rod touching the fixed end screw and the other end facing the micrometer screw but not touching it. Carefully turn the micrometer screw until it just makes contact with the rod. Avoid mechanical backlash (and electrical spark-gap ionization, see below) by always turning the screw *toward* the rod just before reading. Do not force the screw. Record the micrometer setting. Do this three times and take the average as the initial setting. As soon as the initial micrometer is taken, read and record the initial temperature T_0.

   (Some linear expansion apparatus is equipped with an electrical circuit that uses a bell, light, or voltmeter to indicate when contact is made. The averaging process is unnecessary in this case. Also, another type of arrangement uses an optical lever mounted on a vertically standing expansion apparatus to measure ΔL. See Experiment 16 for instructions for the use of an optical lever.)

4. Turn the micrometer screw back from the end of the rod several millimeters to allow for the thermal expansion of the rod with increasing temperature. Light the Bunsen burner and boil the water in the steam generator (which should be about one-half full) so that steam passes through the jacket. The thermometer in the steam jacket should just touch the metal rod.

   Allow steam to pass through the jacket until the thermometer reading stabilizes (several minutes). When equilibrium has been reached, record the thermometer reading. Then carefully advance the micrometer screw until it touches the end of the rod and record the micrometer setting. Do this three times and take the average of the micrometer readings. Turn off the Bunsen burner.

5. Repeat procedures 3 and 4 for the other metal rods. If jacketed rods are exchanged with other lab groups, be careful not to burn yourself in handling the steam jacket.

6. Compute ΔL and ΔT, and find the coefficient of linear expansion for each metal. Compare these α's with the accepted values given in the Appendix by computing the percent errors.

LABORATORY REPORT

DATA TABLE

	Initial length L_0 (mm)	Initial micrometer setting	Final micrometer setting	ΔL	Init. temp. T_0	Final temp. T	ΔT	α meas.	α accepted
1. Type of rod									
2. Type of rod									
3. Type of rod									

Calculations (show work)

Metal	Percent Error
....	....
....	....
....	....

(continued)

QUESTIONS

1. What are the probable sources of error in the experiment?

2. Would the numerical values of the coefficients of linear expansion be the same if the temperatures had been measured in degrees Fahrenheit? Explain and give an example.

3. When a mercury-in-glass thermometer is placed in hot water, the thermometer reading first drops slightly and then rises. Explain why.

4. If flat strips of iron and brass were bonded together and this bimetallic strip heated, what would be observed? (Justify your answer.)

5. Show that the coefficient of volume expansion is approximately equal to 3α.

6. A Pyrex graduated cylinder has a volume of exactly 200 ml at 0°C. If its temperature is increased to 100°C, will its volume increase or decrease? Compute the change in volume.

Name .... Section .... Date ....

Lab Partner(s) ....

EXPERIMENT 22 *Specific Heats of Metals*

ADVANCE STUDY ASSIGNMENT

Read the experiment and answer the following questions.

1. Distinguish between heat capacity and specific heat.

2. Why is the specific heat of water equal to unity?

3. Given that the specific heat of one material is twice that of another, what does this tell you?

4. What is the method of mixtures?

EXPERIMENT 22
Specific Heats of Metals

I. INTRODUCTION

Different substances require different quantities of heat to produce a given temperature change. For example, about three and one-half times as much heat is needed to raise the temperature of 1 kg of iron through a given temperature interval ΔT as is needed to raise the temperature of 1 kg of lead by the same amount. This material behavior is characterized quantitatively by **specific heat,** which is the amount of heat necessary to raise the temperature of a unit mass of a substance by one unit temperature interval, e.g., 1 gram or 1 kilogram of a substance 1 degree Celsius. Thus, in the previous example iron has a greater specific heat than lead.

The specific heat of a material is specific or characteristic for that material. As can be seen from the definition, the specific heat of a given material can be determined by adding a known amount of heat to a known mass of material and noting the corresponding temperature change. It is the purpose of this experiment to determine the specific heats of some common metals by calorimetry methods.

II. EQUIPMENT NEEDED

- Calorimeter
- Boiler and stand
- Bunsen burner and striker
- Two thermometers (0 to 100°C)
- Two kinds of metal (shot form or slugs with attached strings)
- Laboratory balance
- Ice

III. THEORY

The change in temperature ΔT of a substance is proportional to the amount of heat ΔQ added (or removed) from it:

$$\Delta Q \propto \Delta T$$

In equation form, we may write

$$\Delta Q = C\,\Delta T \tag{22-1}$$

where the constant of proportionality C is called the **heat capacity** of the substance.

However, the amount of heat required to change the temperature of an object is also proportional to the mass of the object. Hence, it is convenient to define a specific heat capacity c (or simply specific heat)

$$c = \frac{C}{m} \tag{22-2}$$

which is the heat capacity per unit mass of a substance. Thus, Eq. 22-1 becomes

$$Q = mc\,\Delta T \quad \text{or} \quad c = \frac{\Delta Q}{m\Delta T} \tag{22-3}$$

In common units the specific heat is then the amount of heat (in calories) required to change the temperature of 1 g of a substance 1°C.

The calorie unit of heat is defined as the amount of heat required to raise the temperature of 1 g of water 1°C. By definition, then, water has a specific heat of 1 cal/g-°C.

$$c = \frac{\Delta Q}{m\Delta T} = \frac{1\ \text{cal}}{(1\ \text{g})(1\,^\circ\text{C})} = 1\ \text{cal/g-}^\circ\text{C}$$

[A kilocalorie (kcal) is the unit of heat defined as the amount of heat required to raise the temperature of

1 kg of water by 1 °C. In these units, water has a specific heat of 1 kcal/kg-°C. Your instructor may recommend that you use these units.]

The specific heat of a material can be determined experimentally by measuring the temperature change of a given mass of material produced by a quantity of heat. This is done indirectly by a calorimetry procedure known as the **method of mixtures.** If several substances at various temperatures are brought together, the hotter substances will lose heat and the colder substances will gain heat until all the substances reach a common equilibrium temperature. If the system is insulated so that no heat is lost to the surroundings, then by the conservation of energy, the heat lost is equal to the heat gained.

In this experiment, hot metal is added to water in a calorimeter cup and the mixture is stirred until the system is in thermal equilibrium. The calorimeter insulates the system from losing heat (Fig. 22-1). In mathematical form, we may write

$$\text{heat lost} = \text{heat gained}$$

or

$$\Delta Q_{\text{metal}} = \Delta Q_{\text{water}} + \Delta Q_{\text{cup and stirrer}}$$

and

$$\begin{aligned} m_m c_m (T_m - T_f) &= m_w c_w (T_f - T_w) \\ &\quad + m_{cs} c_{cs} (T_f - T_w) \qquad \textbf{(22-4)} \\ &= (m_w c_w + m_{cs} c_{cs})(T_f - T_w) \end{aligned}$$

Fig. 22-1 Apparatus for specific heat measurements. Metal shot or a solid piece of metal is heated and then placed in an amount of water in a calorimeter, which insulates the system from losing heat.

where T_f is the final intermediate equilibrium temperature of the system. The other subscripts indicate the masses, specific heats, and initial temperatures of the respective components. Hence, Eq. 22-4 may be used to determine the specific heat c_m of the metal if all the other quantities are known.

IV. EXPERIMENTAL PROCEDURE

1. Weigh out 400 to 500 gm (0.4 to 0.5 kg) of one kind of dry metal shot. [Do this by first determining the mass of the empty boiler cup (in which the metal shot is heated) and then adding an appropriate amount of metal shot to the cup and reweighing.]

   Record the mass of the metal m_m in the Laboratory Report. Your instructor may prefer to use a solid piece of metal with a string attached instead of metal shot. In this case it is only necessary to weigh the piece of metal.

2. Insert a thermometer well into the metal shot and place the cup and shot into the boiler and start heating the boiler water.† The boiler should be about half full of water. Keep steam or water from dampening the dry metal by shielding the cup with a cardboard lid (with a hole for the thermometer).

3. While the boiler is heating, determine and record the mass of the inner calorimeter cup (without the ring) and stirrer. Record the total mass m_{cs}. Also, note and record the type metal and specific heat of the cup and stirrer, which is usually stamped on the cup.* (The specific heat may be found in the Appendix if it is not stamped on the cup.)

4. Fill the calorimeter cup about one-half to two-thirds full with cold tap water and weigh the cup, stirrer, and water to determine the mass of the

†If a mercury thermometer is used, special caution must be given. If the thermometer breaks and mercury spills into the hot metal, the mixture should be removed from the room immediately (to an exhaust hood or outdoors). Mercury fumes are *highly* toxic.

*If the cup and stirrer are not of the same material, they must be treated separately, and the last term in Eq. 22-4 becomes $(m_c c_c + m_s c_s)(T_f - T_w)$.

water m_w. (If a solid piece of metal is used, which usually has less mass than the recommended amount of shot, less water should be used so as to obtain an appreciable ΔT temperature change. This may also be the case at high elevations where the temperature of boiling water is substantially less than 100°C).

Place the calorimeter cup with the water and stirrer in the calorimeter jacket and put on the lid with a thermometer that extends into the water.

5. After the water in the boiler boils and the thermometer in the metal has stabilized (allow several minutes), read and record the temperature of the metal T_m.

For good experimental results, T_w should be below room temperature (T_r) by about the same amount that T_f is above room temperature. Call this temperature difference ΔT. Then,

$$\Delta T = T_r - T_w = T_f - T_r$$

If you use these equations to eliminate T_w and T_f from Eq. 22-4, the result is

$$\Delta T = \frac{m_m c_m}{2m_w c_w + 2m_{cs} c_{cs} + m_m c_m} (T_m - T_r)$$

Read and record the room temperature T_r in the data table. Using known and accepted values (from the Appendix), compute the appropriate magnitude of ΔT. Then, adjust the temperature of the inner calorimeter cup and its contents to approximately $T_r - \Delta T$ by placing it into a beaker of ice water. Measure and record the value of T_w.

6. Remove the thermometer from the metal. Then, remove the lid from the calorimeter and quickly, but carefully, lift the cup with the hot metal from the boiler and pour the metal shot into the calorimeter cup with as little splashing as possible so as not to splash out and lose any water. (If a solid piece of metal is used, carefully lower the metal piece into the calorimeter cup by means of the attached string.)

Replace the lid with the thermometer and stir the mixture gently. The thermometer should not touch the metal. While stirring, watch the thermometer and record the temperature when a maximum equilibrium temperature is reached (T_f).

7. Repeat procedures 1 to 6 for another kind of metal sample. Make certain that you use fresh water in the calorimeter cup. (Dump the previous metal shot and water into a strainer in a sink so that it may be dried and used by others doing the experiment at a later time.)

8. Compute the specific heat of each metal using Eq. 22-4. Look up the accepted values in the Appendix table and compute the percent errors.

Name .... Section .... Date ....

Lab Partner(s) ....

EXPERIMENT 22 *Specific Heats of Metals*

LABORATORY REPORT

DATA TABLE

Room temperature T_r ....

Type of metal	Mass of metal, m_m	Mass of calorimeter and stirrer, m_{cs}	Specific heat of calorimeter and stirrer, c_{cs}	Mass of water, m_w	T_m	T_w	T_f

Calculations (show work)

Type of metal	c_m (expt.)	c_m (accepted)	Percent error
....	....	....	....
....	....	....	....

Name .... Section .... Date ....

Lab Partner(s) ....

EXPERIMENT 22 *Specific Heats of Metals*

QUESTIONS

1. The percent errors of your experimental values of the specific heats may be quite large. Identify several sources of experimental error.

2. The specific heat of aluminum is 0.22 cal/g-°C. What is the value of the specific heat in (a) kcal/kg-°C; (b) Btu/lb-°F? (Show your calculations.)

3. If wet shot had been poured into the calorimeter cup, how would the experimental value of the specific heat have been affected?

4. Explain how the unknown temperature of a hot piece of metal may be determined by the method of mixtures. (Assume c_m to be known.)

5. In solar heating applications, heat energy is stored in some medium until it is needed (e.g., to heat a home at night). Should the medium have a high or a low specific heat? Suggest a substance that would be appropriate for use as a heat-storage medium and explain its advantages.

Name .... Section .... Date ....

Lab Partner(s) ....

EXPERIMENT 23 *Heats of Fusion and Vaporization*

ADVANCE STUDY ASSIGNMENT

Read the experiment and answer the following questions.

1. What is latent heat?

2. The heat of vaporization of water is almost seven times its heat of fusion. What does this imply?

3. Why is the water used in the experimental procedures of the heat of fusion and heat of vaporization initially heated and cooled, respectively?

4. Why are the pieces of ice in the heat of fusion procedure dried and handled with a paper towel? Explain the effect on the experimental result if this were not done.

5. What is the purpose of the water trap in the steam line in the heat of vaporization procedure? Explain the effect on the experimental result if it were not used?

EXPERIMENT **23**

Heats of Fusion and Vaporization (optional, Calibration of a Thermometer)

I. INTRODUCTION

When heat is added to a substance, its temperature normally rises. However, when a substance undergoes a change of phase or state, e.g., solid to liquid to gas, the heat energy goes into doing work against the intermolecular forces and is not reflected by a change in the temperature of the substance. This heat energy is called the **heat of fusion** and the **heat of vaporization** for the respective phase changes, which occur at the melting (freezing)-point temperature and boiling-point temperature, respectively. These heats are commonly referred to as **latent heats,** because the heat energy is seemingly hidden or concealed since it is not evidenced by a temperature change. For the inverse processes, when a vapor or gas condenses and a liquid freezes, by the conservation of energy the (latent) heat of vaporization is given up or the (latent) heat of fusion must be extracted.

In this experiment, the heats of fusion and vaporization of water will be determined through the calorimetry method of mixtures.

II. EQUIPMENT NEEDED

- Calorimeter
- Steam generator and stand
- Bunsen burner and striker
- Thermometer (0 to 100° C)
- Rubber hose
- Water trap
- Ice
- Paper towels
- Laboratory balance
- Beaker

III. THEORY

An idealized graph of heat energy versus temperature for a given mass m of water is shown in Fig. 23-1. The sloping phase lines follow the relationship

$$\Delta Q = mc\,\Delta T \qquad \textbf{(23-1)}$$

The slopes of the lines are mc, where c is the specific heat of the particular phase ($c_{\text{water}} = 1.0$ cal/g-°C, and $c_{\text{ice}}=c_{\text{steam}} = 0.5$ cal/g-°C).

At the freezing and boiling points, the addition of certain amounts of heat has no effect on the temperature as indicated by the vertical lines. At these points, the heat energy goes into the work of effecting the phase changes and not into increasing the molecular activity, which would be reflected as a temperature increase.

The **latent heat** L is defined as the amount of heat required to change the phase of a unit mass of a substance (without a change of temperature) and has the units cal/g. Hence, the amount of heat absorbed or given up by a quantity of a substance when it undergoes a change of phase is

$$\Delta Q = mL_i \qquad \textbf{(23-2)}$$

where m is the mass of the substance and L_i the latent heat for the particular phase change.

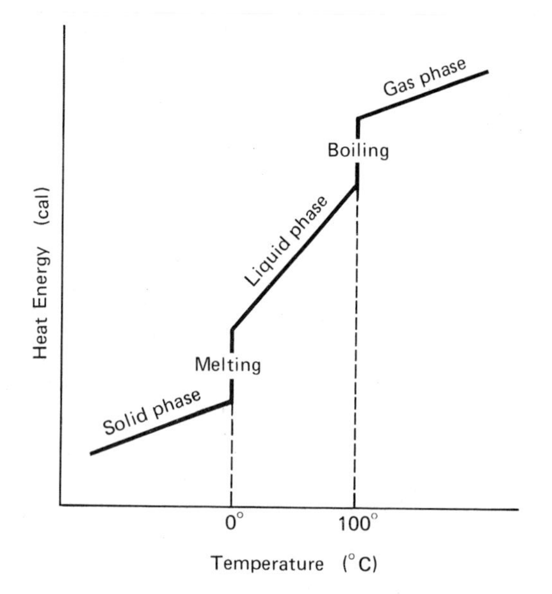

Fig. 23-1 An idealized graph of heat energy versus temperature for a given mass of water.

The latent heats are, of course, characteristic of the substance. For example, the **(latent) heat of fusion** L_f of water is 80 cal/g and the **(latent) heat of vaporization** L_v is 540 cal/g. This means that 80 cal is required to melt 1 g of ice at 0°C (or 80 cal must be removed from 1 g of water at 0°C to freeze it) and that 540 cal is required to convert 1 g of water at 100°C to steam (or 540 cal is released when 1 g of steam at 100°C condenses to water). Other substances have different latent heats.

In this experiment, the heats of fusion and vaporization of water will be measured by a calorimetry procedure called the **method of mixtures.** If several substances at various temperatures are brought together, the hotter substances will lose heat and the colder substances will gain heat until all the substances reach a common equilibrium temperature. This is generally true even if one or more of the substances undergoes a phase change in the process. If the system is insulated and no heat is lost to the surroundings, then by the conservation of energy, the heat lost by the hot substances will equal the heat gained by the colder substances.

A. *Heat of Fusion*

If a quantity of ice of mass m_i at 0°C is added to a sufficient quantity of warm water in a calorimeter cup with a stirrer at T_h, the ice melts and the system comes to equilibrium at some intermediate or final temperature T_f. (The calorimeter insulates the system against heat loss. See Fig. 22-1.) Then, by the conservation of energy,

heat of fusion to melt the ice + heat gained by the ice water

= heat lost by the warm water + heat lost by the calorimeter cup and stirrer

or

$$\Delta Q_{\text{fusion}} + \Delta Q_{\text{ice water}} = \Delta Q_{\substack{\text{warm}\\ \text{water}}} + \Delta Q_{\substack{\text{calorimeter cup}\\ \text{and stirrer}}}$$

and

$$\begin{aligned} m_i L_f + m_i c_w (T_f - 0) &= m_w c_w (T_h - T_f) + m_{cs} c_{cs} (T_h - T_f) \\ &= (m_w c_w + m_{cs} c_{cs})(T_h - T_f) \end{aligned} \qquad \textbf{(23-3)}$$

where the m's and c's are the masses and specific heats, respectively, of the various components as indicated by the subscripts (i = ice, w = water, cs = calorimeter–stirrer). Hence, Eq. 23-3 may be used to determine the heat of fusion L_f of water if all the other quantities are known.

B. *Latent Heat of Vaporization*

If a quantity of steam of mass m_s at 100°C is added to cool water in a calorimeter cup with a stirrer at T_c, the steam condenses and the system comes to equilibrium at some intermediate or final temperature T_f. Assuming no heat loss to the surroundings, by the conservation of energy,

heat of vaporization lost by steam in condensing + heat lost by hot condensed water

= heat gained by the cool water + heat gained by the calorimeter cup and stirrer

or

$$\Delta Q_{\text{vaporization}} + \Delta Q_{\substack{\text{hot}\\ \text{water}}} = \Delta Q_{\substack{\text{cool}\\ \text{water}}} + \Delta Q_{\substack{\text{calorimeter cup}\\ \text{and stirrer}}}$$

and

$$\begin{aligned} m_s L_v + m_s c_w (100 - T_f) &= m_w c_w (T_f - T_c) + m_{cs} c_{cs} (T_f - T_c) \\ &= (m_w c_w + m_{cs} c_{cs})(T_f - T_c) \end{aligned} \qquad \textbf{(23-4)}$$

where the m's and c's are the masses and specific heat, respectively, of the various components as indicated by the subscripts. Hence, Eq. 23-4 may be used to determine the heat of vaporization L_v of water if all the other quantities are known.

IV. EXPERIMENTAL PROCEDURE

A. *Heat of Fusion*

1. Heat some water in the beaker (enough to fill the inner calorimeter cup about half full) to about 10 to 15°C above room temperature. While the water is heating, determine the mass of the inner calorimeter cup (without ring) and stirrer (total mass m_{cs}) on a laboratory balance. Also, note and record the type metal and specific heat of the cup and stirrer, which is usually stamped on the cup.*

2. Fill the inner calorimeter cup about half-full of the warm water and weigh it (with the stirrer) to determine the mass of the water m_w. Place the calorimeter cup with the water and stirrer in the calorimeter jacket and put on the lid with a thermometer that extends into the water. Stir the water gently and record its temperature T_h.

3. Select several small pieces of ice about the size of the end of your thumb and dry them with a paper towel. (It is important that the ice be dry.) Without touching the ice with your bare fingers (use the paper towel), carefully add the pieces of ice to the calorimeter cup one at a time without splashing. (It is good procedure to stir the water and check its temperature again just before adding the ice. If it has changed, take the latter reading as T_h.)

4. Gently stir the water–ice mixture while adding the ice, and add enough ice until the temperature of the mixture is about 10 to 15°C below room temperature. Add the ice more slowly toward the end so that you can better control the final temperature. Continue to stir gently, and read and record the equilibrium temperature T_f when the ice has melted completely.

   Then, weigh and record the mass of the inner calorimeter cup with its contents (water and stirrer) so as to determine the mass of the melted ice water or mass of the ice m_i.

5. Compute the heat of fusion L_f using Eq. 23-3 and compare it with an accepted value $L_f = 80$ cal/g by finding the percent error.

*If the cup and stirrer are not of the same material, they must be treated separately, and the mass term in Eq. 23-3 and 23-4 becomes $(m_w c_w + m_c c_c + m_s c_s)$.

B. *Heat of Vaporization*

6. Set up the steam generator as shown in Fig. 23-2 with the boiler about two-thirds full of water and begin heating the boiler. A water trap is used in the steam line (which should be as short as practically possible) to prevent hot water condensed in the tube from entering the calorimeter. (It is important that the hot water condensed in the tube be prevented from entering the calorimeter—note the hose arrangement in Fig. 23-2.)

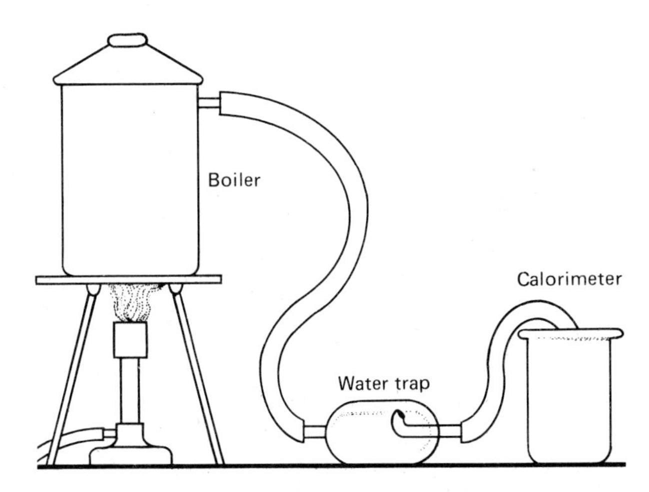

Fig. 23-2 Steam generator arrangement for heat-of-vaporization measurement. The water trap in the steam line prevents hot water condensed in the tube from entering the calorimeter.

7. Fill the inner calorimeter cup about two-thirds full of cool water at about 15°C below room temperature. Either the cool water from the heat-of-fusion experiment or fresh tap water may be used with some ice added to obtain the desired temperature. (Make sure that all the ice has melted, however.)

   Weigh and record the mass of the calorimeter cup with the water and stirrer. Then place the cup in the calorimeter jacket and replace the lid. Gently stir the water and read and record the equilibrium temperature T_c.

8. With the water in the steam generator boiling gently and steam flowing freely from the steam tube (as evidenced by water vapor coming from the tube, recall that steam is invisible), introduce steam into the calorimeter water and stir. (It is good procedure to stir the water and check its temperature again just before introducing the

steam. If it has changed, take the latter reading as T_c).

When the temperature of the water is about 15°C above room temperature, turn off the Bunsen burner and carefully remove the steam line from the calorimeter. Stir gently, and read and record the equilibrium temperature T_f.

Finally, reweigh the inner calorimeter cup and its contents so as to determine the mass of the condensed steam m_s.

9. Compute the heat of vaporization L_v using Eq. 23-4 and compare it with the accepted value $L_v = 540$ cal/g by finding the percent error.

LABORATORY REPORT

Mass of calorimeter cup and stirrer, m_{cs} ....

Specific heat of calorimeter cup and stirrer, c_{cs} ....

A. *Heat of Fusion*

Mass of calorimeter plus water ....

Mass of water, m_w ....

Mass of calorimeter, water, and melted ice ....

Mass of ice, m_i ....

Initial temperature of water, T_h ....

Final equilibrium temperature, T_f ....

Calculations (show work)

Experimental L_f ....

Accepted L_f ....

Percent error ....

Lab Partner(s) ....

EXPERIMENT 23 *Heats of Fusion and Vaporization*

B. *Heat of Vaporization*

Mass of calorimeter plus water ....

Mass of water, m_w ....

Mass of calorimeter, water, and condensed steam ....

Mass of steam, m_s ....

Initial temperature of water, T_c ....

Final equilibrium temperature, T_f ....

Calculations (show work)

Experimental L_v ....

Accepted L_v ....

Percent error ....

QUESTIONS

1. Discuss the most likely sources of error in the experimental procedures.

2. Explain why burns caused by steam at 100°C are more serious than those caused by an equal mass of boiling water at the same temperature.

(continued)

3. How does latent heat figure in the cooling mechanism of our bodies?

4. What is the heat of sublimation? [*Hint:* Dry ice (solid CO_2) sublimes.]

5. A piece of ice with a mass of 30 grams at 0°C is added to 100 ml of water at 20°C. Assuming that no heat is lost to the surroundings, what is the situation when thermal equilibrium is reached? (Ignore the calorimeter or container.)

EXPERIMENT 23 Addendum

Calibration of a Thermometer

I. INTRODUCTION

The latent heats experiment is relatively short and with ice and means to boil water available, you can also quickly calibrate a thermometer, which is quite interesting and instructive.

II. EQUIPMENT NEEDED

- Uncalibrated thermometer (glass)
- Beaker (500 ml)
- Tripod stand and wire gauze pad
- Wax pencil
- Ruler
- Classroom Celsius thermometer

III. THEORY

A thermometer may be calibrated by using two fixed-point temperatures and then making a graduated scale between these points.* For example, the ice point and steam point of water are convenient fixed points (0°C or 32°F and 100°C or 212°F, respectively). If the length between these two points is divided linearly into 100 intervals or degrees, this corresponds with the Celsius temperature scale. Similarly, if the length between the fixed points is divided into 180 intervals or degrees, this corresponds to the Fahrenheit temperature scale.

By analogy, the meter standard was originally referenced to two marks on a metal bar. The length between the marks may be divided into 100 intervals or centimeters. It could be divided into 180 intervals, but this would not be a decimal scale.

IV. EXPERIMENTAL PROCEDURE

1. Start heating a *small* amount of water in the beaker. The water depth in the beaker should be several centimeters, enough to cover the thermometer bulb. (More water may be used, but it is unnecessary and will take longer to boil.)

   While the water is heating, draw an outline of the thermometer on a sheet of paper. The temperature scale will be marked off on the paper and the outline will allow the thermometer to be replaced on the paper in the proper position. (The scales on many thermometers are not engraved on the glass tube of the thermometer, but on an attached frame.)

2. Place the uncalibrated thermometer into the heating water and heat until the water boils vigorously. With a wax pencil, mark the height of the thermometer's liquid column on the stem. Turn off the burner. Remove the thermometer and

*On the SI Kelvin (absolute) scale, the kelvin or interval is defined as 1/273.16 of the temperature of the triple point of water (point where water coexists in three phases, 0.01°C and 4.58 mm Hg). Here the temperature scale is determined by a single fixed point and a defined interval.

place it on the outline, making a mark on the paper at the wax pencil mark of the fixed steam point.

3. Pour the hot water from the beaker into a sink. (Use a paper towel or tongs to transport the hot beaker and pour carefully so as not to burn yourself. Recall that steam has a large latent heat.) Add a small amount of cold water to the beaker (1–2 cm in depth) and enough ice so the beaker is one-third to one-half full.

4. Insert the thermometer into the ice–water mixture and stir. The liquid column of the thermometer should reach a low stationary point after a few minutes. Make a mark on the thermometer stem at this lowest point. (Leave the thermometer in the ice water for another minute or so to make sure the column goes no lower. Adjust the mark accordingly if it does.) In thermal equilibrium, ice and water coexist at the ice point.

5. Remove the thermometer, place it on the outline, and mark the ice point on the paper. Lay the thermometer aside and with a ruler divide the length between the two fixed points into 10 equal intervals (deka-degree or 10-degree intervals). Using the Celsius values for the fixed points, label these marks on one side of the thermometer outline. Then, divide the interval between the 20 and 30 degree marks into 10 equal intervals of degrees. (The same could be done for the other 10-degree intervals, giving a centigrade or 100-degree scale, but this is unnecessary for the purpose of the experiment.)

6. Read the room temperature from the classroom thermometer and mark and record the value on the paper scale. Then, place the uncalibrated thermometer (which should now be at room temperature) on the outline and mark and read the temperature. Compare the temperature reading on your scale with the classroom (accepted) thermometer reading by finding the percent error. Show this calculation on the outline paper.

7. On the other side of the thermometer outline, make a graduated Fahrenheit scale with a fine scale in the vicinity of the room temperature reading. Note the value of your thermometer's room temperature reading on the Fahrenheit scale and record on the paper. Compare the experimental Fahrenheit reading with the experimental Celsius reading by computing T_F from $T_F = (9/5)T_c + 32$ using the experimental value of T_c in the equation. How close did you come?

8. Attach the outline paper to the Laboratory Report for the latent heat portion of the experiment.

Name .... Section .... Date ....

Lab Partner(s) ....

EXPERIMENT 24 *Newton's Law of Cooling: The Time Constant of a Thermometer*

ADVANCE STUDY ASSIGNMENT

Read the experiment and answer the following questions.

1. According to Newton's law of cooling, would a body cool faster in the temperature range 100 to 150°C or 50 to 100°C? Explain.

2. What is meant by an exponential decay function? An exponential growth function?

3. What is a "time constant," and on what does it depend for a cooling process?

4. Is the temperature of a cooling object a linear function of time? Explain in terms of the slope of the function.

5. How is the time constant of a cooling process determined graphically?

EXPERIMENT 24

Newton's Law of Cooling: The Time Constant of a Thermometer

I. INTRODUCTION

All dynamic instruments used to make physical measurements require a time period to respond to the change in conditions which they measure. Some instruments respond very quickly and their response times go unnoticed. However, the times required for some instruments to give "correct" readings are readily observed and measurable. For example, a thermometer requires a finite time to come to thermal equilibrium with its surroundings. This is evidenced by having to allow 2 to 3 min to get a correct reading when measuring one's body temperature with a thermometer. Similarly, a finite time interval is required for a thermometer to cool off from a high temperature reading to room temperature. (A clinical thermometer is prevented from returning to a lower reading by a constriction in its capillary.)

Newton's law of cooling is an empirical observation made by him which states that the rate change of temperature of a body is directly proportional to the difference in temperature between the body and its surroundings, provided that the temperature difference is small.

In this experiment, Newton's law of cooling will be investigated and the time constant of a thermometer will be determined.

II. EQUIPMENT NEEDED

- One metal-stem dial thermometer (0 to 150°C)
- One liquid-in-glass thermometer (0 to 100°C)
- Metal cylinder of thermometer length
- Two stands and clamps
- Bunsen burner and striker
- Laboratory timer or stopwatch
- 1 sheet of Cartesian graph paper

III. THEORY

Newton's law of cooling may be expressed mathematically as

$$\frac{dQ}{dt} = -K\,\Delta T = -K(T - T_r) \qquad \textbf{(24-1)}$$

where dQ/dt is the time rate of cooling, $\Delta T = T - T_r$ the difference in the temperature T and the final temperature T_r (in this experiment room temperature), and K a constant of proportionality. The minus sign indicates that the temperature decreases with time.*

*The heat loss includes contributions from all methods of heat transfer, conduction, convection, and radiation.

A small amount of heat dQ lost by an object over a small temperature interval dT can be expressed in terms of the object's specific heat c:

$$dQ = mc\,dT$$

where m is the mass of the object. Dividing both sides of the equation by a small time interval dt, we have

$$\frac{dQ}{dt} = mc\frac{dT}{dt} \qquad \textbf{(24-2)}$$

Equating Eqs. 24-1 and 24-2 yields

$$\frac{dT}{dt} = \frac{-K}{mc}(T - T_r)$$

By methods of integral calculus, it can be shown that the temperature T of an object at any time t predicted by this equation is given by

$$T - T_r = (T_0 - T_r)e^{-Kt/mc} \qquad \textbf{(24-3)}$$

where the e term is an exponential function ($e = 2.71$), and T_0 is the original or initial temperature of the body at $t = 0$. The quantity $T_0 - T_r$ is therefore a constant.

Equation 24-3 is often written

$$T - T_r = (T_0 - T_r)e^{-t/\tau} \qquad \textbf{(24-4)}$$

where $\tau = mc/K$ is called the **time constant** of the process and can be seen to depend on the mass and specific heat of the object.

The time constant τ has the units of time, so in a time $t = \tau$ (one time constant), $e^{-t/\tau} = e^{-\tau/\tau} = e^{-1} = 0.368$. Thus, the temperature of the object will have fallen to 0.368 (or 36.8 percent) of its original value $(T - T_r)/(T_0 - T_r) = 0.368$, where $(T_0 - T_r)$ is the total temperature drop and $(T - T_r)$ is the remaining temperature interval. For $\tau = 2t$, then $e^{-2} = 0.135$, and so on. A plot of how the temperature of the cooling object varies with time according to Eq. 24-4 is shown in Fig. 24-1.

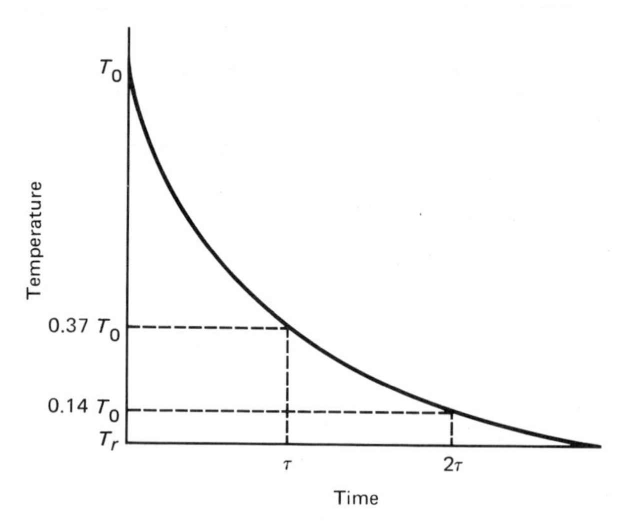

Fig. 24-1 A graph of the temperature of cooling thermometer versus time.

The T versus t plot in Fig. 24-1 is referred to as an **exponential "decay" curve.** The temperature of the object "decays" or decreases exponentially with time as a result of the negative exponent. (An exponential function with a positive exponent would increase or "grow" with time.) As will be noted, the smaller the time constant or the greater the negative exponent, the faster the decay or cooling. It is possible to determine the time constant graphically by using different coordinates. By taking the logarithm (to the base e) of Eq. 24-4, the equation becomes

$$\ln(T - T_r) = -\lambda t + \ln(T_0 - T_r)$$

where $\lambda = 1/\tau$. The equation now has the slope-intercept form of a straight line,

$$y = ax + b$$

and plotting $\ln(T - T_r)$ versus t allows λ to be determined by finding the value to the negative slope of the straight line.

IV. EXPERIMENTAL PROCEDURE

1. Read and record the room temperature T_r of the dial thermometer in Data Table 1. Then clamp the thermometer on the stand and insert the stem into the metal cylinder as shown in Fig. 24-2. Make certain that the stem does not touch the cylinder walls and that the tip of the stem does not extend beyond the end of the cylinder.

2. *Slowly* heat the cylinder by wafting the flame of the Bunsen burner over the length of the cylinder. Do not let the flame touch the thermometer. Heat the cylinder until the thermometer reads about 150°C. Maintain this temperature as closely as possible for a couple of minutes to ensure uniform heating of the thermometer stem. Record this high temperature T_0 in Data Table 1.

3. Remove the Bunsen burner and slide the thermometer out of the cylinder, starting the timer at the same time. Read the thermometer temperature every 30 s for 4 min as it cools in the air and record the readings in Data Table 1.

   During this procedure, make certain that the cylinder is far away from the thermometer so that it will not affect the thermometer's cooling rate. Also, do not disturb the thermometer during the cooling process and avoid any drafts or stirring of air in its vicinity.

4. Repeat procedures 1 to 3 for the liquid-in-glass thermometer with a maximum temperature of about 100°C. Be careful in clamping the thermometer so as not to break it, and be careful not to overheat the thermometer.

5. (a) Compute $(T - T_r)$ and find $\ln(T - T_r)$ as indicated in the data tables (see Appendix B, Table B8).

(b) Plot ln $(T - T_r)$ versus t for each thermometer on the same Cartesian graph. Draw the best straight line that fits the first few data points near $t = 0$ for each set of data and determine the slope and intercept value for each line.

6. From the slope λ of each line, determine the time constant for each thermometer. Also, from the intercepts, compute the initial temperature T_0 of each thermometer. Compare these initial temperature values with the recorded experimental values by computing the percent differences.

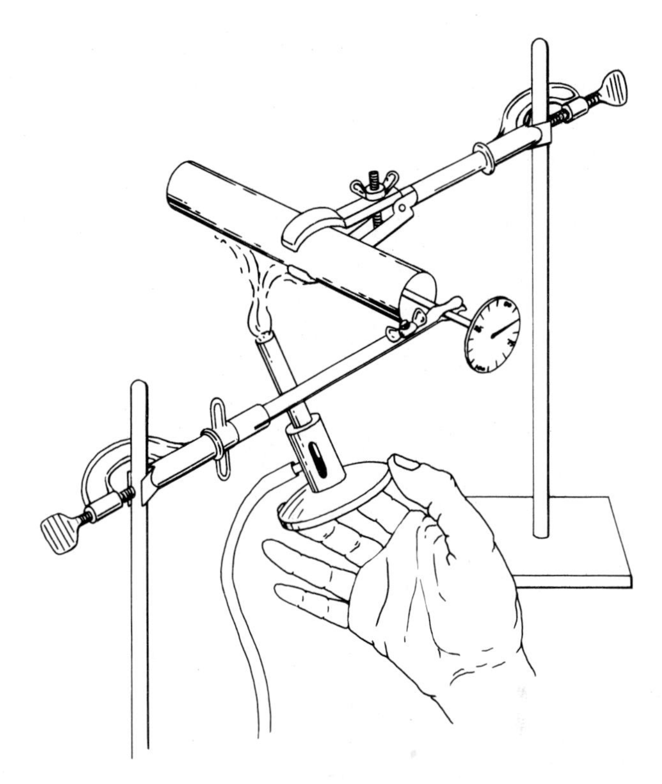

Fig. 24-2 The apparatus setup for Newton's law of cooling.

Name .... Section .... Date ....

Lab Partner(s) ....

EXPERIMENT 24 *Newton's Law of Cooling: The Time Constant of a Thermometer*

LABORATORY REPORT

DATA TABLE 1 DIAL THERMOMETER

T_r ....

Time, t (min)	Temperature, T(°C)	$T - T_r$	$\ln(T - T_r)$
0	(T_0)		

DATA TABLE 2 LIQUID-IN-GLASS THERMOMETER

T_r ....

Time, t (min)	Temperature, T(°C)	$T - T_r$	$\ln(T - T_r)$
0	(T_0)		

Calculations (show work)

	Metal	Glass
Slope	....	....
Time constant	....	....
T_0 (from intercept)	....	....
Percent difference	....	....

Lab Partner(s) ....

EXPERIMENT 24 *Newton's Law of Cooling: The Time Constant of a Thermometer*

QUESTIONS

1. On what experimental parameter(s) do you think the constant K in the exponent of the exponential function depends?

2. Suppose that the cooling of another metal-stem dial thermometer of the same material, but having a larger-diameter stem, is investigated. How would this affect the time constant? Sketch the general form of the cooling curve of such a thermometer on the ln $(T - T_r)$ versus t graph (use a dashed line) and label it accordingly.

3. How would the cooling rate of the metal-stem thermometer be affected if the thermometer were made of a different metal, but still had the same mass?

4. Explain why in general the liquid-in-glass thermometer does not follow a linear relationship for ln $(T - T_r)$ versus t as well as a metal dial thermometer. (Check your graph.)

5. (a) Theoretically, does the T versus t curve (Fig. 24-1) ever reach the horizontal axis? Explain (mathematically) at what time $t = T_r$. (The curve is said to approach the horizontal axis asymptotically.)

(continued)

(b) Practically, after how many time constants would a thermometer be "recovered" or have reached room temperature and be ready to take another reading? (Assume that the difference between the temperature of the thermometer and the temperature of its surroundings has to decrease to less than 1 percent of the initial temperature difference.) How many minutes would this be for the metal thermometer used in the experiment?

Name .... Section .... Date ....

Lab Partner(s) ....

EXPERIMENT 25 *Archimedes' Principle: Buoyancy and Specific Gravity*

ADVANCE STUDY ASSIGNMENT

Read the experiment and answer the following questions.

1. Give the conditions on densities that determine if an object will sink or float in a fluid.

2. Distinguish between specific gravity and density.

3. What does the difference between the weight of an object in air and its measured weight submerged in water equal? How could this be determined otherwise?

4. Why is it important to make certain that no air bubbles adhere to objects during the submerged weighing procedures? How would the experimental results be affected if bubbles were present?

EXPERIMENT **25**

Archimedes' Principle: Buoyancy and Specific Gravity

I. INTRODUCTION

Some objects will float and others will sink in a given fluid—a liquid or a gas. The fact that some objects float means they are "buoyed up" by a force greater than or equal to the weight force of the object. Archimedes (287–212 B.C.), a Greek scientist, deduced that the upward buoyant force acting on a floating object is equal to the weight of the fluid it displaces. Thus, a body will sink if its weight exceeds that of the fluid it displaces.

In this experiment, Archimedes' principle will be studied in the application of determining the densities and specific gravities of solid and liquid samples.

II. EQUIPMENT NEEDED

- Triple-beam pan balance with swing platform (or single-beam double-pan balance with swing platform and set of weights)
- Overflow can (or graduated cylinder and eye dropper)
- Two beakers
- Metal cylinder or irregularly shaped metal object, or metal sinker
- Waxed block of wood
- Saltwater solution or alcohol
- String
- Hydrometer and cylinder

III. THEORY

Archimedes' principle states:

When a body is placed in a fluid, it will be buoyed up by a force equal to the weight of the volume of fluid it displaces.

Archimedes' principle applies to a body wholly or partially immersed in a fluid. The magnitude of the buoyant force depends only on the weight of the fluid displaced by the object.

Archimedes' principle shows that an object

(a) will float in a fluid if the density of the object ρ_o is less than the density of the fluid ρ_f, and

(b) will sink if the object's density is greater than that of the fluid's.

The weight of an object is $w_o = m_o g = \rho_o g V_o$, where V_o is the volume of the object and $\rho_o = m_o/V_o$. Similarly, the weight of the fluid displaced by the object, or the buoyant force, is $F_b = w_f = m_f g = \rho_f g V_f$. If the object is completely submersed in the fluid, then $V_o = V_f$, and dividing one equation by the other,

$$\frac{F_b}{w_o} = \frac{\rho_f}{\rho_o} \quad \text{or} \quad F_b = \left(\frac{\rho_f}{\rho_o}\right) w_o \qquad \textbf{(25-1)}$$

Hence, if $\rho_o < \rho_f$, then $F_b > w_o$ and the object will be buoyed to the surface and will float. If $\rho_o > \rho_f$, then $F_b < w_o$, and the object will sink.

The **specific gravity** of a solid or liquid is defined as the ratio of the weight of a given volume of the substance to an equal volume of water

specific gravity (sp. gr.)

$$= \frac{\text{weight of a substance (of given volume)}}{\text{weight of an equal volume of water}} \quad \textbf{(25-2)}$$

Specific gravity is a density-type designation that uses water as a comparison standard. Since it is a weight ratio, specific gravity has no units, but the numerical value of a substance's specific gravity is the same as the magnitude of its density in cgs units:

$$\text{sp. gr.} = \frac{w_s}{w_w} = \frac{w_s/V_s}{w_w/V_w} = \frac{m_s g/V_s}{m_w g/V_w}$$

$$= \frac{m_s/V_s}{m_w/V_w} = \frac{\rho_s}{\rho_w} \quad \textbf{(25-3)}$$

where the subscripts s and w refer to the substance and water, respectively, and by definition $V_s = V_w$. For practical purposes, the density of water is 1 gm/cm^3 over the liquid temperature range, and in magnitude,

$$\text{sp. gr.} = \frac{\rho_s}{\rho_w} = \frac{\rho_s}{1} = \rho_s \quad \textbf{(25-4)}$$

For example, the density of mercury is 13.6 gm/cm^3 and has a specific gravity of 13.6. A specific gravity of 13.6 indicates that mercury is 13.6 times more dense than water, $\rho_s = (\text{sp. gr.})\rho_w$, or that a sample of mercury will weigh 13.6 times more than an equal volume of water.

In terms of Archimedes' principle, the specific gravity of a *submerged* object is

$$\text{sp. gr.} = \frac{w_o}{w_w} = \frac{w_o}{F_b} \quad \textbf{(25-5)}$$

since $w_w = F_b$. For a heavy object that sinks, if its apparent weight w_o' is measured while it is submerged, we have $w_w = F_b = w_o - w_o'$. Hence,

$$\text{sp. gr.} = \frac{w_o}{w_w} = \frac{w_o}{w_o - w_o'}$$

or in terms of mass which is measured on a balance ($w = mg$),

$$\text{sp. gr.} = \frac{m_o}{m_o - m_o'} \quad \textbf{(25-6)}$$

To find the specific gravity of a solid less dense than water using Archimedes' principle, it is necessary to use another object or sinker of sufficient weight and density to completely submerge the light solid. Using an arrangement as illustrated in Fig. 25-1, $w_1 = w_o + w_s'$ is the measured weight (mass) of the object and the sinker, with only the sinker submerged, and $w_2 = w_o' + w_s'$ is the measured weight when both are submerged. Then $w_1 - w_2 = (w_o + w_s') - (w_o' + w_s') = w_o - w_o'$, or in terms of mass, $m_1 - m_2 = m_o - m_o'$, and the specific gravity can be found from Eq. 25-6.

Fig. 25-1 Arrangement to determine the specific gravity of a solid less dense than water. The weight of the system with only the sinker submerged is measured, then the weight with both the sinker and the light object submerged.

The specific gravity of a liquid can also be found using Archimedes' principle. First, a heavy object is weighed in air (w_o) and then weighed when submerged in liquid (w_o'). Then $(w_o - w_o')_\ell$ is the weight of the volume of liquid the object displaces, by Archimedes' principle. Carrying out a similar procedure for the object in water, $(w_o - w_o')_w$ is the volume of water the object displaces. Then, by the definition of specific gravity (Eq. 25-2),

$$\text{sp. gr.} = \frac{(w_o - w_o')_\ell}{(w_o - w_o')_w} \quad \textbf{(25-7)}$$

You may have been thinking that there are easier ways to determine the density or specific gravity of a solid or liquid. This is true, but the purpose of the experiment is to familiarize you with Archimedes' principle. You may wish to check your experimental results by determining the densities and specific gravities of the solid samples by some other method. The specific gravity of the liquid sample will also be determined using a hydrometer.

IV. EXPERIMENTAL PROCEDURE

A. *Direct Proof of Archimedes' Principle*

1. Weigh the metal sample and record its mass m_o and the type of metal in the Laboratory Report. Also, determine the mass of an empty beaker m_b and record. Fill the overflow can with water and place it on the balance platform. Attach a string to the sample and suspend it from the balance arm as illustrated in Fig. 25-2.*

2. The overflow from the can when the sample is immersed is caught in the beaker. Take a mass reading m'_o of the submerged object. Make certain that no bubbles adhere to the object.

   Then, weigh the beaker and water so as to determine the mass of the displaced water m_w. (If the can does not fit on the balance platform, first suspend and immerse the object in the full overflow can and catch the overflow in the beaker and find m_w. Then attach the sample to the balance arm and suspend it into a beaker of water that will fit on the balance platform to find m'_o.)

3. According to Archimedes' principle, the buoyant force $F_b = m_o g - m'_o g$ should equal the weight of the displaced water

   $$w_w = m_w g \quad \text{or} \quad (m_o - m'_o)\,g = m_w g$$

   Compute the buoyant force and compare it with the weight of the displaced water by finding the percent difference.

B. *Density of a Heavy Solid*

4. Determine the specific gravity and density of the metal sample. This can be computed using the data from part A.

*Note: You may use an alternative method if there is no overflow can available. Attach a string to the sample and place it in a graduated cylinder. Fill the cylinder with water until the sample is completely submerged. Add water (with an eyedropper) until the water level is at a specific reference mark on the cylinder (e.g., 35 ml). Remove the sample, shaking any drops of water back into the cylinder, and weigh the cylinder and water (m_b). Refill the cylinder to the reference mark and weigh it again ($m_w + m_b$). The mass of the "overflow" water is then the difference between these measurements.

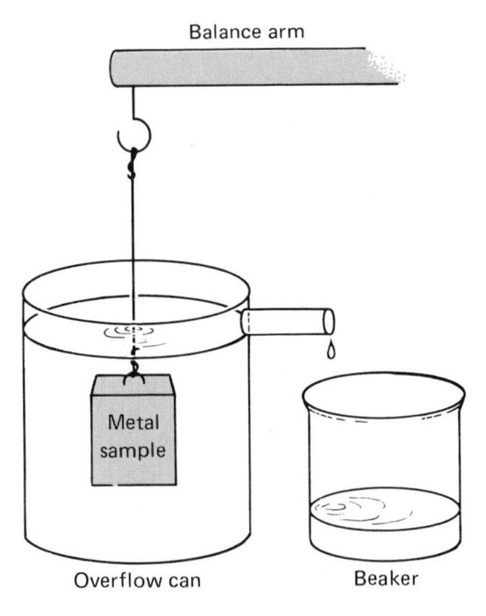

Fig. 25-2 Arrangement for proving Archimedes' principle. (Note the example of refraction at the water-air interface.)

C. *Density of a Light Solid*

5. Determine the specific gravity and density of the wooden block by the procedure described in Section III and illustrated in Fig. 25-1. First, measure the mass of the block alone (in air). Then set up as in Fig. 25-1. Make certain that no air bubbles adhere to the objects during the submerged weighing procedures. [The first weighing procedure may have to be modified to fit the apparatus (i.e., weighing the block and submerged sinker separately).] The block is waxed so that it does not become water logged. Ignore the contribution of the wax film in the procedure.

D. *Density of a Liquid*

6. Determine the specific gravity and density of the liquid provided by the procedure described in Section III. Again, make certain that no air bubbles adhere to the object during the submerged weighing procedures.

7. Determine the specific gravity of the liquid using the hydrometer and cylinder. Compare this value with that found in procedure 6 by computing the percent difference.

Name .... Section .... Date ....

Lab Partner(s) ....

EXPERIMENT 25 *Archimedes' Principle: Buoyancy and Specific Gravity*

LABORATORY REPORT

A. *Direct Proof of Archimedes' Principle*

Type of metal ....

Mass of metal, m_o, in air ....

Mass of beaker, m_b ....

Mass of metal m_o', submerged in water ....

Mass of beaker and displaced water, $m_w + m_b$ ....

Mass of displaced water, m_w ....

Buoyant force ....

Weight of displaced water ....

Percent difference ....

Calculations (show work)

B. *Density of a Heavy Solid ($\rho_s > \rho_w$)*

Calculations (show work)

Specific gravity ....

Density ....

C. *Density of a Light Solid ($\rho_s < \rho_w$)*

Mass of block in air ....

Mass of block and sinker with only sinker submerged ....

Mass of block and sinker with both submerged ....

Specific gravity ....

Density ....

Calculations (show work)

D. *Density of a Liquid*

Mass of object in air ....

Mass of object submerged in liquid ....

Mass of object submerged in water ....

Computed sp. gr. ....

Sp. gr. from hydrometer measurement ....

Percent difference ....

Calculations (show work)

(continued)

QUESTIONS

1. Look up the density of the type of metal of the object used in parts A and B of the procedure, and compare it with the experimental value. Comment on the purity of the metal of the object. (Archimedes developed his principle while working on a similar inquiry. His problem was to determine whether an apparent gold crown had been made with some content of cheaper metal.)

2. What would be the situation of an object immersed in a fluid if the object and the fluid had the same density?

3. Explain how a submarine is caused to submerge and surface without the use of its propulsion propeller.

4. Oil floats on water. What can you say about the density of oil?

5. A block of wood floats in a beaker of water. According to Archimedes' principle, the block experiences an upward buoyant force. If the beaker with the water and floating block is weighed, would the measured weight be less than the sum of the weights of the individual components? Explain.

6. A person can lift 45 kg ($\approx$ 100 lb). Using the experimental value of the specific gravity for the metal object in part B, how many cubic meters of the metal could the person lift (a) in air; (b) in water?

Name .... Section .... Date ....

Lab Partner(s) ....

EXPERIMENT 25 *Archimedes' Principle: Buoyancy and Specific Gravity*

7. Explain the principle and construction of a hydrometer. What is the purpose of the common measurements of the specific gravities of an automobile's radiator coolant and battery electrolyte?

Name .... Section .... Date ....

Lab Partner(s) ....

EXPERIMENT 26 *Fields and Equipotentials*

ADVANCE STUDY ASSIGNMENT

Read the experiment and answer the following questions.

1. What are electric and magnetic fields, and what do they tell you?

2. What are "lines of force"?

3. The electric field begins and ends on what kinds of charges? The magnetic field begins and ends on what kinds of poles?

4. What are equipotentials, and how are they determined experimentally?

5. How are electric field lines and equipotentials oriented relative to each other?

(continued)

6. Does the magnetic field do work on a moving charge? On a magnetic pole? Explain.

EXPERIMENT 26
Fields and Equipotentials

I. INTRODUCTION

When buying groceries, we are often interested in the price per pound. Knowing this, we can determine the price for a given amount of an item. Analogously, it is convenient to know the electric force per unit charge at points in space due to an electric charge configuration or the magnetic force per unit pole or "moving charge." Knowing these, we can easily calculate the electric force or magnetic force an interacting object would experience at different locations.

The electric force per unit charge is called the electric field intensity, or simply the **electric field.** By determining the electric force on a test charge at various points in the vicinity of a charge configuration, the electric field may be "mapped" or represented graphically by lines of force. The English scientist Michael Faraday (1791–1867) introduced the concept of lines of force as an aid in visualizing the magnitude and direction of an electric field.

Similarly, the magnetic force per unit pole is called the magnetic field intensity, or **magnetic field.** In this case, the field is mapped out by using the pole of a magnetic compass.

In this experiment, the concept of fields will be investigated and some electric and magnetic field configurations will be determined experimentally.

II. EQUIPMENT NEEDED

A. *Electric Field*

- Field mapping board and probes
- Conducting sheets with grids*
- Conducting paint*
- Connecting wires
- 1.5-V battery (or 10-V dc source)
- Galvanometer [or vacuum-tube voltmeter (VTVM) with two-point contact field probe]
- Single-throw switch
- 3 sheets of Cartesian graph paper

B. *Magnetic Field*

- 2 bar magnets and one horseshoe magnet
- Iron filings
- 3 sheets of paper or a glass plate
- Small compass
- 3 sheets of Cartesian graph paper or regular paper

III. THEORY

The magnitude of the electrostatic force between two point charges q_1 and q_2 is given by (Coulomb's law)

$$F = \frac{kq_1q_2}{r^2} \qquad \textbf{(26-1)}$$

where r is the distance between the charges and the constant $k = 9 \times 10^9$ N-m²/C². The direction of the force on a charge may be determined by the law of charges: Like charges repel and unlike charges attract.

* Available from Sargent-Welch Scientific Co.

The magnitude of the **electric field** is defined as the electrical force per unit charge, or $E = F/q_0$ (N/C). By convention, the electric field is determined by using a *positive* test charge q_0. In the case of the electric field associated with a single source charge q, the magnitude of the electric field a distance r away from the charge is

$$E = \frac{F}{q_0} = \frac{kq_0q}{q_0r^2} = \frac{kq}{r^2} \qquad \textbf{(26-2)}$$

The direction of the electric field may be determined by the law of charges.

The electric field vectors for several series of radial points from a positive source charge are illustrated in Fig. 26-1(a). Notice that the lengths (magnitudes) of the vectors are smaller the greater the distance from the charge. Why? By drawing lines through the points in the direction of the field vectors, we form lines of force [Fig. 26-1(b)], which give a graphical representation of the electric field. The magnitudes of the electric field are not customarily listed, only the direction of the field lines. However, the closer together the lines of force, the stronger the field.

If a positive charge were released in the vicinity of a stationary positive source charge, it would move along a line of force in the direction indicated (away from the source charge). A negative charge would move along the line of force in the opposite direction. Once the electric field for a particular charge configuration is known, we tend to neglect the charge configuration itself, since the "effect" of the configuration is given by the field.

Since a free charge moves in an electric field by the action of the electric force, we say that work ($W = Fd$) is done by the field in moving charges from one point to another [e.g., A to B in Fig. 26-1(b)]. To move a positive charge from B to A would require work supplied by an external force to move the charge against the electric field (force). The ratio of the work W done to the charge q_0 in moving the charge between two points in an electric field is called the **potential difference ΔV** between the points:

$$\Delta V_{BA} = V_B - V_A = \frac{W}{q_0} \qquad \textbf{(26-3)}$$

(It can be shown that the potential at a particular point a distance r from a source charge q is $V = -kq/r$. See your textbook.)

If a charge is moved along a path at right angles or perpendicular to the field lines, there is no work done ($W = 0$), since there is no force component along the path. Then along such a path (dashed-line paths in Fig. 26-1), $\Delta V = V_B - V_C = W/q_0 = 0$, and $V_C = V_B$. Hence, the potential is constant

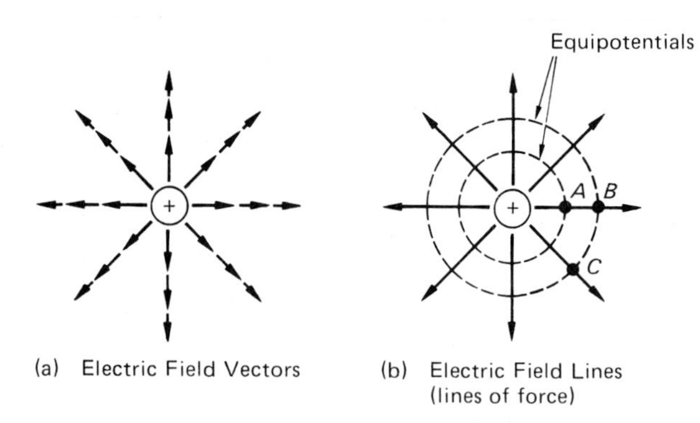

Fig. 26-1 (a) Electric field vectors. (b) Lines of force and equipotentials.

along paths perpendicular to the field lines. Such paths are called **equipotentials** (along an equipotential surface in three dimensions).

An electric field may be mapped experimentally by determining either the field lines (of force) or the equipotential lines. Static electric fields are difficult to measure, and field lines are more easily determined by measuring small electric currents (flow of charges) maintained in a conducting medium between charge configurations in the form of metal electrodes. The steady-state electric field lines closely resemble the static field that a like configuration of static charges would produce. The current is measured in terms of the voltage (potential) difference by a high-impedance voltmeter such as a vacuum-tube voltmeter (VTVM.)

In other instances, equipotentials are determined, and hence the field lines, using a simple galvanometer as a detector. When no current flows between two probe points, as indicated by a zero deflection on the galvanometer, there is no potential difference between the points ($\Delta V = 0$), and the points are on an equipotential.

B. *Magnetic Field*

As in the electric case, a **magnetic field intensity** was originally defined as the magnetic force per unit pole and was given the symbol **H**. The direction of **H** at a particular location is that of the force experienced by a north magnetic pole.

Just as we may map the electric field around an electric charge, we may draw magnetic lines of force around a magnet. A single magnetic pole, or magnetic monopole, has never been observed, so the magnetic field is mapped using the north pole (by convention) of a magnetic dipole, for example, the magnetic needle of a compass. The torque on the compass needle resulting from the magnetic force causes the needle to line up with the field, and the north pole of the compass points in the direction of the field (Fig. 26-2). If the compass is moved in the direction indicated by the north pole, the path of the compass traces out a field line.

We now know that a magnetic field is produced by a moving electric charge or current. For example, the electron motions in the atoms of a permanent magnet produce contributions to the magnetic field of the magnet (along with nonclassical electron spin and nuclear contributions).

Another observation is that an electric charge moving nonparallel to a magnetic field experiences a force. Since the magnetic field is defined at points of space in terms of a magnetic force exerted on a test object, this fact is used to define a **magnetic field** vector **B**. The magnetic field **B** may be thought of as the magnetic force "per unit moving charge." An electric charge moving nonparallel to a magnetic field experiences a force that is perpendicular to the plane defined by the charge's velocity vector and the magnetic field vector ($\mathbf{F} = q\mathbf{v} \times \mathbf{B}$, see textbook for detailed description). The magnetic field **B** (sometimes called the magnetic induction)* is more commonly used than **H**. In space in the vicinity of a magnet, the **B** and **H** fields are in the same direction and differ in magnitude by a constant. In the general mapping of the magnetic fields in this experiment, we will express the **B** field even though the mapping is done with a compass or magnetic pole.

Since the magnetic force on a charged particle moving in a steady magnetic field is always perpendicular to the particle displacement, the magnetic field does no work on the particle. Hence, in a two-dimensional mapping of a magnetic field, the whole plane is an equipotential. However, it is instructive for comparative purposes to draw equipotential lines perpendicular to the field lines as in the electric field case. There would be no work done on a magnetic pole when moved along these equipotential lines. Why?

A common method of demonstrating a magnetic field is to sprinkle iron filings over a paper or glass plate covering a magnet (Fig. 26-3). The iron filings become induced magnets and line up with the field as would a compass needle. This method allows one to quickly visualize the magnetic field configuration.

Fig. 26-2 Magnetic field. The magnetic force causes the compass needle to line up with the field and the north pole of the compass points in the direction of the field. If a compass is moved in the direction indicated by the north pole, the path of the compass traces out a field line.

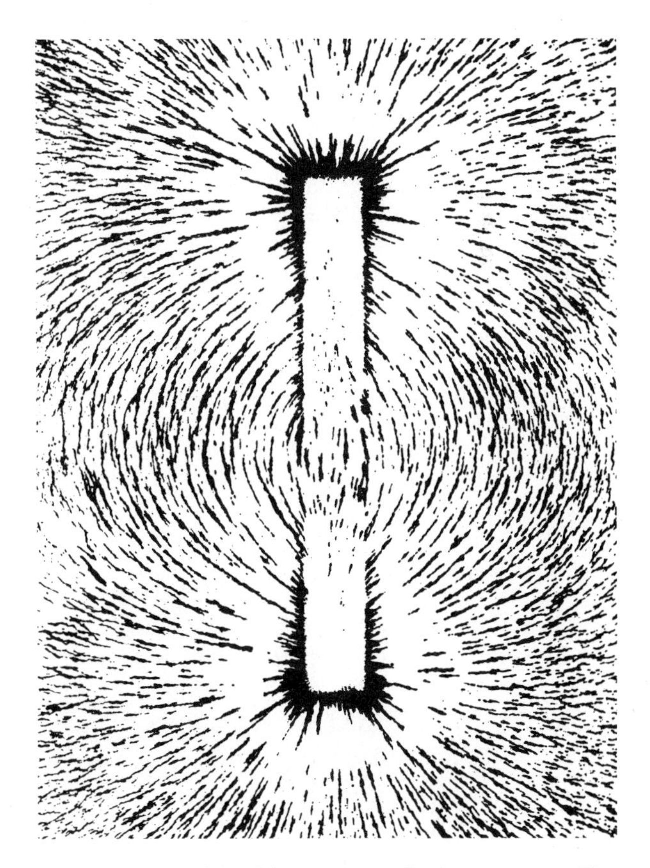

Fig. 26-3 The iron filing pattern of a bar magnet. The iron filings become induced magnets and line up with the field as would a compass needle. (Courtesy *PSCC Physics, 2nd Ed.*, 1965; D. C. Heath and Company with Educational Development Center, Inc., Newton, Mass.)

*The SI unit for the magnetic induction **B** is the tesla (T). Other units are the weber/m^2 (W/m^2) and gauss (G). The units are named after early investigators of magnetic phenomena.

IV. EXPERIMENTAL PROCEDURE

A. *Electric Field*

1. An electric field mapping setup with a galvanometer is shown in Fig. 26-4. The apparatus consists of a flat board on which is placed a sheet of carbonized conducting paper imprinted with a grid. The sheet has an electrode configuration of conducting silver paint, which provides an electric field when connected to a voltage source (e.g., a battery).

   The common electrode configurations ordinarily provided are two dots representing point charges of an electric dipole configuration (Fig. 26-4) and two parallel linear electrodes representing a two-dimensional cross section of a parallel-plate capacitor (on the board in the photo).

2. Draw the electric dipole configuration on a sheet of graph paper to the same scale and coordinates as those of the painted dipole on the imprinted grid on the conducting sheet. Then place the dipole conducting sheet on the board and set the contact terminals firmly on the painted electrode connections. Connect the probes to the galvanometer as shown in the figure. The probes are used to locate points in the field that are at equipotential.

   If a VTVM is used, its field probe should have two contacts mounted about 2 cm apart. Your instructor will give you instructions on the operation of the VTVM. Connect the voltage source of the board terminals (1.5-V battery for galvanometer measurements or 10 V dc from a power supply for VTVM measurements). Place a switch in the circuit and leave it open until you are ready to take measurements (not shown in the figure).

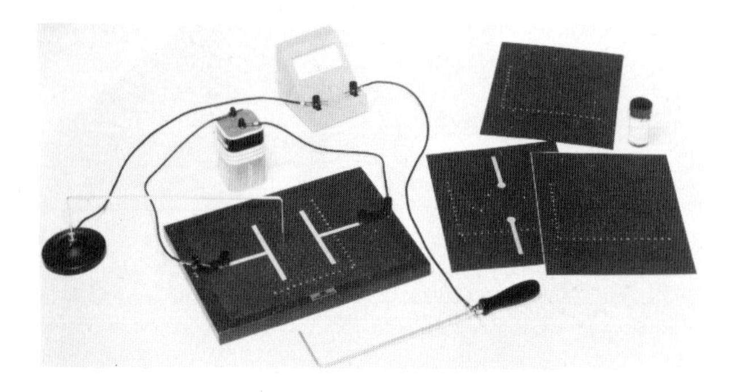

Fig. 26-4 Electric field mapping apparatus. A parallel plate capacitor configuration is on the board and an electric dipole configuration is to the right. (Photo courtesy of Sargent-Welch Scientific Company.)

GALVANOMETER MEASUREMENTS

3. Place the stationary probe on the electric dipole sheet at some general point near the edge of the grid area in the region between the electrodes. The potential at this point will serve as a reference potential. Mark the probe position on your graph-paper map. The movable probe is then used to determine the location of a series of other points having the same potential. When the movable probe is at a point with the same potential as that of the stationary reference probe, no deflection will be observed on the galvanometer.

   The galvanometer is a delicate current-measuring instrument and should be limited to small current values to avoid damage. As a result, the probes should not be allowed to make contact with each other or be brought too close together.

4. Close the switch and place the movable probe on the conducting paper at some location an appreciable distance away from the stationary probe. Move the probe until the galvanometer shows zero deflection (indicating a point of equipotential) and record this point on the graph-paper map. Locate a series of 8 or 10 points of the same potential across the general field region, and draw a dashed-line curve through these points on the graph-paper map.

5. Choose a new location for the reference probe 2 to 3 cm from the previous reference position and locate another series of equipotential points. Continue this procedure until you have mapped the field region. Open the switch. Draw curves perpendicular to the equipotential lines on the graph-paper map to represent the electric field lines. Do not forget to indicate the field direction on the field lines.

6. Repeat the procedure for the parallel linear (plate) electrode configuration. Be sure to investigate the regions around the ends of the plate electrodes.

7. *(Optional)* Your instructor may wish to have you map the electric field for another nonsymmetrical electrode configuration or a configura-

tion of your own choosing. These can be prepared by painting the desired electrode configuration on a conducting sheet with silver paint.

VTVM MEASUREMENTS

8. Close the switch, and with the VTVM set on the 10-V scale, position the negative (−) contact of the field probe near the negative electrode. Using the negative probe point as a pivot, rotate the positive (+) contact around the fixed negative contact until the position with the maximum meter reading is found.

   Record the positions of the probe contacts on the graph-paper map. (The sensitivity of the VTVM may be increased by switching to a lower scale. A midscale reading is desirable. Make sure to rezero the VTVM when scales are switched.)

9. Using the second probe point as a new negative probe point, repeat the procedure to determine another point of maximum meter reading and record. Continue this procedure until the positive electrode is approached. Draw a smooth curve through these points on the graph-paper map.

   Then starting again at a new position near the negative electrode, repeat these procedures for another field line. Trace out four to six field lines in this manner. Do not forget to indicate the field direction on the field lines.

10. Place the negative probe contact near the center of the field region, and rotate the positive contact until a position is found that gives a *zero* meter reading. Record these points on the graph paper with a different symbol than that used for the field lines. Use the second point as a new pivot point as before, and determine a series of null (zero) points. Draw a dashed-line curve through these equipotential points. Determine three to five equipotential lines in this manner.

11. Repeat the procedure for the parallel linear (plate) electrode configuration. Be sure to investigate the regions around the ends of the plate electrodes.

12. *(Optional)* Your instructor may wish to have you map the electric field for another nonsymmetrical electrode configuration or a configuration of your own choosing. These can be prepared by painting the desired electrode configuration on a conducting sheet with silver paint.

B. *Magnetic Field*

13. Covering the magnets with sheets of paper (or glass plate), sprinkle iron filings to obtain an iron filing pattern for each of the arrangements shown in Fig. 26-5. For the bar magnet arrangements, the magnets should be separated by several centimeters, depending on the pole strengths of the magnets. Experiment with this distance so there is enough space between the ends of the magnets to get a good pattern.

14. Sketch the observed magnetic field patterns on Fig. 26-5. After the patterns have been sketched, collect the iron filings on a piece of paper and return them to the filing container (recycling them for someone else's later use). Economy in the laboratory is important.

15. Place the magnets for each arrangement on a piece of graph paper or regular paper. Draw an outline of the magnets for each arrangement on the paper. Using a small compass, trace out (marking on the paper) the magnetic field lines as smooth curves. Draw enough field lines so that the pattern of the magnetic field may be clearly seen. Do not forget to indicate the field direction on the field lines.

16. Draw dashed-line curves perpendicular to the field lines.

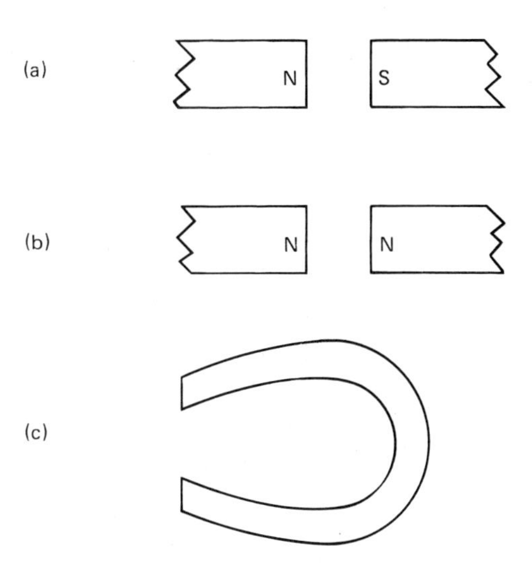

Fig. 26-5

Name .... Section .... Date ....

Lab Partner(s) ....

EXPERIMENT 26 *Fields and Equipotentials*

LABORATORY REPORT

Attach graphs to Laboratory Report.

QUESTIONS

1. Directions of the fields are indicated on field lines. Why are there no directions indicated on equipotential lines?

2. For the dipole configuration, in what region(s) does the electric field have the greatest intensity? Explain how you know from your map, and justify.

3. Comment on the electric field of the parallel plates (a) between the plates, and (b) near the edges of the plates.

4. Sketch the electric field for (a) a negative point charge near a positively charged plate, and (b) two positive point charges.

⊖ ⊕ ⊕

+++++++++++++++++++++++++++++

(a) (b)

5. Compare the electric fields and magnetic fields of the experimental arrangements. Comment on any field similarities and differences.

6. Explain how a gravitational field might be mapped. Sketch the gravitational field for two point masses.

Name .... Section .... Date ....

Lab Partner(s) ....

EXPERIMENT 27 *Ohm's Law*

ADVANCE STUDY ASSIGNMENT

Read the experiment and answer the following questions.

1. According to Ohm's law, how does (a) the current vary with voltage; (b) the voltage vary with resistance; (c) the current vary with resistance?

2. What is an "ohmic" resistance? Are all resistances ohmic in nature?

3. What are the analogs in liquid and electrical circuits?

4. What is the terminal voltage of a battery or power supply equal to in terms of the potential differences or voltage drops across circuit components?

EXPERIMENT 27

Ohm's Law

I. INTRODUCTION

Without doubt, one of the most frequently applied relationships in current electricity is that known as **Ohm's law.** This relationship, discovered by the German physicist Georg Ohm (1787–1854), is basic in the analysis of electrical circuits. Ohm's law applies to many common conductors. For an "ohmic" conductor, the potential difference or voltage drop V across the conductor is proportional to the current I through the conductor. All other conditions remaining unchanged, for a particular voltage and current, Ohm's law is expressed mathematically

$$\frac{V}{I} = R$$

The constant R is called the **resistance** of the conductor and measured in units of ohms.

In this experiment, Ohm's law will be investigated as applied to a circuit component in a simple circuit.

II. EQUIPMENT NEEDED*

- Ammeter (0 to 0.5 A)
- Voltmeter (0 to 10 V dc)
- Decade resistance box (0.1 to 99.9 Ω)
- Rheostat (20 Ω)
- Battery or power supply (6 V)
- Switch
- Connecting wires
- 2 sheets of Cartesian graph paper

III. THEORY

For a wide range of materials used in electrical circuits (e.g., metallic conductors), the relationship of the potential difference or voltage drop V across a circuit component, the current I through the component, and the resistance R of the component is given by **Ohm's law:**

$$V = IR \qquad \textbf{(27-1)}$$

That is, for a given circuit component which has an "ohmic" resistance (obeys Ohm's law), the voltage is linearly proportional to the current.

*The ranges of the equipment are given as examples. These may be varied to apply to available equipment.

To understand the relationships of the quantities in Ohm's law, it is often helpful to consider a liquid current analogy (Fig. 27-1). In a liquid circuit, the force to move the liquid is supplied by a pump. The rate of liquid flow depends on the resistance to the flow (e.g., due to some obstruction in the circuit pipe)—the greater the resistance, the less liquid flow.

Analogously, in an electrical circuit, a voltage source (e.g., a battery or power supply) supplies the (electromotive) force for current flow and the magnitude of the current is determined by the resistance R in the circuit. For a given voltage, the greater the resistance, the less current through the resistance, as may be seen from Ohm's law, $I = V$ (a constant)$/R$.

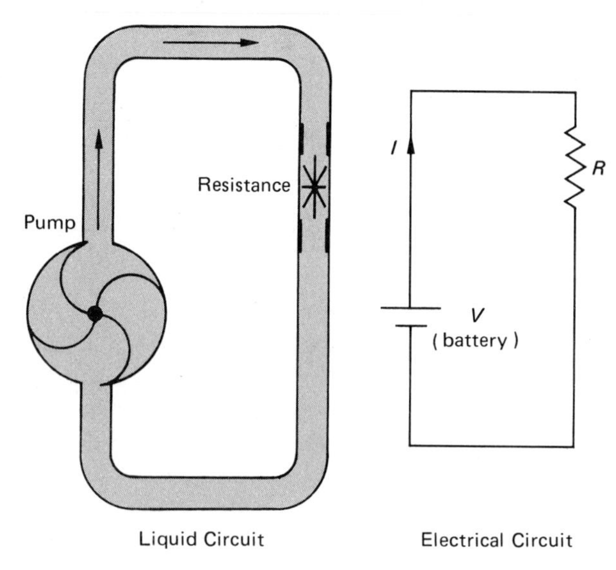

Fig. 27-1 Liquid analogy of an electrical circuit.

Notice that the voltage source supplies a voltage "rise" that is equal to the voltage "drop" across the resistance and is given by $V = IR$ (Ohm's law).

In an electrical circuit with two or more resistances and a single voltage source, Ohm's law may be applied to the entire circuit or to any portion of the circuit. When applied to the entire circuit, the voltage is the terminal input voltage supplied by the voltage source, and the resistance is the total resistance of the circuit. In the case of applying Ohm's law to a particular portion of the circuit, the individual voltage drops, currents, and resistances are used for that part of the circuit.

Consider the circuit diagram shown in Fig. 27-2. The applied voltage is supplied by a power supply or battery. R_h is a rheostat, which is a variable resistor that allows the voltage across the resistance R_s to be varied (sometimes called a voltage divider because it divides the applied voltage across itself and R_s). An ammeter Ⓐ measures the current in the circuit and R_s, and a voltmeter Ⓥ registers the voltage drop across R_s. S is a switch for closing and opening (activating and deactivating) the circuit.

Any component in a circuit that does not generate or supply a voltage acts as a resistance in the circuit. This is true for the connecting wires, the ammeter, and voltmeter. However, the metallic connecting wires and the ammeter have negligibly small resistances, so they do not greatly affect the current. Also, a voltmeter has a high resistance, so very little current flows through the voltmeter. Hence, to good approximations the ammeter registers the current in the circuit and the voltmeter reads the voltage drop across the resistance. These approximations are adequate for most practical applications.

Applying Ohm's law to the portion of the circuit with R_s only, we have

$$V_s = IR_s \quad \textbf{(27-2)}$$

where V_s and I are the voltmeter and ammeter readings, respectively. Notice that the same current I flows through the rheostat R_h and the resistance R_s. The voltage drop across R_h is then

$$V_h = IR_h \quad \textbf{(27-3)}$$

To apply Ohm's law to the entire circuit, we use the fact that the applied voltage "rise" or the terminal voltage V_t of the voltage source must equal the voltage "drops" of the components around the circuit. Then

$$V_t = V_h + V_s$$

or

$$V_t = IR_h + IR_s = I(R_h + R_s) \quad \textbf{(27-4)}$$

From Eq. 27-4, we see that for a constant R_s, the current through this resistance, and hence its voltage drop V_s, can be varied by varying the rheostat resistance R_h (V_t is constant). Similarly, the voltage V_s can be maintained constant when R_s is varied by adjusting R_h.

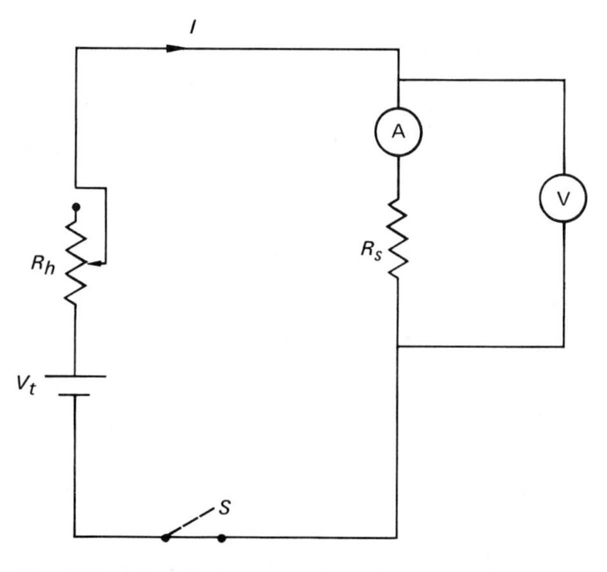

Fig. 27-2 Circuit diagram with voltmeter and ammeter. R_h is a rheostat (a continuously variable resistor).

IV. EXPERIMENTAL PROCEDURE

1. With the voltmeter, measure the terminal voltage of the power supply or battery and record in the Laboratory Report. Start with the voltmeter connection to the largest scale and increase the sensitivity by changing to a smaller scale if necessary. Most common laboratory voltmeters and ammeters have three scale connections and one binding post common to all three scales. It is good practice to initially take measurements with the meters connected to the largest scales. This prevents the instruments from being "pegged" and possibly damaged should the magnitude of the voltage or current exceed the smaller scale limits. A scale setting may be changed for greater sensitivity by moving the connection to a lower scale after the general magnitude and measurement is known.

   Also, attention should be given to the proper polarity (+ and −); otherwise, the meter will be "pegged" into the opposite direction. Connect + to + and − to −.

2. Set up the circuit shown in the circuit diagram (Fig. 27-2) with the switch open. A standard decade resistance box is used for R_s. Set the rheostat resistance R_h for maximum resistance and the value of the R_s to about one-half of the maximum resistance of the decade box. *Have the instructor check the circuit before closing the switch.*

A. *Variation of Current with Voltage (R_s constant)*

3. After the instructor has checked the circuit, close the switch and read the voltage and current on the meters. Open the switch after the readings are taken and record them in Data Table 1. Repeat this procedure for a series of four successively lower rheostat settings along the length of the rheostat. It is convenient for data analysis to adjust the rheostat (after closing the switch) so that evenly spaced and convenient ammeter readings are obtained. The switch should be closed only long enough to obtain the necessary readings. This prevents unnecessary heating in the circuit.

4. Repeat procedure 3 for another value of R_s (about 30 Ω).

5. Plot the results for both resistances on a single V_s versus I_s graph and draw straight lines that best fit the sets of data. Determine the slopes of the lines and compare them with the constant values of R_s of the decade box by computing the percent errors. According to Ohm's law, the respective values should be equal.

B. *Variation of Current and Resistance (V_s constant)*

6. This portion of the experiment uses the same circuit arrangement as before. In this case, the voltage V_s is maintained constant by adjusting the rheostat resistance R_h when the R_s is varied. Initially, set the rheostat near maximum resistance and the resistance R_s of the decade box to about 100 ohms. Record the value of R_s in Data Table 2.

   Close the circuit and read and record the ammeter and voltmeter measurements in Data Table 2. The voltmeter reading is to be taken as the constant voltage V_s. Open the circuit after making the readings.

7. Repeat this procedure for four more successive steps of current by reducing the value of R_s of the decade box. Keep the voltage across R_s constant for each setting by adjusting the rheostat resistance R_h. Do not reduce R_s below 30Ω.

8. Plot the results on a I_s versus $1/R_s$ graph and draw a straight line that best fits the data. (Reciprocal ohms, $1/R$, is commonly given the unit name "mhos.") Determine the slope of the line and compare it with the constant value of V_s by computing the percent error. According to Ohm's law, these values should be equal.

LABORATORY REPORT

A. *Variation of Current with Voltage (R_s constant)*

DATA TABLE 1

Terminal voltage, V_t ....

Reading	Constant R_s		Constant R_s	
	Voltage, V_s	Current, I_s	Voltage, V_s	Current, I_s
1				
2				
3				
4				
5				

Calculations (show work)

Slopes of lines	Percent error from R_s
....	....
....	....

Name .... Section .... Date ....

Lab Partner(s) ....

B. *Variation of Current and Resistance (V_s constant)*

DATA TABLE 2

Constant voltage, V_s ....

Reading	Current, I_s (A)	Resistance, R_s (Ω)	$1/R_s$
1			
2			
3			
4			
5			

Calculations (show work)

Slope of line ....

Percent error from V_s ....

(continued)

QUESTIONS

1. If the switch were kept closed during the procedures and the circuit components heated up, how would this affect the measurements?

2. Devise and draw a circuit using a long, straight wire resistor instead of a decade box that would allow the study of the variation of voltage with resistance (I_s constant). What would a graph of the data from this circuit show?

3. Compute the values of R_h and the voltage drops across this resistance for the two situations in procedure A, reading #1.

Name .... Section .... Date ....

Lab Partner(s) ....

EXPERIMENT 28 *The Potentiometer:* emf *and Terminal Voltage*

ADVANCE STUDY ASSIGNMENT

Read the experiment and answer the following questions.

1. What is an electromotive force? Is it actually a force?

2. Are the emf and the terminal voltage of a battery the same? Are they constant? Explain.

3. Why is a potentiometer more accurate than a voltmeter in measuring emf?

4. What is meant when we say that a potentiometer is "balanced"?

5. Show that a graph of terminal voltage versus the current drawn from a battery has the form of a straight line.

EXPERIMENT **28**

The Potentiometer: *emf* and Terminal Voltage

I. INTRODUCTION

The *potentiometer* is an instrument for the accurate measurement of the potential differences or voltages of electrical sources such as batteries. Unlike the common voltmeter, which measures approximate source potential differences because it draws some current from the source, the potentiometer does not disturb or affect the potential difference of the source. This is accomplished by "balancing" the source to be measured against another "working" source in the circuit. As such, the potentiometer is a valuable instrument for calibrating voltmeters and ammeters as well as in measuring the electromotive force or source potential differences of cells and batteries.

In this experiment, the principle of the potentiometer will be investigated and used in measuring the electromotive forces of battery cells. Also, the internal resistances of the cells will be determined.

II. EQUIPMENT NEEDED

- Slide-wire potentiometer
- Portable galvanometer
- Rheostat (20 to 40 Ω)
- 12-V dc power supply or battery
- Resistor (5 kΩ)
- Two switches, one key, and one single-pole, single-throw (SPST) switch
- Connecting wires
- Ammeter (0 to 3 A)
- Voltmeter (3 to 30 V)
- Two dry cells, 1.5 or 3 V (preferably an old and a fresh cell)
- Standard cell
- 1 sheet of Cartesian graph paper

III. THEORY

An electrical source supplies a potential difference to the circuit to which it is connected. For example, a dry-cell battery transforms chemical energy into electrical energy. (Technically, a battery is a combination of single cells.) When there is no current flowing in the source, the "open circuit" potential difference $\mathscr{E}$ between the terminals is called the **electromotive force (emf).** Electromotive "force" is an unfortunate name because the quantity is not a force but a potential difference or voltage.

Every source of emf has some internal resistance r, which gives rise to an internal voltage drop when a current flows through it. For example, when the terminal voltage of a cell is measured with a voltmeter as in Fig. 28-1, a current I flows in the circuit. By Ohm's law, the terminal potential difference or voltage V as measured by the voltmeter is less than the open circuit emf potential difference of the battery by an *internal* voltage drop Ir due to the internal resistance of the battery, or

$$V = \mathscr{E} - Ir \qquad \textbf{(28-1)}$$

(*terminal voltage* = *emf* − *internal voltage drop*)

As can be seen from the equation, the voltmeter does not read the emf of the cell. However, for many practical situations, the terminal voltage is taken as the emf, since for a high-resistance voltmeter the current I is quite small and the Ir term is

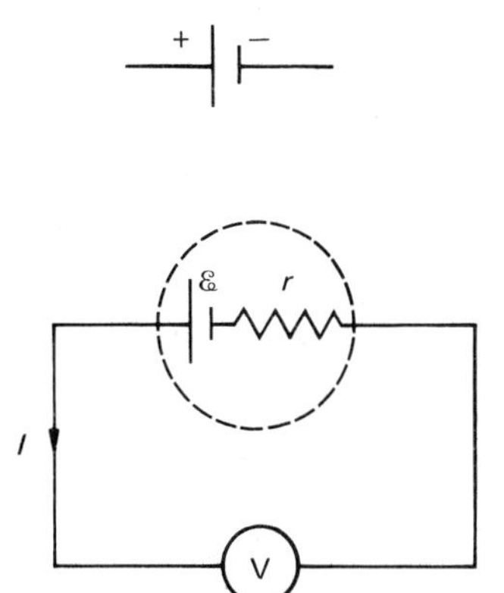

Fig. 28-1 Circuit diagram explicitly showing the internal resistance of a cell.

negligible for common low-resistance cells. Also, as expressed in Eq. 28-1, for a battery cell or other source of constant emf, the terminal voltage decreases linearly as the current drawn from the sources increases. However, a cell of low internal resistance can supply a relatively large current without a great decrease in its terminal voltage.

A simple slide-wire potentiometer circuit is illustrated in Fig. 28-1(a). This circuit may be used to measure the emf of a cell or battery, which requires the condition of zero current through the battery ("open-circuit" condition). The so-called "working source" with an emf of $\mathscr{E}_w$ maintains a constant current I through the resistor.

The resistor is in the form of a wire with a sliding contact key C which allows R_x or L_x to be varied. This is adjusted until the galvanometer (a sensitive current-detecting instrument) has a zero or "null" reading, indicating that no current is flowing through the test cell $\mathscr{E}_x$. As such, $\mathscr{E}_x$ is "balanced" against $\mathscr{E}_w$, and the emf of the test cell is equal to the voltage drop across R_x:

$$\mathscr{E}_x = IR_x \qquad \textbf{(28-2)}$$

In effect, the adjustable resistance acts as a "voltage divider." A portion of the input emf or voltage V_i is tapped off as an output voltage V_o by dividing the resistor [Fig. 28-2(b)]. Since $V_o = IR_o$ and $V_i = IR_i$, $V_o = (R_o/R_i)V_i$. Obviously, if $R_o = R_i$, then $V_o = V_i$.

If the test cell is then replaced by a standard cell of known emf $\mathscr{E}_s$ and the potentiometer is again balanced, then by the same reasoning,

$$\mathscr{E}_s = IR_s \qquad \textbf{(28-3)}$$

The current is the same in both cases, since $\mathscr{E}_w$ and the total resistance R of the wire are constant, $I = \mathscr{E}_w/R$. Then, taking, the ratio of these two equations,

$$\frac{\mathscr{E}_x}{\mathscr{E}_s} = \frac{R_x}{R_s}$$

or

$$\mathscr{E}_x = \left(\frac{R_x}{R_s}\right)\mathscr{E}_s \qquad \textbf{(28-4)}$$

For a uniform slide wire, a length segment of the wire is proportional to the resistance of the segment, and in terms of the lengths L of the two balanced conditions, which are easily measured,

$$\mathscr{E}_x = \left(\frac{L_x}{L_s}\right)\mathscr{E}_s$$

Hence, knowing the emf of the standard cell $\mathscr{E}_s$ and measuring the length segments of the wire for each balanced condition, the emf of the test cell can be computed.

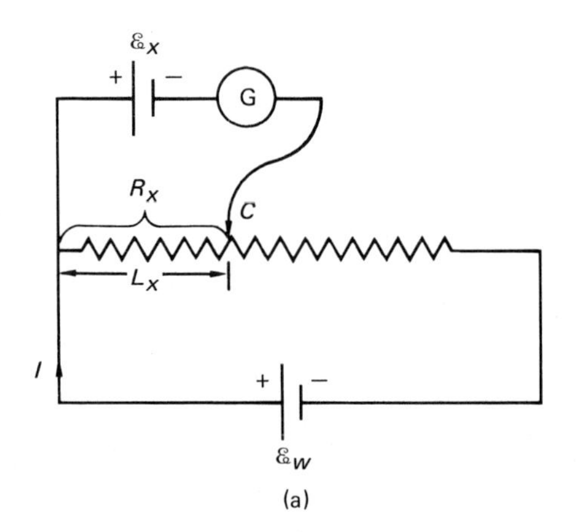

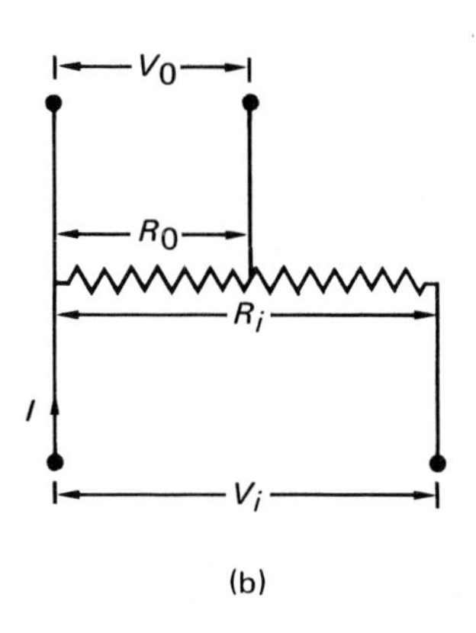

Fig. 28-2 (a) A simple slide-wire potentiometer circuit diagram. (b) The principle of a "voltage divider." See text for description.

IV. EXPERIMENTAL PROCEDURE

A. *Measurement of* emf

1. Set up the potentiometer circuit as shown in Fig. 28-3 with one of the dry cells as $\mathscr{E}_x$. Keep switches S (SPST) and K (Key) open while connections are being made. The resistance R_l (5 kΩ) protects the galvanometer from large initial currents in an unbalanced circuit that could damage the instrument. This protective resistance will be short-circuited during the experiment by closing K. Set the rheostat R_h at its maximum resistance. *Do not plug in the power supply (or connect a battery source) until the instructor has checked the circuit.**

2. After the circuit has been approved, plug in the power supply and close switch S. Depress the contact key C and note the galvanometer deflection. The circuit is balanced when the galvanometer deflection is zero. This is accomplished by sliding the contact key along the wire. *Do not slide the contact key when it is depressed.* This could scrape the wire and make it nonuniform.

   Balance the circuit. Then adjust the rheostat resistance R_h so that when the circuit is balanced again, the length segment L will be about two-thirds to three-fourths of the length of the wire. Note and record the current reading of the ammeter.

*If your galvanometer is equipped with several scale settings, resistor R and switch K may be omitted from the circuit. The galvanometer should be set initially to its least sensitive scale, then changed to a more sensitive scale if applicable. (This is equivalent to closing K.) Your instructor will advise you which scales are more and less sensitive in this case.

3. Close switch K and rebalance the circuit. Keep the current I constant at the value of the previous balanced condition by adjusting the rheostat resistance if necessary. Record the value of this current. The closing of K short-circuits R_1 and greatly increases the sensitivity of the galvanometer. Only a fine adjustment will be needed to balance the circuit. After this is done, open K and deactivate the circuit (unplug the power supply). Measure and record the length segment L of the slide wire.

4. Place the other dry cell in the circuit as $\mathscr{E}_x$ (keeping the same polarity). Activate the circuit and repeat procedure 3.

5. Place the standard cell in the circuit in place of the test dry cell. Record the emf ($\mathscr{E}_s$) of the standard cell (on the cell or provided by the instructor). Activate the circuit and repeat procedure 3. Then, from the experimental data, compute the emf's of the dry cells.

B. *Measurement of Internal Resistance*

6. Set up the circuit as shown in Fig. 28-4 with one of the dry cells. Do not connect one terminal of the cell and have switch S open. Set the rheostat at its maximum resistance. *Have the instructor check the circuit.*

7. After the circuit has been approved, connect the cell terminal and read and record the terminal voltage registered on the voltmeter (switch S still open).

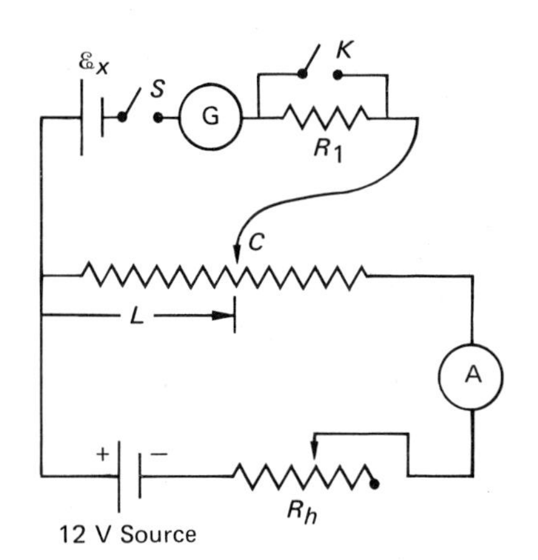

Fig. 28-3 Potentiometer circuit for experimental procedure.

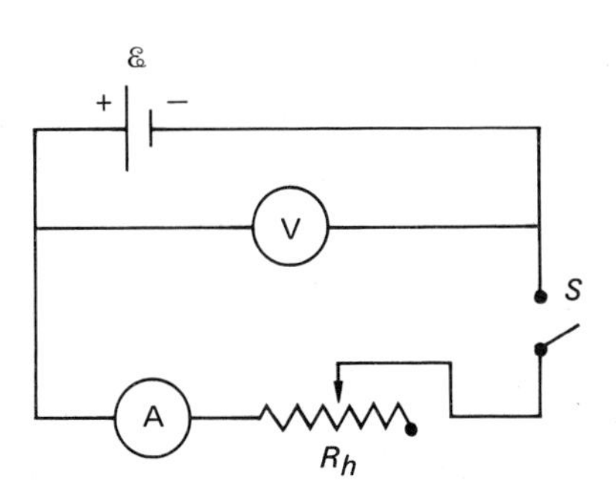

Fig. 28-4 Diagram of circuit for measuring cell internal resistance.

Then close the switch and record the voltmeter and ammeter readings. The ammeter registers the current through the external resistance R_h. Note that the voltage drop across R_h (i.e., $V = IR_h$) is the same as the terminal voltage of the cell.

8. Adjust (reduce) the rheostat resistance in a series of steps, recording the voltmeter and ammeter readings for each step, until the current reaches a value of about 3 A. Then disconnect the cell.

9. Repeat procedures 7 and 8 for the other dry cell.

10. Plot the terminal voltage V versus I (the current delivered to the external resistance) for each cell on the same graph. Draw straight lines that best fit the data with the lines extrapolated to intercept both axes. Determine the (negative) slopes and intercepts of the lines. As can be seen from Eq. 28-1, which has the form of a straight line, the slope of the line is equal to the internal resistance of the cell.

LABORATORY REPORT

A. *Measurement of* emf

Current I ....

Calculations: (show work)

	Slide-wire segment: L(cm)	emf
Cell 1	....	....
Cell 2	....	....
Standard cell	....	....

Name .... Section .... Date ....

Lab Partner(s) ....

EXPERIMENT 28 *The Potentiometer:* emf *and Terminal Voltage*

B. *Measurement of Internal Resistance*

DATA TABLE

Voltage with switch open:

Cell 1 ....

Cell 2 ....

Rheostat step	Cell 1		Cell 2	
	Voltage, V (V)	Current, I (A)	Voltage, V (V)	Current, I (A)
1 $R_{h\max}$				
2 $\approx\frac{1}{2}R_{h\max}$				
3 $\approx\frac{1}{4}R_{h\max}$				
4				
5				
6				
7				
8 short circuit (zero ohms)				

Calculations (show work)

	Cell 1	Cell 2
Slope r	....	....
y intercept	....	....
x intercept	....	....

(continued)

QUESTIONS

1. Compare the terminal voltages of the cells to their emf's by computing the decimal fraction and percent of the terminal voltage of the emf. Give an explanation for the difference in these quantities if any.

2. Were the current capabilities of the dry cells the same? If not, explain why.

3. Interpret the Y- and X-axis intercepts of a graph of V versus I for a cell and explain why the Y intercept corresponds to an "open-circuit" condition and the X intercept to a "short-circuit" condition.

4. Compare the emf's of the cells found in part A with their emf's as determined from the graphical analysis of the data in part B by computing the percent differences.

5. Using known and measured experimental quantities, find the total resistance of the slide wire of the potentiometer.

Name .... Section .... Date ....

Lab Partner(s) ....

EXPERIMENT 29 *The Ammeter and the Voltmeter: Meter Sensitivity*

ADVANCE STUDY ASSIGNMENT

Read the experiment and answer the following questions.

1. What is meant by the current sensitivity of a galvanometer?

2. What is the purpose of the "shunt" resistance in an ammeter? Is the ammeter a high- or low-resistance instrument?

3. How is an ammeter connected in a circuit? Comment on the ammeter reading if it were connected otherwise.

4. What is the purpose of the "multiplier" resistance in a voltmeter? Is the voltmeter a high- or low-resistance instrument.

5. How is a voltmeter connected in a circuit? What would happen if it were connected otherwise?

EXPERIMENT **29**

The Ammeter and Voltmeter: Meter Sensitivity

I. INTRODUCTION

The majority of electrical measurements is concerned with determining the magnitudes of currents and voltages. Electrical currents are commonly measured with ammeters and voltages with voltmeters. These instruments are delicate and may be easily damaged if they are connected improperly in a circuit. Also, the internal resistances of meters may add appreciable error to measurements in some cases. Hence, it is important to understand the basic principles of their operations since you will no doubt be using these instruments several times in various laboratory experiments.

The basic component of both ammeters and voltmeters is the **galvanometer.** The galvanometer is an electromagnetic device capable of detecting very small electrical currents. In this experiment, the characteristics of a galvanometer and the basic construction of the ammeter and voltmeter will be investigated.

II. EQUIPMENT NEEDED*

- Galvanometer
- Standard decade resistance box (0.1 to 99.9 Ω)
- Voltmeter (3 V)
- Rheostat (3500 Ω)
- Resistors (e.g., 1.5 kΩ, 3.0 kΩ, and 15 kΩ, composition type)
- Power supply or battery (6 V)
- Single-pole, single-throw switch
- 1 sheet of Cartesian graph paper

III. THEORY

The basic design of a moving-coil galvanometer is shown in Fig. 29-1 (sometimes called a *D'Arsonval galvanometer,* after the French physicist who invented it around 1882). It consists of a coil of wire on a iron core that is pivoted on bearings between the poles of a permanent magnet. When a current passes through the coil, it experiences a torque and rotates, moving a dial pointer. A balancing countertorque is supplied by control springs. When the coil reaches an equilibrium position, the two opposing torques are equal, and the deflection of the pointer is proportional to the current in the coil.

The scale of a galvanometer is commonly marked with intervals on both sides of a central zero. When the coil current is in one direction, the pointer needle is deflected to the right. If the polarity, and hence the current, are reversed, the needle is deflected to the left. The galvanometer is capable of detecting currents in the microampere (μA) range, and the scale gradations give relative magnitudes of the current.

For absolute current values, the current sensitivity of a specific instrument must be known. The current sensitivity is usually expressed in microamperes per scale division. The number of scale divisions n indicated by the pointer deflection is proportional to the current I_g in the galvanometer coil:

$$I_g \propto n$$

or

$$I_g = kn \qquad \textbf{(29-1)}$$

*The ranges of the equipment are given as examples. These may be varied to apply to available equipment.

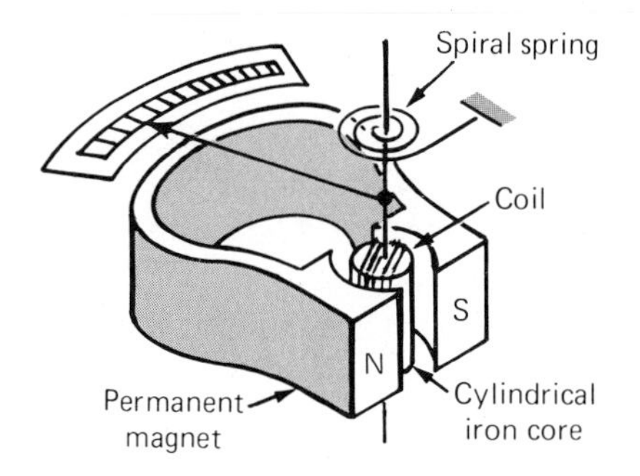

Fig. 29-1 Basic design of a moving-coil (D'Arsonval) galvanometer.

where k is the current sensitivity in microamperes per scale division, $k = I_g/n$. As can be seen from the equation, the smaller k, the greater the sensitivity of the galvanometer (greater deflection for a given current).

☐ **Example 29.1** What current is required for full-scale deflection (50 divisions) of a galvanometer having a current sensitivity of 40 μA per scale division?

Solution With $k = 40\ \mu\text{A/div}$ and $n = 50$ div, the required current is

$$I_g = kn = (40\ \mu\text{A/div})(50\ \text{div}) = 2000\ \mu\text{A} = 2.0 \times 10^{-3}\ \text{A}$$ ☐

The galvanometer coil has a resistance r and the current in the coil is also given by Ohm's law:

$$I_g = \frac{V}{r} \qquad \textbf{(29-2)}$$

where V is the voltage across the galvanometer. As such, a galvanometer could be calibrated in microamps and used as a microammeter. However, this meter would not be useful in circuits in which the current exceeded the microamp range (e.g., 2 mA or 2000 μA in the preceding example). Using such a meter in a circuit with a much larger current than that required for full-scale deflection would "peg" the needle, causing possible damage to the mechanism. Also, a large current would heat the coil as a result of the I^2r losses and would eventually burn out the meter. Thus, a different design is required for a practical ammeter capable of reading current magnitudes in the ampere range.

A. *The dc Ammeter*

To convert a galvanometer to an ammeter capable of reading currents in the ampere range, a small "shunt" resistance R_s is placed in parallel with the galvanometer [Fig. 29-2(a)]. This provides an alternate path whereby part of a large current I can bypass or "shunt" the galvanometer ($I = I_g + I_s$). The circuit in Fig. 29-2(a) constitutes an ammeter (circuit symbol -Ⓐ-) in series with a resistance R, and the ammeter measures the current through R.

Since the voltages across the galvanometer and the shunt resistor are equal, we have

$$V_g = V_s$$

or by Ohm's law,

$$I_g r = I_s R_s$$

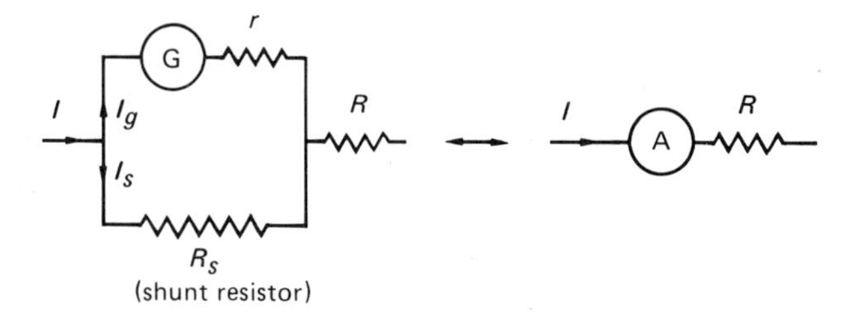

(a) Ammeter

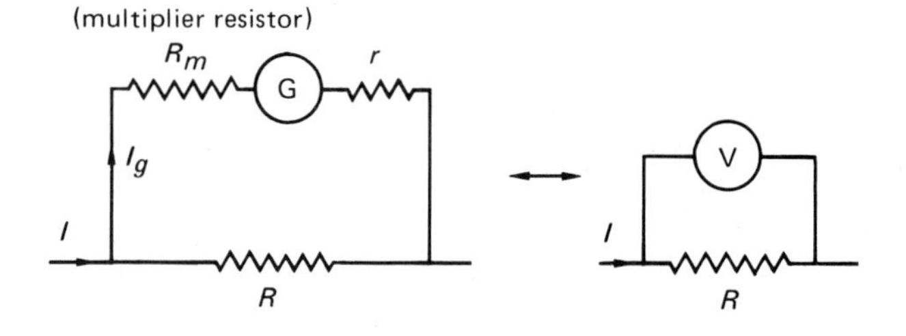

(b) Voltmeter

Fig. 29-2 Circuit diagrams for an ammeter and voltmeter. (a) An ammeter has a small "shunt" resistance R_s in parallel with the galvanometer. (b) A voltmeter has a large multiplier resistance R_m in series with the galvanometer.

Then, in terms of the total current $I = I_g + I_s$,

$$I_g r = (I - I_g)R_s$$

and

$$I_g = \frac{R_s I}{r + R_s} \quad \textbf{(29-3)}$$

Since by Eq. 29-1, $I_g = kn$, we have

$$I_g = kn = \frac{R_s I}{r + R_s} \quad \textbf{(29-4)}$$

☐ **Example 29.2** Suppose that the galvanometer in Example 29.1 has a coil resistance of $r = 100$ Ω and we wish to construct an ammeter with a full-scale reading I_{max} of 3 A. What would be the required value of the shunt resistor?

Solution Given $r = 100\ \Omega$, $I_{max} = 3$ A, and $I_g = 2000\ \mu A = 2.0 \times 10^{-3}$A (from Example 29.1). Then using the general relationship $I_g r = (I - I_g)R_s$,

$$R_s = \frac{I_g r}{I_{max} - I_g} = \frac{(2.0 \times 10^{-3})(100)}{3.000 - 0.002}$$

$$= \frac{0.20}{2.998} = 0.067\ \Omega$$

☐

The resistance of the ammeter, or the total equivalent resistance of the parallel galvanometer and R_s branch, is very small and usually may be considered negligible relative to resistance R. Hence, to a good approximation, the total current I in the circuit is given by

$$I = \frac{V}{R} \quad \textbf{(29-5)}$$

Note: **An ammeter is always connected in line or in series with a circuit component to measure the current flowing through the component. Being a low resistance instrument, if an ammeter were connected in parallel with a circuit component having appreciable resistance, the meter would carry most of the current and could burn out (or hopefully blow a meter fuse).**

B. *The dc Voltmeter*

To convert a galvanometer to a voltmeter that is capable of reading voltages in excess of the microvolt range, a large multiplier resistance R_m is placed in series with the galvanometer [Fig. 29-2(b)]. In this arrangement the voltage drop across the galvanometer branch is $V = V_m + V_g$, and most of the voltage drop across the voltmeter is across the multiplier resistance and not the galvanometer coil. The circuit in Fig. 29-2 constitutes a voltmeter (circuit symbol -Ⓥ-) in parallel with resistance R. Because of the large internal resistance of a voltmeter, it draws little current from the main circuit.

Then, by Ohm's law,

$$V = V_g + V_m$$

$$= I_g r + I_g R_m = I_g(r + R_m)$$

or

$$V = kn(r + R_m) \quad \textbf{(29-6)}$$

which is also equal to the voltage drop across R, since the galvanometer branch and R are in parallel. Hence, by varying the applied voltage, the galvanometer scale can be calibrated in terms of voltage (instead of current).

☐ **Example 29.3** Suppose that we wish to calibrate the galvanometer used in the previous examples as a voltmeter with a full-scale reading V_{max} of 3.0 V. What would be the required value of the series multiplier resistor?

Solution As given previously: $r = 100\ \Omega$ and $I_g = kn = 2.0 \times 10^{-3}$ A. Then, with $V_{max} = 3.0$ V, from Eq. 29-6,

$$R_m = \frac{V_{max} - I_g r}{I_g}$$

$$= \frac{3.0 - (2 \times 10^{-3})(100)}{2.0 \times 10^{-3}} = 1400\ \Omega$$

☐

Note: **A voltmeter is always connected "across" or in parallel with a circuit component to measure the potential difference or voltage drop across the component. Being a relatively high resistance instrument, if a voltmeter were connected in series with a circuit component, it would reduce the current in the circuit (and the voltage drop across the component).**

From the preceding discussion, it should be evident that the critical parameters in calibrating a galvanometer as an ammeter or voltmeter are its current sensitivity k (or its full-scale current $I_g = kn$) and coil resistance r. When these quantities are known, the appropriate resistances for full-scale deflection magnitudes I_{max} and V_{max} can be found from Eqs. 29-4 and 29-6, respectively. Lower values of current and voltage on these scales are directly proportional to the scale divisions n.

IV. EXPERIMENTAL PROCEDURE

1. Examine the resistors. If composition type, the colored bands conform to a color code that gives the resistance value of a resistor. Look up the color code in Appendix A, Table A5, and read and record the value of each resistor. Designate the smallest resistance as R_1 and the consecutively larger values as R_2 and R_m.

A. *Ammeter*

2. Set up the ammeter circuit as shown in Fig. 29-3. Use the decade resistance box as R_s. Set the rheostat resistance R_h at its maximum value. *Have the instructor check the circuit before activating.*

3. After the circuit has been approved, close the switch S, and while watching the meter closely so it does not go off scale, reduce R_h to its minimum value. Next, set R_s so that the galvanometer has a full-scale deflection. Record R_s, the voltmeter reading, and the number of divisions n for full-scale deflection in Data Table 1.

4. Then increase the rheostat resistance R_h for a series of decreasing voltage steps ($\approx$0.5 V) as read on the voltmeter. (Fine adjustments may be made with the power supply if adjustable.) Do not change the value of R_s. Read and record the voltage and the galvanometer scale deflection n for each step. Open the switch after completing the procedure.

5. Compute the current I in the circuit for each voltage setting.

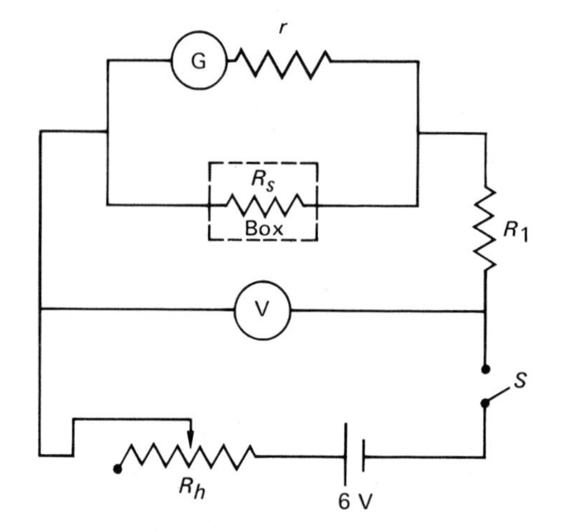

Fig. 29-3 Ammeter circuit for experimental procedure.

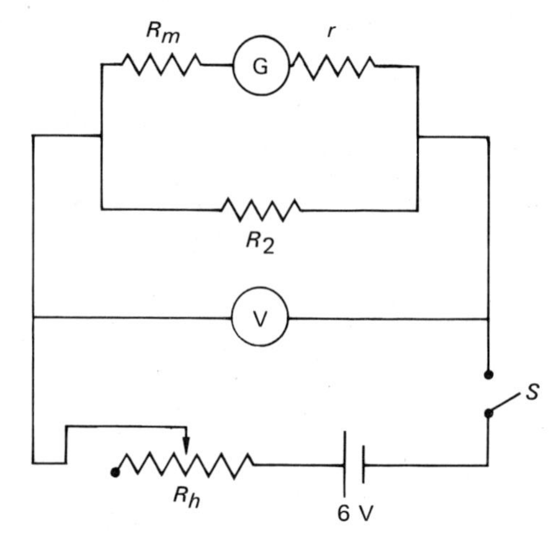

Fig. 29-4 Voltmeter circuit for experimental procedure.

B. *Voltmeter*

6. Set up the voltmeter circuit as shown in Fig. 29-4. Set the rheostat at its maximum resistance. *Have the instructor check the circuit before activating.*

7. After the circuit has been approved, close the switch and record the voltmeter reading and the scale deflection n of the galvanometer in Data Table 2.

8. Then decrease the rheostat resistance R_h for a series of increasing voltage steps ($\approx$0.25 V) as read on the voltmeter. Read and record the voltage and galvanometer scale deflection n for each step. Open the switch after completing the procedure.

9. Plot V versus n and draw the straight line that best fits the data. Determine the slope of the line and record in the data table. Note from Eq. 29-6 that the slope is equal to $k(r + R_m)$. However, since $R_m \gg r$, to a good approximation, slope $\approx kR_m$. Use this approximate formula to compute k.

10. With the determined value of k, calculate the internal resistance r of the galvanometer for three sets of values of I and n from Data Table 1 using Eq. 29-4. Find the average of these values of r and record in the data table.

Name .... Section .... Date ....

Lab Partner(s) ....

EXPERIMENT 29 *The Ammeter and the Voltmeter: Meter Sensitivity*

LABORATORY REPORT

Resistor values: R_1 .... R_m ....

R_s .... R_2 ....

DATA TABLE 1

Voltage, V (V)	Galvanometer deflection, n	$I = V/R_1$ (A)
	(full scale)	

DATA TABLE 2

Voltage, V (V)	Galvanometer deflection, n

Calculations (show work)

Slope (from B) .... $\approx kR_m$ r (average) ....

(continued)

QUESTIONS

1. If the galvanometer used in the experiment is to be incorporated in an ammeter with full-scale readings of (a) 1 A, (b) 3 A, and (c) 30 A, what would be the values of the required shunt resistances?

2. Draw the circuit for a multirange ammeter with the ranges in Question 1.

3. If the galvanometer used in the experiment is to be incorporated in a voltmeter with full-scale readings of (a) 1 V, (b) 3 V, and (c) 30 V, what would be the values of the required multiplier resistances?

4. Draw the circuit for a multirange voltmeter with the ranges in Question 3.

5. The sensitivity of a voltmeter is commonly expressed in *ohms per volt* (Ω/V), which is the total resistance of the meter ($r + R_m$) divided by the full-scale voltage reading. Show that the sensitivity of a voltmeter in Ω/V is related to the current sensitivity of its galvanometer by $\Omega/V = 1/kn$ and compute the Ω/V sensitivity of the experimental voltmeter in the experiment.

Name .... Section .... Date ....

Lab Partner(s) ....

EXPERIMENT 30 *The Measurement of Resistance*

ADVANCE STUDY ASSIGNMENT

Read the experiment and answer the following questions.

A. *Ammeter–Voltmeter Method*

1. When measuring a resistance with an ammeter and voltmeter arrangement, is the resistance given exactly by $R = V/I$? Explain.

2. Comment on the relative magnitudes of the resistances of an ammeter and a voltmeter.

3. Are an ammeter and a voltmeter connected in a series or parallel with a circuit component (a resistance)? Explain.

B. *Wheatstone Bridge Method*

4. Why is the Wheatstone bridge called a "null" instrument?

(continued)

5. When the galvanometer in a Wheatstone bridge circuit shows no deflection, why are the voltages across opposite branches on each side of the galvanometer necessarily equal?

6. For a slide-wire Wheatstone bridge, why should the sliding key not be moved with the key depressed?

EXPERIMENT **30**

The Measurement of Resistance

I. INTRODUCTION

The magnitude of a resistance can be measured by several methods. One common method is to measure the voltage drop V across a resistance in circuit with a voltmeter and the current I through the resistance with an ammeter. By Ohm's law, then, $R = V/I$. However, the ratio of the measured voltage and current does not give the exact value of the resistance, since the resistances of the meters must also be taken into account.

This problem is eliminated when one measures a resistance, or, more properly, compares a resistance with a standard resistance in a Wheatstone bridge circuit [named after Sir Charles Wheatstone (1802–1875)]. In this experiment, the ammeter–voltmeter and the Wheatstone bridge methods of measuring resistances will be investigated.

II. EQUIPMENT NEEDED*

A. *Ammeter–Voltmeter Method*

- Ammeter (0 to 0.5 A)
- Voltmeter (0 to 3 V)
- Rheostat (10 Ω)
- Resistance: decade box or resistors (9.9 Ω and 499 Ω)
- Battery or power supply (3 V)
- Connecting wires

B. *Wheatstone Bridge Method*

- Slide-wire Wheatstone bridge
- Galvanometer
- Standard decade resistance box (0.1 to 99.9 Ω)
- Single-pole, single-throw switch

III. THEORY

A. *Ammeter–Voltmeter Method*

There are two basic arrangements by which resistance is measured with an ammeter and a voltmeter. One circuit is shown in Fig. 30-1. The current I through the resistance R is measured with an ammeter and the potential difference or voltage drop V across the resistance is measured with a voltmeter. Then, by Ohm's law, $R = V/I$.

Strictly, however, this value of the resistance is not altogether correct, since the current registered on the ammeter divides between the resistance R and the voltmeter in parallel. A voltmeter is a high-resistance instrument and draws relatively little current provided that voltmeter resistance R_v is much greater than R (see Experiment 29). Hence, it is more appropriate to write

$$R \simeq \frac{V}{I} \quad \text{if} \quad R_v >> R \qquad \textbf{(30-1)}$$

For more accurate resistance measurement, one must take the resistance of the voltmeter into account. The current drawn by the voltmeter is $I_v = V/R_v$. Since the total current I divides be-

*The ranges of the equipment are given as examples. These may be varied to apply to available equipment.

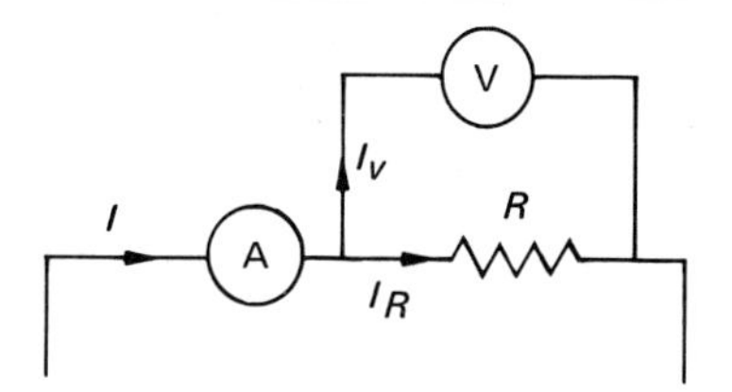

Fig. 30-1 One of the basic current arrangements for measuring resistance with an ammeter and voltmeter. The ammeter measures the sum of the currents through the resistance and voltmeter. Therefore, the true value of R is *greater* than the measured value, if R_{meas} is taken to be V/I.

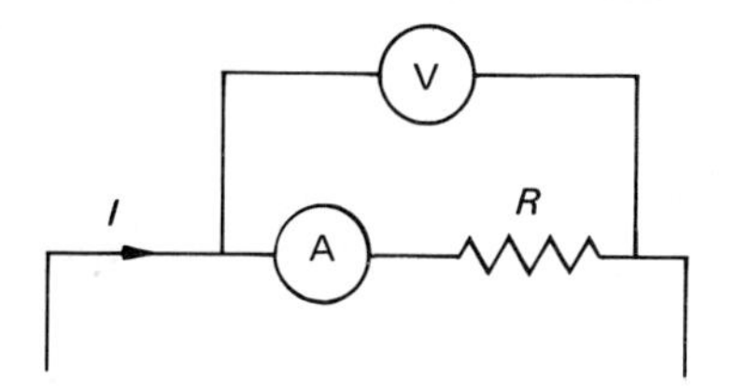

Fig. 30-2 Another basic circuit arrangement for measuring resistance with an ammeter and voltmeter. The ammeter measures the current through R, but the voltmeter measures the voltage across R and the ammeter. Therefore, the true value of R is *less* than the measured value, if R_{meas} is taken to be V/I.

tween the resistance and the voltmeter in the parallel branch, we have

$$I = I_R + I_v$$

or

$$I_R = I - I_v \tag{30-2}$$

where I_R is the true current through the resistance. Then, by Ohm's law,

$$R = \frac{V}{I_R} = \frac{V}{I - I_v} = \frac{V}{I - V/R_v} \tag{30-3}$$

Another possible arrangement for measuring R is shown in the circuit diagram in Fig. 30-2. In this case, the ammeter measures the current through R alone, but now the voltmeter reads the voltage drop across *both* the ammeter and the resistance. Since the ammeter is a low-resistance instrument (Experiment 29), to a good approximation

$$R \simeq \frac{V}{I} \quad \text{if} \quad R_a << R \tag{30-4}$$

where R_a is the resistance of the ammeter. When $R_a << R$, the voltage drop across R_a, that is, $V_a = IR_a$, is small compared to that across R, which is $V_R = IR$.

Taking the voltage drop or the resistance of the ammeter into account, we have

$$V = V_R + V_a = IR + IR_a$$

$$= I(R + R_a) = IR' \tag{30-5}$$

and $R' = R + R_a$. Or, from Eq. 30-5, we may write

$$R = \frac{V}{I} - R_a$$

B. *Wheatstone Bridge Method*

The basic diagram of a Wheatstone bridge circuit is shown in Fig. 30-3. In its simplest form, the bridge circuit consists of four resistors, a battery or voltage source, and a sensitive galvanometer. The values of R_1, R_2, and R_s are all known, and R_x is the unknown resistance. The bridge is "balanced" by adjusting the standard resistance R_s until the galvanometer shows no deflection (indicating no current flow through the galvanometer branch). As a result, the Wheatstone bridge is called a "null" instrument. This is analogous to an ordinary double-pan beam balance, which shows a null or zero reading when there are equal masses on its pans.

Assume that the Wheatstone bridge is balanced so that the galvanometer registers no current. Then points b and c in the circuit are at the same potential; and current I_1 flows through both R_s and R_x and current I_2 flows through both R_1 and R_2. Also, the voltage drop across R_s, V_{ab}, is equal to the voltage drop across R_1, V_{ac}, for a zero galvanometer deflection:

$$V_{ab} = V_{ac}$$

Similarly,

$$V_{bd} = V_{cd} \tag{30-6}$$

Why?

Writing these equations in terms of currents and resistances by Ohm's law,

$$I_1R_x = I_2R_2$$

$$I_1R_s = I_2R_1 \tag{30-7}$$

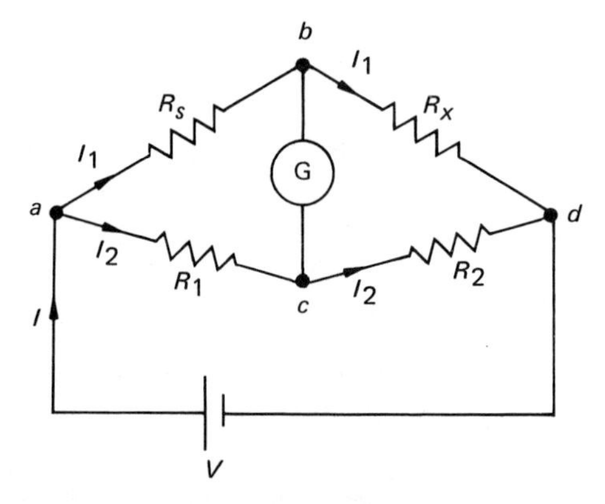

Fig. 30-3 Basic Wheatstone bridge circuit diagram.

Then dividing one equation by the other and solving for R_x yields

$$R_x = \left(\frac{R_2}{R_1}\right) R_s \tag{30-8}$$

Hence, when the bridge is balanced, the unknown resistance R_x can be found in terms of the standard resistance R_s and the ratio R_2/R_1. Notice that the difficulties of the preceding method are eliminated. The Wheatstone bridge in effect compares the unknown resistance R with a standard resistance R_s. Should $R_1 = R_2$, then $R_x = R_s$.

The circuit diagram for a slide-wire form of the Wheatstone bridge is shown in Fig. 30-4 along with a photo of an actual bridge. The line from a to d represents a wire and C is a contact key that slides along the wire so as to divide the wire into different-length segments. The resistances of the segments are proportional to their lengths, so the resistance ratio may be expressed in terms of a length ratio

$$\frac{R_2}{R_1} = \frac{L_2}{L_1} \tag{30-9}$$

Equation 30-8 can then be written in terms of the length ratio

$$R_x = \left(\frac{L_2}{L_1}\right) R_s \tag{30-10}$$

This type of bridge is convenient since the length segments can be easily measured. The resistances R_1 and R_2 of the length segments may be quite small relative to R_x and R_s because the bridge formula depends only on the ratio of R_2/R_1 or L_2/L_1. This fact allows a wire to be used as one side of the bridge.

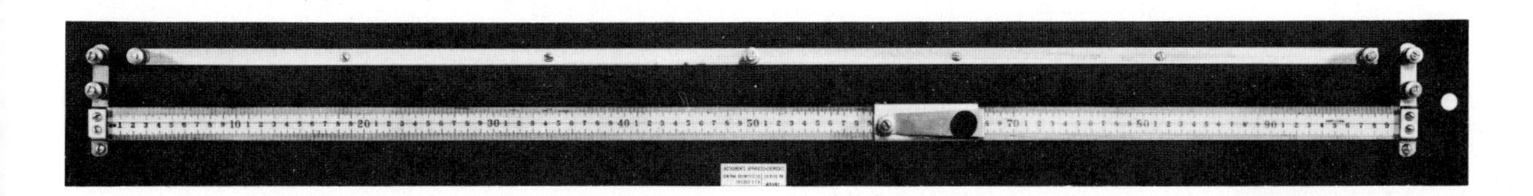

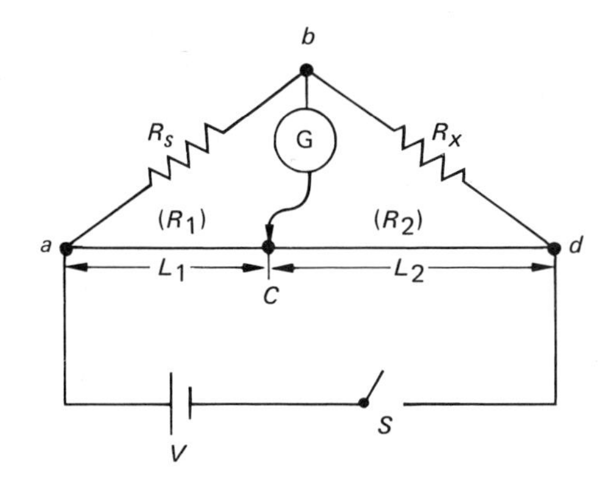

Fig. 30-4 Slide-wire Wheatstone bridge and circuit diagram for resistance measurement. (Photo courtesy of Central Scientific Co., Inc.)

IV. EXPERIMENTAL PROCEDURE

A. *Ammeter–Voltmeter Method*

1. Set up a circuit as shown in Fig. 30-5(a) where R is a small known resistance and R_h is the rheostat. Record the value of R in the data table. Most common meters have three scale connections, with a binding post common to all three scales. It is good practice to initially make connections to the largest scales. This prevents the instruments from being "pegged" and possibly damaged should the magnitudes of the current and voltage exceed the smaller scale limits. The scale setting may be changed for greater sensitivity by moving the connection to a lower scale after the general magnitude of a measurement is known.

   Also, attention should be given to the proper polarity (+ and −). Otherwise, the meter will be "pegged" in the opposite direction. Connect + to + and − to −. However, *do not activate the circuit until your laboratory instructor has checked it.*

2. The current in the circuit is varied by varying the rheostat resistance R_h. Activate the circuit and take three different readings of the ammeter

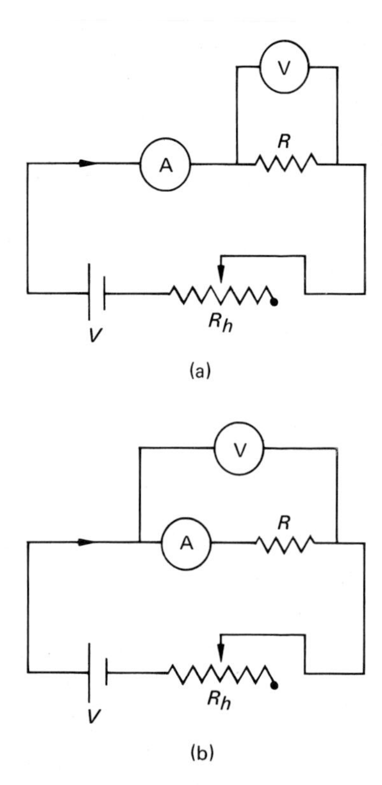

Fig. 30-5 Circuit diagrams for experimental procedure for ammeter–voltmeter method of measuring resistance.

and voltmeter for three different currents. Adjust R_h so that the three currents differ as much as possible. Record the data in Data Table 1 and deactivate the circuit after each of the three readings until the rheostat is set for the next reading.

Also, record the resistance of the voltmeter. The resistance of the meter will be found on the meter face or will be supplied by the instructor. The voltmeter resistance is commonly given as so many ohms per volts, which is the total resistance of the meter divided by the full-scale reading. For example, if the meter has a resistance of 1000 Ω/V and the full-scale reading of a particular range is 3 V, then $R_v = 3\ \text{V}(1000\ \Omega/\text{V}) = 3000\ \Omega$. The resistance in ohms/volt applies to any range setting of the meter.

3. Using Eq. 30-3, compute the value of R for each current setting and find the average value. Compare this with the accepted value by finding the percent error.

4. Set up a circuit as shown in Fig. 30-5(b). This is accomplished by changing only one wire in the previous circuit. Repeat the measurements as in procedure 2 for this circuit.

5. (a) Compute the resistance $R' = V/I$ directly from each set of current and voltage measurements, and find the average value.
   (b) When not taking into account the ammeter resistance, R' would be taken to be the value of the resistance R. Compare the average experimental value of R' with the accepted value of R by finding the percent error.
   (c) Using the values of R and R', compute R_a (Eq. 30-5). Mentally compare the magnitudes of the ammeter and voltmeter resistances.

6. Repeat the previous procedures with a large known resistance.

B. *Wheatstone Bridge Method*

7. Set up a slide-wire Wheatstone bridge circuit as in Fig. 30-4 using the previous small known resistance R as R_x. The wires connecting the resistances and the bridge should be as short as practically possible. The decade resistance box is used for R_s. This should be initially set for a value about equal to R_x. Contact is made to the wire by sliding contact key C. *Do not slide the key along the wire while it is pressed down.* This will scrape the wire, causing it to be nonuniform. Have the instructor check your setup before activating the circuit.

8. Activate the circuit by closing the switch S, and balance the bridge by moving the slide-wire contact. Open the switch and record R_s, L_1, and L_2 in the data table.

9. Repeat procedure 7 for R_s settings of (a) $R_s \approx 10\ R_x$, and (b) $R_s \approx 0.1\ R_x$.

10. Compute the value of R_x for each case and find the average value. Compare this value to the accepted value of R by finding the percent error.

11. Repeat the previous procedures with a large known resistance.

Name .... Section .... Date ....

Lab Partner(s) ....

EXPERIMENT 30 *The Measurement of Resistance*

LABORATORY REPORT

A. *Ammeter–Voltmeter Method*

Accepted value of R ....

DATA TABLE 1

Rheostat setting, R_h	Current, I (A)	Voltage, V (V)	Resistance, R (Ω)
1			
2			
3			
		Average R	

Voltmeter resistance, R_v ....

Percent error of R ....

DATA TABLE 2

Rheostat setting, R_h	Current, I (A)	Voltage, V (V)	$R' = V/I$ (Ω)
1			
2			
3			
		Average R'	

Ammeter resistance, R_a ....

Percent error of R' ....

Calculations (show work)

Accepted value of R ....

DATA TABLE 3

Rheostat setting, R_h	Current, I (A)	Voltage, V (V)	Resistance, R (Ω)
1			
2			
3			
		Average R	

Voltmeter resistance, R_v ....

Percent error of R ....

DATA TABLE 4

Rheostat setting, R_h	Current, I (A)	Voltage, V (V)	$R' = V/I$ (Ω)
1			
2			
3			
		Average R'	

Ammeter resistance, R_a ....

Percent error of R' ....

Calculations (show work)

(continued)

B. *Wheatstone Bridge Method*

DATA TABLE 5

Accepted value of R ....

R_s (Ω)	L_1 (cm)	L_2 (cm)	R (Ω)
		Average R	

Percent error ....

DATA TABLE 6

Accepted value of R ....

R_s (Ω)	L_1 (cm)	L_2 (cm)	R (Ω)
		Average R	

Percent error ....

Calculations (show work)

QUESTIONS

A. *Ammeter–Voltmeter Method*

1. An ideal ammeter would have zero resistance and an ideal voltmeter an infinite resistance. Explain why we would want these ideal cases when using the meters.

2. If in general R were calculated as $R = V/I$, which circuit arrangement in part A of the experiment would have the least error? Explain.

(continued)

3. (a) Prove that the true resistance R is given by

$$R = R'\left(1 - \frac{R_a}{R'}\right)$$

where $R' = V/I$ is the measured resistance as given by the voltmeter and ammeter readings for measurements done by the arrangement in Figs. 30-2 or 30-5(b). Is the true resistance larger or smaller than the apparent resistance?

(b) Prove that the true resistance R is given approximately by

$$R = R'\left(1 + \frac{R'}{R_v}\right)$$

where $R' = V/I$ is the measured resistance as given by the voltmeter and ammeter readings for measurements done by the arrangement in Figs. 30-1 or 30-5(a). *Hint:* By the binomial theorem,

$$\frac{1}{1 - \dfrac{R'}{R_v}} \simeq 1 + \frac{R'}{R_v}$$

Is the true resistance larger or smaller than the apparent resistance?

B. *Wheatstone Bridge Method*

4. Why should the wires connecting the resistances and the bridge be as short as possible?

5. Suppose that the slide-wire on the bridge did not have a uniform cross section. How would this affect your measurements? Was there any experimental evidence of this?

Name .... Section .... Date ....

Lab Partner(s) ....

EXPERIMENT 31 *Resistivity*

ADVANCE STUDY ASSIGNMENT

Read the experiment and answer the following questions.

1. What are the factors affecting the resistance of an electrical conductor?

2. If the length and diameter of a wire conductor were both doubled, would the resistance be the same? Explain.

3. Why is resistivity called a material property?

4. In the experiment, why is it important to close the circuit switch only long enough to obtain meter readings?

EXPERIMENT 31

Resistivity

I. INTRODUCTION

The resistance of an electrical conductor depends on several factors. The physical shape is one factor. The type of conductor material is another, as might be expected. That is, two conductors with the same physical shape, but of different materials, have different resistances. This important material characteristic of resistance is expressed in terms of a quantity called **resistivity.**

Temperature is another factor affecting resistance. However, the temperature dependence of resistance will be investigated in another experiment (Experiment 32). In this experiment, the factors of shape or dimensions and resistivity will be considered.

II. EQUIPMENT NEEDED*

- Ammeter (0 to 0.5 A)
- Voltmeter (0 to 3 V)
- Rheostat (20 Ω)
- Single-pole, single-throw switch
- Battery or power supply (3 V)
- Meter stick
- Micrometer calipers
- Conductor board with wires of various types, lengths, and diameters.† For example, three copper wires of different lengths (1.0, 0.75, and 0.5 m) with the same diameters (No. 30).
- Two copper wires 1.0 m long with different diameters (No. 28 and No. 24).
- Two wires of different materials (constantan and Nichrome) 1.0 m long with the same diameter (No. 30).

III. THEORY

The resistance of an electrical conductor depends on several factors. Consider a wire conductor. The resistance, of course, depends on the *type* of conductor material, and also on (a) the length, (b) the cross-sectional area, and (c) the temperature of the wire. As might be expected the *resistance* of a wire conductor is directly proportional to its length ℓ and inversely proportional to its cross-sectional area A:

$$R \propto \frac{\ell}{A}$$

For example, a 4-m length of wire has twice as much resistance as a 2-m length of the same wire. Also, the larger the cross-sectional area, the greater the current flow (less resistance) for a given voltage. These geometrical conditions are analogous to those for liquid flow in a pipe. The longer the pipe, the more resistance to flow; but the larger the cross-sectional area of the pipe, the smaller the resistance to flow.

The material property of resistance is characterized by the **resistivity** ρ, and for a given temperature,

*The ranges of the equipment are given as examples. These may be varied to apply to available equipment.

†If resistance spools or coils are used, measurements should be made by the Wheatstone bridge method (see Experiment 30).

$$R = \frac{\rho \ell}{A} \qquad \textbf{(31-1)}$$

The resistivity is independent of the shape of the conductor, and rearranging Eq. 31-1,

$$\rho = \frac{RA}{\ell} \qquad \textbf{(31-2)}$$

From the equation, resistivity can be seen to have the units Ω-m or Ω-cm. Common metal conductors have resistivities on the order of 10^{-6} Ω-cm. Another name sometimes used for resistivity is *specific resistance,* indicating that it is specific for a given material.

To determine the resistivities of some materials, a circuit arrangement as illustrated in Fig. 31-1 will be used. The ammeter Ⓐ measures the current I in a wire conductor on the conductor board and the voltmeter Ⓥ registers the voltage drop V across the conductor. Then, the resistance of the wire, by Ohm's law, is $R = V/I$.

Measuring the length ℓ of the wire and its cross-sectional area A (from diameter d measurement, $d/2 = r$ and $A = \pi r^2$), the resistivity of the conductor can be calculated from Eq. 31-2. The rheostat R_h is used to initially limit the current in the circuit so as to protect the meters.

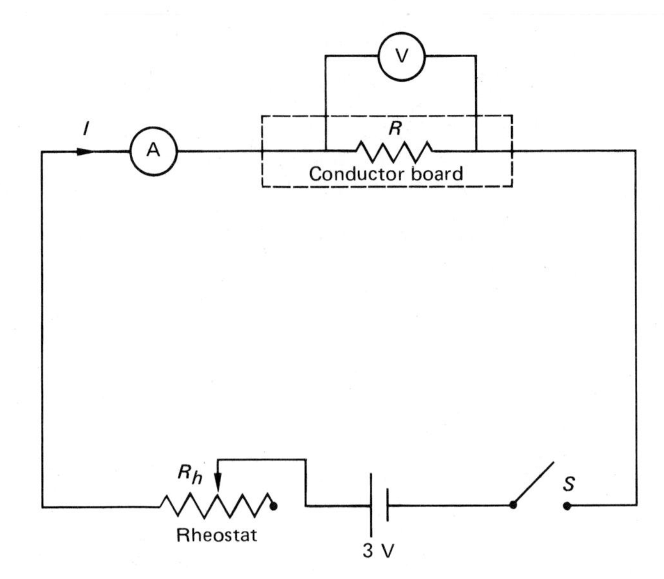

Fig. 31-1 Circuit diagram for experimental procedure to measure resistivity.

IV. EXPERIMENTAL PROCEDURE

1. Set up the circuit shown in Fig. 31-1 with one of the wires on the conductor board in the circuit. Leave the switch S open and set the rheostat at maximum resistance. *Have the instructor check the circuit before activating.**
2. After the circuit has been checked, close the switch and adjust the rheostat until the current in the circuit as indicated on the ammeter is 0.5 A. Read and record the meter values and open the switch as soon as possible to prevent heating and temperature change.
3. Repeat this procedure for all of the other wire conductors with the rheostat initially set at its maximum resistance.
4. Measure the lengths and diameters of the wires and record in the data table.
5. (a) Compute the resistances and cross-sectional areas of the wires, and use these values to determine the resistivities of the materials.
   (b) Find the average resistivity for each material with more than one experimental value.
6. Compare the experimental values of the resistivities with the accepted values listed in Appendix A, Table A6, by computing the percent error.

*For best results, the voltmeter should make contact with the resistance wire (R) about 5 cm *in* from the terminals, not *at* the terminals. Also, the copper wire on the conductor board should be a single piece zig-zagging back and forth across the board so that readings can be taken with various lengths of this wire (e.g., 1, 2, 3, etc., meters).

Name .... Section .... Date ....

Lab Partner(s) ....

EXPERIMENT 31 *Resistivity*

LABORATORY REPORT

DATA TABLE

Wire	Type of material	Voltage, V (V)	Current, I (A)	Resistance, $R = V/I$ (Ω)	Length (cm)	Diameter (cm)	Cross-sectional area (cm^2)	Resistivity (Ω-cm)
1								
2								
3								
4								
5								
6								
7								

Type of material	Accepted value	Expt. value	Percent error
....	....	....	....
....	....	....	....
....	....	....	....

Calculations (show work)

(continued)

QUESTIONS

1. Do the experimental data confirm that the resistance of a conductor is (a) directly proportional to its length and (b) inversely proportional to its cross-sectional area? Support your answer either graphically or mathematically with experimental results.

2. How does the resistance of a wire vary with resistivity?

3. An annealed copper wire (No. 15 AWG gauge) is to be replaced with an aluminum wire with approximately the same length and resistance. What gauge of aluminum wire would be required? (See Appendix A, Table A7, for wire gauges.)

Name .... Section .... Date ....

Lab Partner(s) ....

EXPERIMENT 32 *The Temperature Dependence of Resistance*

ADVANCE STUDY ASSIGNMENT

Read the experiment and answer the following questions.

1. Does the resistance of all substances increase with temperature? Explain.

2. What is the temperature coefficient of resistance, and what are its units?

3. Distinguish between a positive and a negative temperature coefficient of resistance.

4. Are the α of a metal conductor and the β of a thermistor the same? Explain.

5. What are the circuit conditions when a Wheatstone bridge is "balanced"?

EXPERIMENT 32

The Temperature Dependence of Resistance

I. INTRODUCTION

The electrical resistance of all substances varies somewhat with temperature. For pure metals and most alloys, the resistance increases with increasing temperature. However, for some substances, such as carbon and many electrolytes (conducting solutions), the resistance decreases with increasing temperature. Then, too, for some special alloys [e.g., constantan (55 percent Cu–45 percent Ni)] the resistance is virtually independent of temperature over a limited range.

The temperature dependence of resistance for a substance is commonly expressed in terms of its temperature coefficient of resistance, which is the fractional change in the resistance per degree change in temperature. For many electrical applications, it is important to know the temperature coefficient and to take into account the temperature dependence of resistance. In this experiment, this dependence will be investigated and the temperature coefficients of some materials will be determined.

II. EQUIPMENT NEEDED

- Slide-wire Wheatstone bridge assembly (with a 3-V battery and a single-pole, single-throw switch)
- Standard decade resistance box (and an optional standard resistance R_s', depending on the resistance of the decade box)
- Copper coil (and optional constantan or manganese coil)
- Thermistor
- Immersion vessel and stirrer
- Thermometer
- Immersion heater and power source (or Bunsen burner and stand)
- 2 sheets of Cartesian graph paper

III. THEORY

The change in resistance ΔR of a substance is proportional to the change in temperature ΔT. This change in resistance is commonly expressed in terms of the fractional change $\Delta R/R_0$, where R_0 is the initial resistance. For many substances (e.g., metals) the change in resistance is a linear function of temperature, or

$$\frac{\Delta R}{R_0} = \alpha \Delta T \qquad \textbf{(32-1)}$$

where the constant of proportionality α is called the temperature coefficient of resistance and has the units of inverse temperature (1/°C or °C^{-1}).

For the change in temperature $\Delta T = T - T_0$, it is convenient to take as 0°C the initial temperature T_0, and with $\Delta R = R - R_0$, Eq. 32-1 can be written

$$\frac{R - R_0}{R_0} = \alpha T$$

or

$$R = R_0 + R_0 \alpha T = R_0(1 + \alpha T) \qquad \textbf{(32-2)}$$

where R is then the resistance of the conductor at some temperature T (°C) above 0°C, and R_0 is the resistance at $T_0 = 0$°C. The linearity of the temperature dependence is only approximate, but Eq. 32-2 can be used over moderate temperature ranges for all but the most accurate work.

In contrast to pure metals, which have relatively small positive temperature coefficients of resistance (increase in resistance with increase in temperature), thermistors have large negative temperature coefficients (decrease in resistance with an increase in temperature). A thermistor is a thermally sensitive resistor made of semiconducting materials such as oxides of manganese, nickel, and cobalt. Because of the large (negative) temperature coefficients, thermistors are very sensitive to small temperature changes and are used for temperature measurement and in a variety of temperature-sensing applications such as voltage regulation and time-delay switches.

Unlike common metal conductors, the change of resistance with a change of temperature for a thermistor is nonlinear, and the α in Eq. 32-1 is not constant. The temperature dependence of a thermistor is given by an exponential function.

$$R = R_a e^{\beta(1/T - 1/T_a)} \quad \textbf{(32-3)}$$

where R = resistance at a temperature T (in kelvins, K)
R_a = resistance at temperature T_a (K)
T_a = initial temperature (K), near ambient room temperature in the experiment
e = 2.718, the base of natural logarithms
β = *exponential* temperature coefficient of resistance, which has temperature units (K)

In this case, as T increases, the exponential function and hence the resistance R becomes smaller. This expression can be written in terms of the natural logarithm (to the base e) as

$$\ln\left(\frac{R}{R_a}\right) = \beta\left(\frac{1}{T} - \frac{1}{T_a}\right) \quad \textbf{(32-4)}$$

Hence, by plotting $y = \ln (R/R_a)$ versus $x = (1/T - 1/T_a)$ on a Cartesian graph, β is the slope of the line. This, too, is an approximation, but β is reasonably constant for moderate temperature ranges.

The temperature coefficient of resistance of a material can be determined by using an experimental arrangement with a slide-wire Wheatstone bridge circuit, as illustrated in Fig. 32-1. The resistance R_c of a material (coil of wire) when the bridge circuit is balanced is given by

$$R_c = \left(\frac{R_2}{R_1}\right) R_s \quad \text{or} \quad R_c = \left(\frac{L_2}{L_1}\right) R_s \quad \textbf{(32-5)}$$

where R_s is a standard resistance and R_2/R_1 and L_2/L_1 are the ratios of the resistances and lengths of the slide-wire segments, respectively. (See Experiment 30 for the theory of the Wheatstone bridge.) By measuring the resistance of a material at various temperatures, the temperature coefficient can be determined.

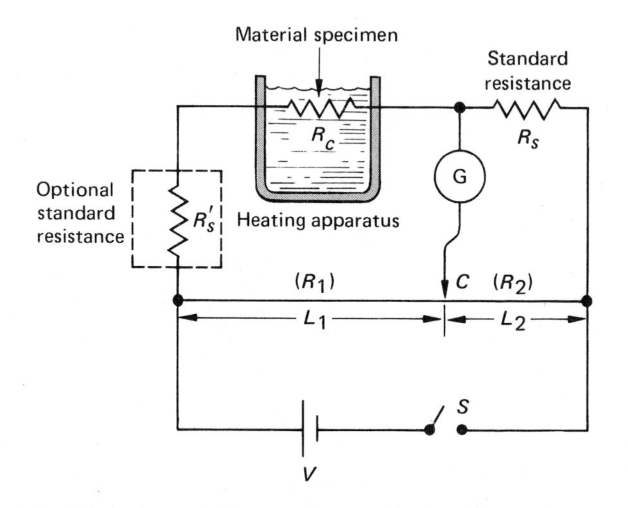

Fig. 32-1 Circuit diagram for experimental procedure to measure the temperature dependence of resistance.

IV. EXPERIMENTAL PROCEDURE

A. *Metal Conductor(s)*

1. Set up the circuit as in Fig. 32-1 with the copper coil in the container of water (near room temperature) and the heating arrangement for the water (immersion heater or Bunsen burner). Place the thermometer in the water. *Have the instructor check your setup.*

2. After your setup has been checked, close the switch and balance the bridge circuit to measure the resistance R_c of the coil at the initial water temperature. The value of the standard resistance R_s should be selected so that the bridge is balanced with the contact key C near the center of the slide-wire. Then with $L_1 \simeq L_2$, we have $R_c \simeq R_s$ (Eq. 32-5). (Because of the small resis-

tance of the copper coil, it may be necessary to place a standard resistance R_s' in series with the coil. This resistance is not placed in the water. Then, the resistance of the coil R_c in terms of the measured *total* resistance R of this bridge arm is $R = R_c + R_s'$, and Eq. 32-5 must be adjusted accordingly.

Record the initial temperature of the water, the magnitude of R_s, and the lengths of the wire segments of the bridge in Data Table 1.

3. *Slowly,* raise the temperature of the water by about 10°C. Stir the water while heating and discontinue heating when the temperature is about 2 degrees below the desired temperature. Continue stirring until a maximum steady temperature is reached. Balance the bridge and record the measurements. Adjust R_s if necessary. Record the temperature and bridge length measurements in the data table.

4. Repeat procedure 3, taking a series of measurements at approximately 10°C temperature intervals until a final temperature of about 90°C is reached.

5. *(Optional)* Repeat the foregoing procedures using the constantan wire coil starting near room temperature. (Data Table 1A.)

6. Compute R_c of the coil(s) at the various temperatures and plot a graph of R_c versus T with a temperature range of 0 to 100°C. Draw the straight line(s) that best fits the data and extrapolate the line(s) to the y axis. Determine the slope and y intercept of the line(s).

   From the slope, find the temperature coefficient of resistance for the specimen(s) and compare with the accepted value found in Appendix A, Table A6, by computing the percent error.

B. *Thermistor*

7. Replace the coil with the thermistor in the bridge circuit and repeat the previous measurement procedures starting at a temperature near room temperature. In this portion of the experiment, exercise great care in order to have temperatures as constant as possible when making resistance measurements, since a thermistor shows considerable variation in resistance with temperature.

8. (a) Find the quantities listed in the second part of Data Table 2. Logarithm values may be found in Appendix B, Table B8.
   (b) Plot a graph of $y = \ln (R/R_a)$ versus $x = (1/T - 1/T_a)\ \mathrm{K}^{-1}$, and draw the straight line that best fits the data.
   (c) Determine the slope of the line, which is the value of β. Compare this to the accepted value provided by the instructor by computing the percent error.

Name .... Section .... Date ....

Lab Partner(s) ....

EXPERIMENT 32 *The Temperature Dependence of Resistance*

LABORATORY REPORT

A. *Metal Conductor*

DATA TABLE 1

Material ....

R'_s (if used) ....

Temperature (°C)	Decade box resistance, R_s (Ω)	L_1 (cm)	L_2 (cm)	Coil resistance, $R = (L_2/L_1)R_s$ (Ω)	$R_c = (L_2/L_1)R_s - R'_s$

Calculations (show work)

Slope, $R_0\alpha$ ....

Intercept, R_0 ....

Experimental α ....

Accepted α ....

Percent error ....

Name .... Section .... Date ....

Lab Partner(s) ....

EXPERIMENT 32 *The Temperature Dependence of Resistance*

DATA TABLE 1A (OPTIONAL)

Material ....

R_s' (if used) ....

Temperature (°C)	Decade box resistance, R_s (Ω)	L_1 (cm)	L_2 (cm)	Coil resistance, R_c (Ω)	$R_c = (L_2/L_1)R_s - R_s'$

Calculations (show work)

Slope, $R_0\alpha$ ....

Intercept, R_0 ....

Experimental α ....

Accepted α ....

Percent error ....

(continued)

B. *Thermistor*

DATA TABLE 2

Temperature, T(°C)	Decade box resistance, R_s (Ω)	L_1 (cm)	L_2 (cm)	Thermistor resistance, $R = (L_2/L_1)R_s$ (Ω)
(T_a)				(R_a)

Temperature (K) $T_K = T_C + 273$	$1/T$	$1/T - 1/T_a$	R/R_a	$\ln (R/R_a)$

Calculations (show work)

Slope, β ....

Accepted α ....

Percent error ....

QUESTIONS

A. *Metal Conductor*

1. What is the value of α for copper in terms of degrees Fahrenheit (1/°F)? Would the resistance be a linear function of temperature on this scale? Explain.

2. Replot the copper data R_c versus T, with a smaller temperature scale extending to −300°C and extrapolate the line to the temperature axis. At what temperature would the resistance go to zero? What are the practical electrical implications for a conductor with zero resistance. [It is interesting to note that the value for many pure metals is roughly the same, approximately $\frac{1}{273}$ or 0.004°C^{-1}. This is the same value of the coefficient of expansion of an ideal gas. Also, in cryogenic (very low temperature) experiments, some metals and alloys do become "superconductors," or have zero resistance near absolute zero.]

3. A coil of copper wire has a resistance of 10.0 Ω and a coil of silver wire has a resistance of 10.1 Ω at 0°C. At what temperature would the resistances of the coils be equal?

B. *Thermistor*

4. Explain why the ambient temperature for T_a for the thermistor cannot be taken as $T_a = 0$°C and the expression for the resistance written $R = R_0 e^{\beta/T}$, where T is in degrees Celsius.

(continued)

5. Assuming that β remained constant, what would be the resistance of the thermistor in the experiment as the temperature approaches 0K?

6. Assume the temperature coefficient of resistance α to be defined over the temperature range $\Delta T = T - T_a$, where $T_a > 273$K (0°C), by $R - R_a = R_a\alpha(T - T_a)$. Show that for a thermistor, α is a function of temperature given by

$$\alpha = \frac{1 - e^{\beta(1/T - 1/T_a)}}{T - T_a}$$

Name .... Section .... Date ....

Lab Partner(s) ....

EXPERIMENT 33 *Resistances in Series and Parallel*

ADVANCE STUDY ASSIGNMENT

Read the experiment and answer the following questions.

1. Explain the difference between series and parallel connections.

2. When resistors are connected in series in a circuit, what are the relationships between the voltage drops across the resistors and the currents through the resistors?

3. When resistors are connected in parallel in a circuit, what are the relationships between the voltage drops across the resistors and the currents through the resistors?

4. Give (draw and explain) a liquid analogy of the series–parallel circuit in part C of the experiment.

5. How would the current divide in a parallel branch of a circuit containing two resistors R_1 and R_2 if (a)$R_1 = R_2$; and (b)$R_1 = 4R_2$?

EXPERIMENT **33**

Resistances in Series and Parallel

I. INTRODUCTION

The circuit components of simple circuits are connected in series and/or parallel arrangements. Each component may be represented as a resistance to the flow of current in the circuit. In computing the voltage and current requirements of the circuit (or part of the circuit), it is necessary to know the equivalent resistances of the series and parallel arrangements.

In this experiment, the circuit characteristics of resistors in series and parallel will be investigated. A particular circuit will first be analyzed theoretically and then the theoretical predictions will be checked experimentally.

II. EQUIPMENT NEEDED*

- Battery or power supply (3 V)
- Ammeter (0 to 500 mA)
- Voltmeter (0 to 3 V)
- Single-pole, single-throw switch
- Four resistors (10 Ω, 20 Ω, 100 Ω, and 10 kΩ, composition type, 1 W)
- Connecting wires

III. THEORY

A. *Series Resistance*

Resistors are said to be connected in **series** when they are connected as in Fig. 33-1. (The resistors are connected in line or "head to tail," so to speak, although there is no distinction between the connecting ends of a resistor.) When connected to a voltage source V and the switch closed, the source supplies a current I to the circuit. By the conservation of charge, this current I flows through each resistor. However, the voltage drop across each resistor is not equal to V, but the sum of the voltage drop is:

$$V = V_1 + V_2 + V_3 \tag{33-1}$$

In an analogous liquid-gravity circuit (Fig. 33-1), a pump, corresponding to the voltage source, raises the liquid a distance h. The liquid then falls or "drops" through three paddle-wheel "resistors" and the distances h_1, h_2, and h_3. The liquid rise supplied by the pump is equal to the sum of the liquid "drops," $h = h_1 + h_2 + h_3$. Analogously, the voltage "rise" supplied by the source is equal to the sum of the voltage drops across the resistors (Eq. 33-1).

The voltage drop across each resistor is given by Ohm's law (e.g., $V_1 = IR_1$) and Eq. 33-1 may be written

$$\begin{aligned} V &= V_1 + V_2 + V_3 \\ &= IR_1 + IR_2 + IR_3 \\ &= I(R_1 + R_2 + R_3) \end{aligned} \tag{33-2}$$

For a voltage across a single resistance R_s in a circuit, $V = IR_s$, and by comparison,

$$R_s = R_1 + R_2 + R_3 \tag{33-3}$$

* The ranges of the equipment are given as examples. These may be varied to apply to available equipment.

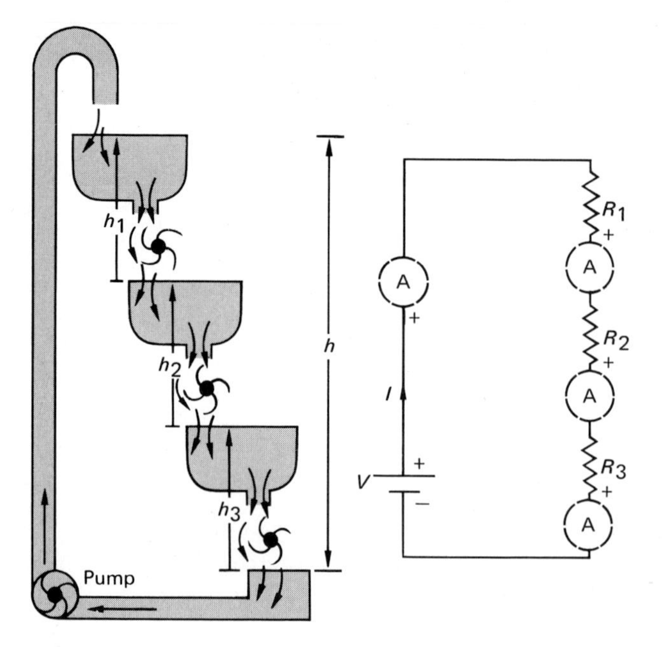

Fig. 33-1 Liquid analogy and circuit diagram for resistors in series.

where R_s is the equivalent resistance of the resistors in series. That is, the three resistors in series could be replaced by a single resistor with a value of R_s and the same current I would flow in the circuit.

B. *Parallel Resistance*

Resistors are said to be connected in **parallel** when connected as in Fig. 33-2. (In this arrangement, all the "heads" are connected together, as are all of the "tails.") The voltage drops across all the resistors are the same and equal to the voltage V of the source. However, the current I from the source divides among the resistors such that

$$I = I_1 + I_2 + I_3 \tag{33-4}$$

In the liquid circuit analogy (Fig. 33-2), the height h the pump raises the liquid is equal to the distance the liquid "drops" through each paddle-wheel "resistor." The liquid flow coming into the junction of the parallel arrangement divides among the three pipe paths, analogously to the current dividing in the electrical circuit.

The current in an electrical parallel circuit divides according to the magnitudes of the resistances in the parallel branches—the smaller the resistance of a given branch, the greater the current through that branch. The current through each resistor is given by Ohm's law (e.g., $I_1 = V/R_1$) and Eq. 33-4 may be written

$$I = I_1 + I_2 + I_3 = \frac{V}{R_1} + \frac{V}{R_2} + \frac{V}{R_3}$$
$$= V\left(\frac{1}{R_1} + \frac{1}{R_2} + \frac{1}{R_3}\right) \tag{33-5}$$

For the current through a single resistance R_p in a circuit, $I = V/R_p$, and by comparison

$$\frac{1}{R_p} = \frac{1}{R_1} + \frac{1}{R_2} + \frac{1}{R_3} \tag{33-6}$$

where R_p is the equivalent resistance of the resistors in parallel. That is, the three resistors in parallel could be replaced by a single resistor with a value of R_p and the same current I would flow in the circuit.

The previous developments for equivalent resistances may be extended to any number of resistors (i.e., $R_s = R_1 + R_2 + R_3 + R_4 + \cdots$ and $1/R_1 + 1/R_2 + 1/R_3 + 1/R_4 + \cdots$). In many instances, two resistors are connected in parallel in a circuit, and

$$\frac{1}{R_p} = \frac{1}{R_1} + \frac{1}{R_2}$$

or

$$R_p = \frac{R_1 R_2}{R_1 + R_2} \tag{33-7}$$

This particular form of R_p for two resistors may be more convenient for calculations than the reciprocal form. Also, in a circuit with three resistors in parallel, the equivalent resistance of two of the resistors can be found by Eq. 33-7, and then the equation may be applied again to the equivalent resistance and the other resistance in parallel to find the total equivalent resistance of the three parallel resistors. However, if your calculator has a $1/x$ function, the reciprocal form is easier to use.

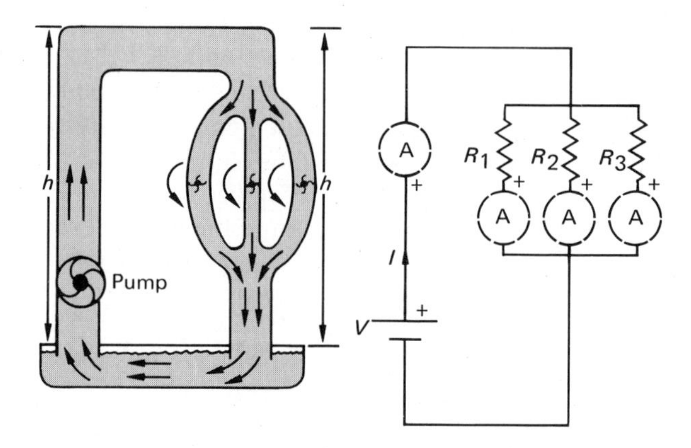

Fig. 33-2 Liquid analogy and circuit diagram for resistors in parallel.

Consider the circuit in Fig. 33-3. To find the equivalent resistance of this series–parallel circuit, one first "collapses" the parallel branch into a single equivalent resistance, which is given by Eq. 33-7. This equivalent resistance is in series with R_1 and the total equivalent resistance R of the circuit is $R = R_1 + R_p$.

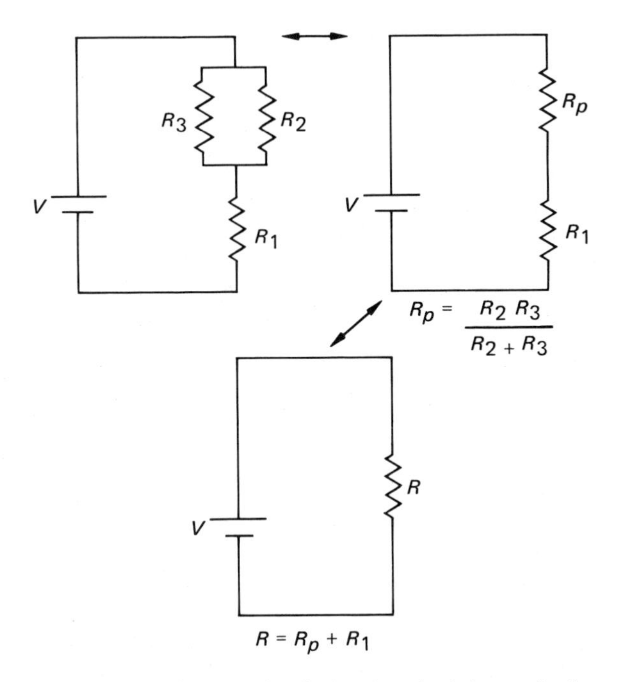

Fig. 33-3 Equivalent circuits used to find the equivalent resistance of a series–parallel circuit.

Note that the voltage drops across R_1 and R_2 are the same, and by Ohm's law,

$$I_1R_1 = I_2R_2$$

or

$$\frac{I_1}{I_2} = \frac{R_2}{R_1} \quad \textbf{(33-8)}$$

Thus, the ratio of the resistances gives the relative magnitudes of the currents in the resistors.

☐ **Example 33.1** Given two resistors, R_1 and R_2, with $R_2 = 2R_1$ in parallel in a circuit. What fraction of the current I from the voltage source goes through each resistor?

Solution With $R_2 = 2R_1$, or $R_2/R_1 = 2$, by Eq. 33-8

$$I_1 = \left(\frac{R_2}{R_1}\right)I_2 = 2I_2$$

Since $I = I_1 + I_2$, we have

$$I = I_1 + I_2 = 2I_2 + I_2 = 3I_2$$

or

$$I_2 = \frac{I}{3}$$

Hence, the current divides such that one-third goes through R_2 and two-thirds goes through R_1. ☐

IV. EXPERIMENTAL PROCEDURE

1. Examine the resistors. The colored bands conform to a color code that gives the value of a resistor. Look up the color code in Appendix A, Table A5 and read and record the value of each resistor. Designate the smallest resistance as R_1, and consecutively larger values as R_2, R_3, and R_4.

2. In the following procedures, you will be asked to theoretically compute various quantities for a given circuit arrangement. The quantities are then determined by actual circuit measurements and the calculated and experimental results compared. Before initially activating each circuit arrangement, *have the circuit checked by the instructor, unless otherwise instructed.*

A. *Resistors in Series*

3. If R_1, R_2, and R_3 are connected in a circuit in series as in Fig. 33-1, (a) what is the current flowing in the circuit; (b) what is the voltage drop across each resistor? Show your calculations in the Laboratory Report. Ask your instructor for the voltage of the source if this is not known.

4. Set up the actual circuit with a switch and the ammeter in the circuit next to the current source. A convenient way to check a circuit to see if it is properly connected is to trace the path of current flow (with your finger) through the circuit. Do this for the circuit under considera-

tion to make sure that the current will flow through each circuit component in series. Remember, an ammeter is *always* connected in series, and for proper polarity, + connected to +.

Close the switch (after having the circuit checked by the instructor) and read and record the ammeter value of the current (I) from the voltage source. (If using a variable power supply, adjust voltage if required, to 3.0 volts with circuit closed. Otherwise, the voltage drop in the power supply may yield significantly less than 3 V to the external circuit.)

Open the switch and move the ammeter in the circuit to a position "after" a resistor [i.e., on the opposite side of the resistor from the voltage source so as to measure the current flowing through (from) the resistor]. The ammeter positions are shown in Fig. 33-1. Carry out this procedure for each resistor and record the currents in the data table. Leave the switch closed only while readings are being taken.

5. Remove the ammeter from the circuit, and with the voltmeter, measure and record the voltage drop across each resistor and across all three resistors as a group. Remember, a voltmeter is *always* connected in parallel or "across" a circuit element to measure its voltage drop.

6. Compare the experimentally measured values with the theoretically computed values by finding the percent error. (Use the theoretical values as the accepted values.)

B. *Resistors in Parallel*

7. If R_1, R_2, and R_3 are connected in a circuit in parallel as in Fig. 33-2, (a) how much current (I) is supplied by the source; (b) what is the current flowing through each resistor? Show your calculations in the Laboratory Report.

8. Set up the actual circuit. Check the circuit arrangement by tracing the current from the source through the circuit to see that it divides into three parallel branches at the junction of the resistors and comes together again at the opposite junction. Then measure and record the voltage drops across each resistor and across all three resistors as a group. As before and throughout the remaining procedures, leave the switch closed only while readings are being taken.

9. Remove the voltmeter and connect the ammeter between the source and resistors so as to measure the current I supplied by the source. Measure and record this current. Then measure the current through each resistor by connecting the meter between a given resistor and one of the common junctions. The ammeter positions are shown in Fig. 33-2.

10. Compare the theoretical and experimental values by computing the percent errors.

11. *(Optional)* Repeat procedures 7 to 10 with R_2 replaced by R_4.

C. *Resistors in Series–Parallel*

12. If R_1 is connected in series with R_2 and R_3 in parallel (Fig. 33-3):
    (a) How much current is supplied by the source?
    (b) What is the voltage drop across R_1?
    (c) What is the voltage drop across R_2 and R_3?
    (d) What is the voltage drop across all three resistors?
    (e) What are the currents through R_2 and R_3?

13. Set up the actual circuit and trace the current flow to check the circuit. With the voltmeter and ammeter, measure and record the calculated quantities. You need not compute the percent errors in this case. However, make a mental comparison to satisfy yourself that the measured quantities agree with the computed values within experimental error.

Name .... Section .... Date ....

Lab Partner(s) ....

EXPERIMENT 33 *Resistances in Series and Parallel*

LABORATORY REPORT

Resistor values R_1 .... R_3 ....

R_2 .... R_4 ....

A. *Resistors in Series*

Calculations (show work)

Current I ....

Voltage drops across resistors

V_1 ....

V_2 ....

V_3 ....

Experimental measurements

				Percent error
I	....			....
I_1	....	V_1	....	....
I_2	....	V_2	....	....
I_3	....	V_3	....	....
		V	....	

B. *Resistors in Parallel*

Calculations (show work)

Current I ....

Current through resistors

I_1 ....

I_2 ....

I_3 ....

(continued)

Experimental measurements

				Percent error
		I	....	....
V_1	....	I_1	....	....
V_2	....	I_2	....	....
V_3	....	I_3	....	....

(Optional Procedure)

Calculations (show work)

Current I ....

Current through resistors

I_1 ....

I_2 ....

I_3 ....

Experimental measurements

				Percent error
		I	....	....
V_2	....	I_2	....	....
V_3	....	I_3	....	....

C. *Resistors in Series–Parallel*

Calculations (show work)

Current I ....

Voltage drops

V_1 ....

$V_2 = V_3$ ....

Currents

I_2 ....

I_3 ....

Experimental measurements

I ....

V_1 ....

$V_2 = V_3$ ....

I_2 ....

I_3 ....

Name .... Section .... Date ....

Lab Partner(s) ....

EXPERIMENT 33 *Resistances in Series and Parallel*

QUESTIONS

1. Discuss the sources of error in the experiment.

2. Suppose that the resistors in the various circuit diagrams represented the resistances of light bulbs in such circuits. When a light bulb "burns out," the circuit is open through that particular component, i.e., R is infinite. What would happen to the rest of the bulbs if:

   (a) R_2 burned out in the circuit in part A?

   (b) R_1 burned out in the circuit in part B?

   (c) Then R_3 also burned out in the circuit in part B?

   (d) R_3 burned out in the circuit in part C?

   (e) Then R_1 also burned out in the circuit in part C?

(continued)

3. Explain the effect of replacing R_2 with R_4 in procedure 11. (Explain theoretically even if procedure 11 of the experiment was not done.)

4. Given the four resistors in the experiment, how many possible different resistance values could be obtained by using one or more of the resistors? [List the specific combinations (e.g., R_1 and R_2 in series).]

Name .... Section .... Date ....

Lab Partner(s) ....

EXPERIMENT 34 *Multiloop Circuits: Kirchhoff's Rules*

ADVANCE STUDY ASSIGNMENT

Read the experiment and answer the following questions.

1. Do Kirchhoff's rules represent any new physical principles in the sense that Ohm's law does? Explain.

2. (a) What is a junction?

   (b) Distinguish between a branch and a loop.

3. A household wiring circuit consists of a voltage source (ac power from the electric company) connected across many loads (resistances) in parallel.
   (a) Draw a circuit diagram showing the voltage source and three loads, e.g., a light bulb, a clock, and a TV set.

   (b) The diagram has several loops. Are Kirchhoff's rules needed to analyze the current through each load? Explain.

(continued)

4. (a) The direction one goes around a circuit loop makes no difference in the loop theorem equation obtained for the loop. Show this explicitly by going around the loops for the circuit in Example 34.1 in the opposite directions.

(b) Apply the loop theorem to loop 3 in Fig. 34-1 and show that it is redundant or unnecessary with the inside loop equations (Example 34.1).

EXPERIMENT **34***

Multiloop Circuits: Kirchhoff's Rules

I. INTRODUCTION

The analysis of electrical circuits is the first step toward understanding their operation. By "analysis" is meant the process of calculating how the electrical currents in a circuit depend on the values of the voltage sources (or vice versa). For our discussion, we will refer to any voltage source as a battery with a terminal or "operating" voltage V.

Many electrical circuits can be analyzed by using nothing more than Ohm's law (Experiment 27). The simplest situation consists of one battery (V) and one resistor (R) connected in a single closed loop (Fig. 27-1). In this case, $I = V/R$. Several batteries may appear in series, or several resistors (Experiment 33) may appear in series. In these cases, a combination can be represented by a single equivalent element, and then Ohm's law can be used. Situations that look more complicated include the potentiometer (Fig. 28-3) and the Wheatstone bridge (Fig. 30-3). These circuits contain more than one loop. However, the loops are independent, that is, there is no current flow between the loops when the bridge is balanced. So even these circuits can be analyzed with Ohm's law in this condition.

The more general electrical circuit contains several loops, with batteries and currents shared among the loops. Such a general circuit cannot be analyzed directly by using Ohm's law; however, it can be analyzed using Kirchhoff's rules, or laws as they are sometimes called, [named after Gustav Kirchhoff (1824–1887), the German physicist who developed them]. In this experiment, we will investigate, use, and verify Kirchoff's rules in analyzing multiloop circuits.

II. EQUIPMENT NEEDED

- Ammeter (0 to 10/100/1000 mA)
- Voltmeter (0 to 5/25 V)
- Two batteries or voltage supplies (6 V and 12 V)
- Two single-pole, single-throw switches
- Composition resistors, 2-watt rating (100 Ω, 150 Ω, 220 Ω, 330 Ω, 470 Ω, 680 Ω, 1000 Ω)
- Connecting wires

Note: Items may be varied to apply to available equipment.

III. THEORY

A simple multiloop circuit is shown in Fig. 34-1, which will be used to illustrate the principles of Kirchhoff's rules and the terminology involved. Terminology definitions vary among textbooks, even though the principles remain the same. Therefore, it is important to define terms carefully as they will be used.

A **junction** is a point in a circuit at which three or more connecting wires are joined together, or a point where the current divides or comes together in

*Contributed in part by Professor I. L. Fischer, Bergen Community College.

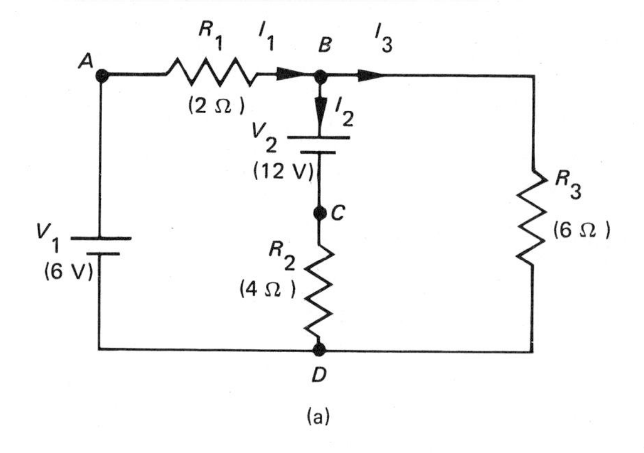

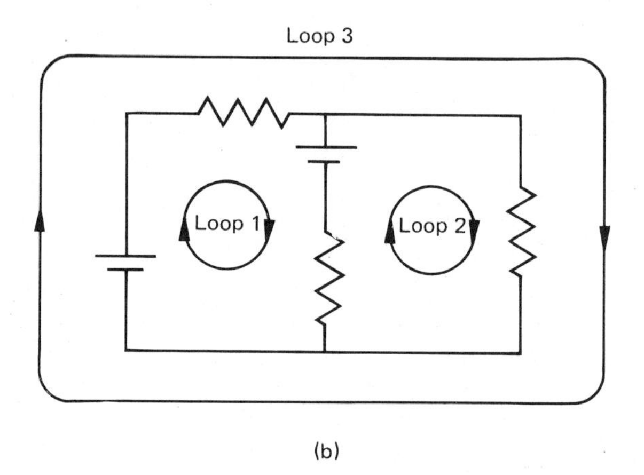

Fig. 34-1 (a) By Kirchhoff's junction theorem, the sum of the currents at a junction is zero, i.e., the current in equals the current out, or $I_1 = I_2 + I_3$ at junction B. (b) The circuit has three loops about which the sum of the voltage changes is zero (Kirchhoff's loop theorem).

a circuit. For example, points B and D in Fig. 34-1(a) are junctions. A **branch** is a path connecting two junctions, and it may contain one element, or two or more elements. In Fig. 34-1(a) there are three branches connecting junctions B and D. These are the left branch BAD, the center branch BCD, and the right branch BD.

A loop is a closed path of two or more branches. There are three loops in the circuit in Fig. 34-1 as shown in diagram (b)—two inside loops (loops 1 and 2) and one outside loop (loop 3). Notice that each loop in this case is a closed path of two branches.

Kirchhoff's Rules

These rules do not represent any new physical principles. They embody two fundamental conservation laws: conservation of electrical charge and conservation of (electrical) energy.

A current flows in each branch of a circuit. In Fig. 34-1(a) these are labeled as I_1, I_2, and I_3. At a junction, by the conservation of electrical charge, the current (or currents) into a junction is equal to the currents leaving the junction. For example, in Fig. 34-1(a),

$$\underset{\text{(current in}}{I_1} \underset{=}{=} \underset{\text{current out)}}{I_2 + I_3}$$

By the conservation of electrical charge, this means that charge cannot "pile up" or "vanish" at a junction. This current equation may be written

$$I_1 - I_2 - I_3 = 0 \qquad \textbf{(34-1)}$$

Of course, we do not generally know whether or not a particular current flows into or out of a junction by looking at a multiloop circuit diagram. We simply assign labels and assume the directions of the branch current flows at a particular junction. If these assumptions are wrong, we will soon find out from the mathematics, as will be shown in a following example. Notice that once the branch current directions are assigned at one junction, the currents at a common branch junction are fixed; for example in Fig. 34-1(a) at junction D, $I_2 + I_3 = I_1$ (current(s) in = current out).

Equation 34-1 may be written in mathematical notation as

$$\sum_i I_i = 0 \qquad \textbf{(34-1)}$$

which is a mathematical statement of **Kirchhoff's first rule or junction theorem:**

The algebraic sum of the currents at any junction is zero.

In a simple single-loop circuit as in Experiment 27, Fig. 27-1, it is easy to see that by the conservation of energy the voltage "drop" across the resistor must be equal to the voltage "rise" of the battery,* i.e.,

$$V_{\text{battery}} = V_{\text{resistor}}$$

where the voltage drop across the resistor is by Ohm's law equal to IR, i.e., $V_{\text{resistor}} = IR$. The conservation law holds for any loop in a multiloop circuit, although in this case there may be more than one battery and more than one resistor in a particular loop. Similar to the summation of the currents in the first rule, we may write for the voltages **Kirchhoff's second rule or loop theorem:**

$$\sum_i V_i = 0 \qquad \textbf{(34-2)}$$

*Here we take the terminal or "operating" voltage of the battery instead of the emf ($\mathscr{E}$). The terminal or "operating" voltage of the battery is $V = \mathscr{E} - Ir$, where r is the internal resistance of the battery (not usually known) and Ir is the internal voltage "drop" of the battery (see Experiment 28).

or

The algebraic sum of the voltage changes around a closed loop is zero.

Since one may go around a circuit loop in either a clockwise or counterclockwise direction, it is important to establish a sign convention for voltage changes. For example, if we went around a loop in one direction and crossed a resistor, this might be a voltage drop (depending on the current flow). However, if we went around the loop in the opposite direction, we would have a voltage "rise" in terms of potential.

We will use the convention illustrated in Fig. 34-2. The voltage change of a battery is taken as positive when the battery's "positive" terminal is in the direction of the motion (a voltage "rise"), and negative if the battery's "negative" terminal is in the direction of the motion when traversed in going around a loop. *Note* that the assigned branch currents have nothing to do with determining the voltage change of a battery, only the direction one goes around a loop or through a battery.

The voltage change across a resistor, on the other hand, involves the direction of the assigned current through the resistor. The voltage change across a resistor is taken to be negative if it is traversed in the direction of the assigned current (a voltage "drop"), and positive if traversed in the opposite direction.

The sign convention allows you to go around a loop either clockwise or counterclockwise at your choice. Going in opposite directions merely makes all the signs opposite and Eq. 34-2 is the same.

Kirchhoff's rules may be used in circuit analysis in several ways. We will consider two methods.

Branch (Current) Method

First, label a current for each branch in the circuit. This is done by a current arrowhead, which also indicates the current direction, and is most conveniently done at a junction as in Fig. 34-1 at junction B. Kirchhoff's first rule applies at any junction. Remember, the current directions are arbitrary, but there must be at least one current in and one current out. Why? Then, draw loops so that every branch is in at least one loop. This is again shown for the circuit in Fig. 34-1, which has three loops. Again, the direction of a loop is arbitrary because of our sign convention.

With this done, current equations are written for each junction according to Kirchhoff's junction theorem (rule 1) that gives different equations. In general, this gives a set of equations that includes all branch currents. For the simple circuit in Fig. 34-1,

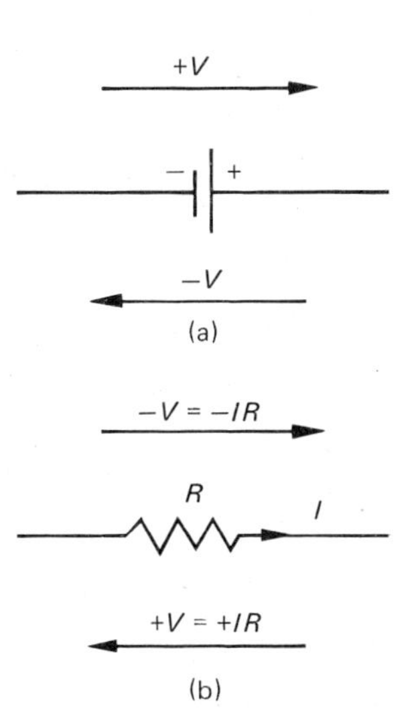

Fig. 34-2 Sign convention for Kirchhoff's rules. (a) When traversing a loop and passing through a battery, the voltage change is taken to be positive when the battery's positive terminal is in the direction of motion, and negative when the motion is in the direction of the battery's negative terminal. (b) When passing through a resistor, the voltage change across the resistor is taken to be negative if it is traversed in the direction of the assigned current flow (a voltage "drop"), and positive if traversed in the opposite direction.

this is one equation, since the sum of the currents at junction D is the same as that at junction B.

Then, Kirchhoff's loop theorem (rule 2) is applied to the circuit loops. This gives additional equations that form a set of N equations with N unknowns, which can be solved for the unknowns. There may be more loops than necessary. Only the number of loops that include all branches is needed.

To illustrate this method, the circuit in Fig. 34-1 is analyzed in the following example.

☐ **Example 34.1** Apply Kirchhoff's rules and the above sign convention to the circuit shown in Fig. 34-1 and find the value of the current in each branch.

Solution By rule 1 (junction theorem),

$$I_1 = I_2 + I_3 \tag{34-3}$$

with directions as assigned in the figure.

Going around loop 1 as indicated in the figure, with Kirchhoff's second rule (loop theorem) and our sign convention, we have starting at battery 1,

$$V_1 - I_1R_1 - V_2 - I_2R_2 = 0$$

or with known values

$$6 - I_1(2) - 12 - I_2(4) = 0$$

and
$$I_1 + 2I_2 = -3 \tag{34-4}$$

Similarly, around loop 2 starting at battery 2,

$$V_2 - I_3R_3 + I_2R_2 = 0$$

or
$$12 - I_3(6) + I_2(4) = 0$$

and
$$3I_3 - 2I_2 = 6 \qquad \textbf{(34-5)}$$

Equations 34-3, 34-4, and 34-5 constitute a set of three equations with three unknowns from which the values of I_1, I_2, and I_3 can be found. Solving these equations for the currents, one obtains

$$I_1 = -\tfrac{3}{11}\text{ A}$$
$$I_2 = -\tfrac{15}{11}\text{ A}$$
$$I_3 = \tfrac{12}{11}\text{ A}$$

The negative values of I_1 and I_2 indicate that the wrong directions were assumed for these currents. In the actual circuit, I_2 would flow into junction B, and I_1 would flow out of the junction as well as I_3 as assumed. Hence, if we had guessed correctly, we would have written for junction B,

$$I_2 = I_1 + I_3$$

$$\underset{\text{(current in)}}{\tfrac{15}{11}\text{ A}} = \underset{\text{(current out)}}{\tfrac{3}{11}\text{ A} + \tfrac{12}{11}\text{ A}}$$

and $\Sigma I_i = 0$ as required by rule 1.

In looking at Fig. 34-1 more carefully, one might have surmised this. Battery 2 (12 V) has twice the voltage as battery 1 (6 V) and it would have been a good guess that (conventional) current would flow out of battery 2 toward junction B. If battery 1 were a rechargable battery, it would be recharging in the circuit. Why?

Notice that loop 3 was not used to solve the problem. This loop would have provided another redundant equation with the other two loop equations. However, loop 3 could have been used with one of the other loops to solve the problem. □

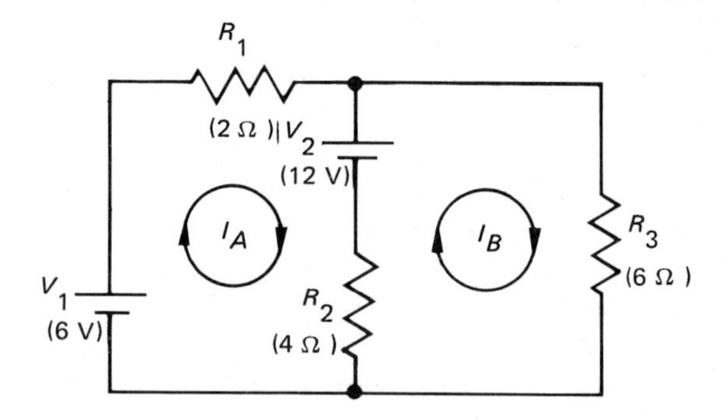

Fig. 34-3 Loop method. In this method, the loops are taken as "current loops" with a particular current assigned to each loop.

Loop (Current) Method

This method is similar to the previous branch method, but some consider it to be simpler mathematically. Loops are drawn as before so that every branch is in at least one loop. Then, each loop is labeled as if it were a current as shown in Fig. 34-3 for the circuit in Fig. 34-1. The "current loop" direction is arbitrary as in the previous method for loop directions.

Kirchhoff's loop theorem (rule 2) is then used to write an equation for each loop applying the sign convention. For the circuit in Fig. 34-3, there are two loop equations for the "loop currents" I_A and I_B. These are starting at V_1 and V_2, respectively,

$$\text{(loop } A\text{)} \quad V_1 - I_AR_1 - V_2 - I_AR_2 + I_BR_2 = 0$$
$$\text{(loop } B\text{)} \quad V_2 - I_BR_3 - I_BR_2 + I_AR_2 = 0$$

Note that you have to take into account all the "loop currents" that affect each loop. That is, when going around a loop and for a branch resistor through which two "loop currents" flow, the voltage changes for both currents must be considered for the resistor as for R_2 in this case.

Traversing the current loops produces two equations for this circuit with two unknowns, I_A and I_B. (Note that in the branch method analysis in Example 34.1, the loop equations had three unknowns.) Putting in the known values, the equations may be solved simultaneously to find the value of each unknown:

$$\text{(loop } A\text{)} \quad 6 - I_A(2) - 12 - I_A(4) + I_B(4) = 0$$
$$\text{(loop } B\text{)} \quad 12 - I_B(6) - I_B(4) + I_A(4) = 0$$

which reduce to:

$$-3I_A + 2I_B = 3$$
$$2I_A - 5I_B = -6$$

Solving for I_A and I_B, we find

$$I_A = -\tfrac{3}{11}\text{ A}$$
$$I_B = \tfrac{12}{11}\text{ A}$$

The computed values of the "loop currents" may then be utilized in a straightforward way to find the actual branch currents. Kirchhoff's junction theorem is applied to the various junctions in the circuit and the branch currents are compared with the "loop currents." For example, compare the branch currents at junction B as in Fig. 34-1 and the "loop currents" in Fig. 34-3. (The branch currents may be drawn directly on the current loop diagram.) It should be evident by comparison that:

$$I_1 = I_A = -\tfrac{3}{11}\text{ A}$$
$$I_2 = I_A - I_B = -\tfrac{3}{11}\text{ A} - \tfrac{12}{11}\text{ A} = -\tfrac{15}{11}\text{ A}$$
$$I_3 = I_B = \tfrac{12}{11}\text{ A}$$

which are the same values obtained for the circuit by the branch method.

IV. EXPERIMENTAL PROCEDURE

1. Examine the resistors. The colored bands on composition resistors conform to a color code that gives the resistance value of the resistor. Look up the color code in Appendix A, Table 5A, to identify each resistor. Note that the actual resistance value may vary according to the tolerance indicated by the last band (gold $\pm 5\%$, silver $\pm 10\%$, no band $\pm 20\%$).

2. Connect the two-loop circuit as shown in Fig. 34-4. If you are using variable power supplies, adjust each power supply as closely as possible to the values specified in the figure. Leave the switches open until the circuit has been checked by the instructor.

   Note: Lay out the circuit on your table exactly as shown in the diagram. This will help prevent errors and will facilitate your measurements.

3. After your circuit has been checked, close the switches and measure the "operating" value of each battery (V_1 and V_2) by temporarily connecting the voltmeter across a battery. Record these operating values in the data table.

   Caution: To avoid damage to the voltmeter, always start with the meter on its least-sensitive scale. Increase the sensitivity of the meter only as needed for accurate measurement, and remember to return the meter to its least-sensitive scale before proceeding.

4. Temporarily open the switches. Insert the ammeter in series with one of the branches. Close the switches, measure and record the branch current, then open the switches.

   Caution: Observe the same precautions described under procedure 3 to avoid damage to the ammeter. Also, if the ammeter deflects downscale (below zero), open the switches before reversing the polarity of the meter.

5. Repeat procedure 4 for each of the branches.

6. Compute the theoretical values of each branch current for this circuit. In the analysis, use the measured values of the batteries V_1 and V_2 and the labeled values of the resistors (procedure 1). Compare the measured values of the branch currents with the computed theoretical values by finding the percent error.

7. Connect the three-loop circuit as shown in Fig. 34-5. Repeat procedures 3 through 6 for this circuit. The instructor may wish to provide you with a different circuit to investigate. In Fig. 34-5, why is there no current indicated between the two connection points on the bottom wire?

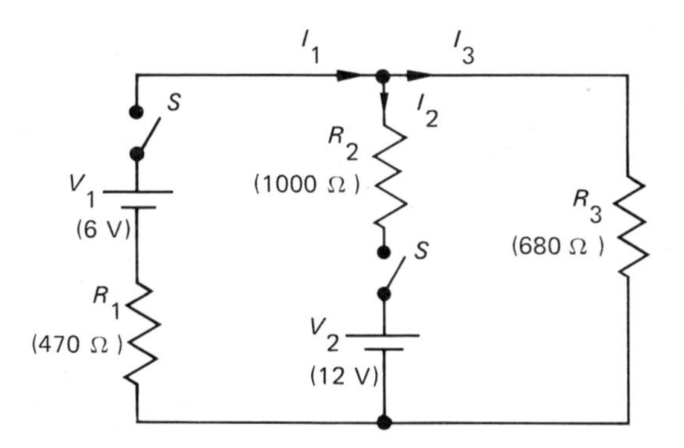

Fig. 34-4 Experimental two-loop circuit.

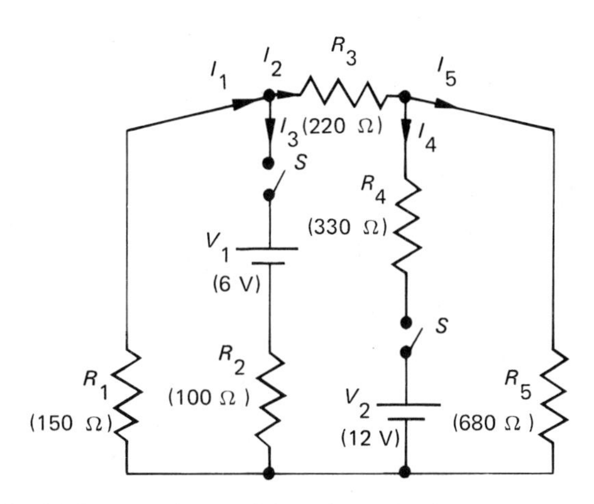

Fig. 34-5 Experimental three-loop circuit.

Name .... Section .... Date ....

Lab Partner(s) ....

LABORATORY REPORT

Two-loop circuit (Fig. 34-4)

R_1 ....

R_2 ....

R_3 ....

	Meas.	Theor.	Percent error
V_1		X	X
V_2		X	X
I_1			
I_2			
I_3			

Calculations (show work)

Three-loop circuit (Fig. 34-5)

R_1 .... R_4 ....

R_2 .... R_5 ....

R_3 ....

	Meas.	Theor.	Percent error	
V_1		X	X	volts
V_2		X	X	volts
I_1				mA
I_2				mA
I_3				mA
I_4				mA
I_5				mA

Calculations (show work)

Lab Partner(s) ....

EXPERIMENT 34 *Multiloop Circuits: Kirchhoff's Rules*

QUESTIONS

1. A typical ammeter has a resistance of 22 ohms when it is set to its 10-mA scale, a resistance of 2.7 ohms on its 100-mA scale, and a resistance of 0.7 ohms on its 1000-mA (1-A) scale.
   (a) What percent error might the ammeter resistance cause in the circuit of Fig. 34-4 when reading the current through R_3? (*Hint:* The fractional error is approximately the ratio of the meter resistance and the circuit resistance.)

   (b) How does this error compare with the resistor tolerance noted in procedure 1 of the experiment?

   (c) If all the resistors were 1/100 times the values shown in the figure, the currents would be 100 times greater. What percent error would now be introduced by the ammeter?

2. The potentiometer (Fig. 28-3) and the Wheatstone bridge (Fig. 30-3) each have several loops.
   (a) Assume the potentiometer or bridge is balanced, i.e., the galvanometer reads zero. Are Kirchhoff's rules needed to calculate the currents through the resistors? Explain. Could Kirchhoff's rules be used?

   (b) Assume the galvanometer does not read zero. Are Kirchhoff's rules now needed? Explain.

3. Draw a diagram showing the junctions and loops for the analysis of an unbalanced Wheatstone bridge (Experiment 30). How many simultaneous equations would be needed for analysis?

(continued)

4. Two batteries with emf's of 1.5 V and 1.2 V and internal resistances of 0.5 ohm and 0.8 ohm, respectively, are connected in parallel.
   (a) Determine the current they would deliver to an external resistance of 2 ohms.

   (b) Would it be more efficient to use only one battery? If so, which one? Explain.

Name .... Section .... Date ....

Lab Partner(s) ....

EXPERIMENT 35 *Joule Heat*

ADVANCE STUDY ASSIGNMENT

Read the experiment and answer the following questions.

1. Explain (a) joule heat; (b) I^2R losses.

2. What is the difference between joule heat and power?

3. Given two different resistances, how does the joule heat of the resistances vary if they are connected in a circuit (a) in series; and (b) in parallel?

4. In the experiment, why isn't the final temperature of the system read at the same time that the power supply is unplugged and the timer stopped?

5. Suppose that oil were used in the experiment instead of water. Would (a) the joule heat and (b) the temperature rise be the same? Explain.

EXPERIMENT **35**

Joule Heat

I. INTRODUCTION

Whenever an electrical current flows in a conductor, some electrical energy is converted into heat energy. For a given current I, the energy conversion is greater in a conductor of greater resistance. This is analogous to the conversion of mechanical energy into heat energy due to frictional resistance. The heat generated in an electrical circuit is commonly referred to as **joule heat,** after James Prescott Joule (1818–1889), the English scientist who investigated the conversion of electrical energy to heat (and also the mechanical equivalent of heat).

In many electrical applications, such as electrical motors, the joule heat is an undesirable loss of energy. However, in other applications, such as toasters and electrical heaters, electrical energy is purposefully converted to heat energy. In this experiment, the heating effect of an electrical current and the "electrical equivalent of heat" will be investigated.

II. EQUIPMENT NEEDED

- Electrocalorimeter (immersion heater and calorimeter)
- Power supply or battery (12 V)
- Ammeter (0 to 3 A)
- Voltmeter (30 V)
- Rheostat (40 Ω)
- Connecting wires
- Thermometer
- Stopwatch or laboratory timer
- Laboratory balance

III. THEORY

The work W done (or energy expended) per unit charge in moving a charge q from one point to another is the potential difference or voltage V

$$V = \frac{W}{q}$$

or

$$W = qV \tag{35-1}$$

The time rate of flow of charge is described in terms of current I and

$$I = \frac{q}{t} \tag{35-2}$$

Hence, Eq. 35-1 may be written

$$W = qV = IVt \tag{35-3}$$

which represents the work done or the energy expended in a circuit in a time t. This is often expressed as work or energy per time or power P,

$$P = \frac{W}{t} = IV \tag{35-4}$$

The expended energy can be written in terms of the resistance R of the circuit or a particular circuit element by Ohm's law, $V = IR$. Using this relationship, Eq. 35-3 has the various forms

$$W = IVt = I^2Rt = \frac{V^2t}{R} \tag{35-5}$$

The electrical energy expended is manifested as heat energy and is commonly called joule heat or I^2R

losses, I^2R being the power or energy expended per time.

Equation 35-5 shows how the joule heat varies with resistance:

1. For a constant current I, the joule heat is directly proportional to the resistance, I^2R.
2. For a constant voltage V, the joule heat is inversely proportional to the resistance, V^2/R.

The energy expended in an electrical circuit as given by Eq. 35-5 is in the units of joules. The relationship (conversion factor) between joules and heat units in calories was established by James Joule from mechanical considerations—the *mechanical equivalent of heat.* You may recall that in his mechanical experiment Joule had a descending weight turn a paddle wheel in a liquid. He then correlated the mechanical (gravitational) potential energy lost by the descending weight to the heat generated in the liquid. The result was 1 cal = 4.18 J. A similar electrical experiment may be done to determine the "electrical equivalent of heat." By the conservation of energy, the heat equivalents of mechanical and electrical energy are the same (i.e., 1 cal = 4.18 J).

Experimentally, the amount of electrical joule heat generated in a circuit element of resistance R is measured by calorimetry methods. If a current is passed through a resistance (immersion heater) in a calorimeter with water in an arrangement as illustrated in Fig. 35-1, by the conservation of energy the electrical energy expended in the resistance is equal to the heat energy (joule heat) Q gained by the system:

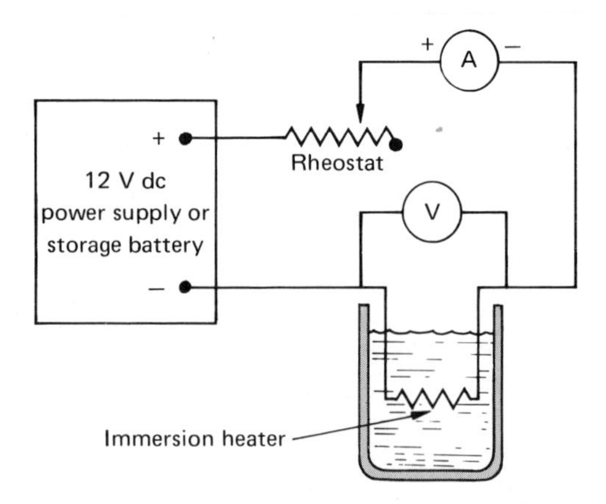

Fig. 35-1 Circuit diagram for experimental procedure to measure joule heat.

$$\text{electrical energy expended} = \text{heat gained}$$

$$W = Q$$

or

$$IVt = (m_w c_w + m_{\text{cal}} c_{\text{cal}} + m_{\text{coil}} c_{\text{coil}})(T_f - T_i) \quad \textbf{(35-6)}$$

where the m's and c's are the masses and specific heats of the water, calorimeter cup, and immersion coil, respectively, as indicated by the subscripts. T_f and T_i are the final and initial temperatures of the system, respectively. (See Experiment 22 for the detailed theory of calorimetry procedure.)

IV. EXPERIMENTAL PROCEDURE

1. Determine and record in the Laboratory Report the masses of the inner calorimeter cup (without ring) and the coil of the immersion heater. (The latter may be supplied by the instructor if the coil is permanently mounted.)

   Also, record the types of materials and their specific heats. (The type of material and specific heat of the calorimeter cup is usually stamped on the cup. For the coil, usually copper, a table of specific heats is given in the Appendix.)

2. Fill the calorimeter cup about two-thirds full of cool tap water several degrees below room temperature. (The cup should be filled to a level so that the immersion heater will be completely covered when immersed later.) Determine and record the mass of the calorimeter cup with the water.

3. Place the immersion heater in the calorimeter cup and set up the circuit as illustrated in Fig. 35-1 with the rheostat set at its maximum resistance. Make certain that the heating coil is completely immersed. If not, add more water and reweigh the cup and water as in procedure 2. *Do not plug in the power supply (or connect battery) until the circuit has been checked by the instructor.*

4. After the circuit has been checked, plug in the power supply set to 10 to 12 V. Adjust the rheostat until a constant current between 2 and 3 A flows in the circuit as indicated on the ammeter. (If a variable power supply is used, it may also be used to make fine current adjustments.) Then unplug the power supply. This procedure should be done as quickly as possible to avoid heating the water.

5. Stir the water in the calorimeter with the immersion coil or thermometer, and measure and record its equilibrium temperature (T_i). Then plug in the power supply and at the same time start the stopwatch or laboratory timer. Immediately read and record the initial ammeter and voltmeter readings. As time goes on, keep the current constant by varying the rheostat (and/or the power supply). Record the voltage every minute. Stir the water frequently.

6. When the temperature of the water (and calorimeter system) is 10 to 15°C above the initial temperature, simultaneously unplug the power supply and stop the timer at the time of a particular minute interval reading. Continue stirring until a maximum temperature is reached and record this temperature (T_f).

7. Compute the electrical energy expended in the coil (in joules) from the electrical and time readings. Use the average value of the voltage readings as the effective voltage across the coil.

8. (a) Compute the heat energy (in calories) gained by the calorimeter system.
   (b) Then taking the ratio of the electrical and heat energy results, find the "electrical equivalent of heat." Compare this to the value of the mechanical equivalent of heat by computing the percent error.

9. If time permits (ask instructor), repeat the experiment and use the average value of the experimental results in determining the percent error.

LABORATORY REPORT

	Mass	Material	Specific heat
Calorimeter cup	....	....	....
Immersion coil	....	....	....
Calorimeter cup and water	....		
Water	....	....	....

DATA TABLE

Time (s)	Voltage, V (V)	Current, I (A)	Temperature (°C)
0			
			T_i ....
			T_f ....

Average voltage ....

(continued)

Calculations (show work)

Electrical energy expended ....

Heat energy gained ....

Ratio of results ....

Percent error ....

QUESTIONS

1. What are the major sources of error in the experiment?

2. Why was it necessary to make adjustments to maintain a constant current in the circuit?

3. If the cost of electricity is 12 cents per kWh, what was the cost of the electricity used in performing the experiment?

4. A piece of No. 24 gauge (AWG) copper wire is three times the length of a piece of No. 18 gauge copper wire. If the same voltage is applied to the wires, which wire would have more joule heat and how many times more? (*Hint:* See Experiment 31 and Appendix A, Table A7.)

5. Do heating appliances such as electric blankets and toasters have high-resistance or low-resistance elements? Explain.

Name .... Section .... Date ....

Lab Partner(s) ....

EXPERIMENT 36 *The Measurement of Capacitance: Bridge Method*

ADVANCE STUDY ASSIGNMENT

Read the experiment and answer the following questions.

1. What is capacitance?

2. How is the capacitance bridge balanced, and what are the potential differences across the circuit components in this condition? Compare the capacitance bridge to a Wheatstone Bridge.

3. If the applied ac voltage varies with time (ac audio oscillator has a sinusoidal voltage), why can capacitance be measured through the charges on the capacitor plates?

4. Is the equivalent capacitance of series and parallel combinations of capacitors the same? Explain.

5. Could the experiment be performed using a dc power supply instead of the ac audio oscillator source? Explain.

EXPERIMENT 36

The Measurement of Capacitance: Bridge Method

I. INTRODUCTION

A **capacitor** generally consists of two metal plates separated by an insulator. (The term "condenser" was formerly used, but capacitor is now preferred.) As such, a capacitor does not allow the conduction of a constant current in a dc circuit. However, charge is conducted until a charge Q builds up on the plates, and

$$Q = CV \tag{36-1}$$

where V is the applied voltage. The constant C is called the **capacitance** and has the unit of farad (F). It expresses the charge on a capacitor for a given potential difference or voltage, $C = Q/V$. In an ac circuit, the charge alternately builds and discharges on the plates so that there is an effective current in the circuit. This allows the measurement of capacitance in a bridge circuit.

In this experiment, the capacitance of capacitors and that of various combinations of capacitors will be measured.

II. EQUIPMENT NEEDED

- Standard capacitor (e.g., 0.5 μF)
- Two standard decade resistance boxes (0 to 999.9 Ω)
- Two unknown capacitors (e.g., Mylar capacitors wrapped in masking tape to mask values)
- Telephone receiver
- Audio oscillator (ac source)
- Connecting wires

III. THEORY

A capacitance bridge as illustrated in Fig. 36-1 is similar to a Wheatstone bridge for resistance measurements (see Experiment 30). The circuit symbol for a capacitor is ⊣⋲. (The curved line helps avoid confusion with the battery symbol ⊣⊢. An older circuit symbol for a capacitor is ⊣⊢.)

For a capacitance bridge, an alternating voltage source is used instead of a dc power supply or battery, and a telephone receiver T is used in place of a galvanometer to determine the balanced condition of the bridge circuit. When the circuit is "balanced," the hum in the telephone receiver is a minimum, indicating that the voltage drop between points B and D is zero.

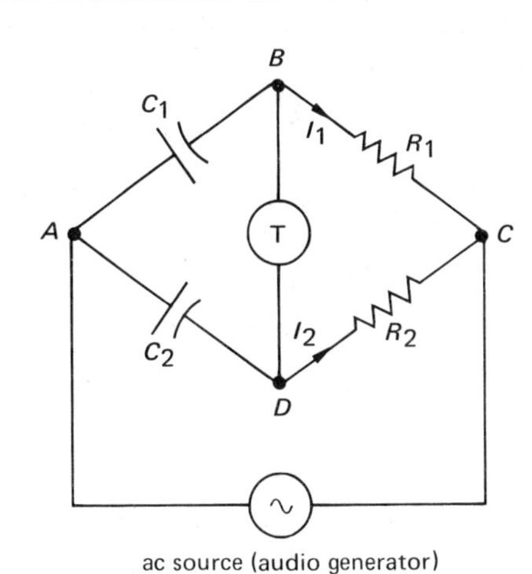

Fig. 36-1 Circuit diagram for a capacitance bridge.

In the balanced condition, the voltages across the capacitors C_1 and C_2 must then be equal, and

$$Q_1 = C_1V \quad \text{and} \quad Q_2 = C_2V$$

or

$$\frac{Q_1}{Q_2} = \frac{C_1}{C_2} \tag{36-2}$$

(Even though the applied voltage changes with time, the potential difference across the plates of the capacitors must be the same at all instants of time.) The charges on the capacitors are proportional to the currents through the respective branches of the circuit, and

$$\frac{Q_1}{Q_2} = \frac{I_1}{I_2} = \frac{C_1}{C_2} \tag{36-3}$$

Similarly, the voltages across the resistances must be equal at any time, and by Ohm's law, $V = IR$,

$$I_1R_1 = I_2R_2$$

or

$$\frac{I_1}{I_2} = \frac{R_2}{R_1} \tag{36-4}$$

Then, combining Eqs. 36-3 and 36-4,

$$\frac{C_1}{C_2} = \frac{R_2}{R_1}$$

$$C_1 = \left(\frac{R_2}{R_1}\right)C_2 \tag{36-5}$$

Hence, if C_1 is an unknown capacitance C_x and C_2 is a standard capacitance C_s, with known resistances R_1 and R_2, the value of the unknown capacitance is given by

$$C_x = \left(\frac{R_2}{R_1}\right)C_s \tag{36-6}$$

Capacitors in Series and Parallel

Suppose that capacitors are connected in series as illustrated in Fig. 36-2(a). When charged, each capacitor has a charge Q. (Actually, there is a charge $+Q$ on one plate and an equal charge $-Q$ on the other plate, but we say that the capacitor has a charge Q.) Then the potential difference across C_1 is $V_1 = Q/C_1$, with similar equations for C_2 and C_3.

The total potential difference across the capacitors is equal to the sum of the individual potential differences:

$$V = V_1 + V_2 + V_3 = \frac{Q}{C_1} + \frac{Q}{C_2} + \frac{Q}{C_3}$$

or

$$V = Q\left(\frac{1}{C_1} + \frac{1}{C_2} + \frac{1}{C_3}\right) \tag{36-7}$$

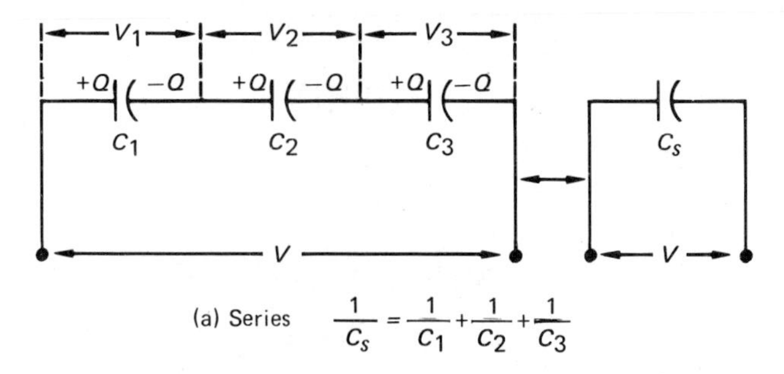

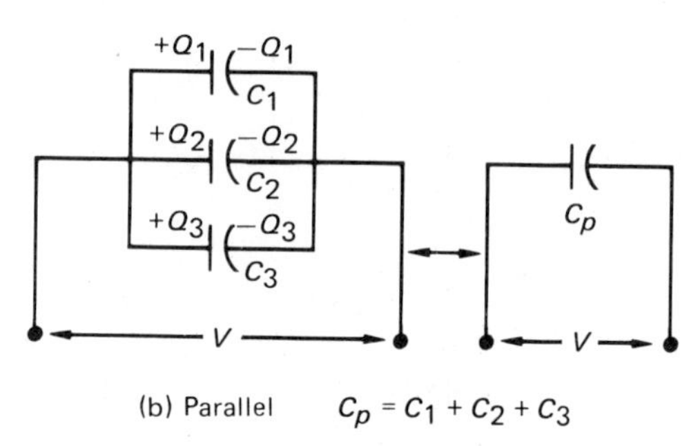

Fig. 36-2 Circuit diagrams for capacitors in (a) series and (b) parallel, and their single equivalent capacitance circuits.

The relationship of an equivalent capacitance C_s of the series combination in a circuit would have the form $V = Q/C_s$, and by comparison

$$\frac{1}{C_s} = \frac{1}{C_1} + \frac{1}{C_2} + \frac{1}{C_3} \tag{36-8}$$

That is, a single capacitor with a capacitance C_s is equivalent to and could replace the three capacitors in series. This development can be extended to any number of capacitors in series.

If the capacitors are connected in parallel as illustrated in Fig. 36-2(b), the potential difference is the same across all of the capacitors, and the charge on a capacitor depends on its capacitance. For example, for C_1, $Q_1 = C_1V$, with similar equations for C_2 and C_3. The total charge Q on the capacitors is equal to the sum of the charges on the individual capacitors:

$$Q = Q_1 + Q_2 + Q_3 = C_1V + C_2V + C_3V$$

or

$$Q = (C_1 + C_2 + C_3)V \tag{36-9}$$

The relationship for an equivalent capacitance C_P of the parallel combination in a circuit would have the form $Q = C_pV$, and by comparison

$$C_p = C_1 + C_2 + C_3 \tag{36-10}$$

This is, a single capacitor with a capacitance C_p is equivalent to and could replace the three capacitors in parallel. This development can be extended to any number of capacitors in parallel.

Name .... Section .... Date ....

Lab Partner(s) ....

EXPERIMENT 36 *The Measurement of Capacitance: Bridge Method*

IV. EXPERIMENTAL PROCEDURE

1. Set up the circuit arrangement as shown in Fig. 36-1 with the decade boxes as R_1 and R_2, and one of the unknown capacitors and the standard capacitor as C_1 and C_2, respectively. Record the capacitance of the standard capacitance in the Laboratory Report. Set one of the resistance boxes at 200 Ω. Set the audio oscillator between 2000 and 3000 Hz. *Do not turn on the audio oscillator until the instructor has checked the circuit.*

2. After the circuit has been approved, turn on the audio oscillator. Leaving the one decade box resistance (R_1) set at 200 Ω, adjust the resistance of the other box until the sound heard in the telephone receiver is a minimum. Record the values of R_1 and R_2 in the data table.

3. Replace the test capacitor with the other unknown capacitor and repeat procedure 2. When the experiment is completed, turn off the oscillator.

4. Compute the capacitances of the two unknown capacitors. Remove the masking tape from the capacitors and read their marked values or ask the instructor for the capacitor values. Compare the experimental values with the rated values by computing the percent errors.

Capacitors in Series and Parallel

5. Compute the equivalent capacitances of the two capacitors connected (a) in series; and (b) in parallel. Use the known (labeled) values of C_1 and C_2.

6. Connect the capacitors in series and measure the capacitance of the combination in the capacitance bridge as in procedure 2.

7. Connect the capacitors in parallel and measure the capacitance of the combination in the capacitance bridge.

8. Compare the measured values with the computed values by finding the percent differences.

LABORATORY REPORT

	Unknown capacitor 1	Unknown capacitor 2	Standard capacitor C_s
R_1	....	....	....
R_2	....	....	....
Experimental capacitance	....	....	....
Actual capacitance	....	....	....
Percent error	....	....	....

Calculations (show work)

(continued)

Capacitors in Series and Parallel

Calculations (show work)

	C_s	C_p
Computed value (from known values)	....	....
R_1	....	....
R_2	....	....
Experimental value	....	....
Percent difference	....	....

QUESTIONS

1. If a 1.0-μF capacitor is charged by a 6-V dc voltage source, what is the magnitude of the charge on the capacitor plates when it is fully charged?

2. What is the magnitude of the charge on each of the capacitors in the circuit shown in Fig. 36-Q2?

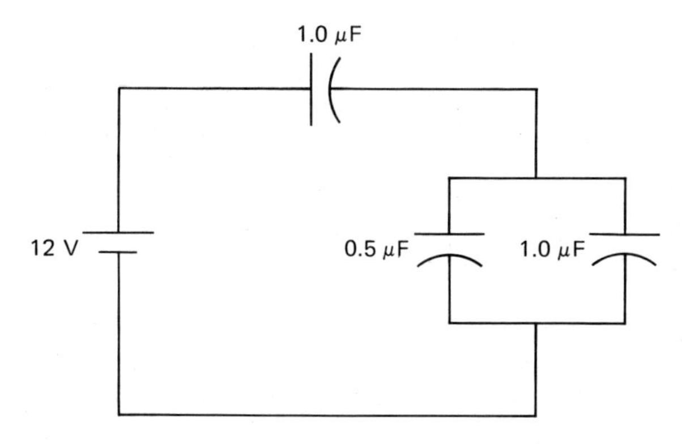

Fig. 36-Q2

3. Given three capacitors, 0.25 μF, 0.5 μF, and 1.0 μF, how many different values of capacitance can be obtained using one or more of the capacitors, and what are these values?

Lab Partner(s) ....

EXPERIMENT 37 *The RC Time Constant*

ADVANCE STUDY ASSIGNMENT

Read the experiment and answer the following questions.

1. What is an *RC* time constant?

2. When an *RC* circuit is connected to a dc source, what is the voltage on a capacitor after one time constant when (a) charging from zero voltage; and (b) discharging from a fully charged condition?

3. If the resistance in a capacitor circuit is increased, would the charging time of the capacitor increase or decrease? Explain.

4. Can the voltage across a capacitor be measured with a common voltmeter? Explain.

5. What is an "infinite resistance" voltmeter?

EXPERIMENT 37

The *RC* Time Constant

I. INTRODUCTION

When a capacitor is connected to a dc power supply or battery, charge builds up on the capacitor plates and the potential difference or voltage across the plates increases until it equals the voltage of the source. At any time, the charge Q of the capacitor is related to the voltage across the capacitor plates by $Q = CV$, where C is the capacitance of the capacitor in farads (F). The rate of voltage rise depends on the capacitance of the capacitor and the resistance in the circuit. Similarly, when a charged capacitor is discharged, the rate of voltage "decay" depends on the same parameters.

Both the charging and discharge times of a capacitor are characterized by a quantity called the **time constant** τ, which is the product of the capacitance C and the resistance R, that is, $\tau = RC$. In this experiment, the time constants and the charging and discharging characteristics of capacitors will be investigated.

II. EQUIPMENT NEEDED

- Two capacitors (0.5-μF and 1.0-μF Mylar or other high-quality capacitors)
- Two resistors (50 MΩ and 80 MΩ, or enough resistors to make equivalent series resistances)
- Power supply or battery (12 V)
- Infinite resistance (electrostatic) voltmeter (or VTVM)
- Single-pole, double-throw switch
- Connecting wires
- Laboratory timer
- 2 sheets of Cartesian graph paper

III. THEORY

When a capacitor is charged through a resistor by a dc voltage source (the single-pole, double-throw switch in position a in Fig. 37-1), the charge on the capacitor and the voltage across the capacitor increase with time. The voltage V as a function of time is given by

$$V = V_0(1 - e^{-t/RC}) \qquad \textbf{(37-1)}$$

where the exponential $e = 2.718$ is the base of natural logarithms and V_0 is the voltage of the source. The curve of the exponential rise of the voltage with time during the charging process is illustrated in Fig. 37-2. The quantity $\tau = RC$ is called the **time constant** of the circuit.

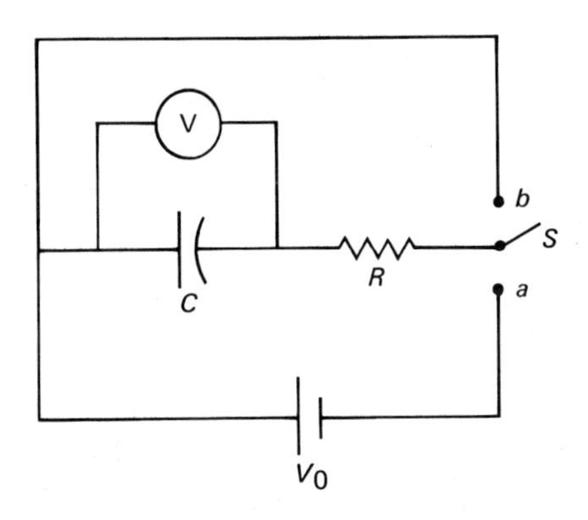

Fig. 37-1 Circuit for charging (S in position a) and discharging (S in position b) a capacitor through a resistor.

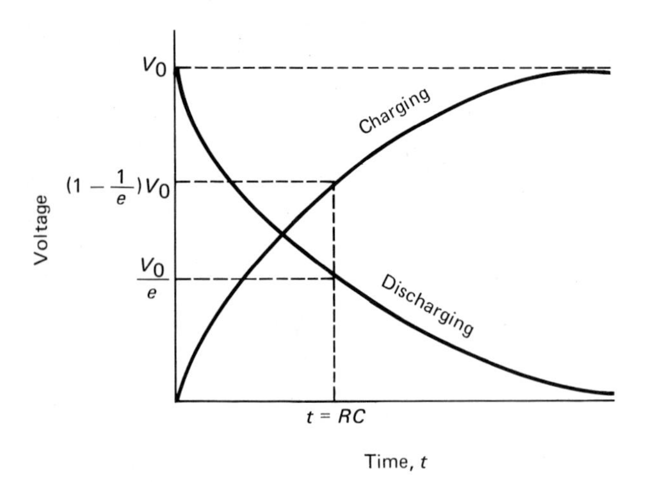

Fig. 37-2 A graph illustrating voltage versus time for a charging and discharging capacitor. The "steepness" of the curves depends on the time constant RC.

After a time $t = \tau = RC$ (one time constant), the voltage across the capacitor has grown to a value of $(1 - 1/e)$ of V_0 [i.e., $V = V_0\ (1 - e^{-RC/RC}) = V_0(1 - e^{-1}) = V_0(1 - 1/e) = 0.63\ V_0$].

When the fully charged capacitor is discharged through a resistor (switch in position b in Fig. 37-1), the voltage across (and the charge on) the capacitor "decays" or decreases with time according to the equation

$$V = V_0 e^{-t/RC} \qquad \textbf{(37-2)}$$

The exponential decay of the voltage with time is also illustrated in Fig. 37-2. After a time $t = \tau = RC$ (one time constant), the voltage across the capacitor has decreased to a value of $1/e$ of V_0 (i.e., $V = V_0e^{-t/RC} = V_0e^{-RC/RC} = V_0e^{-1} = V_0/e = 0.37\ V_0$).

From Eq. 37-1, we obtain

$$(V_0 - V) = V_0 e^{-t/RC}$$

and taking the natural logarithm gives

$$\ln (V_0 - V) = \frac{-t}{RC} + \ln V_0 \qquad \textbf{(37-3)}$$

From Eq. 37-2,

$$\ln V = \ln V_0 - \frac{t}{RC}$$

or

$$\ln \left(\frac{V_0}{V}\right) = \frac{t}{RC} \qquad \textbf{(37-4)}$$

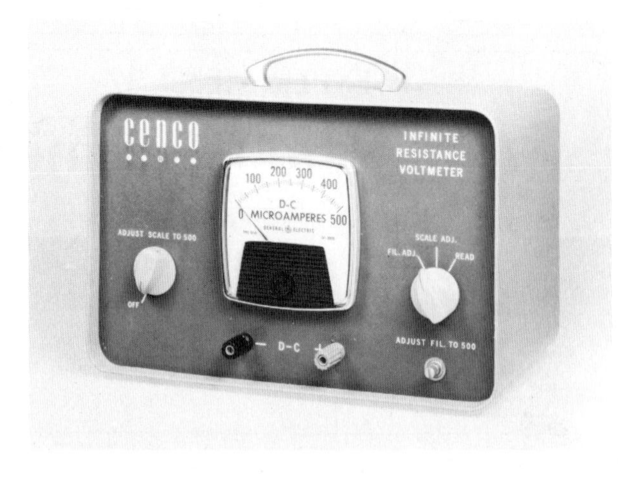

Fig. 37-3 An electrostatic or infinite resistance voltmeter. (Photo courtesy of Central Scientific Co.)

Both of these equations have the form of the equation of a straight line, $y = ax + b$, with slopes of magnitude $1/RC$ (Eq. 37-3, negative slope; Eq. 37-4, positive slope). Hence, the time constant can be found from the slopes of the graphs of $\ln (V_0 - V)$ versus t and/or $\ln (V_0/V)$ versus t.

The potential difference or voltage across a capacitor is difficult to measure with a common voltmeter since the capacitor quickly loses charge though the voltmeter. One method to measure this voltage is by using an *electrostatic* or *infinite resistance voltmeter* (Fig. 37-3). The basic component of this instrument is a variable air capacitor. The movable plates vary according to the electrostatic force between the charge on the fixed and variable plates, which is related to the voltage across the plates or terminals of the voltmeter. The instrument scale is calibrated in volts. Because an air capacitor presents an open circuit or infinite resistance in a dc circuit, an electrostatic voltmeter is sometimes called an infinite resistance voltmeter. [An alternate method is the use of another high-resistance instrument, the vacuum tube voltmeter (VTVM). If this is used your instructor will explain its operation.]

IV. EXPERIMENTAL PROCEDURE

1. Set up the circuit as shown in Fig. 37-1 with the capacitor of smaller capacitance and resistor of larger resistance. It is often necessary to use a series combination of resistors to obtain the large resistances required in the experiment. The resistance of a resistor may be determined from the colored bands on the resistor. (See Appendix A, Table A5, for the resistor color code.) Record the value of the capacitance C_1 and the resistance R_1 in Data Table 1. Also prepare the laboratory timer for time measurements. *Have the instructor check the circuit before closing the switch.*

2. Close the switch to position *a* and note the voltage rise of the capacitor on the voltmeter. When the capacitor is fully charged, move the switch to position *b* and note the voltage decrease as the capacitor discharges. In the following procedures, the voltage is read as a function of time. You should try trial time runs to become familiar with the procedures.

3. Simultaneously close the switch to position *a* and start the timer. Read and record the capacitor voltage at small time intervals (e.g., 5 s) until the capacitor is fully charged (V_0). This should be done with two persons working together. However, if necessary, the switch may be opened (and the timer stopped) to stop the charging process after a given interval without appreciable error if a high-quality, low-leakage capacitor is used.

4. After the capacitor is fully charged, open the switch to the neutral position and reset the timer. Then, simultaneously close the switch to position *b* and start the timer. Read and record the decreasing voltage at small time intervals. Open the switch when the capacitor is discharged.

5. Replace R_1 and C_1 with R_2 and C_2 (smaller resistance and larger capacitance), and repeat procedures 3 and 4.

6. Compute the quantities $(V_0 - V)$ and (V_0/V) for the charging and discharging processes, respectively. Then, using Appendix B, Table B8 or a calculator, find the value of $\ln(V_0 - V)$ and $\ln(V_0/V)$.

7. On a Cartesian graph, plot $\ln(V_0 - V)$ versus t for both sets of data, and $\ln(V_0/V)$ vs t for both sets of data on another graph. Draw the straight lines that best fit the data and determine the slope of each line. Record the slopes in the data table. Compute the time constants from the average slope values.
8. Compute $\tau_1 = R_1C_1$ and $\tau_2 = R_2C_2$ from the known resistance and capacitance values and compare with the experimental values by finding the percent errors.

Name .... Section .... Date ....

Lab Partner(s) ....

EXPERIMENT 37 *The RC Time Constant*

LABORATORY REPORT

DATA TABLE 1

C_1 ....

R_1 ....

Charging				Discharging			
V (V)	t (s)	$V_0 - V$	$\ln(V_0 - V)$	V (V)	t (s)	V_0/V	$\ln(V_0/V)$
				V_0			
V_0							

Calculations (show work)

Slope ....

R_1C_1 (from slope) ....

R_1C_1 (from known values) ....

Percent error ....

Name .... Section .... Date ....

Lab Partner(s) ....

EXPERIMENT 37 *The RC Time Constant*

DATA TABLE 2

C_2 ....

R_2 ....

Charging				Discharging			
V (V)	t (s)	$V_0 - V$	$\ln(V_0 - V)$	V (V)	t (s)	V_0/V	$\ln(V_0/V)$
				V_0			
V_0							

Calculations (show work)

Slope ....

R_2C_2 (from slope) ....

R_2C_2 (from known values) ....

Percent error ....

(continued)

QUESTIONS

1. Show that the magnitude of the charge on a capacitor is given by $Q = Q_0(1 - e^{-t/\tau})$ and $Q = Q_0e^{-t/\tau}$ for charging and discharging, respectively.

2. What is the voltage across a capacitor after a time of two time constants when (a) charging from zero voltage, and (b) discharging from a fully charged condition?

3. A 2.0-μF capacitor in a circuit in series with a resistance of 1.0 MΩ is charged with a 6.0-V battery. How long would it take to charge the capacitor to three-fourths of its maximum voltage?

Name .... Section .... Date ....

Lab Partner(s) ....

EXPERIMENT 38 *Introduction to the Oscilloscope*

ADVANCE STUDY ASSIGNMENT

Read the experiment and answer the following questions.

1. What is a CRT?

2. What is the purpose of the deflection plates in the oscilloscope CRT?

3. How is a graph of voltage versus time obtained on the oscilloscope screen?

4. What is the condition for a stationary trace of an ac voltage input signal?

5. What are Lissajous figures, and how are they produced?

EXPERIMENT **38**

Introduction to the Oscilloscope

I. INTRODUCTION

The cathode-ray oscilloscope is one of the most versatile laboratory instruments for studying ac circuits. Having long been quite common in the physics laboratory, the oscilloscope is finding increasing uses and applications in biology and medical fields. In its most basic application, the oscilloscope is used to display a graph of voltage versus time on its screen. The signal may be the voltage across a component in an electrical circuit or that generated by a nerve impulse or a heartbeat.

In this experiment, an introduction to the basic principles of the oscilloscope is presented, and you will operate an oscilloscope so as to become familiar with its controls and characteristics.

II. EQUIPMENT NEEDED

- Cathode-ray oscilloscope (student model)
- Audio-signal function generator (sine and square waves)
- (60-Hz sine wave source, or second generator if oscilloscope does not have internal line input)
- Connecting wires

III. THEORY

The basic component of an oscilloscope (or "scope") is a cathode-ray tube (CRT) or electron-beam tube (Fig. 38-1). The name "cathode-ray tube" comes from early experiments with gas discharge tubes in which "rays" coming from the cathode or negative electrode were observed. The beam of electrons (cathode rays) in a CRT is formed by an "electron gun," in which electrons thermally emitted from a cathode filament are accelerated through a potential difference of several thousand volts and focused into a beam. The electron beam strikes a fluorescent screen coated with a phosphor that emits visible light and a spot of light is seen on the screen.

The CRT is also equipped with sets of vertical and horizontal deflection plates. If no voltage signals are applied to the deflection plates, the beam is undeflected and strikes the center of the screen. However, if a voltage signal is applied to the horizontal deflection plates, the electron beam will experience a force and be deflected horizontally.

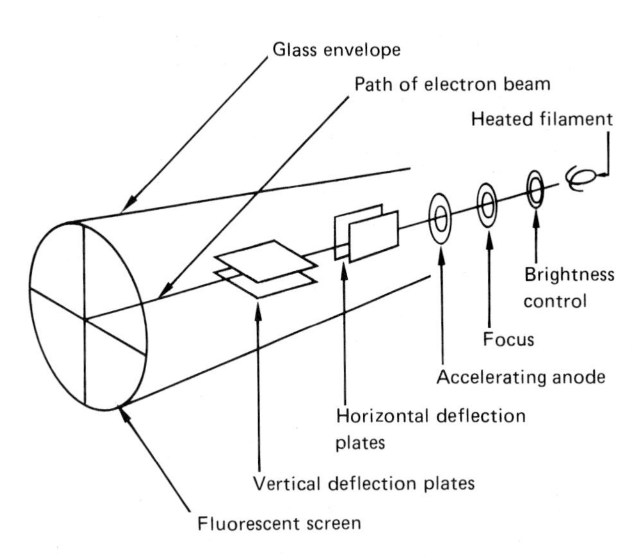

Fig. 38-1 The basic components of an oscilloscope cathode ray tube (CRT).

A constant dc voltage will deflect the beam spot on the screen a fixed distance. An ac voltage, on the other hand, will deflect the beam back and forth, since the polarity is continually changing. If the frequency of the ac voltage signal is large enough, the beam spot traces out an observable continuous horizontal line. This is due to the relatively slow decay of the brightness of the fluorescent screen after each excitation and the persistence of vision in the human eye.

Similarly, a voltage signal applied to the vertical deflection plates causes the beam to move vertically. In either case, the magnitude of the deflection of the beam spot from the center of the screen is proportional to the magnitude of the voltage applied to the deflection plates. As such, the cathode-ray oscilloscope is an extremely fast X-Y plotter that is capable of plotting an input signal versus time or another signal.

In ac voltage applications, it is usually desired to display the voltage on the screen as a function of time (i.e., a graph of voltage versus time). The signal to be studied is applied to the vertical deflection plates. The oscilloscope has an internal vertical amplifier or gain to amplify weak input signals. A horizontal linear time axis is obtained if the beam spot moves horizontally (left to right) with a constant speed.

The time axis is generated by applying a deflecting voltage to the horizontal plates, which increases linearly with time (e.g., a voltage signal with a "sawtooth" wave form, as illustrated in Fig. 38-2). As the voltage increases uniformly, the spot sweeps uniformly (with constant speed) across the screen from left to right. When the voltage suddenly drops to zero, the beam flies back to its initial position and begins another horizontal sweep with the "triggering" of the next sawtooth.

With a slow sweep rate, the spot can be observed moving periodically from left to right. However, with a fast sweep rate, a continuous trace is observed, as the eye cannot follow the motion. The oscilloscope is equipped with an internal variable sawtooth generator to supply such horizontal signals with no signal on the horizontal plates.

Suppose that a sinusoidal voltage having the form $V = V_0 \sin 2\pi ft$ is applied to the vertical plates. As discussed previously, the beam would move up and down and trace out a vertical line on the screen. However, if a sawtooth voltage signal is applied to the horizontal plates with the same sweep-rate frequency as the frequency f of the vertical sinusoidal voltage signal, the beam will sweep from left to right while it moves vertically up and down (Fig. 38-3). The combined motions of the beam spot then trace out a graph of the voltage of the applied vertical signal with time, and a sine wave is seen on the screen.

A. *Frequency Measurements*

If the sine-wave pattern retraces itself and appears to stand still, we say that the signals are synchronized. In this case, the frequency of the sinusoidal voltage input is equal to the sweep rate, which is the number of horizontal repetitions (cycles) of the beam spot per second. That is, there is one vertical oscillation for each horizontal sweep.

It follows that if the sinusoidal voltage frequency is twice that of the sweep rate, then two sine-wave cycles would appear on the screen, and so on. For example, if the sweep rate frequency is 100 Hz (cycles per second) and two cycles of a sine wave appear on the screen, the frequency input is 200 Hz. In general, a stationary pattern results when the ratio

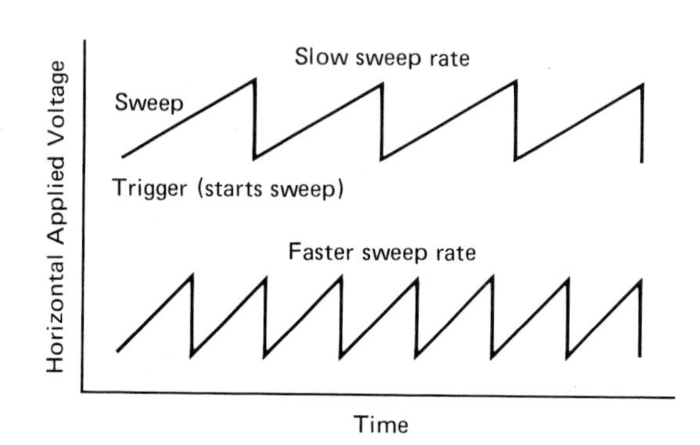

Fig. 38-2 Sawtooth voltage functions that supply different oscilloscope sweep rates.

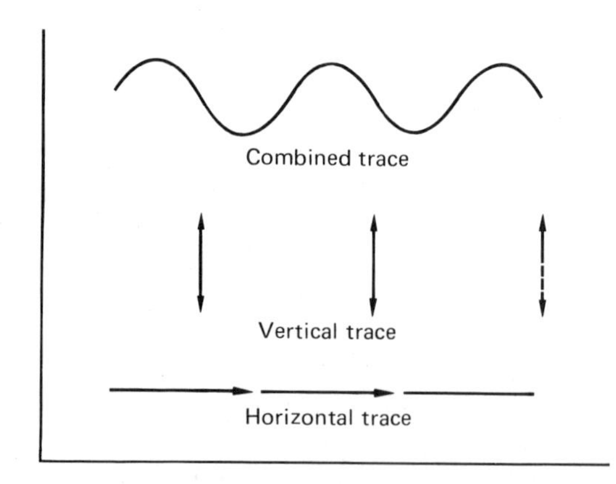

Fig. 38-3 The combination of a horizontal sweep signal and a vertical sinusoidal signal produces a sine-wave trace on an oscilloscope screen.

of the sweep-rate frequency and the frequency of the input-voltage signal is an integral or half-integral.

The sweep time of the trace across the screen is the reciprocal of the sweep rate frequency. For example, if the sweep-rate frequency is 100 Hz, the sweep time is $\frac{1}{100} = 0.01$ s. Knowing the sweep time, the frequency of a vertical input voltage can be determined by dividing the number of stationary-wave cycles observed on the screen by the sweep time.

Applied to the previous example with a sweep-rate frequency of 100 Hz and 2 cycles observed on the screen, the frequency of the input voltage is $f = \frac{2}{0.01} = 200$ Hz.

The frequency of a sinusoidal voltage can also be determined by comparing it with another calibrated sinusoidal signal. The simplest oscilloscope pattern results when the two sinusoidal signals have the same frequency and are in phase, for example, have the form

$$x = A \sin 2\pi ft$$

and

$$y = A \sin 2\pi ft \tag{38-1}$$

A diagonal line is observed on the screen (Fig. 38-4). The pattern observed when the frequency of the y signal is three times that of the x signal is also shown in Fig. 38-4. Such patterns are called *Lissajous figures.* As before, a stationary pattern results when the frequency of the X-axis signal and the frequency of the Y-axis signal are integral or half-integral.

Suppose that two sinusoidal signals have the same frequency but different phases, for example of the form

$$x = A \sin 2\pi ft$$

and

$$y = A \sin (2\pi ft - \delta) \tag{38-2}$$

where δ is the phase angle or the phase difference. The phase angle of the x signal is equal to zero. Then an ellipse will be traced out on the screen (Fig. 38-5). If the phase difference is 90°, the pattern is a circle.

B. *Voltage Measurements*

In addition to frequency measurements, the oscilloscope can be used as a voltmeter to read the peak-to-peak voltages of ac signals as well as dc voltages. The voltage readings are read directly from a plastic screen or grid attached to the face of the oscilloscope tube. Recall that the vertical deflection of the beam is proportional to the voltage of the signal applied to the vertical deflection plates.

To calibrate the oscilloscope screen, a voltage signal of known magnitude is used. For ac measurements, an internal or external reference voltage is applied to the vertical input of the scope with the horizontal sweep turned off. The resulting vertical line represents the peak-to-peak voltage of the input

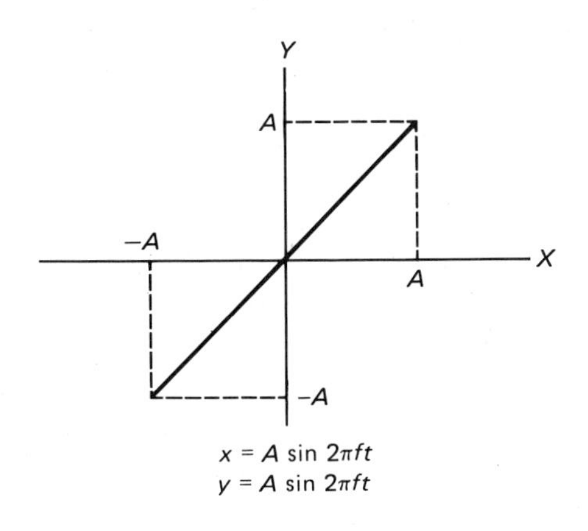

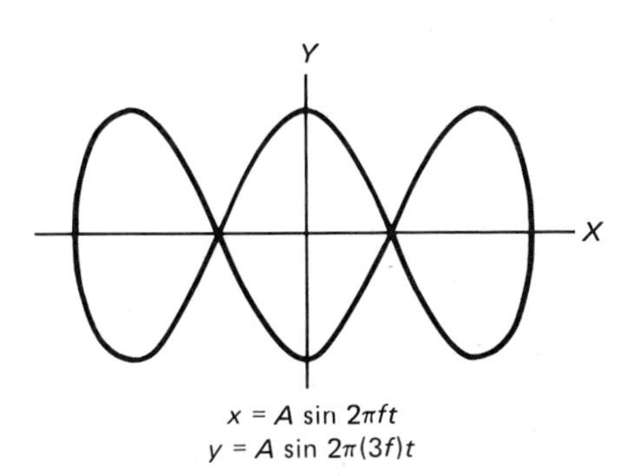

Fig. 38-4 Lissajous figures for different x and y signals.

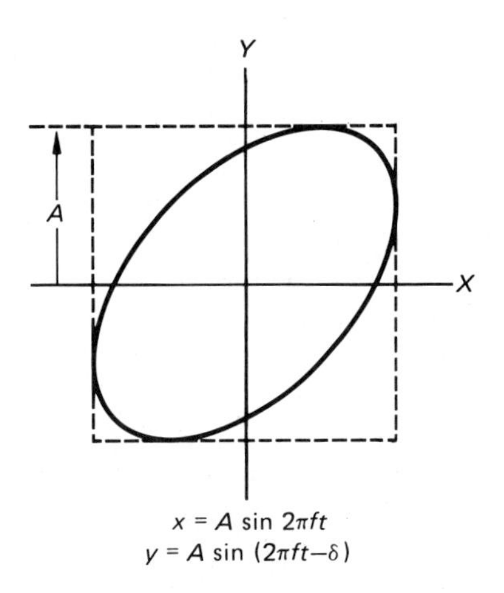

Fig. 38-5 The Lissajous figure for signals having the same frequency but different phases is an ellipse.

signal. For example, suppose the reference voltage is 5 volts peak-to-peak. By adjusting the vertical height control or gain so the vertical trace line corresponds to 5 scale divisions, the calibration is 5 volts/5 divisions or 1 volt/div. The vertical height control is left in this position after calibration. (Why?) Most scopes have internal voltage calibrations in several voltage steps or values.

The peak-to-peak height of the trace of an ac input signal is then measured and the peak-to-peak voltage computed from the calibration. Dividing by two gives the peak voltage, and multiplying by $1/\sqrt{2} = 0.707$ gives the root-mean-square (rms) value of the voltage.

There is a large variety of oscilloscopes and the controls of each model and make cannot be discussed here. All oscilloscopes have some operating controls in common, however, some have controls that others lack, depending on the purpose and sophistication of the scope. Also, controls and connectors are not always found in the same location or in the same form on all oscilloscopes. A typical type oscilloscope (B+K, Model 1466A) found in an introductory physics lab is shown in Fig. 38-6(a).

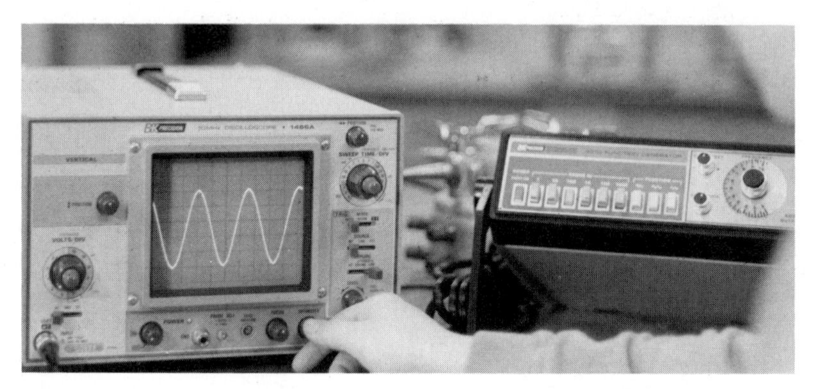

Fig. 38-6 (a) A typical student model oscilloscope. (b) A sine wave on oscilloscope from a function generator input. (Photo in (a) courtesy of B+K Precision, Dynascan Corporation.)

EXPERIMENTAL PROCEDURE

1. Locate and familiarize yourself with the following general controls and connections on your oscilloscope. The operating manual of your instrument should be consulted for specific and detailed explanations of the operating controls and for initial control settings.

DISPLAY CONTROLS

(a) INTENSITY—adjusts the rate of electron emission of the cathode, and hence the brightness of the spot.
(b) FOCUS—adjusts the sharpness of the spot.
(c) ASTIG (astigmatism)—adjusts vertical and horizontal positions of trace to same position. If a well-defined trace cannot be obtained with focus control, it may be necessary to adjust the astigmatism control.
(d) SCALE LIGHT (optional)—adjusts a lamp that lights the edges of the screen scale.
(e) POWER SWITCH (OFF-ON)—turns the oscilloscope on and off. This switch is often on the same shaft as one of the preceding controls.

POSITION CONTROLS

(a) HORIZONTAL POSITION (H-POS)—adjusts the horizontal position of the spot or trace on the screen.
(b) VERTICAL POSITION (V-POS)—adjusts the vertical position of the spot or trace on the screen.
(c) HORIZONTAL GAIN—provides continuous adjustment of the gain of the horizontal amplifier (i.e., amplifies or "magnifies" the trace).
(d) VERTICAL GAIN—provides continuous adjustment of the gain of the vertical amplifier (i.e., amplifies or "magnifies" the trace).
(e) VOLTS/DIV or VERTICAL ATTENUATOR—provides step adjustment of vertical sensitivity in calibrations of volts per grid division or volts/cm. [On uncalibrated scopes, at the "1" position the amplitude of the signal

voltage is unchanged (no attenuation). At the "10" or "10:1" position, the amplitude is attenuated by a factor of 10, and so on.]

(f) VARIABLE VOLTS/DIV (VERTICAL GAIN)—provides fine control of vertical sensitivity. In CAL (calibrated) position, the VERTICAL ATTENUATOR is calibrated. May also serve as vertical gain in some operations. (On uncalibrated scopes, the VERTICAL GAIN provides fine control of vertical sensitivity between steps of VERTICAL ATTENUATOR settings, usually in steps of 10.)

SWEEP TIME AND TRIGGERING MODE

(a) SWEEP TIME/DIV or SWEEP (HOR) SELECTOR—selects calibrated horizontal sweep rates in terms of time per grid division. On instruments with sweep selector frequency controls, the switch positions are calibrated in terms of frequency rather than time. For example, set on the 100- to 1000-Hz range, the sweep time is between $\frac{1}{100}$ and $\frac{1}{1000}$ of a second ($t = 1/f$). Consult instrument Instruction Manual for operation of any additional sweep selections.

(b) VARIABLE TIME/DIV (*X* or HORIZONTAL GAIN or SWEEP VERNIER)—provides fine sweep time adjustment. In CAL (calibrated) position, the sweep time is calibrated in time/div steps. (On uncalibrated scopes, horizontal gain adjustment is provided by a HORIZONTAL GAIN control.)

(c) TRIG MODE—three-position switch that selects triggering mode.
   - AUTO—triggered sweep operation when trigger signal is present and automatically generates sweep in absence of trigger signal.
   - NORM—normal triggered sweep operation. No trace unless proper trigger signal is applied.
   - *X-Y*—vertical input signal produces vertical (*Y*-axis) deflection, EXT input signal produces horizontal (*X*-axis) deflection.

(d) SOURCE or SYNC (HOR) SELECTOR—selects the triggering source (i.e., source that determines *when* spot is triggered or starts sweeping across screen).
   - INT—wave form being observed is used as sync trigger.
   - LINE—sweep is triggered by line voltage or in step with (60 Hz) line frequency.
   - EXT—sweep is triggered by signal applied to EXT jack.

EXTERNAL CONNECTIONS

(a) VERTICAL INPUT—applies an external signal to the vertical amplifier. The lower terminal is usually grounded (GND) to the instrument case. (Sometimes there are two terminals above ground for balanced inputs. The lower terminal must be grounded for grounded-side input and is usually gauged to the ground terminal by a connector. See the Instruction Manual for the instrument.

(b) HORIZONTAL INPUT and/or EXT TRIG —applies an external signal to the horizontal amplifier. May also be input terminal for external trigger signal.

2. Consult the instrument Instruction Manual for initial starting procedure and control settings. Make the appropriate control adjustments. Turn on oscilloscope. A trace should appear on the screen.

   Adjust the INTENSITY and FOCUS. *Never advance the intensity control to the point where an excessively bright spot or trace appears on the screen. A bright spot can burn the screen and decrease its useful life.* Adjust the HORIZONTAL POSITION and VERTICAL POSITION controls so that the spot is in the center of the screen. If you have trouble obtaining a spot, ask the instructor for assistance.

3. Calibrated Scopes: With a low SWEEP TIME/DIV setting, adjust the VARIABLE TIME control and note the effect.

   Uncalibrated Scopes: Advance the HORIZONTAL GAIN and note the trace. (The horizontal trace may have an unavoidable stray sinusoidal "ripple.") Return the gain to zero.

   The following general procedures are divided into Calibrated Scope and Uncalibrated Scope sections, with the latter being assumed to have an internal voltage calibration signal and an internal 60-Hz line horizontal input.

A. *Time and Frequency Measurements*

CALIBRATED SCOPES

1. Set the TRIG MODE to NORM, the SOURCE SWITCH to INT, and the SWEEP TIME variable control to CAL. Connect the function generator sine-wave output to the VERTICAL (*X*) INPUT of the scope. Turn on the function generator and set the generator frequency at 90 Hz. Adjust the function generator amplitude control

so that the sine-wave pattern is almost full-scale on the screen. (Switch the TRIG MODE momentarily to X-Y. What do you observe and why?)

2. Adjust the SWEEP TIME/DIV control so that a wave pattern with two peaks appears on the screen. Read the number of divisions for one full sine-wave cycle and record in the Laboratory Report.

3. (a) Compute the time period (T) of one cycle of the wave pattern using the calibrated SWEEP TIME/DIV setting and record.
   (b) Compute the sine-wave frequency ($f = 1/T$) and compare with the function generator setting by computing the percent difference.

4. Repeat the preceding procedures for a function generator output of 300 Hz.

5. With the generator output still at 300 Hz, adjust the SWEEP TIME/DIV control in various steps and note the relationship of the number of wave cycles to the sweep time/div. Can you explain? (See Question 1 at the end of the experiment.)

6. Set the SOURCE SWITCH to LINE. Adjust the function generator frequency to 60 Hz so a stationary pattern appears on the screen. This matches the generator frequency to the relatively stable 60-Hz line frequency, which is more accurate than the calibration markings on the function generator. Compare the generator frequency setting to the line frequency by finding the percent error.

7. Adjust the calibrated SWEEP TIME/DIV control until one or more full sine-wave cycles appear on the screen. Then, compute the frequency of the sine wave appearing on the scope as before.

UNCALIBRATED SCOPES

1. Set the VERTICAL ATTENUATOR (or CALIBRATION control) to calibration. (This introduces a 60-Hz sinusoidal calibrated voltage signal to the vertical amplifier by which the vertical voltage scale may be calibrated—so many volts per centimeter on the screen scale.) Adjust VERTICAL GAIN and note effect. With the SWEEP SELECTOR set on the 10- to 100-Hz range, advance the HORIZONTAL GAIN and note the trace.

2. Adjust the SWEEP VERNIER to 60 Hz and note the trace. The sweep-rate frequency is now equal to the vertical voltage frequency. What should be observed, and why? Adjust the SWEEP VERNIER so that exactly one sine-wave cycle appears on the screen. This matches the sweep-rate frequency to the line frequency (60 Hz). The sweep-knob setting may be slightly off.

3. Set the HORIZONTAL SELECTOR to 60 Hz (or LINE) and note the pattern. Can you explain what appears? Adjust the HORIZONTAL and VERTICAL GAINS and note the pattern. Are the input voltages exactly in phase?

4. Leave the SWEEP VERNIER set at the matched 60-Hz setting, set the HOR SELECTOR to + or −, and the VERTICAL GAIN and VERTICAL ATTENUATOR at their lowest settings. Connect the function generator sine-wave output to the V-INPUT terminals. Turn on the function generator and set the generator frequency at 60 Hz. Adjust the generator amplifier (wave amplitude) so that the sine wave is almost full scale on the oscilloscope screen. Advance the VERTICAL ATTENUATOR to the 10 (or 0.1) setting (or the VOLTS/DIV control) and note the attenuation. Advance the VERTICAL GAIN until the pattern is about half-scale on the screen.

5. Finely adjust the frequency of the function generator so that the pattern is as stationary as possible. Record the generator frequency setting in the Laboratory Report. The sweep frequency, which has been matched to the relatively stable 60-Hz line frequency, is much more accurate than the calibration markings on the function generator and may be used to calibrate the function generator.

   Compare the generator frequency to the line-matched 60-Hz sweep frequency by finding the percent error.

6. Set the function generator frequency to 30 Hz. (Make a fine adjustment of the generator frequency to obtain a stationary pattern.) Note and record the number of whole and one-half fractional parts of complete sine-wave cycles. Also make a sketch of the pattern.

   Advance the generator frequency and find several frequencies that produce stationary patterns. Record the generator frequency and the number of sine-wave cycles for each stationary pattern. Round off the generator frequencies to

the nearest whole-10 Hz (e.g., 30 Hz, etc.) to take into account the error in the generator frequency calibration markings. Advance the HORIZONTAL GAIN to spread out the pattern if necessary.
Compute the ratio of the generator frequency and the line frequency (60 Hz) for each case.

7. Set the function generator for a frequency of 200 Hz and adjust the oscilloscope controls until a stationary pattern of one sine-wave cycle is obtained on the screen. Note the sweep rate or frequency of the signal indicated by the oscilloscope controls. Set the function generator to 2 kHz. How many cycles are there now? Adjust for one cycle. Do you see how the sweep rate can be calibrated from a generator?

B. *Voltage Measurement*

CALIBRATED SCOPES

1. (a) With the full sine-wave cycles on the screen (procedure 7), record the VOLTS/DIV control setting (with variable control at CAL).
   (b) Read the number of peak-to-peak divisions for the height of the wave pattern on the screen.
   (c) Compute the rms voltage of the wave.

UNCALIBRATED SCOPES

1. Set the VERTICAL ATTENUATOR (or CALIBRATION control) to calibration. This introduces a 60-Hz sinusoidal calibrated voltage signal to the vertical amplifier by which the vertical voltage scale may be calibrated—so many volts per centimeter on the screen scale. Consult the Instruction Manual or the instructor for the magnitude of the internal calibration signal.
   (a) Using the VERT GAIN, adjust the voltage signal so as to extend over several vertical scale divisions and compute the volts per division calibration. (Leave the VERTICAL GAIN set at this position throughout the rest of this procedure.)
   (b) With a 60-Hz input from the function generator, obtain a stationary wave pattern on the screen.
   (c) Read the number of peak-to-peak divisions for the height of the wave pattern.
   (d) Compute the rms voltage of the wave.

C. *Lissajous Figures*

CALIBRATED SCOPES

1. Set TRIG MODE to *X-Y* and apply a 60-Hz sine wave to the HORIZONTAL or *X* INPUT. A second function generator may be used. (Do not exceed the voltage limitation of the oscilloscope in any case.)
   With either the VERTICAL or HORIZONTAL INPUT set at 60 Hz, adjust the other input frequency to 30 Hz, making a fine adjustment of this generator frequency to obtain a stationary pattern. (If two generators are used, adjust the generator amplitude control(s) so the signals have equal amplitudes. This is done by alternately adjusting the signal amplitude with the other generator switched off.) Observe the pattern on the screen.

2. Adjust the generator frequency to 60 Hz and obtain a stationary wave pattern. Note that by slight adjustment you can vary the pattern between a straight line or a circle. Can you explain why?

3. Continue to increase the generator frequency and observe the various stationary and moving patterns. Record in the data table.

UNCALIBRATED SCOPES

1. Set the HORIZONTAL SELECTOR to 60 Hz (or LINE). Repeat the series of generator frequency settings as in the previous procedure 6 and observe the patterns, adjusting the frequency so a pattern is as stationary as possible. Record the frequencies and patterns in the data table. In particular, at the 60-Hz generator setting, notice that by slight adjustment of the frequency control that you can vary the pattern between a straight line and a circle. Can you explain this behavior and the Lissajous figures?

General: Calibrated and Uncalibrated Scopes

You should now be getting a feeling for the oscilloscope controls and operations. Adjust the various control knobs so as to better understand their functions. Also, connect the square-wave output of the function generator to the VERTICAL INPUT and investigate this wave form.

Name .... Section .... Date ....

Lab Partner(s) ....

EXPERIMENT 38 *Introduction to the Oscilloscope*

LABORATORY REPORT

Calibrated Scopes

A. *Time and Frequency Measurements*

Generator frequency 90 Hz

Calculations (show work)

No. divisions ....

Time Sweep/Div ....

Period ....

Frequency ....

Percent difference ....

Generator frequency 300 Hz

Calculations (show work)

No. divisions ....

Time Sweep/Div ....

Period ....

Frequency ....

Percent difference ....

Source—Line (60 Hz)

Generator frequency setting ....

Percent error ....

No. divisions ....

Time Sweep/Div ....

Period ....

Frequency ....

Calculations (show work)

Lab Partner(s) ....

EXPERIMENT 38 *Introduction to the Oscilloscope*

B. *Voltage Measurement*

Calculations (show work)

Volts/div ....

No. of divisions (peak-to-peak) ....

Rms voltage ....

Uncalibrated Scopes

A. Time and Frequency Measurements

Function generator frequency ....

Line frequency ....

Percent error ....

Stationary Wave Patterns Generator frequency	Sine cycles (number and sketch)	Ratio of generator frequency and line frequency
30 Hz	....	....
....	....	....
....	....	....
....	....	....
....	....	....
....	....	....

B. *Voltage Measurement*

Calculations (show work)

Volts/div ....

No. of divisions (peak-to-peak) ....

Rms voltage ....

Lissajous Figures

X-input frequency	Y-input frequency	Sketches of patterns
....	....	
....	....	
....	....	
....	....	
....	....	
....	....	

(continued)

QUESTIONS

Calibrated Scopes

1. How does the number of wave cycles seen on a screen for a fixed input frequency vary with the SWEEP TIME/DIV control setting? Why?

2. Explain why the 60-Hz Lissajous figure could be varied between a straight line and a circle.

Uncalibrated Scopes

1. What is the sweep time of a trace with a sweep rate of 60 Hz?

2. In the stationary wave patterns procedure, compare the number of wave cycles and the frequency ratio. Explain any similarity.

3. Explain why the 60-Hz Lissajous figure could be varied between a straight line and a circle.

Name .... Section .... Date ....

Lab Partner(s) ....

EXPERIMENT 39 *The RC Circuit: Oscilloscope Study*

ADVANCE STUDY ASSIGNMENT

Read the experiment and answer the following questions.

1. Compare the voltages across a capacitor in dc and ac *RC* circuits.

2. How is the time base of the horizontal oscilloscope trace determined?

3. Is the time base affected by changing (a) the horizontal and vertical gains or (b) the sweep frequency? Explain.

4. Explain how the time constant of an *RC* circuit is determined from a stationary oscilloscope pattern.

EXPERIMENT **39**

The RC Circuit: Oscilloscope Study

I. INTRODUCTION

The oscilloscope can be used to study many ac circuit characteristics. The screen display of voltage versus time allows a variety of measurements. In particular, in an **RC (resistance–capacitance) circuit,** the charging of the capacitor can be visually observed. And using the horizontal time scale, the time constant of the charging process can be readily determined.

In this experiment, the oscilloscope will be used to determine the time constant of an *RC* circuit as the capacitor is continually charged and discharged by an ac signal voltage.

II. EQUIPMENT NEEDED

- Function generator (square wave)
- Oscilloscope
- Two capacitors (0.01 μF and 0.1 μF) (or substitution capacitor box)
- Two known resistors (10 kΩ and 100 kΩ) and one unknown resistor of similar range (the unknown resistor may be wrapped in masking tape to conceal its value)
- Connecting wires

III. THEORY

The voltage across a capacitor in an *RC* circuit increases according to the formula

$$V = V_0(1 - e^{-t/RC}) \qquad \textbf{(39-1)}$$

where e is the base of the natural logarithm (2.718), V_0 is the maximum voltage amplitude, R the resistance in the circuit, C the capacitance, and the product RC the **time constant** of the circuit. (See the Theory section in Experiment 37.) Taking $V = 0$ at $t = 0$, after a time of one time constant, $t = RC$, the voltage is

$$V = V_0(1 - e^{-RC/RC}) = V_0(1 - e^{-1}) = V_0(0.63)$$

or

$$\frac{V}{V_0} = 0.63 \qquad \textbf{(39-2)}$$

That is, the voltage across the capacitor is 0.63 (or 63 percent) of its maximum value (Fig. 39-1).

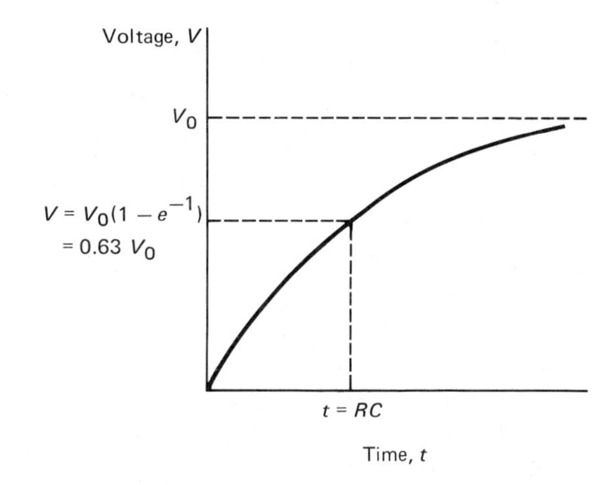

Fig. 39-1 A graph of voltage versus time for a capacitor charging in an *RC* circuit. In a time $t = RC$, the capacitor charges to 63 percent of its maximum voltage.

For a dc voltage source, the capacitor voltage further increases to V_0 and maintains this voltage unless discharged. However, for an ac voltage source, the capacitor voltage increases and decreases as the voltage of the applied signal alternately increases and decreases. For example, suppose that a square-wave ac signal as illustrated in Fig. 39-2 is applied to the circuit. The voltage across the capacitor increases according to Eq. 39-1 and then decreases according to the relationship

$$V = V_0 e^{-t/RC} \tag{39-3}$$

On an oscilloscope, the time base or the magnitude of the horizontal time axis is determined by the SWEEP TIME/DIV or the sweep frequency on the SWEEP SELECTOR and VERNIER. For example, for the latter, if the sweep frequency is 60 Hz, the time base is $\frac{1}{60}$ s (time base = 1/sweep frequency). Then, if a stationary pattern with two wave cycles is on the screen, the time or period of one cycle is $\frac{1}{120}$ s.

The time constant of an RC circuit can be determined from a stationary oscilloscope pattern of the capacitor voltage versus time. This is done by finding the fraction of the horizontal distance (time) needed for the trace to reach 0.63 V_0 with respect to the distance (time) for one full cycle. During the time the capacitor voltage is increasing, the voltage approaches V_0. Although the voltage decreases during the next half-cycle, it is assumed that if the voltage were still increasing, it would essentially reach V_0 during the time for one cycle. Hence, the fractional horizontal scale distance to reach 0.63 V_0 times the cycle period time is equal to one time constant RC.

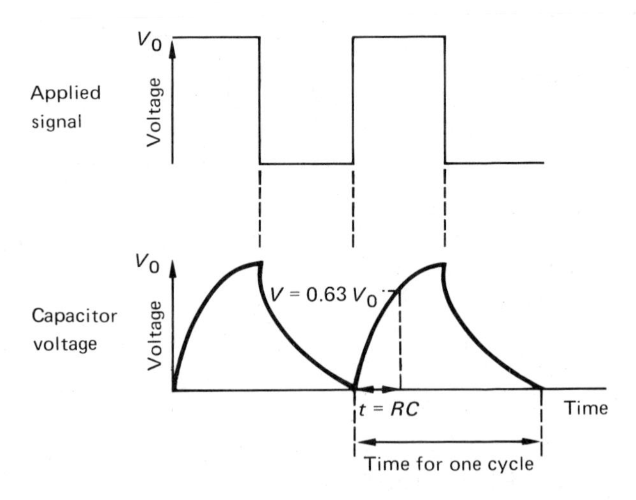

Fig. 39-2 When a "square"-wave signal is applied to a capacitor in an RC circuit, the capacitor periodically charges and discharges.

Example 39.1 It is estimated that the capacitor voltage reaches 0.63 V_0 in two horizontal scale divisions when the trace for one cycle covers 10 scale divisions. If the time for one cycle is $\frac{1}{120}$ s, what is the time constant?

Solution The fraction of the cycle time for the voltage to reach $0.63V_0$ is 2 div/10 div = 0.2, which for a cycle time of $\frac{1}{120}$ s is a time of

$$t = RC = 0.2 \times \tfrac{1}{120}\ \text{s} = 0.0017\ \text{s}$$

IV. EXPERIMENTAL PROCEDURE

1. Turn on the oscilloscope and function generator. Set the function generator frequency to 100 Hz and the wave amplitude near maximum. Connect the square-wave output of the function generator directly to the vertical input terminals of the oscilloscope. Obtain stationary traces of (a) one cycle and (b) two cycles of the square-wave pattern on the oscilloscope screen, using the proper oscilloscope techniques learned in Experiment 38.

2. Note and record in the Laboratory Report the SWEEP TIME/DIV or the sweep frequency of the oscilloscope for the two-cycle pattern. Compute the time base of the trace and the time for one cycle. As long as the sweep frequency is not changed, this trace can be used as a time base.

3. Then, set up the circuit as shown in Fig. 39-3, with $R = R_1 = 10\ \text{k}\Omega$ and $C = C_1 = 0.1\ \mu\text{F}$. *Have the instructor check the circuit before attaching the final lead to the oscilloscope.*

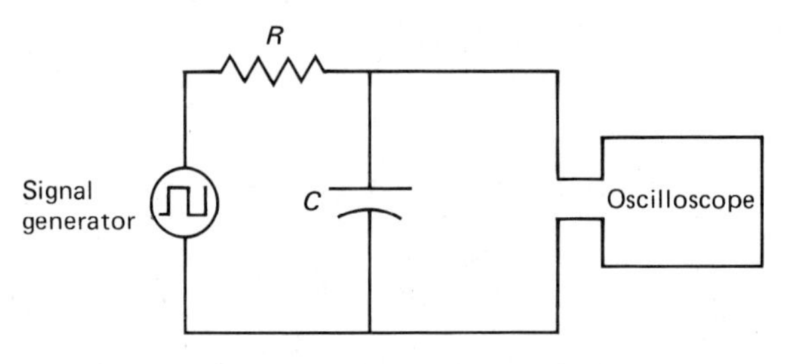

Fig. 39-3 Circuit diagram for experimental procedure for studying RC circuits.

4. Close the oscilloscope circuit by attaching the lead wire and note the pattern on the screen. A stationary two-cycle pattern should be observed. If the pattern is not stationary, adjust the sweep time or frequency and recompute the time base and time for one cycle. Make the appropriate gain adjustments if necessary. This does not change the time base.

5. From the stationary pattern on the screen, determine the time for the capacitor to charge to 0.63 V_0 by finding the fractional horizontal scale distance with respect to the distance for one full cycle. Use the cycle beginning at the center of the screen. Record the scale distance measurements in the data table.

6. Repeat procedure 5 with $R = R_2 = 100$ kΩ and $C = C_2 = 0.01$ μF.

7. Repeat procedure 5 using the unknown resistance as R and either C_1 or C_2 as C, whichever is more appropriate.

8. (a) Compute the time constants of the first two RC combinations, using the known R and C values and compare with the experimentally determined values by finding the percent errors.
   (b) Using the experimental time constant of the third measurement and the known capacitance value, compute the value of the unknown resistance.
   (c) Compare this to the accepted value (read from the unwrapped resistor or provided by the instructor) by computing the percent error.

9. Change the function generator frequency to 220 Hz and experimentally find the time constants of two more RC combinations not previously used. Change the SWEEP TIME/DIV or the sweep frequency of the oscilloscope if necessary. (Do not forget to compute the new time base if oscilloscope has sweep frequency.) Compare the experimental and computed values.

LABORATORY REPORT

Data

Signal generator frequency	100 Hz	220 Hz
Oscilloscope sweep frequency (if applicable)	....	....
Time base (or Sweep Time/Div)	....	....
Time for one (of two) cycles	....	....

(continued)

DATA TABLE

	R (kΩ)	C (μF)	Scale divisions for $0.63V_0$	Scale divisions for one cycle	Fractional scale divisions for $0.63V_0$	Expt. RC (Ω-F)	Computed RC (Ω-F)	Percent error
Case 1								
Case 2								
Case 3 (unknown R)							(computed R)	
Case 4								
Case 5								

Calculations (show work)

QUESTIONS

1. Based on the results of cases 1 and 2, under what conditions are the charging times of different RC circuits the same?

2. Is the charge of the capacitor zero at the beginning of each cycle? Explain in terms of the voltage increase.

3. How could the value of a unknown capacitance be determined using the experimental procedures? Show explicitly by assuming a value for an experimentally determined time constant.

Name .... Section .... Date ....

Lab Partner(s) ....

EXPERIMENT 40 *Phase Measurements and Resonance in ac Circuits*

ADVANCE STUDY ASSIGNMENT

Read the experiment and answer the following questions.

1. What is meant by the phase difference between the voltage and current in an ac circuit? Is there such a phase difference in dc circuits?

2. What is the effect of (a) capacitive reactance and (b) inductive reactance, and how does each affect the phase?

3. What is impedance, and is a circuit with impedance capacitive or inductive?

4. State the condition for resonance in a series *RLC* circuit, and describe the circuit effect of resonance.

5. Explain how the phase difference between voltage and current can be measured with a single-beam oscilloscope.

EXPERIMENT **40**

Phase Measurements and Resonance in ac Circuits

I. INTRODUCTION

In **ac (alternating current) circuits,** the current in the circuit alternates in direction. For an applied sinusoidal voltage, the current varies sinusoidally with time with the voltage. However, the current is not always exactly with the voltage in time. The current may "lead" or "lag" the voltage, depending on the circuit elements. That is, the current and voltage are out of phase.

In this experiment, the phase properties of particular circuit elements will be studied, together with the important condition of resonance. In this condition, the current and voltage are in phase and the current in the circuit is a maximum. This means that there is maximum power transfer from the voltage source to the circuit.

II. EQUIPMENT NEEDED

- Audio signal generator (sine wave)
- Oscilloscope (with horizontal and vertical amplifiers in phase or completely out of phase)
- Standard decade resistance box (10 to 9990 Ω)
- Two capacitors (0.1 μF and 0.5 μF, or capacitor substitution box)
- Air-core inductor (250 mH)
- Connecting wires

III. THEORY

When a dc (direct current) voltage source, such as a battery, is applied to an electrical circuit, the resulting current flows only in one direction, as the polarity (+ and −) of the voltage remains constant. For an ac (alternating current) voltage source, however, the polarity of the voltage alternates with time and the direction of the current flow in the circuit alternates with the same frequency as that of the voltage source.

One of the most commonly used ac voltages is one that varies sinusoidally with time. This may be generally described by the equation

$$V(t) = V_m \sin 2\pi ft \qquad \textbf{(40-1)}$$

where V_m is the maximum voltage amplitude and f is the frequency of the source. The angle $\theta = 2\pi ft = \omega t$ is called the **phase angle.**

The current in an ac circuit may or may not be in phase with the voltage, depending on the nature of the components in the circuit. In any case, if the applied voltage is sinusoidal, the current I will also be sinusoidal, and as a function of time may be expressed

$$I(t) = I_m \sin (2\pi ft - \phi) \qquad \textbf{(40-2)}$$

where I_m is the maximum amplitude of the current and ϕ the **phase constant.** In Eq. 40-1 the phase constant of the voltage has been assumed to be zero for simplicity. In this case, ϕ is also the phase difference between the applied voltage and the resulting current. (More generally, $\phi = \phi_I - \phi_V$ and is the angle between I_m and V_m, as illustrated in Fig. 40-1.)

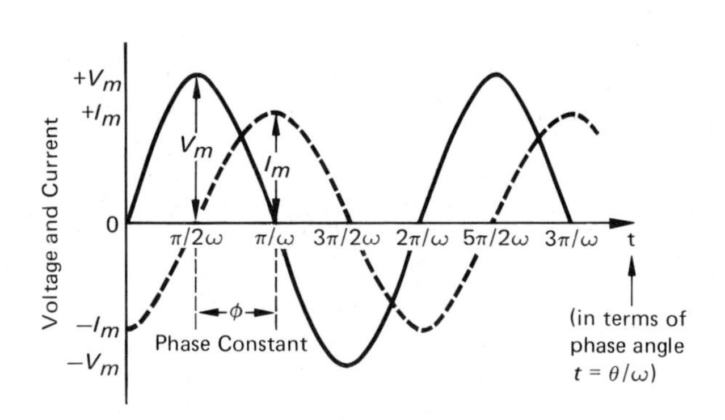

Fig. 40-1 ac Voltage and current. Here, the voltage leads the current by a constant phase difference ϕ.

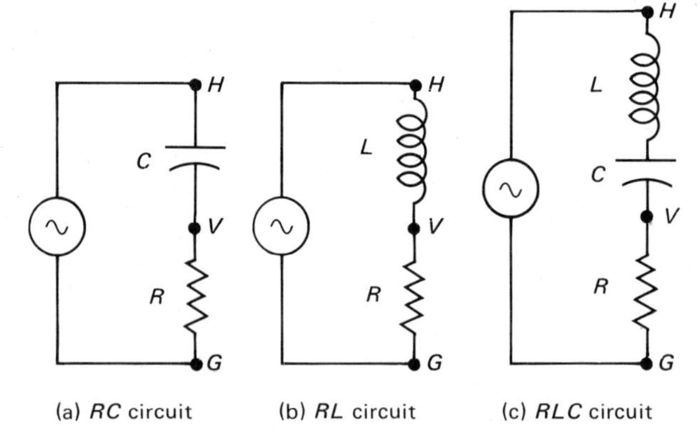

Fig. 40-2 Circuit diagrams for (a) *RC*, (b) *RL*, and (c) *RLC* series circuits.

The phase constant ϕ in Eq. 40-2 can be either positive or negative. When it is positive, the maximum value of the current I_m is reached at a later time than the maximum value of the voltage V_m (Fig. 40-1). In this case, we say that the voltage *leads* the current or the current *lags* behind the voltage by a phase difference ϕ. Similarly, if ϕ is negative, the current leads and the voltage lags.

When there is a capacitive element in the circuit [an *RC* circuit, Fig. 40-2(a)], the alternate charging and discharging of the capacitor opposes the current flow. This opposition is expressed as **capacitive reactance** X_C and

$$X_C = \frac{1}{2\pi fC} \qquad \textbf{(40-3)}$$

where C is the value of capacitance (in farads, F). The unit of X_C is ohms.

Similarly, when there is an inductive element in an ac circuit [an *RL* circuit, Fig. 40-2(b)], the self-induced counter emf in the induction coil opposes the current. The **inductive reactance** X_L is given by

$$X_L = 2\pi fL \qquad \textbf{(40-4)}$$

where L is the inductance of the coil (in henrys, H). The unit of X_L is ohms.

Many ac circuits have both capacitive and inductive reactance elements. The combined opposition to the current flow of resistive and reactive elements in a series circuit as shown in Fig. 40-2(c) (a series *RLC* circuit) is expressed in terms of the **impedance** Z of the circuit, which is given by

$$Z = [R^2 + (X_L - X_C)^2]^{1/2}$$

$$= [R^2 + (2\pi fL - \frac{1}{2\pi fC})^2]^{1/2} \qquad \textbf{(40-5)}$$

The unit of Z is also ohms. (R is the total resistance of the circuit. In general, it is assumed that the resistance of the induction coil is negligible compared with the resistor element.)

For the *RLC* series circuit, it can be shown that the phase relation of the voltage and current is given by

$$\tan \phi = \frac{X_L - X_C}{R} \qquad \textbf{(40-6)}$$

This relationship is often represented in a phasor diagram, in which the resistance and reactances are added like vectors (Fig. 40-3). Note that the angle ϕ is either positive or negative, depending on whether the inductive or capacitive reactance is greater. If X_L is greater than X_C, ϕ is positive and the current lags behind the applied voltage in time. The circuit is then said to be inductive. Similarly, if X_C is greater than X_L, ϕ is negative and the current leads the voltage. In this case, the circuit is said to be capacitive. [Equation 40-6 can be applied to single reactance circuits as in Fig. 40-2(a) and (b) by letting the appropriate reactance be zero.]

A common way to remember the phase relationship in inductive and capacitive circuits is by the phrase

ELI the *ICE* man

Here E is equivalent to the voltage (V). In an inductive circuit (L), the voltage leads the current as indicated by E "leading" I in *ELI* (see Fig. 40-1). Similarly, in a capacitive circuit (C), the current leads the voltage, *ICE*.

Since the voltage and current are continually changing in an ac circuit, it is convenient to consider effective or time-average values of the voltage and current. These root-mean-square (rms) values are given by $V = V_m/\sqrt{2}$ and $I = I_m/\sqrt{2}$, where V_m and I_m are the maximum or "peak" voltage and current, respectively. The rms values of V and I are the

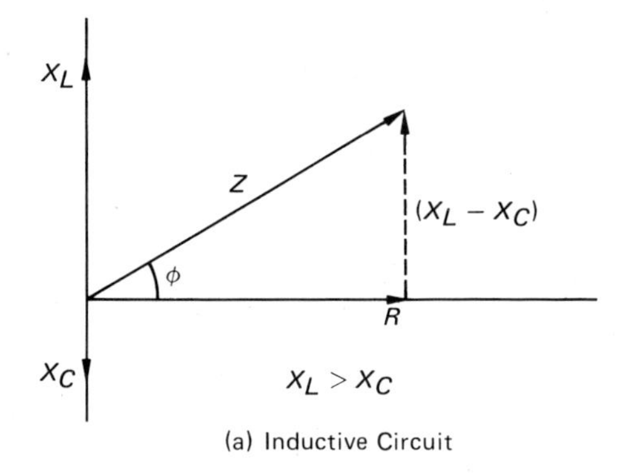

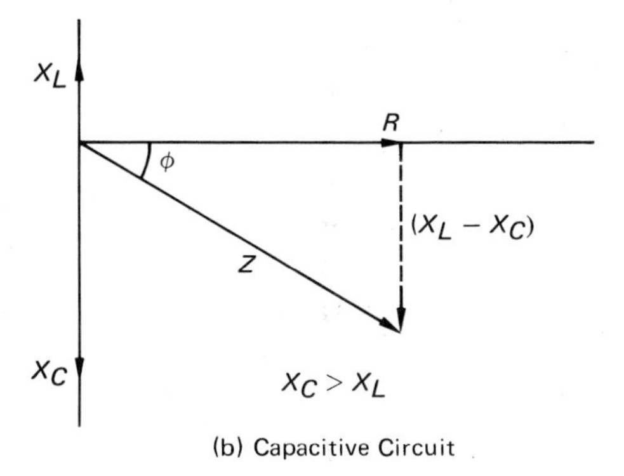

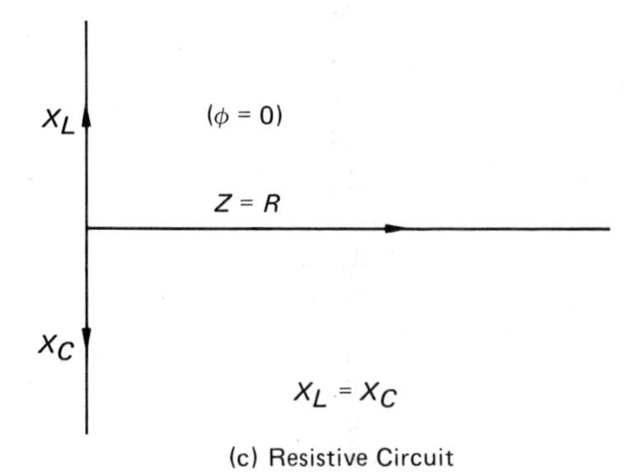

Fig. 40-3 Phasor diagrams for (a) an inductive circuit, (b) a capacitive circuit, and (c) a resistive circuit. In a phasor diagram, resistances and reactances are added like vectors.

values read on most ac voltmeters and ammeters. An Ohm's law type relationship holds between these values and the impedance Z:

$$V = IZ \tag{40-7}$$

and it follows that

$$V_m = I_m Z$$

Thus, for a given applied voltage, the smaller the impedance of a circuit, the greater the current in the circuit ($I = V/Z$). Notice in the reactance term in Eq. 40-5 that there is a minus sign and that the individual reactances are reciprocally frequency-dependent. As a result, for given L and C values, the total reactance can be zero for a particular frequency when $X_L = X_C$. That is,

$$2\pi fL - \frac{1}{2\pi fC} = 0 \tag{40-8}$$

and solving for f,

$$f_r = \frac{1}{2\pi\sqrt{LC}} \tag{40-9}$$

where f_r is called the **resonance frequency.** In this condition, the circuit is said to be in resonance. The impedance is then equal to the resistance in the circuit, $Z = R$, and the circuit is resistive [Fig. 40-2(c)].

Since the impedance is a minimum at the resonance condition, the maximum current flows in the circuit from the voltage source, and maximum power $P = I^2Z = I^2R$. For fixed values of L and C, resonance occurs at the particular resonance frequency f_r given by Eq. 40-9. However, notice from Eq. 40-8 that for a given source frequency, resonance can also be obtained by varying L and/or C in the circuit.

☐ **Example 40-1** A 10 (rms)-V, 1000-Hz voltage is applied to a series RLC circuit with $R = 1\,\text{k}\Omega$ and $L = 250$ mH. (a) What value of capacitance would put the circuit in resonance? (b) How much current would flow in the circuit in this condition?

Solution Given: $V = 10$(rms) V, $f = 10^3$ Hz, $R = 1\,\text{k}\Omega = 1000\ \Omega$, and $L = 250$ mH $= 0.250$ H. (a) To have f_r be 10^3 Hz, the required capacitance is given by

$$f_r = \frac{1}{2\pi\sqrt{LC}}$$

or

$$C = \frac{1}{L4\pi^2 f_r^2} = \frac{1}{(0.250)(4\pi^2)(10^3)^2}$$
$$= 0.1 \times 10^{-6}\ \text{F} = 0.1\ \mu\text{F}$$

(b) Since $Z = R$ for resonance,

$$I = \frac{V}{Z} = \frac{V}{R} = \frac{10}{10^3} = 0.01\ \text{A (rms)}$$ ☐

In this experiment, it is desired to measure the phase difference between the applied voltage and current in ac circuits, and to investigate resonance condition in a series (or parallel) RLC circuit using an oscilloscope. Notice from Eq. 40-6 that for the resonance condition, $X_L = X_C$, the voltage and current are in phase ($\phi = 0$), since $\tan \phi = 0$ (and $\tan 0 = 0$). To obtain a double-trace graph of V and I as in Fig. 40-1 requires a double-beam oscilloscope or special electronics, which are usually not available. However, phase measurements can be made with an ordinary oscilloscope by the following method.

Suppose that different voltage signals are applied to the horizontal and vertical inputs of an oscilloscope. If the ratio of the horizontal and vertical input frequencies is an integral or half-integral (see Experiment 38), then a stationary elliptical pattern such as in Fig. 40-4 is observed. Assume that the applied voltages have the forms

$$x = A \sin 2\pi ft$$
$$y = B \sin (2\pi ft - \phi) \qquad \textbf{(40-10)}$$

where A and B are the amplitudes. [Compare with Eqs. 40-1 and 40-2 that give the voltage $V(t)$ and circuit current $I(t)$, respectively.] Note that the y intercepts, $+b$ and $-b$, occur when $x = 0$, or when $2\pi ft = 0$ (since $\sin 0 = 0$). Then from the second equation, $b = B \sin(-\phi)$. Hence,

$$\sin \phi = \pm \frac{b}{B}$$

and similarly

$$\sin \phi = \pm \frac{a}{A} \qquad \textbf{(40-11)}$$

You should be able to prove and understand that if $A = B$ and $\phi = 90°$, the trace would be a circle. Also, if $A = B$ and $\phi = 0°$, the trace is a straight line.

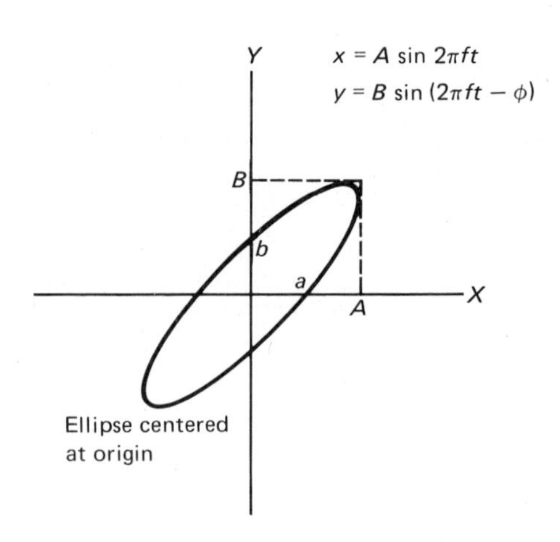

Fig. 40-4 The elliptical pattern used to measure the phase angle difference.

To measure the phase angle difference of the voltage and current in a circuit, the voltage signal from the signal generator is applied to the horizontal input. For example, for the circuits in Fig. 40-2, connections are made to points H and G, where G is to the horizontal ground terminal. The voltage input signal has the form of Eq. 40-1.

The current signal is applied to the vertical input with connections made to points V and G, where G is to ground. This is actually a voltage input from across the resistor in the circuit. However, it is proportional to and in phase with the current through the resistor (and therefore through the entire series circuit). Hence, the phase angle difference between the voltage and current can be determined from the shape of the resulting elliptical oscilloscope pattern, as described above.

IV. EXPERIMENTAL PROCEDURE

1. Turn on the oscilloscope and signal generator. It is necessary for the horizontal and vertical amplifiers of the oscilloscope to be in phase (or completely out of phase, $\phi = 180°$) to measure the phase differences. Check to see if this is the case by connecting the signal generator output to both the horizontal and vertical inputs of the oscilloscope.

   Set the generator frequency at 1000 Hz with the wave amplitude near maximum. If the horizontal and vertical amplifiers are in phase, the resulting trace is a diagonal straight line through the first and third quadrants. If the amplifiers are completely out of phase, the diagonal line will be through the second and fourth quadrants. In either case, phase measurements can be made. Vary the frequency of the generator to see if the phase difference of the amplifiers remains reasonably constant, in particular over the range 100 to 1000 Hz.

A. *Capacitive Circuit*

2. Set up a circuit as in Fig. 40-2(a). Use the standard decade resistance box as R set at 500 Ω and a capacitance of about 0.5 μF. Set the signal generator at a frequency of 1000 Hz. Connect

point G to one of the oscilloscope ground terminals and points V and H to the vertical and horizontal inputs, respectively. The SWEEP (HOR) SELECTOR should be on EXT.

Obtain an elliptical pattern on the screen of appropriate size for measurement by adjusting the HORIZONTAL and VERTICAL GAINS. Adjust the INTENSITY and FOCUS controls to obtain a sharp pattern.

3. Then vary (a) the generator frequency, (b) R, and (c) C (if a substitution box is used) one at a time and observe the effects.

4. Using the parameter values suggested in procedure 2, make measurements on the shape of the elliptical trace required to determine the phase difference angle ϕ. (Measure $2b$ and $2B$ or $2a$ and $2A$ for convenience, $\sin b/B = \sin 2b/2B$.) Record the measurements in the Laboratory Report.

5. Compute the phase-difference angle (Eq. 40-11) and the value of R (Eq. 40-6). Compare the experimental and known values of R by finding the percent error.

B. *Inductive Circuit*

6. Replace the capacitor with the inductor [Fig. 40-2(b)] and make $R = 2000\ \Omega$. Repeat part A for this arrangement. Alternatively, if the value of L is unknown, use the known values of R and f to compute L.

C. *RLC Circuit Resonance*

7. Set up a circuit as in Fig. 40-2(c) with $R = 500\ \Omega$ and $C = 0.5\ \mu$F. Vary the frequency of the signal generator and observe what happens. Determine, as accurately as possible, the frequency at which the circuit is in resonance. Record the resonance frequency in the data table. Using the known values of L and C, compute the resonance frequency and compare to the measured value by finding the percent difference.

8. Repeat procedure 7 with the 0.5-μF capacitance replaced by a 0.1-μF capacitance.

9. Change the scope SWEEP TIME SELECTOR from EXT to a suitable SWEEP rate. Observe the magnitude of the current $I(t)$ as the generator frequency is varied around the resonance frequency f_r, with the circuit as shown in Fig. 40-2(c). What happens to I at f_r? How sharp is the effect? Calculate and record

$$\frac{I\,(\text{at } f_r)}{I\,(\text{at } 2f_r \text{ or } 0.5f_r)}$$

(Use number of scale divisions for I. This is proportional to actual value.)

10. Rearrange the circuit for *anti-resonance* as shown in Fig. 40-5, and repeat procedure 9. Now what happens to I at f_r? How sharp is the effect? Calculate and record

$$\frac{I\,(\text{at } 2f_r \text{ or } 0.5f_r)}{I\,(\text{at } f_r)}$$

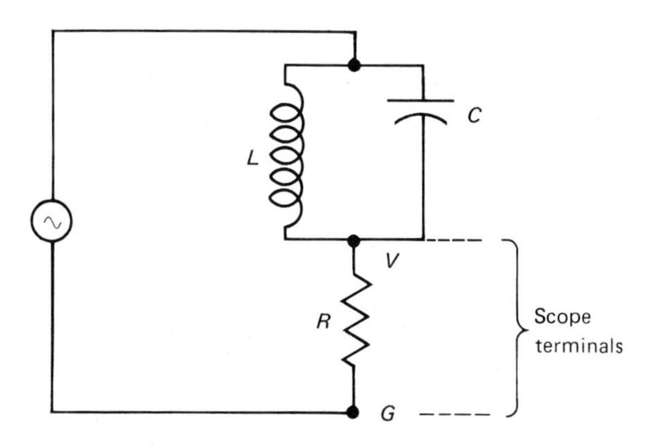

Fig. 40-5 Parallel (anti-resonance) LC circuit.

Name .... Section .... Date ....

Lab Partner(s) ....

EXPERIMENT 40 *Phase Measurements and Resonance in ac Circuits*

LABORATORY REPORT

A. *Capacitive Circuit*

R ....

C ....

Calculations (show work)

Generator frequency ....

$2a$ or $2b$ ....

$2A$ or $2B$ ....

Computed ϕ ....

Computed R ....

Percent error ....

B. *Inductive Circuit*

R ....

L ....

Calculations (show work)

Generator frequency ....

$2a$ or $2b$ ....

$2A$ or $2B$ ....

Computed ϕ ....

Computed R ....

Percent error ....

C. *RLC Circuit Resonance*

1. R .... Measured resonance frequency ....

   L .... Computed resonance frequency ....

   C .... Percent error ....

2. R .... Measured resonance frequency ....

   L .... Computed resonance frequency ....

   C .... Percent error ....

Describe current behavior at $f = f_r$.

Divisions at $f = f_r$ .... Sharpness ratio ....

Divisions at $f = 2f_r$ ....

3. Parallel (anti-resonance) LC circuit

   R .... Measured resonance frequency ....

   L .... Computed resonance frequency ....

   C .... Percent error ....

Describe current behavior at $f = f_r$.

Divisions at $f = f_r$ .... Sharpness ratio ....

Divisions at $f = 2f_r$ ....

(continued)

QUESTIONS

1. Prove mathematically that the oscilloscope pattern for horizontal and vertical input signals of the same frequency and amplitude (a) is a circle when the phase difference of the signals is 90°; (b) is a straight line for the resonance condition.

2. In tracing out an elliptical pattern on the screen, the oscilloscope beam moves in either a clockwise or a counterclockwise sense.
   (a) Describe the motions of the beam in terms of voltage and current lead or lag and the capacitive and inductive nature of the circuit.

   (b) Which way would the beam move in the *RLC* circuit in the experiment if the generator frequency were (1) greater than f_r, and (2) less than f_r?

3. For a series *RLC* circuit driven by a voltage signal of a given frequency other than f_r and a fixed R, how should L and C be varied within finite ranges so that the maximum current flows in the circuit?

4. Suppose that the resistance of the inductor coil were not negligible compared with the resistance in an *RL* or *RLC* circuit. How would this affect the impedance?

5. Why is a parallel *LC* circuit called an "anti-resonance" circuit?

Name .... Section .... Date ....

Lab Partner(s) ....

EXPERIMENT 41 *Electromagnetic Induction*

ADVANCE STUDY ASSIGNMENT

Read the experiment and answer the following questions.

1. If a straight wire in the plane of this paper and parallel to this typed sentence carries a current from left to right, indicate the directions of the magnetic field on both sides of the wire. (Make a sketch.)

2. What does magnetic flux measure?

3. Explain Faraday's law of induction.

4. Explain how Lenz's law gives the direction of the induced current in a wire when the change of flux is known.

5. What is magnetic permeability?

EXPERIMENT 41
Electromagnetic Induction

I. INTRODUCTION

Electromagnetic induction is one of the most important and applied relationships between electricity and magnetism. A classic discovery was made by the Danish physicist Hans Oersted in 1820 when he noticed the deflection of the needle of a compass near a current-carrying wire. This led to the conclusion that a magnetic field was produced by an electrical current.

A reasonable question to ask is whether the reverse is possible. That is, can a magnetic field produce an electrical current or potential difference in a conductor? The English scientist Michael Faraday investigated this question and found that the answer was yes, under certain conditions. This phenomenon is expressed in Faraday's famous law of induction, which is the basis of generation of electrical power.

II. EQUIPMENT NEEDED

- Pair of cylindrical coils (many turns on secondary relative to primary)
- Iron and brass or aluminum-core rods
- Two bar magnets of different pole strengths
- Low-voltage dc power supply or dry cell
- Magnetic compass
- Knife switch
- Portable galvanometer
- Metric ruler or meter stick
- Connecting wires
- 1 sheet of Cartesian graph paper

III. THEORY

Oersted discovered that a magnetic field is associated with an electrical current. He further established a relationship between the direction of the current and the direction of the magnetic field. The magnetic field lines are found to be closed circles around the current-carrying wire, as may be illustrated with a compass. The **direction of the magnetic field** is given by a **right-hand rule** (Fig. 41-1):

If a current-carrying wire is grasped with the right hand with the thumb extended in the direction of the *conventional* current, the curled fingers will indicate the circular sense of the magnetic field line.

The direction of the magnetic field at any point is then tangential to the circle. (Recall that the direction of the conventional current is in the direction that *positive* charges would flow in a circuit.)

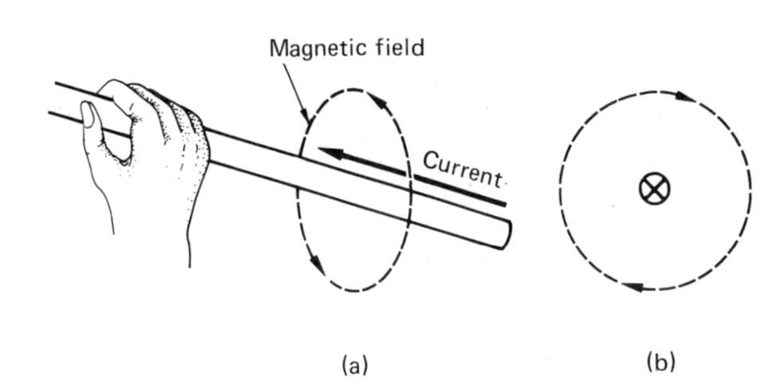

Fig. 41-1 The right-hand rule for determining the direction of the magnetic field. (a) When the thumb of the right hand points in the direction of the conventional current, the fingers curl in the circular sense of the magnetic field lines. (b) Looking along the wire in the direction of the current.

Faraday investigated the possible reverse effect of a current being produced by a magnetic field in the vicinity of a wire. No effect was found with a stationary magnet or magnetic field. However, it was discovered that a current is induced in a wire when there is a relative motion between the wire and the magnet. Thus, electromagnetic induction involves a time-varying magnetic field, for example, when a magnet is moved toward or away from a loop of wire. Since magnetic induction depends on *relative* motion, the same effect could be produced by moving the coil toward or away from a stationary magnet.

Investigations led Faraday to the conclusion that the important factor in electromagnetic induction was the time rate of change of the magnetic field B through a loop. The "total" magnetic field through a loop of wire can be characterized by what is called **magnetic flux** Φ:

$$\Phi = \mathbf{B} \cdot \mathbf{A} = BA \cos \theta \qquad \textbf{(41-1)}$$

where A is the cross-sectional area of the loop and θ is the angle between a normal to the plane of the loop and the magnetic field. Hence, the time rate of change of the magnetic field, or the number of field lines, perpendicular to and through a loop is given by the time rate of change of the magnetic flux, $\Delta\Phi/\Delta t = (\Delta B/\Delta t)A$.

The result of this work was **Faraday's law of induction,** which relates the induced voltage or emf in a wire to the time rate of change of flux:

$$V = -\frac{\Delta\Phi}{\Delta t} \qquad \textbf{(41-2)}$$

where V is the average value of the induced voltage over the time interval Δt. (By Ohm's law, the average induced current is $I = V/R$, where R is the resistance of the circuit.) Note that

$$\frac{\Delta\Phi}{\Delta t} = \left(\frac{\Delta B}{\Delta t}\right)A + B\left(\frac{\Delta A}{\Delta t}\right) \qquad \textbf{(41-3)}$$

That is, a flux change can be due to a change in the magnetic field through a loop of constant area, $(\Delta B/\Delta t)A$, and/or due to a constant magnetic field and a change in the area of the loop.

In either case, the number or density of field lines through the loop changes. The latter effect is commonly obtained by rotating a loop of wire in a constant magnetic field so that the "effective" area of the loop exposed to the field, and hence the flux, changes. In this experiment, we are concerned only with the effect of the first term in the equation.

The negative sign in Eq. 41-2 expresses another important law of electromagnetic induction, **Lenz's law,** which gives the direction of the induced current:

An induced current is in such a direction that its effects oppose the change that produces it.

Essentially, this means the induced current gives rise to a magnetic field that opposes the change in the original magnetic field. If this were not the case and the magnetic field arising from the induced current augmented the original field (i.e., was in the same direction as the original field), the induced field would increase the flux, which would increase the current, which would increase the flux, and so on. This would give rise to a something-for-nothing situation that violates the conservation of energy.

Another way to produce a time-varying magnetic field, and an induced voltage in a stationary wire loop, is to vary the current in a current-carrying loop (Fig. 41-2). When the switch in the circuit is closed, the current in the loop goes from zero to a constant value in a short time. During this time, however, the magnetic field associated with the current also increases or changes with time.

The magnetic flux through the adjacent loop then changes with time and an induced current momentarily flows in the loop, as indicated by the galvanometer deflection. The induced current goes to zero when the current in the battery circuit has a constant value. (Why?) Similarly, when the switch is opened, the magnetic field decreases and a current flows momentarily in the detector loop in the opposite direction to that of the first induced current. (Why?)

If there are a number of loops N in the detector circuit, the flux change through each loop contributes to the induced current or voltage, and Faraday's law becomes

$$V = -N\frac{\Delta\Phi}{\Delta t} \qquad \textbf{(41-4)}$$

Similarly, the magnitude of the magnetic field (number of field lines) would be increased if there were a number of loops in the battery circuit. If loops of wire are wound in a tight helix so as to form a coil, the arrangement is called a **solenoid.**

In a current-carrying solenoid, it can be shown that the magnetic field near and along the axis of the solenoid is given by

$$B = \mu_0 n I \qquad \textbf{(41-5)}$$

where n is the linear turn density of the coil (i.e., the number of turns per unit length N/L, where L is the length of the coil). The constant μ_0 is called the **permeability of free space** (vacuum or approximately air), and indicates that the solenoid has an air core.

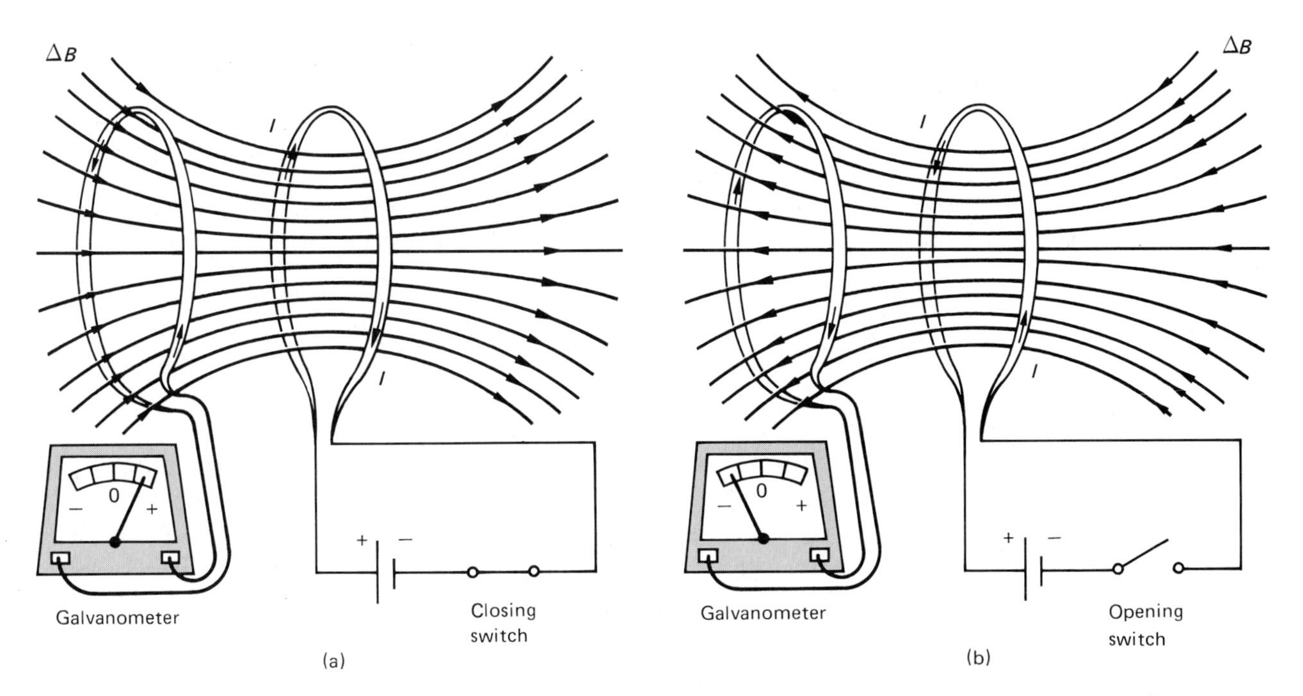

Fig. 41-2 (a) A time-varying field is produced when the switch in a circuit is closed and the current goes from zero to some steady value. The time-varying field builds up through a nearby loop and induces a current in the loop. (b) A similar, reverse situation occurs when the switch is opened.

If a material with a magnetic permeability μ is used as a solenoid core, μ_0 is replaced by μ in Eq. 41-5. The permeability expresses the magnetic characteristics of a material. Some materials with large permeabilities can be used to increase greatly the magnetic field in a current-carrying solenoid.

IV. EXPERIMENTAL PROCEDURE

1. In this experiment, it is important to know the direction of the induced current in a circuit. This is related to the positive and negative deflections of the galvanometer. To establish the direction of the galvanometer deflection to a known current direction, connect one terminal of the dry cell (or dc power supply at 1.5 to 3.0 V) to one terminal of the galvanometer and the other source terminal *through a large resistance* to the other galavanometer terminal (Fig. 41-3). Use yourself as the large resistance. (Really get into the circuit, so to speak.)

   From the known polarity of the source, relate the galvanometer deflection to the direction of current flow. For conventional current (assuming positive charge carriers), the current would flow *from* the positive source terminal. Galvanometer deflections to the right are usually labeled as positive and deflections to the left as negative. It is convenient to have a conventional current entering the positive galvanometer terminal to give a positive deflection.

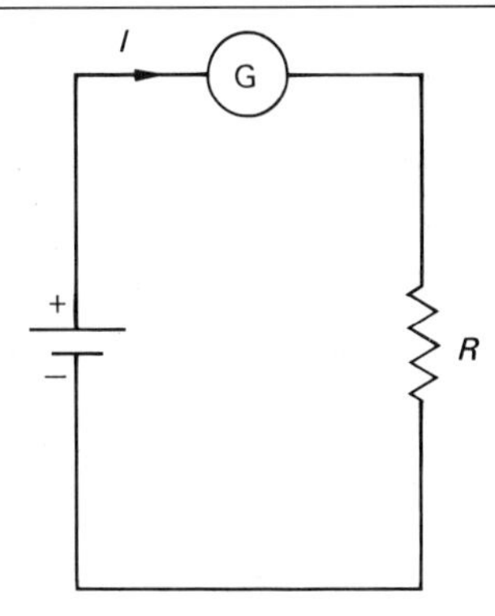

Fig. 41-3 Circuit for determining the direction of galvanometer deflection with respect to direction of current flow.

A.

2. Connect the galvanometer to the terminals of the secondary coil (the larger coil with the greater number of turns) as shown in Fig. 41-4. Use the compass to determine the relative strengths of the bar magnets. Then, using the stronger magnet,

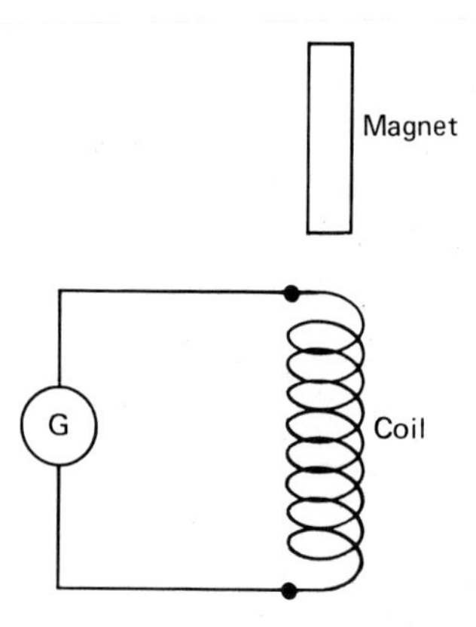

Fig. 41-4 Experimental setup for determining induced current due to a moving magnet.

move the magnet in and out of the coil, noting and recording the effects (relative magnitude and direction of the galvanometer deflection with (a) the speed at which the magnet is moved; and (b) the change of the magnet's polarity. Also note the effect when the magnet is stationary in the coil.

3. Repeat procedure 2 with the other magnet, and draw conclusions based on the experimental data.

B.

4. Set up the primary coil circuit as shown in Fig. 41-5 with the knife switch *S* open. (A secondary coil circuit is not needed in parts B and C.) Close the switch and with the compass, investigate the magnetic field around the coil. Make a sketch of the field pattern in the Laboratory Report.

C.

5. Open the switch and insert the stronger bar magnet into the primary coil almost the full length of the magnet. Close the switch and slowly remove the magnet from the coil. Note, record, and explain any observed effects.

D.

6. Open the switch and insert the primary coil into the secondary coil, which is connected to the galvanometer (Fig. 41-5). Close and open the switch, noting and recording the magnitude and direction of the galvanometer deflection in each case.

7. Repeat procedure 6 with each of the two metal cores. Draw conclusions based on the experimental data.

E.

8. Measure and record the length of the primary coil. With the switch open, insert the primary coil with the iron core completely into the secondary coil. Make a series of observations of the magnitudes of the deflections as the switch is opened and closed, withdrawing the primary coil 1 cm between the observations. Record the length of the primary coil still inside the secondary in each case.

   Find the average magnitude of the plus and minus deflections for each observation, and plot a graph of the average deflection magnitude versus the length of the primary inside the secondary. Use an abcissa scale of decreasing length. Interpret the results of the data.

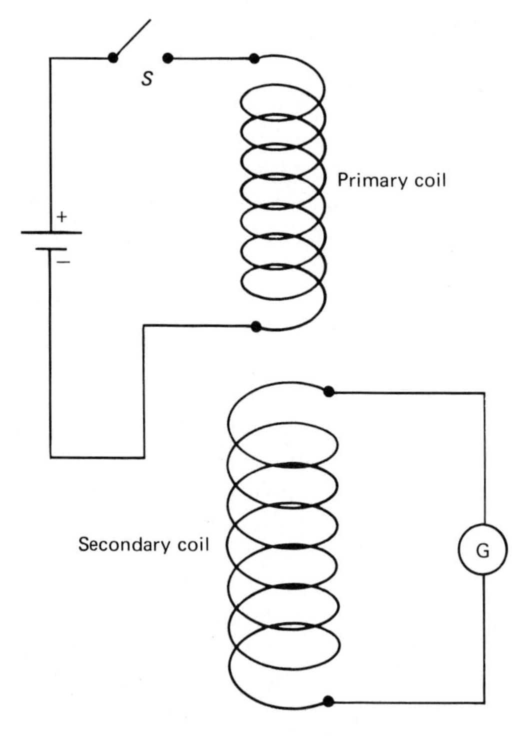

Fig. 41-5 Experimental setup for investigating electromagnetic induction.

Name .... Section .... Date ....

Lab Partner(s) ....

EXPERIMENT 41 *Electromagnetic Induction*

LABORATORY REPORT

A.

	Maximum deflection (scale divisions)	Current direction (+ or −)
Magnet 1		
Motion toward coil		
v_1	....	....
$v_2 > v_1$	....	....
Motion away from coil		
v_1	....	....
$v_2 > v_1$	....	....
Magnet 2		
Motion toward coil		
v_1	....	....
$v_2 > v_1$	....	....
Motion away from coil		
v_1	....	....
$v_2 > v_1$	....	....

Conclusions

B. *(Sketch)*

C. *(Observation and explanation)*

(continued)

D.

	Galvanometer deflection (scale divisions)	Secondary current direction (+ or −)
Air core		
switch closed	....	....
switch opened	....	....
Iron core		
switch closed	....	....
switch opened	....	....
.... core		
switch closed	....	....
switch opened	....	....

Conclusions

E.

DATA TABLE

Primary coil length L ....

Length of primary in secondary	Switch-closed deflection	Switch-opened deflection	Average deflection

Name .... Section .... Date ....

Lab Partner(s) ....

EXPERIMENT 41 *Electromagnetic Induction*

Interpretation of Data

QUESTIONS

1. Suppose that a bar magnet is dropped through a horizontal loop of wire connected to a galvanometer. Explain what would be observed on the galvanometer, and why.

2. Describe the change of flux through and the induced current in a loop of wire rotated in a uniform magnetic field. (Rotational axis of loop perpendicular to field.)

Name .... Section .... Date ....

Lab Partner(s) ....

EXPERIMENT 42 *Rectification: Semiconductor Diodes*

ADVANCE STUDY ASSIGNMENT

Read the experiment and answer the following questions.

1. What is a semiconducting material?

2. Distinguish between P-type and N-type semiconductors.

3. What is a semiconductor diode?

4. What is rectification?

5. How is a diode used to produce half-wave rectification? Explain in terms of voltage bias.

6. What is the frequency of the principal ac component of the rectified voltage from (a) a half-wave rectifier, and (b) a full-wave rectifier? Assume the original ac has a frequency of 60 Hz.

EXPERIMENT **42**

Rectification: Semiconductor Diodes

I. INTRODUCTION

Rectification is the conversion of an alternating current (ac) to direct current (dc). A battery delivers direct current, since the polarity of the battery voltage is fixed and never changes or alternates. On the other hand, commercial electricity, as in the home, is of the alternating type. That is, the polarity of the voltage (and the direction of the current) periodically changes with time.

In many applications, direct current is required, and, if an alternating source is to be used, the current must first be rectified. For example, a battery charger used to charge an automobile battery rectifies ac to dc. In the past, rectification was done exclusively with vacuum-tube diodes. Vacuum tubes have now been replaced for the most part by solid-state semiconductor diodes because of economy and compactness. In this experiment, the characteristics and rectification properties of semiconductor diodes will be investigated.

II. EQUIPMENT NEEDED

- Four signal diodes (e.g., IN 60)
- Resistor of known value (e.g., 1 kΩ)
- Oscilloscope
- Sine-wave generator
- Connecting wires
- 1 sheet of Cartesian graph paper
- Filter components: 10-Ω resistor, 30-μF capacitor (> 30 VDC rating)

III. THEORY

A **semiconductor** is a material that has a conductivity between that of good conductors (e.g., metals) and good insulators (e.g., glass and plastics). Common semiconductor materials used in solid-state electronic applications include germanium, silicon, and selenium. These crystalline materials in pure form are really poor conductors. They are made semiconducting by **doping** or adding impurity atoms.

A. *N-Type Semiconductors*

If the impurity atoms have more valence electrons than the atoms of the host material, there will be extra electrons within the material for conduction. Such negatively doped materials are called **N-type semiconductors.** The charge carriers in N-type materials are the *negative* excess electrons.

B. *P-Type Semiconductors*

If the impurity atoms have fewer valence electrons than the atoms of the host material, the material will be deficient in electrons. When a voltage is applied across the material, electrons from the host atoms move to fill the electron vacancies or "holes" created by the impurity atoms. The relative effect is the motion of the holes in the direction opposite that of the electrons. The moving holes take on the character of positive charges Such positively doped materials are called **P-type semiconductors.** The "charge" carriers in P-type materials are considered to be the *"positive"* holes.

C. *The Diode*

The boundary of two adjoining P-type and N-type materials is called a **P-N junction** (Fig. 42-1). At the junction, there is some diffusion of the materials and some recombination of electrons filling holes occurs. However, an electrostatic potential barrier is developed across the junction by those carriers that do not recombine. This barrier prevents the further diffusion motion of holes and electrons across the junction.

The P-N junction arrangement forms a **semiconducting diode.** When a battery is connected to the

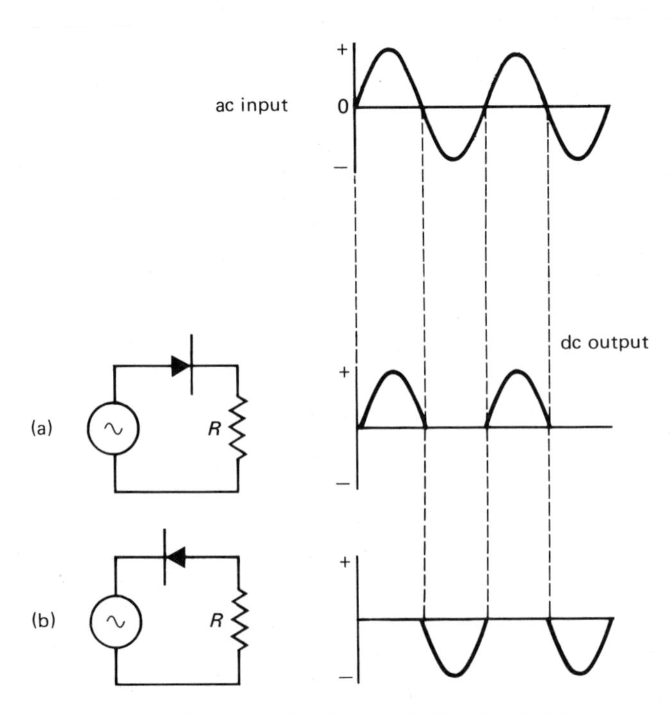

Fig. 42-2 Diode rectification. Diode circuit (a) conducts when the applied voltage is positive, and diode circuit (b) conducts when the applied voltage is negative.

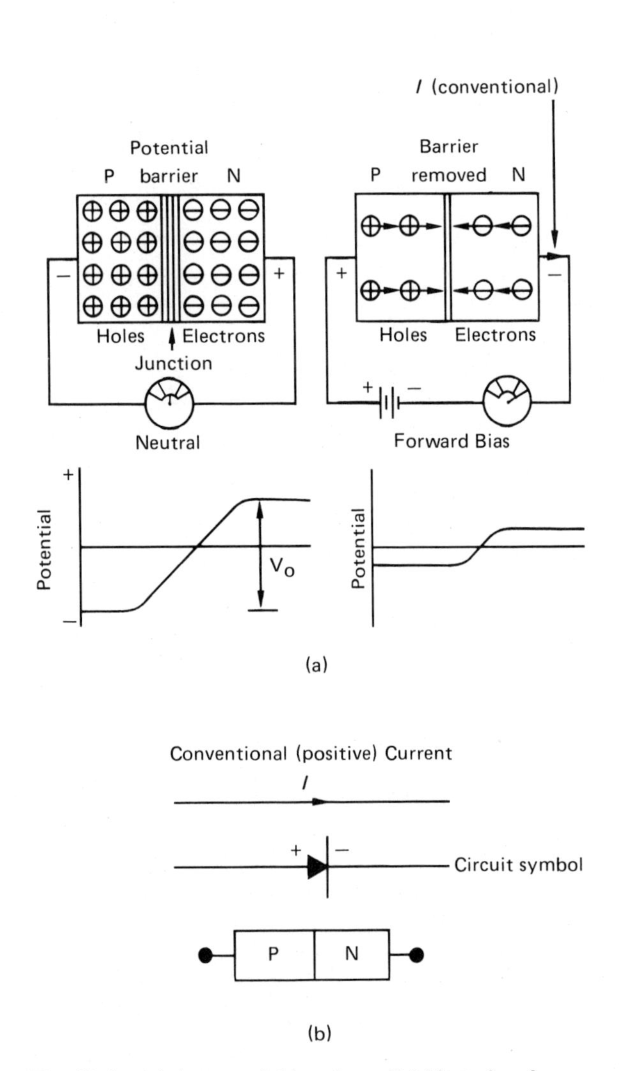

Fig. 42-1 (a) A potential barrier at P-N junction for neutral and forward-biased conditions. (b) Circuit symbol for semiconductor diode.

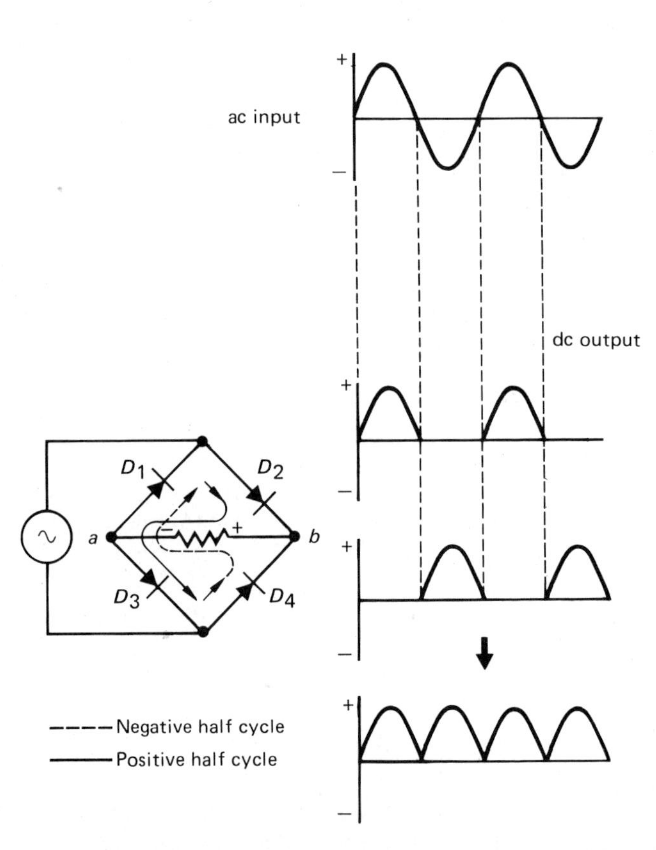

Fig. 42-3 Bridge rectifier circuit for full-wave rectification. See the text for a description.

diode, with the N-side negative and the P-side positive, the polarity is opposite to that of the electrostatic barrier at the junction. Hence, the barrier is diminished and current flows (Fig. 42-1). When the voltage is applied in this manner, the diode is said to be *biased* in the *forward* direction. By reversing the polarity, the junction barrier is increased and (ideally) no current flows. In this case, the diode is said to have a *reverse bias.* The circuit symbol for the diode shows the direction a conventional current would flow through the diode.

Because of these properties, the diode acts like a current "valve" that passes current only in one direction. When an alternating voltage is applied to a diode in series with a resistor, rectification is obtained (Fig. 42-2). In diagram (a), when the instantaneous voltage is positive, the diode conducts (is ON or forward-biased), and there is a voltage across the resistor. When the instantaneous voltage becomes negative, the diode is OFF (or reverse-biased) and does not conduct. Similarly, by reversing the direction of the diode, only the negative portion of the input voltage signal appears across the resistor. Each circuit gives what is called **half-wave rectification.**

Full-wave rectification can be obtained by using a bridge rectifier circuit as illustrated in Fig. 42-3. In operation, D_2 and D_3 are ON during the positive half-cycle, and D_1 and D_4 are ON during the negative half-cycle. Because both sets of diodes feed into the same end of the resistor, the output wave form is all positive. The negative half-cycles of the input signal are effectively inverted.

Filtering may be used to convert the half-wave or full-wave rectified ac into a voltage that more nearly resembles the steady dc voltage of a battery. Such filters are called *low-pass* filters because they allow the low-frequency (dc) component of the rectified voltage to pass through the load. The undesirable high-frequency component is either blocked (by an inductor) or shunted to ground (by a capacitor), or both. The simplest kind of low-pass filter is the $R'-C'$ network shown in Fig. 42-6. It is placed between the diode rectifiers and the load (R).

IV. EXPERIMENTAL PROCEDURE

A. *Half-wave Rectification*

1. Set up the circuit as shown in Fig. 42-4(a) with the oscilloscope vertical input at position A. (A band on the diode is commonly used to mark the cathode or negative end of the diode. Consult your instructor if you are not certain.)

   Set the sine-wave generator frequency to 60 Hz and adjust the output voltage so that the height of the wave form observed on the oscilloscope occupies one-half or more of the screen. Measure the ac voltage at the generator. For safety, it should be less than 20 V. Adjust the oscilloscope sweep selector so that a stationary pattern is obtained.

   Sketch the observed wave form in the Laboratory Report.

2. Move the oscilloscope lead to position B and observe the oscilloscope pattern. Sketch the wave form and explain the difference between that observed with the lead in position A.

3. Set the oscilloscope for calibrated voltage readings (V/cm) on the vertical axis of the display grid. (The calibration procedure varies with different models of oscilloscopes. Ask your instruc-

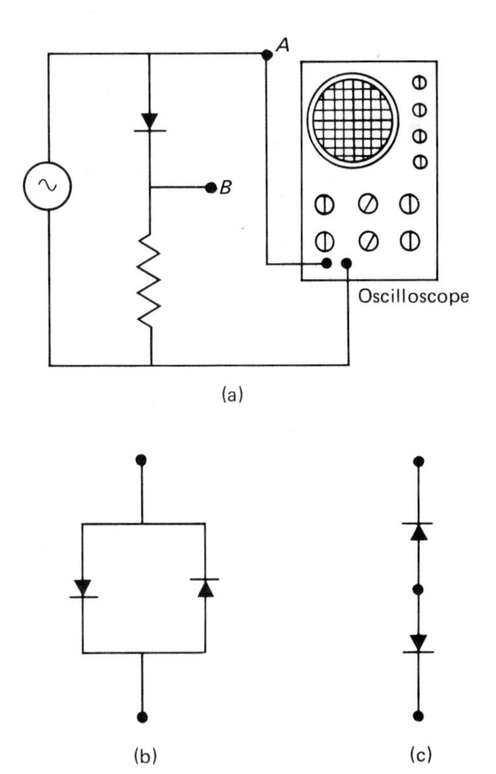

Fig. 42-4 Circuit diagrams for experimental procedure for half-wave rectification and forward and reverse resistances.

tor for assistance if necessary. For uncalibrated oscilloscopes, measure the relative voltage in centimeters.)

Measure and record the voltage at positions *A* and *B* for five or six different voltage settings on the signal generator. It is convenient to vary V_B in grid-mark steps.

4. Plot V_A versus V_B to determine if the resistance of the diode is linear.

B. *Forward and Reverse Resistances*

5. Replace the single diode in the circuit with the parallel diode arrangement shown in Fig. 42-4(b). Sketch the wave form of V_B in the Laboratory Report and explain it in terms of diode characteristics.

6. Measure the voltages V_A and V_B for a fixed generator voltage setting and record. Using the known value of the resistor, compute the forward resistance of the diodes. (*Hint:* Consider the circuit as two resistances in series.)

7. Repeat procedures 5 and 6 for the series diode arrangement shown in Fig. 42-4(c). Compute or estimate the reverse resistance of the diodes.

C. *Full-wave Rectification*

8. Set up the bridge rectifier circuit as shown in Fig. 42-5. Notice that this is equivalent to the circuit in Fig. 42-3 with the oscilloscope leads connected to points *a* and *b*. Sketch the observed wave form and explain it.

D. *Filtering*

9. Add the filter circuit as shown in Fig. 42-6, with $R' = 10\ \Omega$, $C' = 30\ \mu F$, and *R* the same as before (1 kΩ). Sketch the observed wave form. Indicate on the sketch the peak-to-peak amplitude of the ac component of the wave form. Compare this with the corresponding measurement in procedure 8 (without the filter).

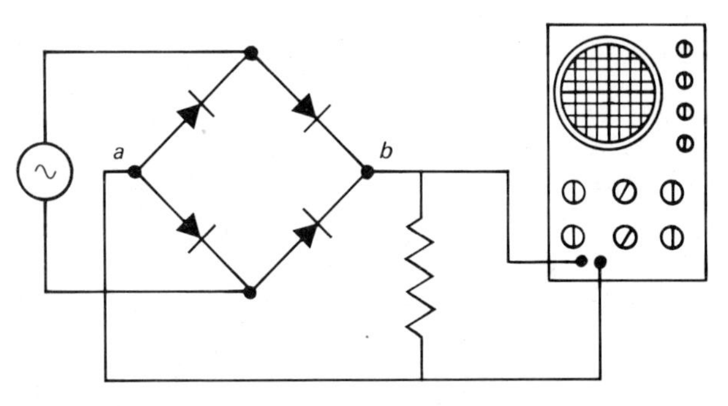

Fig. 42-5 Circuit diagram for experimental procedure for full-wave rectification.

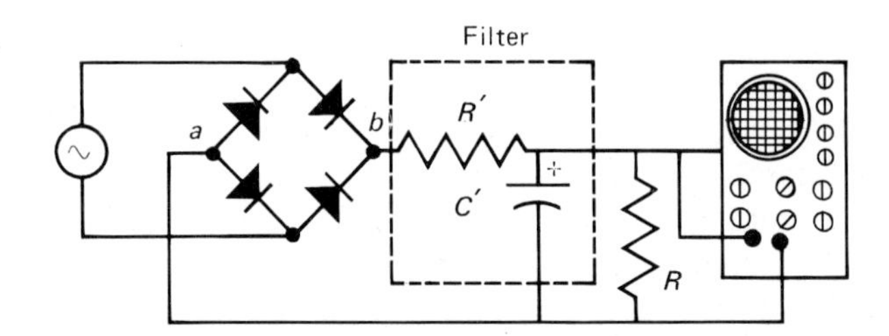

Fig. 42-6 Circuit diagram for filtering procedure.

Name .... Section .... Date ....

Lab Partner(s) ....

EXPERIMENT 42 *Rectification: Semiconductor Diodes*

LABORATORY REPORT

A. *Half-wave Rectification*

Observed Wave Forms

(1) Position *A*

(2) Position *B*

Explanation

V_A	V_B
....	....
....	....
....	....
....	....
....	....
....	....
....	....

B. *Forward and Reverse Resistance*

Parallel Arrangement Wave Form

(continued)

Explanation

V_A .... V_B ....

Resistor R (known) ....

Calculations (show and explain in detail)

Forward diode resistance ....

Series Arrangement Wave Form

Explanation

V_A .... V_B ....

Reverse diode resistance ....

C. *Full-wave Rectification*

Wave Form (V_B) (indicate peak-to-peak ac amplitude)

Explanation

D. *Filtering*

Wave Form (indicate peak-to-peak ac amplitude)

Improvement due to filtering:

$$\frac{\text{Filtered ac amplitude}}{\text{Unfiltered ac amplitude}} = \text{..................................}$$

(continued)

QUESTIONS

1. How does the graph of V_A versus V_B determine whether or not the forward diode resistance is linear? That is, what is the relationship of V_A and V_B to the diode resistance?

2. What is the purpose of the resistor in the rectifying circuits?

3. Another method of producing full-wave rectification is through the use of the center-tapped transformer and diode circuit shown in Fig. 42-Q3. Explain how the rectification occurs.

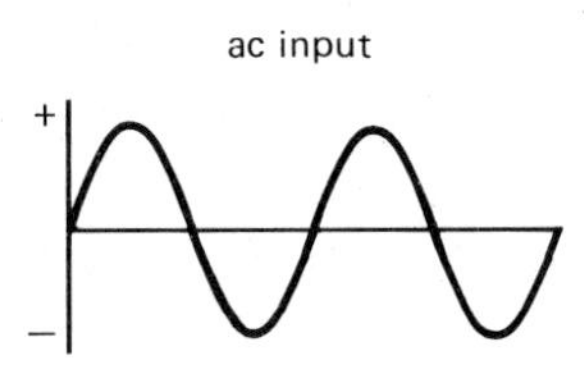

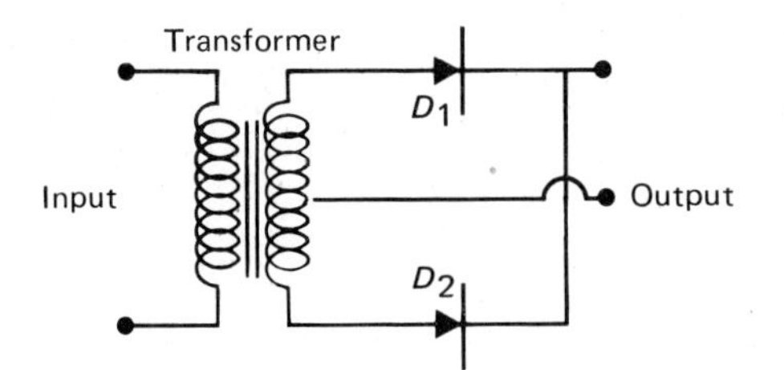

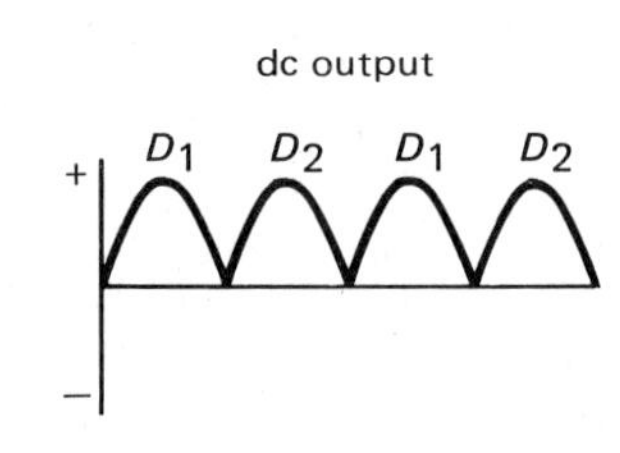

Fig. 42-Q3

4. Another low-pass filter makes use of an inductor (L') and a capacitor (C'). If an inductor L' were placed in series with the rectifier and the load in place of R' in Fig. 42-6, how would you connect the capacitor C' in the filter box to give low-pass filtering? (*Hint:* Think in terms of impedance and frequency.)

Name .... Section .... Date ....

Lab Partner(s) ....

EXPERIMENT 43 *Transistor Characteristics*

ADVANCE STUDY ASSIGNMENT

Read the experiment and answer the following questions.

1. Give several reasons why transistors make modern computers desirable and popular in comparison to early vacuum tube computers.

2. List the chief ways in which (a) transistors are related to diodes, and (b) PNP and NPN transistors differ.

3. List the changes that would have been made in Fig. 43-5 if a PNP transistor were used instead of an NPN transistor.

4. Describe the function of resistors R_1 and R_2 in Fig. 43-5.

5. Define what is meant by the current gain for a transistor, and give typical values for the current gain for (a) the common emitter configuration and (b) the common base configuration.

EXPERIMENT **43***

Transistor Characteristics

I. INTRODUCTION

The **transistor** is a solid-state device that is used to strengthen (amplify) electrical signals. It can be used to amplify information signals, from the very low levels that are present during transmission to the much higher levels that are needed at an output such as an audio speaker. A transistor can also be used as a switch to control many watts of output power, while using only milliwatts of control (input) power.

Early electrical control devices included mechanical relays and vacuum tubes. These devices had the disadvantages of being bulky, unreliable, and inefficient. (Relays wear out after a limited number of operations. Vacuum tubes need large amounts of filament heating power in order to function, and the heat causes rapid degradation of the device besides being wasteful.) The transistor is compact, reliable, and efficient. It has replaced the older devices in all but very special applications. Modern integrated circuits combine thousands of individual transistors on a single silicon chip that is smaller than a fingernail.

In this experiment, the electrical characteristics of a transistor will be investigated. A transistor circuit will be constructed to demonstrate the ability of a transistor to amplify an audio signal and to switch power.

II. EQUIPMENT NEEDED

- NPN transistor (150 mW, 20 V minimum)
- Composition resistors (1.5 kΩ, 15 kΩ, 22 kΩ, 33 kΩ, 47 kΩ, 68 kΩ)
- Two variable power supplies, 0–12 V, metered
- dc milliammeter, multiscale, 1/10 mA
- dc microammeter, multiscale, 100/1000 μA
- dc voltmeter, 5/25 V
- 1 sheet of Cartesian graph paper

(Optional)

- Resistors (0.1 kΩ, 1.0 kΩ)
- Two capacitors (1 μF)
- Audio generator
- Oscilloscope
- Diode
- High-resistance earphones

Note: These items may be varied to apply to available equipment. For example, a metered variable power supply may be replaced by a 1-kilohm potentiometer and a 25-VDC meter across two 6-volt batteries.

III. THEORY

A. *The Diode*

(See Experiment 42 for an introduction to semiconductor material, doping, P-type and N-type carriers, the P-N junction, and diode biasing.)

The semiconductor diode is an almost perfect rectifier. When the diode is forward biased ($V > 0$ in Fig. 43-1), large currents result from very small voltages. Typically, $I = +50$ mA when $V = +0.7$ V. When the diode is reverse biased ($V < 0$), very little current flows even for a large magnitude of voltage. Typically, $I = -0.01$ mA with $V = -10$ V. Therefore, very little power is dissipated or wasted in the diode itself when it is "on" (forward biased), and even less when it is "off" (reverse biased). For example, the *typical* semiconductor diode dissipates

*Contributed by Professor I. L. Fischer, Bergen Community College.

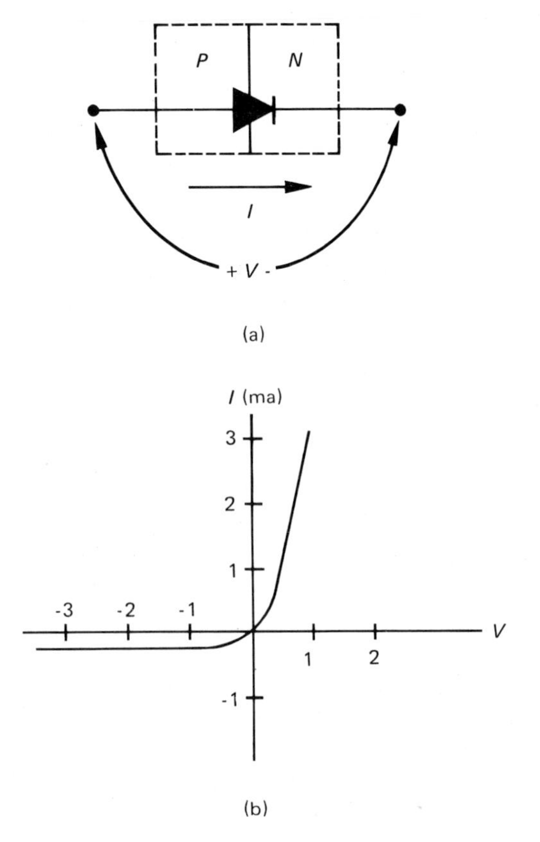

Fig. 43-1 The semiconductor junction diode. (a) Semiconductor doping, circuit symbol, and directions for voltage and (conventional) current. (b) Characteristic voltage-current curve. Forward bias corresponds to $V > 0$ and reverse bias to $V < 0$.

power $P = IV = (50 \text{ mA})(0.7 \text{ V}) = 35$ mW in the forward-biased condition and $P = IV = (-0.01 \text{ mA})(-10 \text{ V}) = 0.1$ mW when reverse biased. And yet, a power of about $P_{\text{max}} = I_{\text{max}}V_{\text{max}} = (50 \text{ mA})(10 \text{ V}) = 500$ mW is being "handled" by the diode.

Dissipation in this diode is clearly less than 10 percent of the load power being handled. If there were some way of turning the diode on and off, it would become an efficient control switch.

B. *The Transistor*

A transistor behaves much like a pair of back-to-back diodes with a common middle region, the **base.** This configuration enables us to achieve the desired control action. The outer sections of the transistor are called the **emitter** and the **collector.***

*Some transistors are actually symmetrical so that it doesn't matter which outer section is used as a collector or emitter, i.e., the circuit connections to the emitter and collector may be interchanged with virtually no change in performance. Most transistors, however, are not symmetrical; the doping levels of the emitter and collector are different for special reasons.

Semiconductor doping and circuit symbols for NPN and PNP transistors are shown in Fig. 43-2.

Consider the NPN transistor. Its "normal" biasing is shown in Fig. 43-3. The emitter-base diode is forward biased by the low-voltage battery E_{eb}.† The collector-base diode is reverse biased by the high-voltage battery E_{cb}.

Electrons are injected into the emitter from the negative terminal of E_{eb} and travel across the emitter-base diode junction into the base region. The base region is lightly doped to keep the injected electrons from recombining with the holes (positive charges) in this region. Thus, most of the injected electrons travel across the base region and are swept up by the positive collector. From the collector, they return through the load resistor (R_c) and E_{cb} to E_{eb} and ultimately to the emitter.

If the emitter-base diode is reverse biased instead of being forward biased, no electrons are injected into the emitter and almost no load current (I_c) will flow. Thus a large load current can be controlled by a small voltage change at the emitter-base junction.

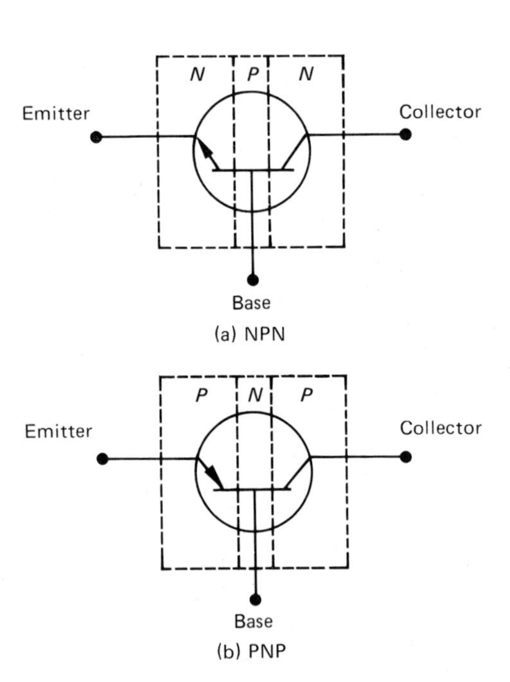

Fig. 43-2 The transistor semiconductor sections (doping) and circuit symbols. The emitter arrow indicates the direction of conventional current flow when the transistor is biased normally.

†It is common to use the symbol E for battery voltage. This should not be confused with the electrical field, which is also commonly represented by the symbol E.

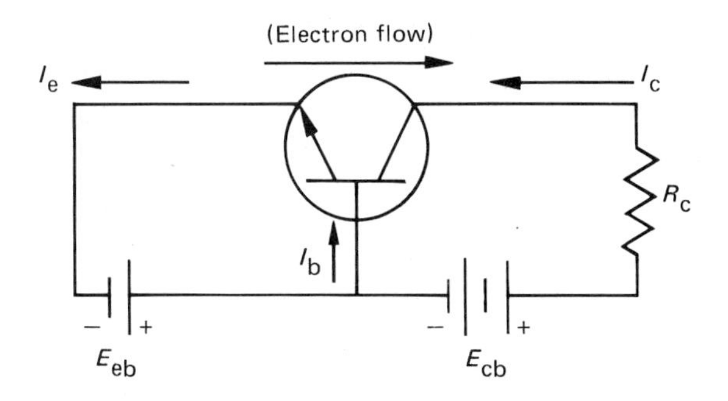

Fig. 43-3 Normal biasing of NPN transistor in the common base configuration. Current arrows generally refer to direction of conventional current flow. Electron flow direction is added for illustration.

C. *Amplification by a Transistor*

A very small amount of base current is needed to account for the small amount of recombination that does occur in the base region. This small base current effectively controls the much larger collector current. Hence, there is current gain or amplification from the base to the collector ($I_c > I_b$). Fig. 43-3(b) shows the current gain for a typical NPN transistor in its common emitter configuration. Note that a change of 0.04 mA of base current causes a change in the collector current of about 2.0 mA, yielding a current gain of about $I_c/I_b = 50$.

There is also considerable voltage gain in this configuration. The emitter-base diode is forward biased, so it has a low resistance and requires very little voltage to produce I_b. Conversely, the collector-base diode is reversed biased, has high resistance, and will yield a large voltage change in response to I_c. Thus, the voltage gain V_c/V_b is even greater than the current gain.

D. *Measurement of Current Gain*

In this experiment we will use the basic circuit shown in Fig. 43-4 to plot the current gain characteristics of an NPN transistor in its common emitter configuration. The variable dc power supply E_c provides a collector current I_c through the load resistor R_c. A milliammeter is used to measure I_c as shown. Similarly, the base current (I_b) loop includes E_b, R_b, and a microammeter.

The available ammeters may not be sufficiently sensitive to measure the small values of I_b. In that case, I_b may be calculated from the approximate equation

$$I_b = \frac{(E_b - V_{b_0})}{R_b} \tag{43-1}$$

where V_{b_0} is the "threshold" value of the transistor's base-emitter voltage V_b. (The voltage V_{b_0} is the value of V_b at which base current just begins to flow.) Equation 43-1 assumes that the value of V_b remains nearly constant at V_{b_0}. This is a reasonable approximation in our experimental procedure.

E. *Transistor Amplification (optional)*

In order to demonstrate transistor amplification in a realistic situation, we must add some components to

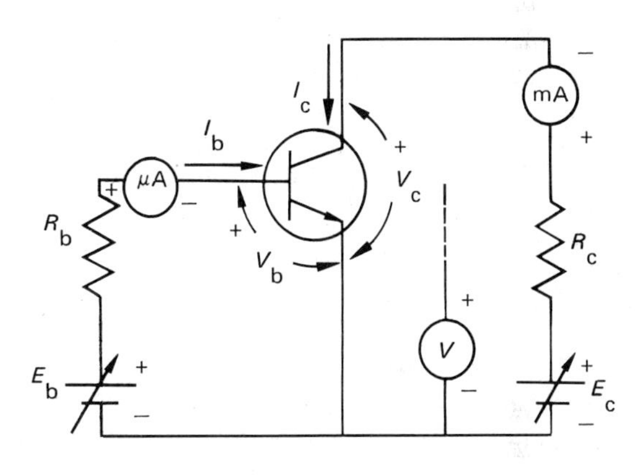

(a) Experimental circuit

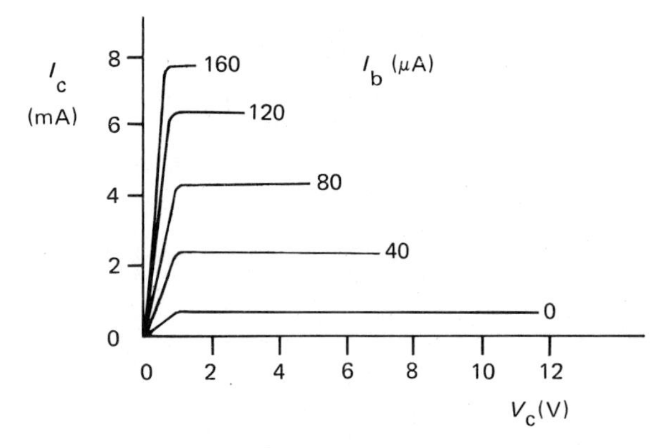

(b) Characteristic curves

Fig. 43-4 Normal biasing of NPN transistor in the common emitter configuration.

the basic circuit as shown in Fig. 43-5. Capacitors C_1 and C_2 allow us to connect an audio generator and earphones to the transistor without disturbing the transistor's dc bias. Diode D_1 protects the transistor against damage that might be caused by excessive reverse voltage across the base-emitter junction. Resistors R_1 and R_2 reduce the output of the generator to a usable level.

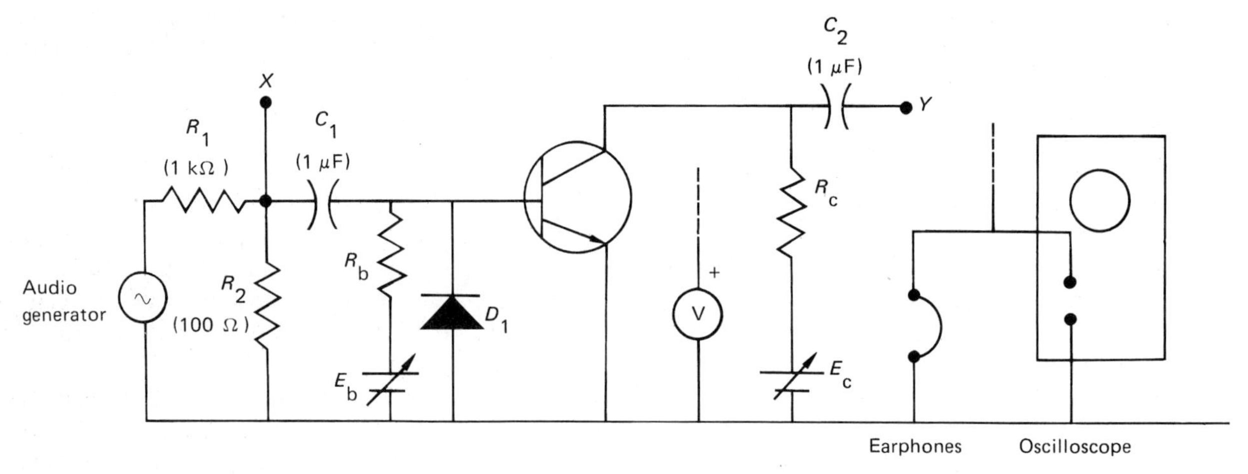

Fig. 43-5 Experimental circuit for demonstrating transistor amplification.

IV. EXPERIMENTAL PROCEDURE

A. *Transistor Current Gain*

1. Set up the circuit as shown in Fig. 43-4(a). Leave E_b and E_c set to *zero* until the circuit has been checked by your instructor. In the figure:
   (a) E_b and E_c are variable power supplies, 0–12 VDC. Each supply should include a voltmeter to constantly monitor the supply's output voltage.
   (b) (mA) is a multiscale milliammeter for measuring I_c. (μA) is a multiscale microammeter for measuring I_b. (See discussion in Theory section for Eq. 43-1.)
   (c) (V) is a multiscale voltmeter for measuring V_c or V_b. Begin with the positive lead of this voltmeter connected to the transistor collector to measure V_c.
   (d) Resistor values are $R_c = 1.5$ kΩ and $R_b = 15$ kΩ. R_b probably will be changed later. Identify the resistances by the colored bands on the resistor (see Appendix A, Table A5).

Note: Lay out the circuit on your table exactly as shown in the diagram. This will help to prevent errors and will facilitate your measurements.

Caution: Always begin with meters set to their least sensitive scale to avoid damage resulting from overloads. Increase sensitivity only as needed for a measurement and return the meter to its safest sensitivity setting immediately afterwards.

2. Prepare a sheet of graph paper of a plot of I_c versus V_c. Label and scale the axes as shown in Fig. 43-4(b).

   Plot data directly on the graph paper. This is much faster than recording the data and recopying it later.

3. With $E_b = 0$, slowly increase E_c from 0 to 12 volts.

   Caution: Do not let E_c exceed 12 V. Higher voltages may damage the transistor as a result of excessive power.

   Measure V_c and I_c at several points along the way, and plot these points on the graph. Do not be surprised if I_c seems to remain at zero. (Why?) If $I_c > 1$ mA, your transistor may be faulty—notify the instructor. Connect the data points with a smooth curve and label this curve $I_b = 0$.

4. With $E_c = 12$ V, slowly increase E_b until you get a small but measurable change in I_c (about 0.1 mA).
   (a) Temporarily transfer the positive lead of the voltmeter from the collector to the base of

the transistor in order to measure V_{b_0}, the *threshold* value of V_b. Record the value of V_{b_0} in the data table.

(b) Further increase E_b until V_c drops to 1.0 V. (Remember to make sure that the ammeters do not overload.) Measure I_b and V_b. These are the base current and voltage values needed to drive the transistor nearly into *saturation* (fully on). Record these values as I_{b_1} and V_{b_1} in the data table.

5. You are going to plot characteristic curves for values of I_b that increase in *steps* of I_b. The optimum size of I_b depends on the current gain of your transistor. A transistor with a high current gain should have a small I_b (and a large value of R_b). Calculate the optimum values of I_b and R_b for your transistor as follows:

$$I_b = \frac{I_{b_1}}{5} \quad \text{and} \quad R_b = \frac{(10\ \text{V})}{I_{b_1}}$$

Select the closest convenient value or I_b and R_b, and record your selections in the data table.

Reduce both power supplies to zero volts and change R_b to your selected value.

6. Adjust I_b for the first step ($I_b = I_b$). Slowly vary E_c through its range of 0 to 12 V. Remember not to exceed $E_c = 12$ V. Measure V_c and I_c at several points. Plot these points and connect with a smooth curve. Label the curve with the value of I_b.

7. Increase I_b in steps of I_b, plotting the characteristic curve for each value of I_b. Plot at least six curves. Make sure that your final curve has a low V_c (less than 1.0 V) even with E_c at its maximum value of 12 V.

B. *The Transistor as an Audio Amplifier (optional)*

8. Remove the ammeters from the experimental circuit and add components R_1, R_2, C_1, C_2, and D_1 as shown in Fig. 43-5. Leave E_b, E_c, and the audio generator output set to *zero* until the circuit has been checked.

The earphone/oscilloscope will be connected to either point X (the *input* of the transistor amplifier) or to point Y (the *output*). Begin with them connected to Y. The phones give an audible indication of the ac voltage levels in the circuit, while the scope produces a visual indication. They may be used together as shown, or either one may be used alone.

9. Set E_c to 12 V. Adjust E_b until V_c is about 5 V. Then remove the voltmeter.

10. Set the audio generator for a frequency of about 400 Hz sine wave. Slowly increase the generator output until a substantial sound is heard in the phones, and/or a substantial deflection is observed on the scope. Adjust the scope for a stationary full-screen pattern. Record the output level indicated by the scope (in divisions or in volts peak-to-peak).

11. Transfer the phones/scope to the input (X). Note the level of sound in the phones. Record the input level indicated by the scope. Calculate and record the voltage gain (output/input) for this circuit.

C. *The Transistor as a Power Switch (optional)*

12. Return the scope to the output (Y). Slowly increase the generator signal amplitude until the scope indicates that the amplifier is *clipping* (flattening) the top or bottom of the signal. This is called distortion.

Caution: If the sound in the phones becomes excessively loud before distortion is visible, disconnect the phones immediately. Otherwise, they may be damaged.

13. Continue increasing the signal amplitude until the output is well clipped at both top and bottom. The amplifier is now operating as a power switch, i.e., turning the signal power fully on and off during each cycle.

Disconnect the phones. Record the output level indicated by the scope in volts peak-to-peak.

Name .... Section .... Date ....

Lab Partner(s) ....

EXPERIMENT 43 *Transistor Characteristics*

LABORATORY REPORT

A. *Transistor Characteristics*

Threshold value of V_b: V_{b_0} ....

Saturation values of I_b and V_b: I_{b_1} .... V_{b_1} ....

Selected values of I_b and R_b: I_b .... R_b ....

Attach graph showing the characteristic curves of your transistor.

B. *Transistor Amplifier (optional)*

Output level ....

Input level ....

Voltage gain ....

C. *Transistor Switch (optional)*

Output level ....

Lab Partner(s) ....

EXPERIMENT 43 *Transistor Characteristics*

QUESTIONS

1. (a) Use your graph of transistor characteristic curves to calculate the current gain of the transistor (I_c/I_b along a vertical line representing a constant V_c) in three regions: (a) $V_c = 2$ V and $I_c = 1$ mA, (b) $V_c = 2$ V and $I_c = 6$ mA, and (c) $V_c = 9$ V and $I_c = 1$ mA.

   (b) How did the current gain depend on collector current? How much did they vary? (Calculate the percent difference between the highest and lowest values in regions *a, b,* and *c*.)

2. Use your measurements of V_{b_0} and V_{b_1} to calculate the voltage gain of the transistor as an amplifier as follows: (a) Compute the input voltage change, $\Delta V_b = V_{b_1} - V_{b_0}$; and (b) compute the corresponding output voltage change ΔV_c. (c) The voltage gain is $\Delta V_c/\Delta V_b$. Does the value of ΔV_b justify the appproximation used in Eq. 43-1?

3. Use your graph to calculate the efficiency of the transistor as a power switch as follows. Refer to your curve of largest I_b. Measure V_c and I_c at the highest point on that curve corresponding to $E_c = 12$ V. (a) Compute the output power being dissipated or wasted by the transistor, $P_o = I_c V_c$. (b) Compute the input power needed to turn the transistor "on," $P_i = I_b V_{b_0}$. (c) Compute the load power, $P_\ell = I_c E_c$. (d) Compute the efficiency $= 1 - \dfrac{(P_o + P_i)}{P_\ell}$, and express as a percent.

(continued)

(Optional procedures)

4. In procedure 11, you measured the voltage gain of the transistor as an amplifier. (a) How does the measurement compare with the calculation in Question 2? Which do you think is more accurate? Why?

   (b) How does the voltage gain compare with the current gain (Question 1). Explain why there is such a difference in magnitude between these two "gains."

5. In procedure 13, you measured the peak-to-peak output of the transistor as a power switch. How does this output compare with the value of the voltage supply E_c? Explain why this is so.

Name .... Section .... Date ....

Lab Partner(s) ....

EXPERIMENT 44 *Reflection and Refraction*

ADVANCE STUDY ASSIGNMENT

Read the experiment and answer the following questions.

1. What is the law of reflection, and does it apply to all reflecting surfaces?

2. Distinguish between regular and irregular reflection. Give an example of each.

3. Why is light refracted when it passes from one medium into an optically different medium?

4. Show by Snell's law that if the speed of light is less in a particular medium, then a light ray is bent toward the normal when entering that medium. What happens if the speed of light is greater in the medium?

5. What is the difference between the relative index of refraction and the absolute index of refraction? Explain why we can experimentally determine the absolute index of refraction relatively accurately using air as a medium.

EXPERIMENT **44**

Reflection and Refraction

I. INTRODUCTION

Reflection and refraction are two commonly observed optical properties of light. The reflection of light from smooth and polished surfaces, such as ponds of water and mirrors, allows us to view the images of objects. Also, when light passes from one medium into another, it is bent or refracted. As a result, a stick in a pond or a spoon in a glass of water appears to be bent.

As part of geometrical optics, these phenomena are explained by the behavior of light rays. Through ray tracing, the physical laws of reflection and refraction can be conveniently investigated in the laboratory. In this experiment a plane mirror and a glass plate are used to study these laws and the parameters used in describing the reflection and refraction of light.

II. EQUIPMENT NEEDED

- Pins
- Pin board (cardboard or poster board suffices)
- Sheets of white paper ($8\frac{1}{2} \times 11$ in.)
- Ruler and protractor
- Short candle (less than 5 cm)
- Rectangular mirror (and holder if available)
- Thick glass plate (approximately 8×10 cm)

Note: Ray boxes may be used if available.

III. THEORY

A. *Reflection*

When light strikes the surface of a material, some light is usually reflected. The reflection of light rays from a plane surface like a glass plate or a plane mirror is described by the **law of reflection:**

The angle of incidence is equal to the angle of reflection (i.e., $\theta_i = \theta_r$).

These angles are measured from a line perpendicular or normal to the reflecting surface at the point of incidence (Fig. 44-1).

The rays from an object reflected by a smooth plane surface appear to come from an image behind the surface, as shown in the figure. From equal triangles it can be seen that the image distance d_i from the reflecting surface is the same as the object distance d_o. Such reflection is called **regular or specular reflection.** The law of reflection applies to any reflecting surface. If the surface is relatively rough, like the paper of this page, the reflection will become diffused or mixed, so that no image of the source or object will be produced. This type of reflection is called **irregular or diffuse reflection.**

B. *Refraction*

When light passes from one medium into an optically different medium at an angle other than normal to the surface, it is "bent" or undergoes a

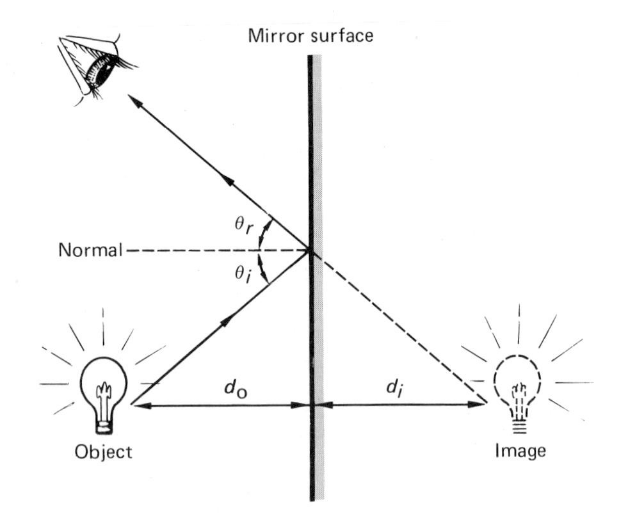

Fig. 44-1 The law of reflection, $\theta_i = \theta_r$.

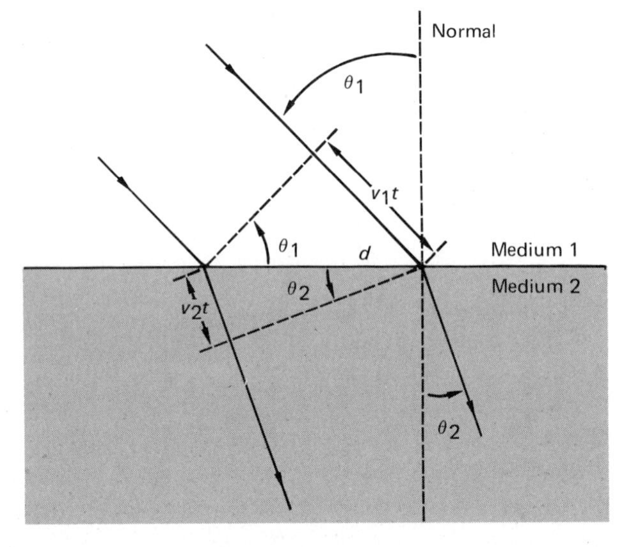

Fig. 44-2 The refraction of two parallel rays. When medium 2 is more optically dense than medium 1, $v_2 < v_1$, and the rays are bent toward the normal as shown here. If $v_2 > v_1$, the rays are bent away from the normal.

change in direction, as illustrated in Fig. 44-2 for two parallel light rays in a beam of light. This is due to the different velocities of light in the different media. In the case of refraction, θ_1 is the angle of incidence and θ_2 is the angle of refraction.

From the geometry of Fig. 44-2, we have

$$\sin\theta_1 = \frac{v_1 t}{d} \quad \text{and} \quad \sin\theta_2 = \frac{v_2 t}{d}$$

or

$$\frac{\sin\theta_1}{\sin\theta_2} = \frac{v_1}{v_2} = n_{12} \qquad \textbf{(44-1)}$$

where the ratio of the velocities n_{12} is called the **relative index of refraction,** and Eq. 44-1 is known as **Snell's law.** If $v_2 < v_1$ (as in Fig. 44-2), the rays are bent toward the normal in the second medium. And if $v_2 > v_1$, the rays are bent away from the normal (e.g., reversed rays in Fig. 44-2 with medium 2 taken as medium 1).

For light traveling initially in vacuum or approximately in air, the relative index of refraction is called the **absolute index of refraction** or simply the **index of refraction,** and

$$n = \frac{c}{v} \qquad \textbf{(44-2)}$$

where c is the speed of light in vacuum and v the speed of light in the medium. Hence, the index of refraction of vacuum is $n = c/c = 1$, and for air, $n \simeq c/c = 1$. For water, $n = 1.33$.

Snell's law can then be written

$$\frac{\sin\theta_1}{\sin\theta_2} = \frac{v_1}{v_2} = \frac{c/n_1}{c/n_2} = \frac{n_2}{n_1}$$

or

$$n_1 \sin\theta_1 = n_2 \sin\theta_2 \qquad \textbf{(44-3)}$$

where n_1 and n_2 are the indices of refraction of the first and second medium, respectively.

IV. EXPERIMENTAL PROCEDURE

A. *Reflection*

GLASS PLATE AS A MIRROR

1. Place a sheet of white paper on the table. As illustrated in Fig. 44-3, draw a line where the candle (or object) will be placed. The line should be drawn parallel to the edge of the width of the page and about 3 to 4 cm from the edge of the paper. Make a mark near the center of the line and place the candle on the mark.

   Put the glass plate near the center of the paper as shown in the figure. With the length of the plate parallel to the candle line, draw a line along the edge of the glass plate (side toward the candle). Light the candle, and looking *directly over the candle* with your eye as in position 1 in Fig. 44-3, you will observe an image of the candle in the glass plate. The glass plate reflects light and serves as a mirror.

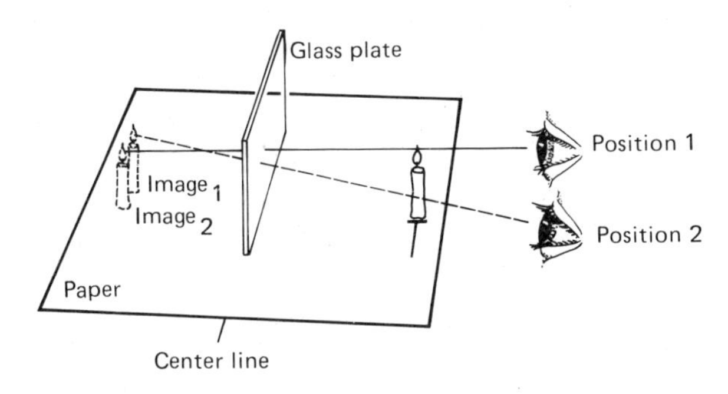

Fig. 44-3 Arrangement for experimental procedure for using a glass plate as a mirror.

2. Observing from a side position 2 (observe the top of the flame), you will see two images, one nearer than the other. Can you explain why?

   Place a pin near the glass plate so that it is aligned (in the line of sight) with the front or nearer image of the candle. Place another pin closer to the edge of the paper so that both pins and the candle image are aligned. Mark the locations of the pins.

   Repeat this procedure viewing from a position on the other side of the candle.

3. Remove the equipment from the paper. Draw straight lines through the respective pair of pin points extending from the candle line through the glass plate line. The lines will intersect on the opposite side of the plate line at the location of the candle image.

   Draw lines from the actual candle mark to the points of intersection of the previously drawn lines and the plate line. These lines from the candle (mark) to the glass plate line and back to the observation positions are ray tracings of reflected light rays.

4. Draw normal lines to the glass plate line at the points of intersection of the ray lines. Label and measure the angles of incidence θ_i and reflection θ_r. Record the data in the data table.

   Also, measure the perpendicular distances from the glass plate line to the candle mark (the object distance d_o) and to the candle image position (the image distance d_i). Compute the percent differences of the quantities indicated in the data table.

PLANE MIRROR

5. (a) Place the mirror near the center of a sheet of paper similar to the glass plate used previously. (The mirror may be propped up by some means or a holder used if available.) Draw a line along the silvered side of the mirror. Then lay an object pin about 10 cm in front of the mirror and parallel to its length (Fig. 44-4). Mark the locations of the ends of the object pin on the paper with a pencil.
   (b) Stick a reference pin R in the board to one side of the object pin and near the edge of the paper, as illustrated in Fig. 44-4, and mark its location.
   (c) Placing another pin nearer the mirror, visually align this pin and the reference pin with the head of the object pin's image in the mirror. Mark the position of this pin and label with an "H." Then, move this pin over and align it and the reference pin with the "tail" of the image pin. Mark this location and label with a "T."
   (d) Repeat this procedure on the opposite side of the object pin with another reference pin.

6. Remove the equipment from the paper and draw straight lines from the reference points through each of the H and T locations and the mirror line. The H lines and T lines will intersect and define the locations of the head and tail of the pin image, respectively.

   Draw a line between the line intersections (the length of the pin image). Measure the length of this line and the length of the object pin and record. Also, measure the object distance d_o and the image distance d_i from the mirror line and record.

   Compute the percent differences of the respective measured quantities.

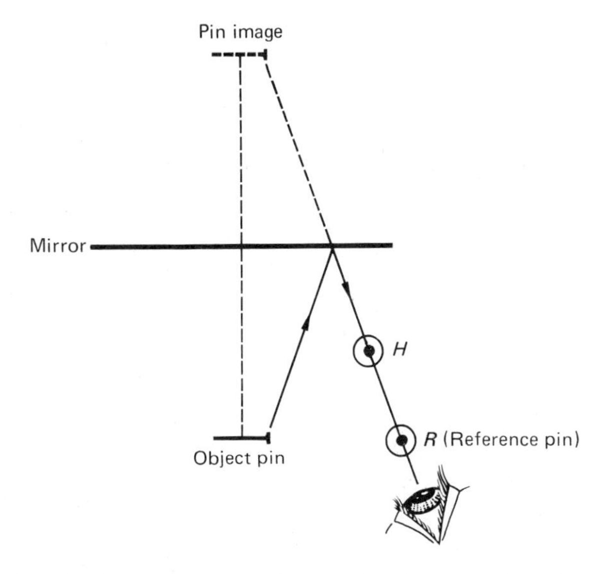

Fig. 44-4 Arrangement for experimental procedure for a plane mirror.

ROTATION OF A MIRROR

7. Place the mirror near the center of a sheet of paper (as described above) and draw a line along the length of the silvered side of the mirror. Measure so as to find the center of the line and mark that location. Stick two pins (*A* and *B*) in the board to one side and in front of and in line with the center of the mirror as in Fig. 44-5. Viewing the aligned images of these pins from the other side of the page, place two more pins (*C* and *D*) in alignment. Label the locations of the pins.

8. Leaving pins *A* and *B* in place, rotate the mirror a small but measurable angle θ (approximately 10 to 15°) about its center point, and draw a line along the silvered side of the mirror. Align two pins (*E* and *F*) with the aligned images of *A* and *B* and mark and label the locations of *E* and *F*.

9. Remove the equipment from the paper and draw the incident ray and the two reflected rays. Measure the angle of rotation of the mirror θ and the angle of deflection ϕ between the two reflected rays and record in the Laboratory Report.

   Double θ and compute the percent difference between 2θ and ϕ. Make a conclusion about the relationship of the angle of rotation of a mirror and the angle of deflection of a ray.

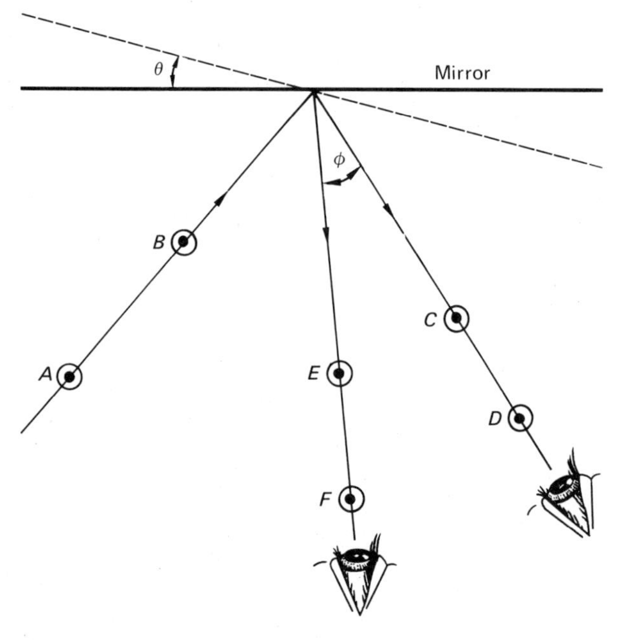

Fig. 44-5 The experimental procedure for rotation of a mirror.

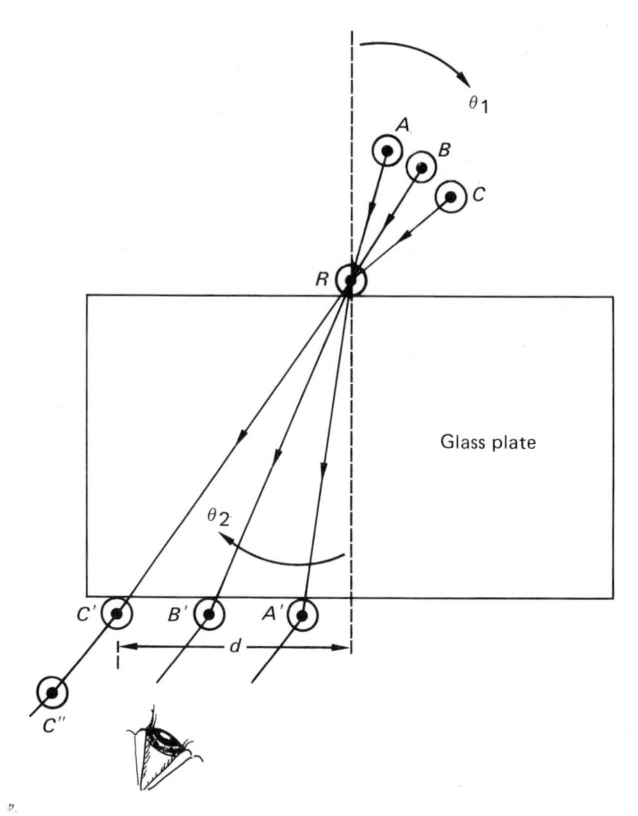

Fig. 44-6 The experimental procedure for determining the index of refraction of a glass plate.

B. *Refraction*

INDEX OF REFRACTION OF A GLASS PLATE

10. Lay the glass plate in the center of a sheet of paper and outline its shape with a pencil (Fig. 44-6). Draw a line normal to one of the sides of the plate and place a pin (*R*) at the intersection of this line and the face of the plate. Measure an angle of θ_1 of 15° relative to this line and place a pin (*A*) about 6 to 8 cm from the plate at this angle.

    Then sighting through the opposite edge of the plate, place a pin (*A'*) adjacent to the face of the plate so that it is aligned with *R* and *A*. Mark and label the locations of the pins and repeat this procedure with pins (*B* and *C*) at angles of 30° and 45°, respectively. For the 45°-angle case, align an additional pin (*C''*, Fig. 44-6).

11. Trace the various rays and measure and record θ_1 and θ_2 for each case. Also measure and record the displacement *d* of ray *C'C''* from the normal. Using Eq. 44-3, compute the index of refraction of the glass. Compare the average experimental value of the index of refraction with the general range of the index of refraction of glass ($n = 1.5 \sim 1.7$, depending on type).

Name .... Section .... Date ....

Lab Partner(s) ....

EXPERIMENT 44 *Reflection and Refraction*

LABORATORY REPORT

A. *Reflection*

Glass Plate as a Mirror

	θ_i	θ_r		
Ray 1	....	....	d_o	
Ray 2	....	....	d_i	

Percent differences between θ_i and θ_r

Ray 1 ....

Ray 2 ....

Percent difference between d_o and d_i ....

Plane Mirror

Length of pin .... d_o ....

Length of image .... d_i ....

Percent difference between pin length and image length ....

Percent difference between d_o and d_i ....

Rotation of a Mirror

Angle of rotation, θ .... 2θ ....

Angle of deflection of ray, ϕ ....

Percent difference between ϕ and 2θ ....

B. *Refraction*

Index of Refraction of a Glass Plate

Calculations (show work)

	θ_1	θ_2	Computed n
Ray ARA'	....	....	....
Ray BRB'	....	....	....
Ray CRC'	....	....	....

Average n ....

General range of the index of refraction of glass..................................

Displacement d of ray $C'C''$ ....

(continued)

QUESTIONS

1. (a) Why are two images seen in the glass plate when it is viewed from position 2 in part A of the procedure? Why is only one image seen when it is viewed from position 1? (b) Explain why reflections are easily seen at night in a window pane from inside the house, but during the day they are not.

2. Based on your experimental data, draw conclusions (a) about the relationship of the distance of the object in front of a plane mirror and the distance of its image "behind" the mirror; and (b) the image magnification (i.e., how much bigger the image is than the object).

3. Explain the situation shown in Fig. 44-Q3. Is someone dumb enough to hold the palm of his hand over a burning candle? (*Hint:* The author's hand extends inside the sliding glass-windowed door of a laboratory cabinet.)

Fig. 44-Q3 (Photo courtesy Gerald Taylor)

4. Prove mathematically that when a plane mirror is rotated an angle θ about an axis through its center (part A of the procedure), the angle of deflection of a light ray ϕ is equal to 2θ. Draw a diagram and show the work involved in your proof. Attach additional sheet if necessary.

5. Referring to the situation in Fig. 44-6 and using θ_1 for ray CR, show theoretically that ray $C'C''$ is parallel and compute the displacement d of the ray passing through the glass plate. Compare this with the measured experimental displacement.

6. Using the experimentally determined n for the glass plate, compute the speed of light in the glass plate.

Name .... Section .... Date ....

Lab Partner(s) ....

EXPERIMENT 45 *Spherical Mirrors and Lenses*

ADVANCE STUDY ASSIGNMENT

Read the experiment and answer the following questions.

1. Distinguish between concave and convex spherical mirrors.

2. What is the difference between a real image and a virtual image?

3. Distinguish between diverging and converging lenses.

4. What is the characteristic of the focal point for spherical mirrors and lenses with regard to the term "focal"?

5. If an object is placed 15 cm in front of a concave mirror with a radius of curvature of 20 cm, what are the image characteristics? (Show work.)

EXPERIMENT **45**

Spherical Mirrors and Lenses

I. INTRODUCTION

Mirrors and lenses are familiar objects that are used daily. The most common mirror is a plane mirror, the type we look into every morning to see our image. However, spherical mirrors have many common applications. For example, convex spherical mirrors are used in stores to monitor aisles and merchandise and concave spherical mirrors are used as flashlight reflectors.

Mirrors reflect light, whereas **lenses** transmit light. Spherical lenses are used to converge and focus light (convex spherical lenses) and to diverge light (concave spherical lenses). Many of us wear lenses in the form of eyeglasses. Cameras and projectors use lens systems to magnify and form images on screens.

In this experiment, the fundamental properties of spherical mirrors and lenses will be investigated so as to learn the parameters that govern their use.

II. EQUIPMENT NEEDED

- Spherical metal mirror (or individual concave and convex spherical mirrors)
- Convex lens (focal length 10 to 20 cm)
- Concave lens
- Meter stick optical bench (or precision bench) with lens holder, screen, and screen holder (white cardboard can serve as the screen)
- Light source: candle and candle holder, or electric light source with object arrow

III. THEORY

A. *Spherical Mirrors*

A **spherical mirror** is a section of a sphere and is characterized by a center of curvature (Fig. 45-1). The distance from the center of curvature to the vertex of the mirror along the optic axis is called the **radius of curvature** R. The focal point is midway between R and the vertex, and the **focal length** f is one-half the radius of curvature:

$$f = \frac{R}{2} \tag{45-1}$$

If the reflecting surface is on the inside of the spherical section, the mirror is said to be **concave.** For a **convex** mirror, the reflecting surface is on the outside of the spherical section.

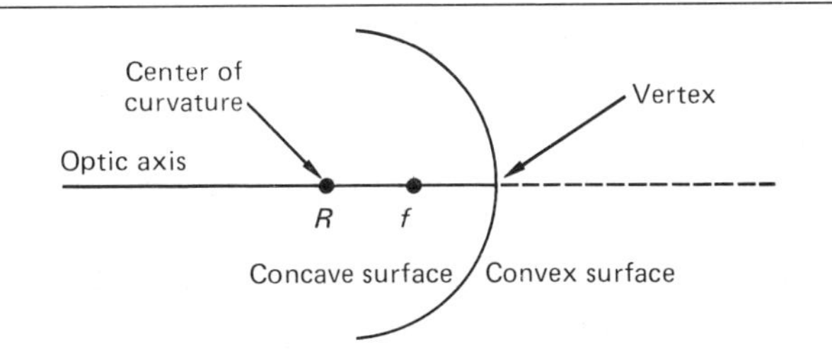

Fig. 45-1 The parameters of spherical mirror surfaces.

The characteristics of the images formed by spherical mirrors can be determined either graphically or analytically. Examples of the graphical ray method are shown in the ray diagrams in Fig. 45-2. A ray from an object through the center of curvature R, called the **chief ray** (1), is reflected back through R. A ray parallel to the optic axis, called a **parallel ray** (2), is reflected through the focal point f. The inter-

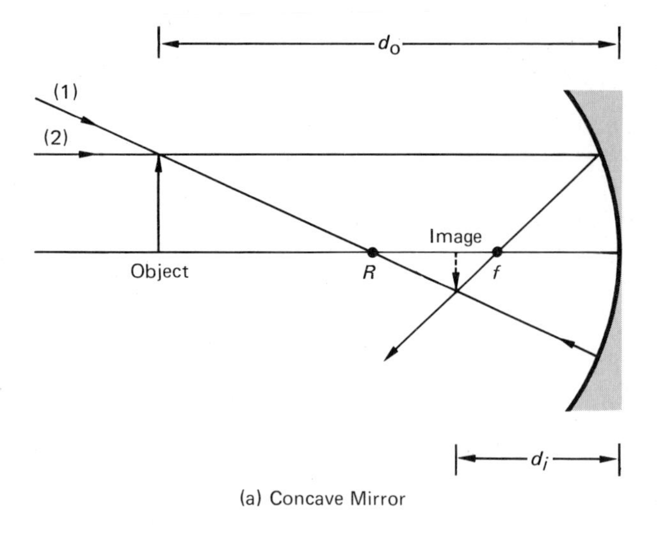

(a) Concave Mirror

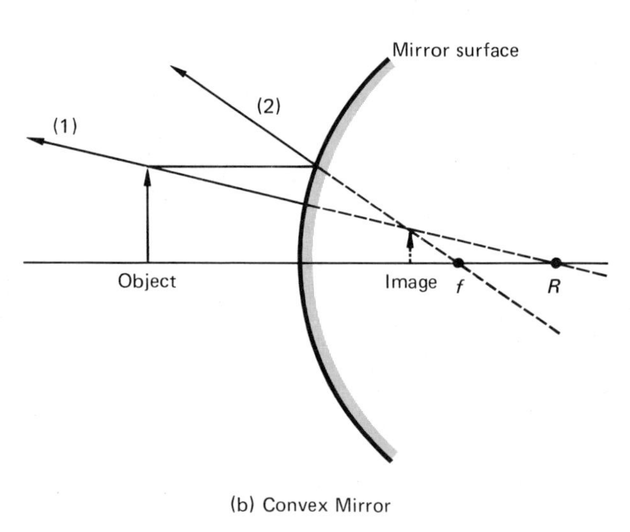

(b) Convex Mirror

Fig. 45-2 Examples of the ray method for determining the image characteristics for (a) a concave spherical mirror and (b) a convex spherical mirror.

section of these rays defines the location of the image plane. For a convex mirror, the rays appear to go through R and f [Fig. 45-2(b)]. A third ray, the focal ray through f that is reflected parallel to the optic axis, may be drawn, but it is not needed.

A concave mirror is called a **converging mirror** because rays parallel to the optic axis converge at the focal point. Similarly, a convex mirror is called a **diverging mirror** because the rays parallel to the optic axis appear to diverge from the focal point.

If the image is formed on the same side of the mirror as the object, the image is said to be a **real image** and can be observed on a screen. An image formed "behind" the mirror is called a **virtual image** and obviously cannot be observed on a screen. An image is described by the following characteristics: (1) real or virtual, (2) upright (erect) or inverted (relative to the object orientation), and (3) magnified or reduced. In Fig. 45-2(a) the image is real, inverted, and reduced; and in Fig. 45-2(b) the image is virtual, upright, and reduced.

The distance from the object to the vertex along the optic axis, d_o, is called the **object distance,** and the distance from the vertex to the image is the **image distance** d_i. Knowing the focal length f of the mirror, the position of the image d_i can be found from the **mirror equation,**

$$\frac{1}{d_o} + \frac{1}{d_i} = \frac{1}{f} \tag{45-2}$$

In the case of a concave mirror, the focal length is taken to be positive (+), and for a convex mirror the focal length is taken to be negative (−). The object distance d_o is taken to be positive in either case. The sign convention is: If d_i is positive, the image is real; and if d_i is negative, the image is virtual. The **magnification factor** M is given by

$$M = -\frac{d_i}{d_o} \tag{45-3}$$

If M is positive (d_i negative), the image is upright; and if M is negative (d_i positive), the image is inverted. The sign convention is summarized in Table 45.1.

☐ **Example 45.1** An object is placed 45 cm in front of a concave mirror with a focal length of 15 cm [corresponding to the case in Fig. 45-2(a)]. Determine the image characteristics analytically.

Solution With d_o = 45 cm and f = 15 cm,

$$\frac{1}{45} + \frac{1}{d_i} = \frac{1}{15} = \frac{3}{45}$$

By inspection, $1/d_i = 2/45$ or $d_i = 45/2 = 22.5$ cm. Then

$$M = -\frac{d_i}{d_o} = -\frac{22.5}{45} = -\frac{1}{2}$$

Thus, the image is real (positive d_i), inverted (negative M), and reduced by a factor of $\frac{1}{2}$ (i.e., one-half as tall as the object). ☐

B. *Spherical Lenses*

The shapes of biconvex and biconcave spherical lenses are illustrated in Fig. 45-3. A radius of curvature R is defined for each spherical surface. A focal length f is defined for a lens. R and f are located along the principal axis. A convex lens is called a

Table 45.1 Sign Convention for Spherical Mirrors and Lenses

Concave mirror,	convex lens:	f positive	d_o always positive
Convex mirror,	concave lens:	f negative	(in this experiment)*
d_i	Image	M	Image
+	Real	+	Upright
−	Virtual	−	Inverted

*In some cases of lenses combinations, d_o may be negative when an image is used as an object. See Experiment 46.

converging lens because rays parallel to the principal axis converge at the focal point. A concave lens is called a **diverging lens** because rays parallel to the principal axis appear to diverge from the focal point.

Similar to spherical mirrors, the characteristics of the images formed by spherical lenses can be determined graphically or analytically. The chief (1) and parallel (2) rays for the graphical method are illustrated in the ray diagrams in Fig. 45-4. In the case of lenses, the chief ray through the center of a lens is undeviated. A ray parallel to the principal axis is refracted such that it goes through the focal point on transmission through the lens.

In the case of a concave lens, the ray appears to have passed through the focal point on the object side of the lens. If the image is formed on the side of the lens opposite to that of the object, it is real and can be observed on a screen. However, if the image is on the same side of the lens as the object, it is virtual and cannot be seen on a screen.

The lens equation and magnification factor for analytically determining the image characteristics are identical to those for spherical mirrors (Eqs. 45-2 and 45-3). The sign convention is also similar (see Table 45.1). It should be noted that this lens equation applies only to *thin* lenses.

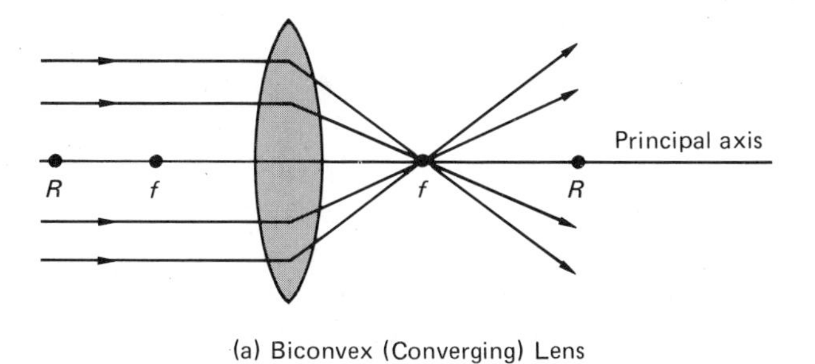

(a) Biconvex (Converging) Lens

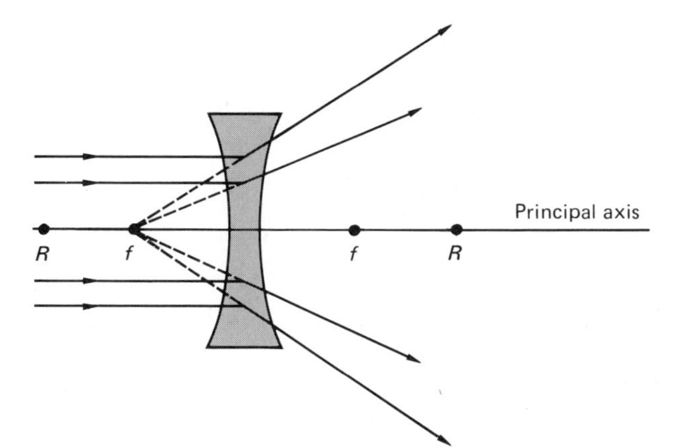

(b) Biconcave (Diverging) Lens

Fig. 45-3 (a) A biconvex (converging) lens. (b) A biconcave (diverging) lens.

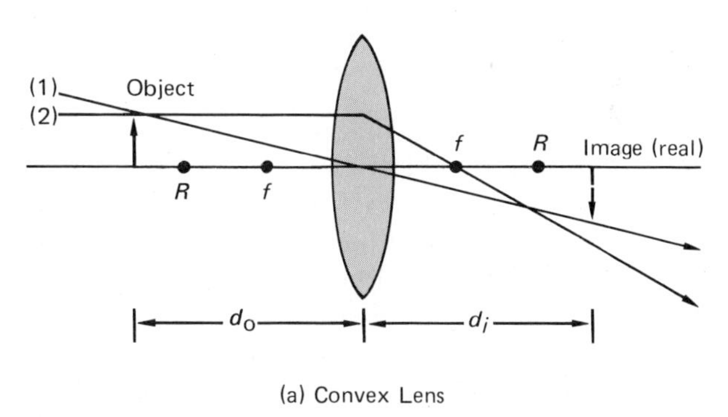

(a) Convex Lens

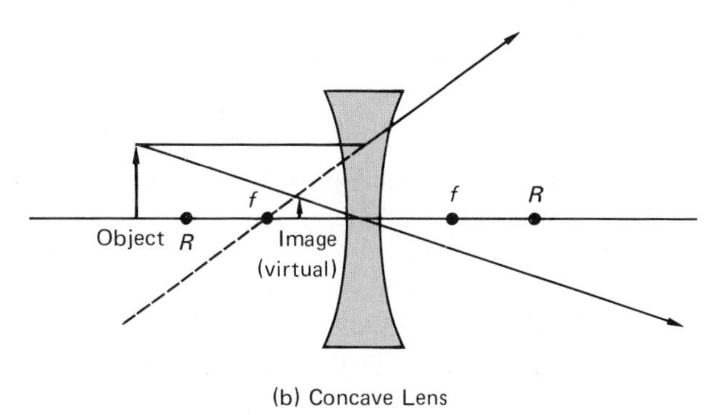

(b) Concave Lens

Fig. 45-4 Examples of the ray method for determining the image characteristics for (a) a convex lens and (b) a concave lens.

☐ **Example 45.2** An object is placed 30 cm from a biconcave lens with a focal length of 10 cm [corresponding to the case in Fig. 45-4(b)]. Determine the image characteristics analytically.

Solution With $d_o = 30$ cm and $f = -10$ cm (negative by convention for a concave lens),

$$\frac{1}{30} + \frac{1}{d_i} = -\frac{1}{10}$$

or

$$\frac{1}{30} + \frac{1}{d_i} = -\frac{3}{30}$$

By inspection,

$$\frac{1}{d_i} = -\frac{4}{30} \quad \text{or} \quad d_i = -\frac{30}{4} = -7.5 \text{ cm}$$

Then,

$$M = -\frac{d_i}{d_o} = \frac{-(-30/4)}{30} = +\tfrac{1}{4}$$

Thus, the image is virtual (negative d_i), upright (positive M), and reduced by a factor of $\frac{1}{4}$. ☐

However, the relationship between the focal length and the radius of curvature for a spherical lens is not as simple as for that of a spherical mirror (Eq. 45-1). For a lens, the focal length is given by what is known as the lensmaker's equation:

$$\frac{1}{f} = (n - 1)\left(\frac{1}{R_1} + \frac{1}{R_2}\right) \qquad \textbf{(45-4)}$$

where n is the index of refraction for the lens material and the R's are taken as positive for *convex* surfaces. For glass with $n = 1.5$ and symmetrical converging lenses ($R_1 = R$ and $R_2 = R$), this equation yields $f = R$. In drawing diagrams, we will assume that $n = 2$ for convenience, so that $f = R/2$ for lenses, as for mirrors. Keep in mind, however, that the focal length of a lens depends in general on the value of n as well as the R's. In computations, the experimentally determined value of f will be used.

IV. EXPERIMENTAL PROCEDURE

A. *Spherical Mirrors*

CONCAVE MIRROR

1. (a) Construct a ray diagram for a concave mirror with an object located at its focal point. (Drawing provided in the Laboratory Report.) It should be observed from the diagram that the reflected rays are parallel. In this case we say that the rays "converge" at infinity or that the image is formed at infinity. Inversely, rays coming from an object at infinity converge to form an image at the focal point or in the focal plane (plane perpendicular to the optic axis).
   (b) Construct a ray diagram with several rays parallel to the optic axis to show they converge at f.
   (c) Using the spherical-mirror equation, determine the image distance for an object at infinity (∞).

2. This focal property allows the experimental determination of the focal length of the mirror. An object a great distance from the mirror is essentially at infinity relative,to the dimensions of the mirror. Take the mirror and screen to a window. Holding the mirror in one hand and the screen in the other, adjust the distance of the screen from the mirror until the image of some outside distant object is observed on the screen (hence a real image).*

   Measure the distance from the mirror vertex to the screen (f) and record in the Laboratory Report. Repeat this procedure twice and take the average of the three measurements as the focal length of the mirror.

3. *Case 1:* $d_o > R$.
   (a) Sketch a ray diagram for an object at a distance slightly beyond R (i.e., $d_o > R$) and note the image characteristics.
   (b) Set this situation up on the optical bench as illustrated in Fig. 45-5, with the object placed several centimeters beyond the radius of curvature (known from f determination in procedure 2, $R = 2f$). Measure and record the object distance d_o. (Data Table 1).

   It is usually convenient to hold the mirror manually and adjust the object distance by moving the mirror rather than the object light source. Move the screen along the side

*If a window is not available or it is a dark day, use procedure 4 to experimentally determine f. In this case, show first that $d_i = d_o = R$ and $M = 1$. Then, finding d_i, $f = R/2$.

(a)

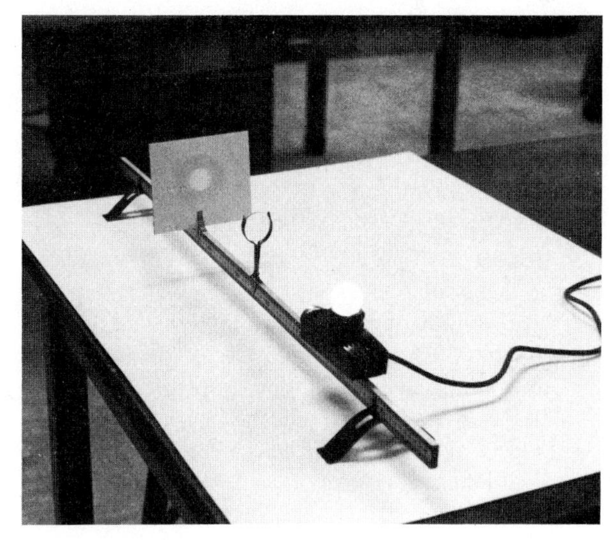

(b)

Fig. 45-5 Arrangements for experimental procedures for (a) spherical mirrors and (b) spherical lens.

of the optical bench until an image is observed on the screen. This is best observed in a darkened room. The mirror may have to be turned slightly to direct the rays toward the screen.

(c) Estimate the magnification factor M and measure and record the image distance d_i.

(d) Using the mirror equation, compute the image distance and the magnification factor.

(e) Compare the computed value of d_i with the experimental value by computing the percent difference.

4. *Case 2:* $d_o = R$. Repeat procedure 3 for this case.

5. *Case 3:* $f < d_o < R$. Repeat procedure 3 for this case.

6. *Case 4:* $d_o < f$. Repeat procedure 3 for this case.

CONVEX MIRROR

7. Sketch ray diagrams for objects at (1) $d_o > R$, (2) $f < d_o < R$, and (3) $d_o < f$, and draw conclusions about the characteristics of the image of a convex mirror. Experimentally verify that the image of a convex mirror is virtual (i.e., try to locate the image on the screen).

B. *Spherical Lenses*

CONVEX LENS

8. (a) Sketch a ray diagram for a convex lens with the object at its focal point. Similar to the concave mirror (procedure 1), the image is formed at infinity.

   (b) Using the lens equation, determine the image characteristics for an object at infinity.

   (c) Experimentally determine the focal length of the lens by a procedure similar to that used for the concave mirror. (The lens may be placed in a lens holder and mounted on a meter stick.)*

9. Repeat the four cases for the lens as was done for the concave mirror in procedures 3 to 6 with R replaced by $2f$ (see Fig. 45-5). It is initially instructive to continuously move the lens toward the object light source (decreasing d_o) from a $d_o > 2f$ and observe the image on the screen, which also must be moved continuously to obtain a sharp image. In particular, notice the change in the size of the image as d_o approaches f.

**Note:* In general for a lens, $f \neq R/2$. However, it can be shown for the case of $d_i = d_o$ that $d_o = 2f$. See Question 4 at the end of the experiment.

CONCAVE LENS

10. Repeat the procedures carried out for the convex mirror in procedure 7 for the concave lens with R replaced by $2f$.

11. It is possible to experimentally determine the focal length of a concave lens by placing it in contact with a convex lens so as to form a lens combination. The combination will form a real image. If two lenses of focal lengths f_1 and f_2 are placed in contact, the lens combination has a focal length f_c given by

$$\frac{1}{f_c} = \frac{1}{f_1} + \frac{1}{f_2} \qquad \textbf{(45-5)}$$

Place the concave lens in contact with the convex lens (convex surface to concave surface) in a lens holder and determine the focal length of the lens combination f_c by finding the image of a distant object as in procedure 8. Record in the Laboratory Report.

Using Eq. 45-5 with the focal length of the convex lens determined in procedure 8, compute the focal length of the concave lens.

LABORATORY REPORT

A. *Spherical Mirrors*

Concave Mirror: Ray diagrams

$d_o = f$

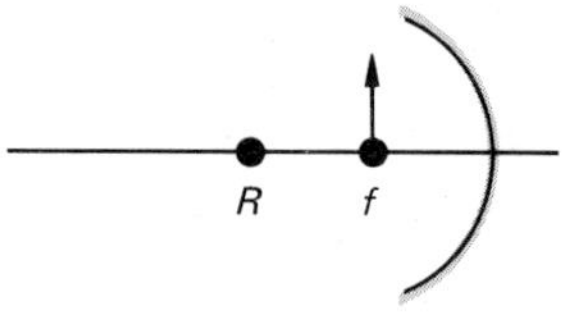

R f

$f < d_o < R$

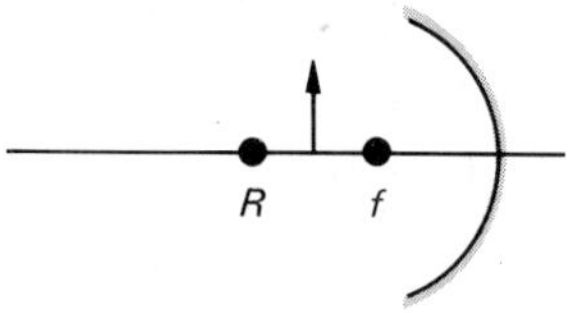

R f

$d_o > R$

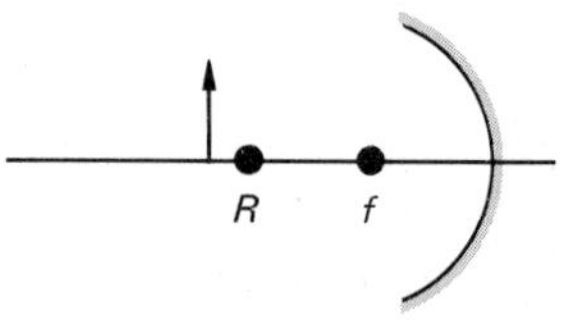

R f

$d_o < f$

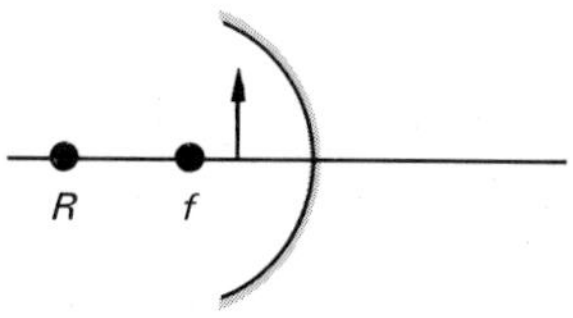

R f

$d_o = R$

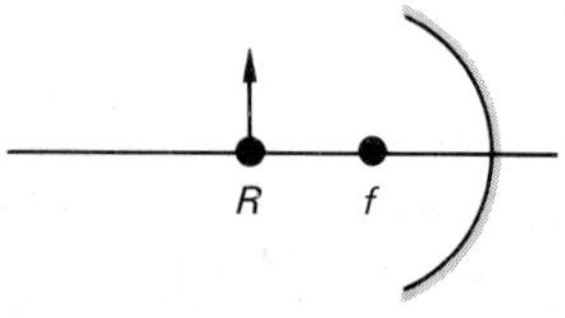

R f

Name .... Section .... Date ....

Lab Partner(s) ....

EXPERIMENT 45 *Spherical Mirrors and Lenses*

Calculation of d_i for object at ∞

Experimental focal length, f ....

....

....

Average ....

DATA TABLE 1

	Experimental			Computed		d_i percent difference
	d_o (cm)	d_i (cm)	M factor (estimated)	d_i (cm)	M	
$d_o > R$						
$d_o = R$						
$f < d_o < R$						
$d_o < f$						

Convex mirror: Ray diagrams

$d_0 > R$

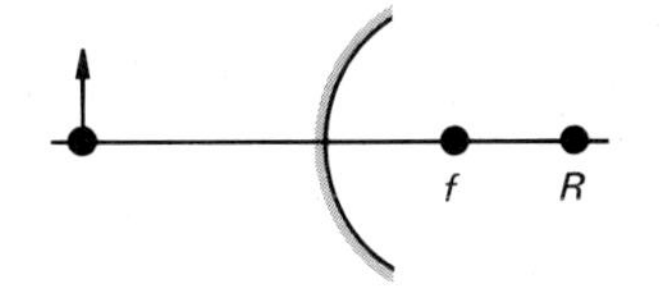

$f < d_0 < R$

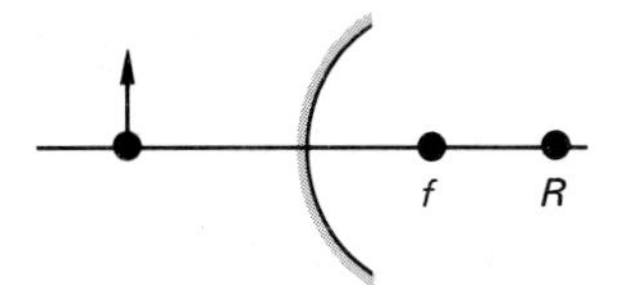

$d_0 < f$

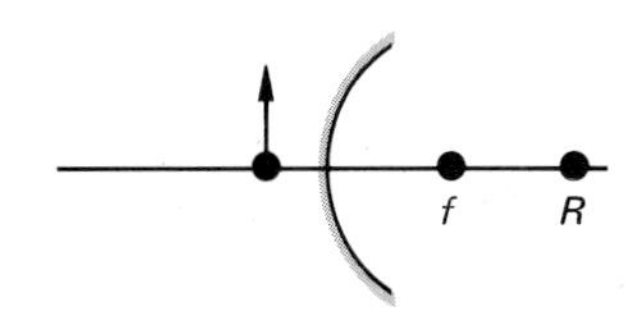

Conclusions

(continued)

B. *Spherical Lenses*

Convex lens: Ray diagrams

$d_o = f$

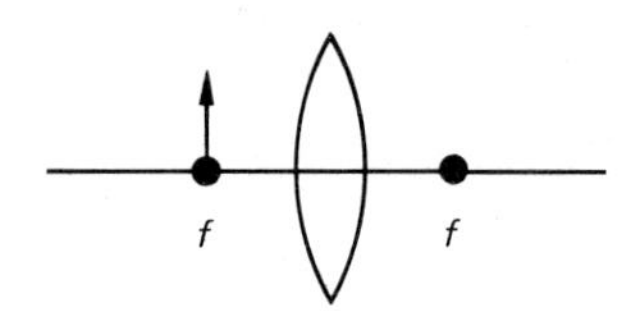

$f < d_o < 2f$

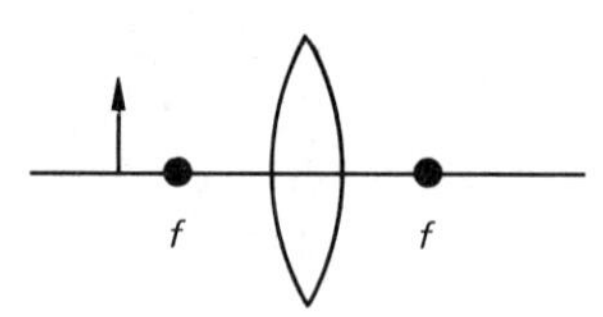

$d_o > 2f$

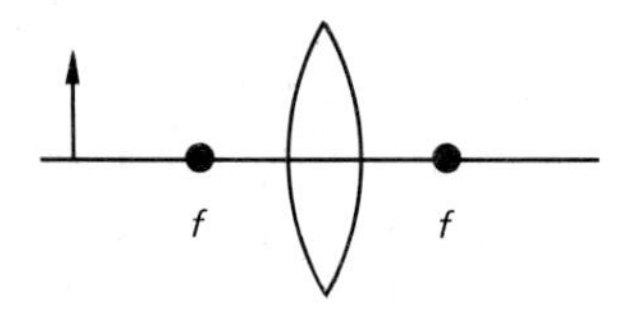

$d_o < f$

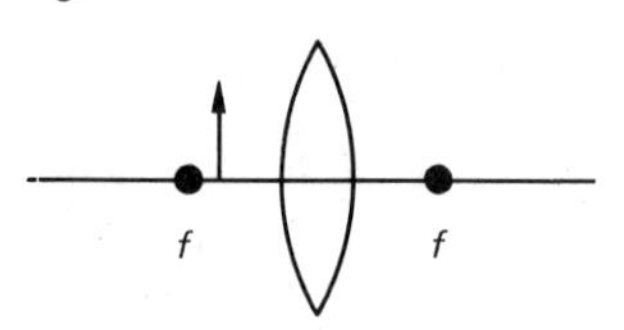

$d_o = 2f$

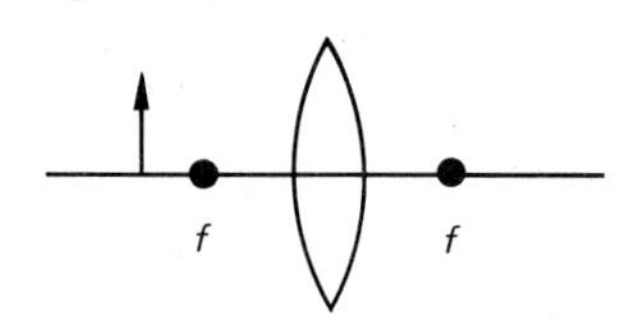

Calculation of d_i for object at ∞

Experimental focal length, f ....

....

....

Average ....

DATA TABLE 2

	Experimental			Computed		d_i percent difference
	d_o (cm)	d_i (cm)	M factor (estimated)	d_i (cm)	M	
$d_o > 2f$						
$d_o = 2f$						
$f < d_o < 2f$						
$d_o < f$						

Concave lens: Ray diagrams

$d_o > 2f$

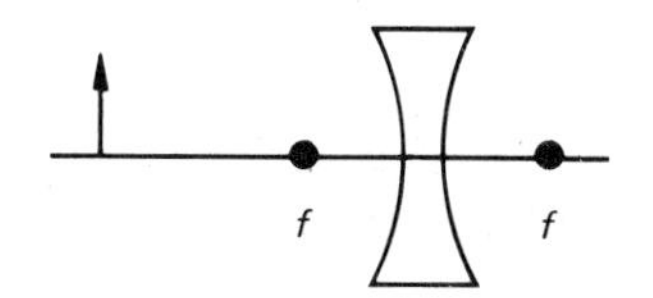

$f < d_o < 2f$

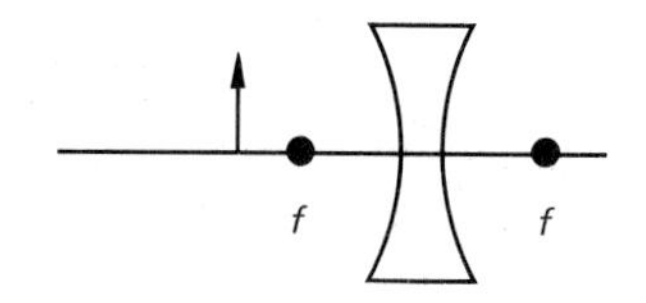

$d_o < f$

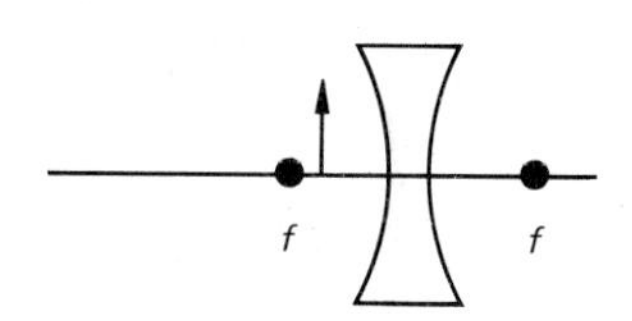

Conclusions

(continued)

QUESTIONS

1. A plane mirror essentially has a radius of curvature of infinity. Using the mirror equation, show that (a) the image of a plane mirror is always virtual; (b) the image is "behind" the mirror the same distance as the object is in front of the mirror; and (c) the image is always upright.

2. Show that the magnification factor for a mirror or lens $M = d_i/d_o$ (sign convention omitted) is the lateral magnification or the ratio of the height (lateral size) of the image to that of the object. (*Hint:* Draw a ray diagram.)

3. Explain what characteristics make convex spherical mirrors applicable for store monitoring and concave spherical mirrors applicable as flashlight reflectors.

4. Prove that for a converging lens for the case $d_i = d_o$ that $d_i = d_o = 2f$.

5. Using the experimental value of f for the converging lens and an $n = 1.5$, compute the radius of curvature of the lens' surfaces using the lensmaker's equation. (b) Taking $f = R/2$, compare this computed value of f with the experimental value. (c) The index of refraction of the lens could have a different value (n of glass varies generally from 1.5–1.7). Would this make a difference?

Name .... Section .... Date ....

Lab Partner(s) ....

EXPERIMENT 46 *Optical Instruments: The Microscope and the Telescope*

ADVANCE STUDY ASSIGNMENT

Read the experiment and answer the following questions.

1. Distinguish between angular magnification and lateral magnification.

2. What is the advantage of the eyepiece in a compound microscope?

3. What is the difference between reflecting and refracting telescopes?

4. Distinguish between astronomical and terrestrial telescopes.

5. How is the greatest magnification achieved for a telescope?

6. Explain how the focal lengths of the lenses are determined in the experiment. (Draw a ray diagram for illustration.)

EXPERIMENT **46**

Optical Instruments: The Microscope and the Telescope

I. INTRODUCTION

Optical instruments exist in abundance for a variety of purposes. In general, we use optical instruments to improve and to extend our visual observations. Two such common instruments are the microscope and telescope, which are used to view objects and their details that may not be visible to the unaided eye.

Although some optical instruments are extremely complex, they can be analyzed in terms of their individual components, usually mirrors and lenses. In this experiment, investigations will be limited to the study of lenses and simple lens systems that illustrate the basic principles of telescopes and microscopes.

II. EQUIPMENT NEEDED

- Two biconvex of converging lenses with relatively short and long focal lengths (e.g., 10 cm and 30 cm)
- One biconcave or diverging lens ($f \simeq 15$ cm)
- Meter stick optical bench (or precision bench) with three lens holders, screen, and a screen holder (white cardboard can serve as the screen)
- Light source object: candle and candle holder or electric light source with object arrow
- Ruler
- Magnification scales (1) Microscope: white cardboard screen with horizontal pencil lines at regular intervals (e.g., 0.5 cm) (2) Telescope: sheet of white paper with heavy black lines (felt pen) drawn horizontally at regular intervals (e.g., 5 cm)
- Tape (to fix telescope scale on wall or cabinet)

III. THEORY

A. *Microscopes*

THE MAGNIFYING GLASS OR SIMPLE MICROSCOPE

The **magnifying glass (simple microscope)** is simply a biconvex or converging lens of relatively short focal length. When an object is viewed through the lens, the observer sees an *enlarged,* upright, virtual image, as illustrated in Fig. 46-1. (The virtual image acts as an object for the lens of the eye, which forms a real image on the retina or "screen" of the eye.)

The magnification of an object *viewed through* a magnifying glass is expressed in terms of the angular magnification or magnifying power $m = \theta_i/\theta_o$ (not to be confused with the lateral magnification $M = d_i/d_o$; see Fig. 46-1). The angular magnification is defined as the ratio of the *angular* size of the object viewed through the lens θ_i to the angular size of the object seen without the lens θ_o. (See textbook for further explanation.)

When viewing small objects without a lens, the most enlarged, distinct version occurs for the normal eye when the object is brought about 20–25 cm from the eye, the near point of the eye.* When viewing through a magnifying glass, adjustment is made until the virtual image of the lens (which acts

*A person's near point gradually recedes with age as the crystalline lens of the eye loses its ability to adjust its contour and focal length. This is why older people hold reading materials farther away.

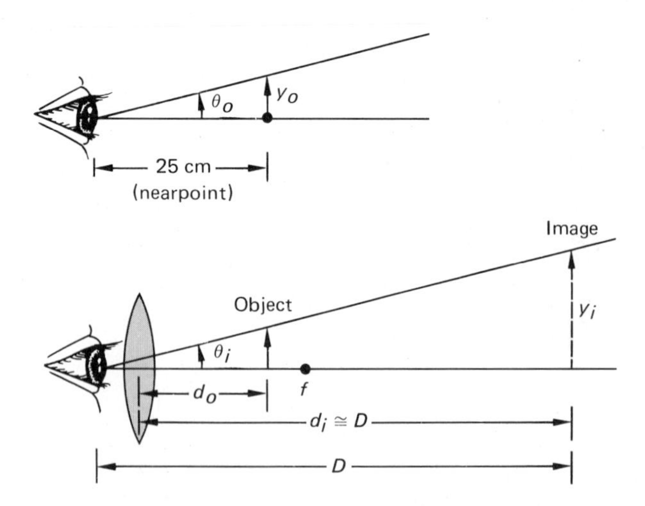

Fig. 46-1 Angular magnification is defined as $m = \theta_i/\theta_o$, where θ_o is the angular size of the object seen without the magnifying glass and θ_i is the angular size of the image viewed through the glass. The distance of the image from the viewer is $D \simeq -d_i$. Also, when $d_o = f$, then $D = \infty$.

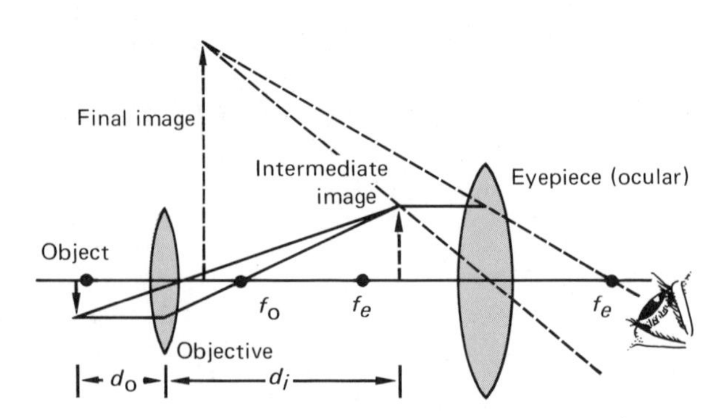

Fig. 46-2 The basic components of the compound microscope.

as an object for the eye) is about 25 cm from the eye or lens, with the eye being near the lens. Under these conditions, the magnifying power can be shown to be (see textbook)

$$m = \frac{25}{f} + \frac{25}{D} \qquad \textbf{(46-1)}$$

where f is the focal length of the converging (biconvex) lens and D is the distance of the image from the eye (or lens with the eye near the lens). See Fig. 46-1. Maximum magnification is obtained when the image is at the near point (taken to be 25 cm), or $D = 25$ cm; hence,

$$m = \frac{25}{f} + 1 \qquad \textbf{(46-2)}$$

where m is the maximum angular magnification for distinct vision.

THE COMPOUND MICROSCOPE

To obtain greater magnification than that provided by a simple microscope, a compound microscope is used. Basically, the **simple compound microscope** consists of a pair of converging lenses, each of which contributes to the magnification (Fig. 46-2). A converging lens with a relatively short focal length f_o, called the **objective,** produces a real image just inside the focal point of the second lens, called the **eyepiece (or ocular),** which has a longer focal length f_e. The convergent eyepiece forms a magnified virtual image that is viewed by the observer. The objective acts as a projector and the eyepiece is used as a simple microscope to view the projected image.

The total magnification M_t of the compound microscope lens system is given by the product of the lateral magnification of the objective, $M_o = d_i/d_o$ (minus sign omitted), and the angular magnification of the eyepiece, $m_e = 25/f_e + 1$ (maximum magnification for distinct vision with the eyepiece close to the eye Eq. 46-2):

$$M_t = M_o m_e = \left(\frac{d_i}{d_o}\right)\left(\frac{25}{f_e} + 1\right) \qquad \textbf{(46-3)}$$

where d_i is the image distance of the intermediate image formed by the objective and d_o is the object distance from the objective.

B. *Telescopes*

There are two basic types of telescopes, reflecting and refracting. A **reflecting telescope** uses a concave mirror to form the image of a distant object, which is then magnified by a lens eyepiece. The **refracting telescope,** which is the subject of this experiment, uses a converging lens instead of a mirror to form the image of a distant object. The differences between the two types of refracting telescopes, astronomical and terrestrial, depend on the type of lens used for the eyepiece.

THE ASTRONOMICAL TELESCOPE

The **astronomical telescope** uses a large converging lens of long focal length as an objective and a converging lens with a relative short focal length as a movable eyepiece (Fig. 46-3).

Parallel rays from a distant object form an intermediate image at the focal point f_o of the objective

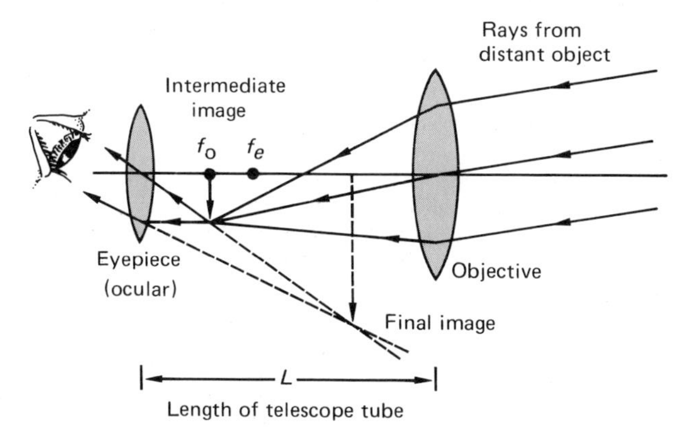

Fig. 46-3 The basic components of the astronomical telescope.

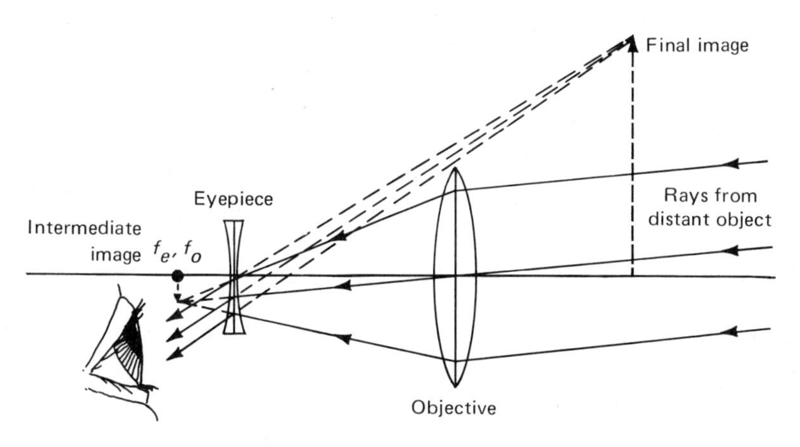

Fig. 46-4 The basic components of the terrestrial or Galilean telescope.

lens. This image is magnified by the eyepiece, which is moved so that the intermediate image is at or slightly inside the focal point of the eyepiece f_e. With the image at f_e, the light rays emerging from the eyepiece are parallel and the final image viewed through the eyepiece is at infinity. As can be seen from Fig. 46-3, the length of the telescope with the intermediate image at f_e would be $f_e + f_o$.

The magnifying power of a telescope focused for a final image at infinity is given by

$$m = \frac{f_o}{f_e} \tag{46-4}$$

Hence, to achieve the greatest magnification, the focal length of the objective should be made as long as possible and the focal length of the eyepiece as short as possible. Notice that the final image of this type of telescope, which is commonly used in astronomical work, is inverted. This poses no problem for the astronomer, however; in viewing objects on earth, it is more convenient to have upright images.

THE TERRESTRIAL (GALILEAN) TELESCOPE

In a **terrestrial telescope,** the viewed image is upright. This may be accomplished by several methods. One type of terrestrial telescope was used by Galileo in 1609. The Galilean telescope has a converging lens as an objective but a diverging lens as an eyepiece (Fig. 46-4).

The eyepiece is positioned so that the image of the converging objective lens is formed at the focal point of the diverging eyepiece lens and acts as a virtual object for this lens. [A *virtual* object is an image that is not formed before the rays are intercepted by the lens (i.e., the rays pass through the second lens). In using the lens equation for the second lens, the object distance of a virtual object is taken to be negative.]

As a result, the viewer sees a magnified, upright, virtual image. "Opera glasses" are a small version of the Galilean telescope. The magnification power is given by the same relationship as for the astronomical telescope (Eq. 46-4).

IV. EXPERIMENTAL PROCEDURE

A. *Microscopes*

1. Your instructor will give you or ask you to determine the focal lengths of the three lenses. Measurement procedures are as follows:
   (a) For the two converging lenses, use one or both of the methods described in Experiment 45, viz., (i) the image of a distant outdoors object is formed on a screen at the focal point, and (ii) in using an optical bench, when the object distance (d_o) of a light source (e.g., a candle) is equal to the image distance (d_i) of an image formed on a screen, then $d_i = d_o = 2f$.
   (b) To determine the focal length of the diverging (biconcave) lens, use the method described in Experiment 45 with the diverging lens and one of the converging lenses of known focal length in a lens holder in combination (contact). Determine the focal length of the lens combination f_c and compute the focal length of the diverging lens from the formula $1/f_c = 1/f_1 + 1/f_2$, where f_1 and f_2 are the focal lengths of the individual lenses in combination.

THE MAGNIFYING GLASS OR SIMPLE MICROSCOPE

2. Mount the longer-focus converging lens in a lens holder and place it on the optical bench near one end. Mount the screen with the scale on the bench, and with the eye close to the lens, adjust the distance between the lens and the screen until the most distinct image of the scale lines is seen. (A mark such as a cross or a letter may be made on the screen in the scale region to focus on, instead of the scale lines.)

   Then hold a ruler vertically and a little to one side of the screen scale. With both eyes open, view the image of the screen scale through the lens with one eye and the ruler with the other eye (not through the lens). By adjusting your head (sideways), with a little practice you will see the magnified screen scale superimposed on the unmagnified ruler scale.

   Estimate how many centimeters on the ruler one or more of the magnified scale divisions covers. The ratio of the magnified scale interval y_i to the unmagnified ruler scale interval y_o is the magnifying power of the lens (i.e., $m = \theta_i/\theta_o = y_i/y_o$). Record the lengths of the intervals and find m. Using Eq. 46-2 with the focal length of the lens, compute m and compare to the experimental value by finding the percent difference.

THE COMPOUND MICROSCOPE

3. Using the longer-focus converging lens as an eyepiece and the shorter-focus converging lens as an objective, mount the eyepiece near one end of the optical bench, the objective farther on, and the microscope scale screen beyond this. Make sure that f_o and f_e (Fig. 46-2) are separated by at least 10 cm. Increasing this separation distance will increase the magnification. However, don't use too much magnification or the screen scale may appear distorted.

   Viewing through the eyepiece with the eye close to the lens, adjust the position of the objective until the magnified image of the screen scale is seen most distinctly. Then, holding a ruler vertically near the optical bench, about 25 cm from the eye end, determine the magnification M_t of the compound microscope by the method used for the simple microscope in procedure 2.

4. Without moving the lenses, replace the screen with a light source at the exact position of the screen location in procedure 3. Then, using the screen, find the position of the intermediate image formed by the objective lens (Fig. 46-2), which can be seen on the screen since it is a real image. Record the object and image distances and compute the magnification of the microscope from Eq. 46-3 using the magnification of the eyepiece (m_e) found in procedure 2.

   Compare this to the experimental value by computing the percent difference.

B. *Telescopes*

THE ASTRONOMICAL TELESCOPE

5. Construct an astronomical telescope using the shorter-focus converging lens as an eyepiece and the longer-focus converging lens as an objective (Fig. 46-3). With the eyepiece mounted near one end of the optical bench and the eye close to the lens, view some distant object through the lens system of the telescope, adjusting the position of the objective until a distinct image is seen. Is the image upright or inverted?

6. Fix the telescope magnification scale to a distant (across the room) wall or cabinet by means of a piece of tape at a level so that the scale can be viewed through the telescope while it is sitting on the lab table. Adjust the telescope objective so

that the magnified scale is seen through the telescope.

Once the distinct image of a distant object is seen through the telescope, measure and record the distance between the objective and eyepiece, and compare this distance to the sum of the focal lengths of the lenses.

With both eyes open, view the image of the scale through the telescope with one eye and look directly at the scale with the other (not through the telescope), similar to the procedure used for the microscope. The magnified scale image will be seen superimposed on the unmagnified scale. Estimate the number of scale divisions as viewed directly (y_o) that correspond to one or more of the magnified scale divisions (y_i). Find the magnification from the ratio of these lengths or number of scale divisions. Compute the magnification from Eq. 46-4 and compare with the experimental value by calculating the percent difference.

THE TERRESTRIAL (GALILEAN) TELESCOPE

7. Using the same objective lens as in the astronomical telescope, construct a Galilean terrestrial telescope using the divergent lens as an eyepiece. Repeat the steps as done in procedure 6. Is the image upright or inverted?

LABORATORY REPORT

Focal Lengths of Lenses

Converging biconvex lenses

f

....

....

Diverging biconcave lens

f_c

....

f

....

Calculations (show work)

The Magnifying Glass or Simple Microscope

y_i .... m ....

y_o ....

Computed value (Eq. 46-2) of m ....

Percent difference ....

The Compound Microscope

y_i .... M_t ....

y_o ....

d_i .... M_o ....

d_o ....

Computed value (Eq. 46-3) of M_t ....

Percent difference ....

(continued)

The Astronomical Telescope

Distance between objective and eyepiece ....

$f_o + f_e$ ....

Magnification

y_i ....

y_o ....

m ....

Computed value (Eq. 46-4) of m ....

Percent difference ....

The Galilean Telescope

y_i ....

y_o ....

m ....

Computed value (Eq. 46-4) of m ....

Percent difference ....

QUESTIONS

1. The (biconvex) crystalline lens of the eye adjusts in shape and focal length to form images on the retina. If an object is at the near point, what is the focal length of the crystalline lens if the spherical chamber of the eyeball is 1.5 cm in diameter? What is the size of the image on the retina?

2. Show that the ratio of the magnified scale interval y_i to the unmagnified scale interval y_o is the angular magnification (i.e., $m = \theta_i/\theta_o$).

3. According to Eq. 46-2, the magnifying power of a magnifying glass could be made quite large by using a small-focal-length lens. However, magnifying glasses are generally limited to $2\times$ or $3\times$ magnification. Explain why.

4. How could an astronomical telescope be converted to a terrestrial telescope using a third converging lens?

Name .... Section .... Date ....

Lab Partner(s) ....

EXPERIMENT 47 *Polarized Light*

ADVANCE STUDY ASSIGNMENT

Read the experiment and answer the following questions.

1. Is the plane of polarization of a polarizing polymer sheet in the same direction as the molecular chain orientation? Explain.

2. What is the condition for optimum polarization by reflection? Is the polarization angle the same for every material? Explain.

3. Describe two methods by which light can be polarized by refraction.

4. Why is sky light partially polarized and why does it appear blue?

5. What is meant by optical activity?

EXPERIMENT **47**

Polarized Light

I. INTRODUCTION

When we speak of polarized light, Polaroid sunglasses usually come to mind, as this is one of the most common applications of polarization. However, few people understand how such sunglasses reduce "glare." Since the human eye cannot distinguish between polarized light and unpolarized light, we are not normally aware of the many instances of polarized light around us. Bees, on the other hand, with their many faceted eyes, can detect polarized light and use scattered polarized sunlight in navigating.

Although the human eye cannot detect polarized light, with a little help polarization can be studied and experimentally investigated. This is the purpose of this experiment, and we will also learn how light is polarized.

II. EQUIPMENT NEEDED

- 3 polarizing sheets
- Polarizing sunglasses
- Exposure meter and sheet of Cartesian graph paper (optional)
- Lamp (and converging lens for parallel beam if needed)
- Protractor
- 6–8 glass microscope slides
- Glass plate
- Tripod stand (open ring top)
- Calcite crystal
- Mica sheet
- Cellophane tape
- Lucite or other plastic pieces (e.g., U-shaped, hook, or hollow triangle)
- LCD (student's wrist watch or hand calculator)

III. THEORY

Light, like all electromagnetic radiation, by wave theory is a **transverse wave.** That is, the directions of the vibrating electric and magnetic field vectors are at right angles to the direction of propagation as illustrated schematically in Fig. 47-1. (If the vector vibrations were parallel to the direction of propagation, light would be a longitudinal wave.) The phenomenon of polarization is firm evidence that light is a transverse wave.

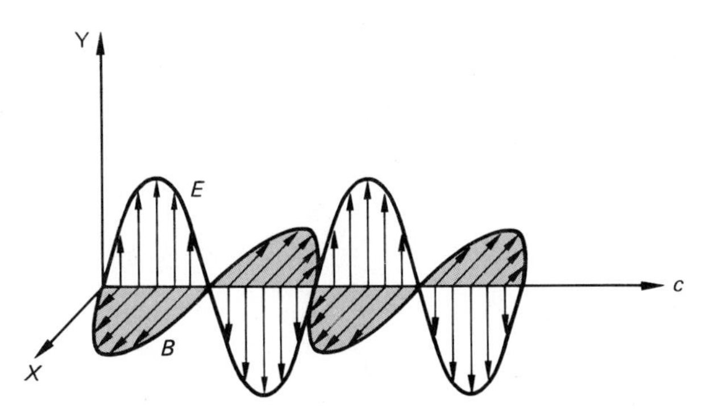

Fig. 47-1 An illustration of an electromagnetic wave. The electric and magnetic field vectors (E and B) vibrate at right angles to each other and to the direction of propagation.

Polarization refers to the orientation of the vibrating vectors of electromagnetic radiation. Light from an ordinary light source consists of a large number of waves emitted by the atoms or molecules of the source. Each atom produces a wave with its own orientation of the E (and B) vibration corresponding to the direction of the atomic vibration. However, with numerous atoms, all directions are possible. The result is that the emitted light is **unpolarized,** or the vibration vectors are randomly oriented with all directions equally probable. This is represented schematically as shown in Fig. 47-2(a), which views a superposition of waves along the axis of propagation.

If for some reason the light vectors become preferentially oriented, we say it is partially polarized [Fig. 47-2(b)]. Should there be only one direction of vibration for the E vectors [Fig. 47-2(c)], the light is then **linearly polarized.** This is sometimes called **plane polarized** or simply **polarized.** The direction of the E vector vibration defines the plane or direction of polarization. The polarization of light may be effected by several means: (a) selective absorption, (b) reflection, (c) refraction, and (d) scattering. Let's take a look at these.

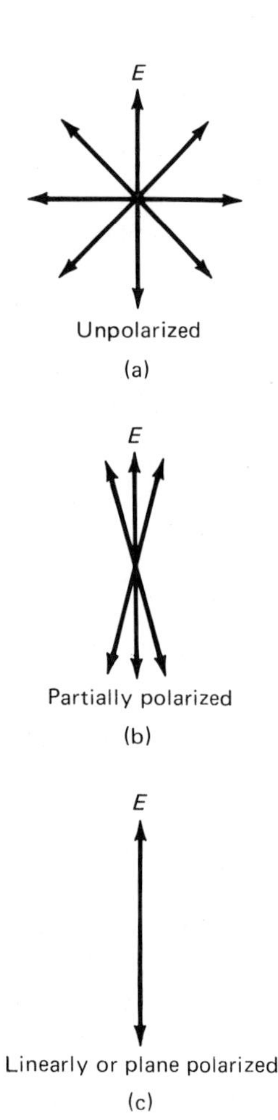

Fig. 47-2 Polarization of light as denoted by the E vector viewed along the line of propagation.

A. *Polarization by Selective Absorption*

A substance that has the property of transmitting light with the electric field vector in only one direction (linearly polarized light) is said to be **dichroic.** This is the principle of the most common technique for obtaining polarized light. In 1938, E. H. Land discovered a material, known by the commercial name of Polaroid, that polarizes light through selective absorption by oriented long-chain polymer molecules. This is in the form of a thin sheet in which the molecular alignment is produced by stretching during fabrication. With proper treatment, the outer (valence) molecular electrons can move along the oriented chains. As a result, the molecules readily absorb light with E vectors parallel to the oriented chains and transmit light with the E vectors perpendicular to the chains.

The direction perpendicular to the oriented molecular chains is commonly called the **transmission axis, plane of polarization,** or **polarization direction.** Hence, when unpolarized light falls on a polarizing sheet (polarizer), polarized light is transmitted. This is illustrated in Fig. 47-3. Various crystals are dichroic. In fact, an early version of Polaroid film developed by Land around 1930 consisted of herapathite (quinine sulfate periodide) crystals imbedded in a polymer sheet.

The polarization of light may be analyzed (detected) by means of another polarizer, which acts as an analyzer (Fig. 47-3). The component of the E vector parallel to the transmission axis of the analyzer is $E_o \cos\theta$. Since the intensity varies as the square of the amplitude, the transmitted intensity of light through the analyzer is

$$I = I_o \cos^2\theta \qquad \textbf{(47-1)}$$

where I_o is the intensity of the polarized light incident on the analyzer. If $\theta = 90°$, we have a condition of cross-polarizers and no light is transmitted through the analyzer.

B. *Polarization by Reflection*

Polarized light may also be obtained by reflection. The degree of polarization depends on the angle of incidence. For angles of incidence of 0° and 90° (grazing and normal angles), the reflected light is unpolarized. However, for intermediate angles, the light is polarized to some extent. Complete polarization occurs at an optimum angle called the **polarization angle** θ_p (Fig. 47-4). This occurs when the reflected and refracted beams are 90° apart and is specific for a given material.

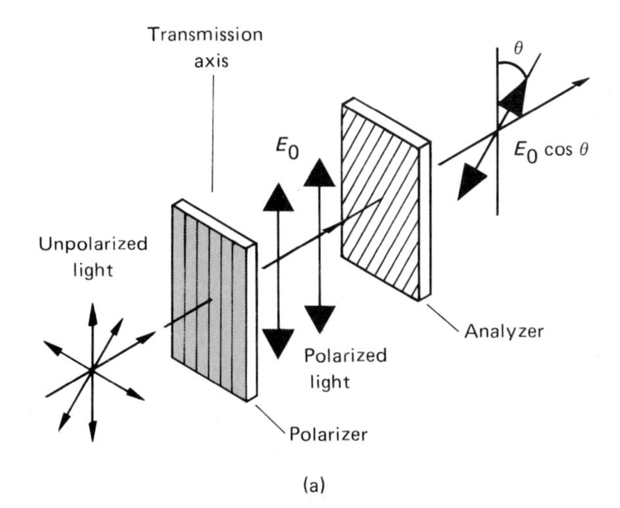

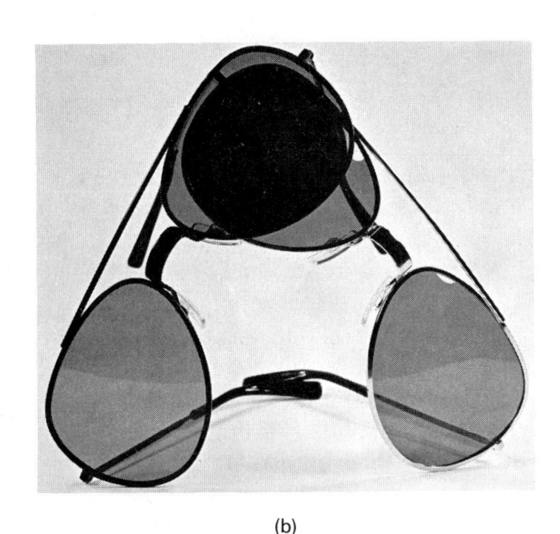

(b)

Fig. 47-3 Polarizer and analyzer. (a) The transmission axis (or plane of polarization or polarization direction) is perpendicular to the oriented molecular chains. When the transmission axes of a polarizer and analyzer are not parallel, less light is transmitted. (b) For "crossed polaroids" (θ =90°), little or no light is transmitted as shown for "crossed" polarizing sunglass lenses. (Photo courtesy *An Introduction to Physical Science,* 4th ed., by Shipman, et al. Lexington, Mass.: D.C. Heath and Company, 1983.)

Referring to Fig. 47-4, when $\theta_1 = \theta_p$, we have $\theta_1 + 90° + \theta_2 = 180°$. Then, $\theta_1 + \theta_2 = 90°$, or $\theta_2 = 90° - \theta_1$. By Snell's law (Experiment 44),

$$\frac{\sin \theta_1}{\sin \theta_2} = n$$

where n is the index of refraction and $\sin \theta_2 = \sin(90° - \theta_1) = \cos \theta_1$. Thus,

$$\frac{\sin \theta_1}{\sin \theta_2} = \frac{\sin \theta_1}{\cos \theta_1} = \tan \theta_1 = n$$

or

$$\tan \theta_p = n \qquad \textbf{(47-2)}$$

This expression is sometimes called **Brewster's law,*** and θ_p is called Brewster's angle.

□ **Example 47-1** A glass plate has an index of refraction of 1.48. What is the angle of polarization for the plate?

Solution With $n = 1.48$,

$$\theta_p = \tan^{-1}(1.48) = 56°$$

□

Notice from Fig. 47-4 that the reflected beam is horizontally polarized. Sunlight reflected from water, metallic surfaces (e.g., from a car), etc., is partially polarized; and if the surface is horizontal, the reflected light has a strong horizontal component. This fact is used in polarizing sunglasses. The transmission axis of the lenses is oriented vertically so as to absorb the reflected horizontal component and reduce the glare or intensity.

Also notice that the refracted beam is partially polarized. By using a stack of glass plates, the transmitted beam would become almost linearly polarized.

C. *Polarization by Refraction*

In a sense, the preceding case of a polarized transmitted beam from a stack of glass plates might be thought of as polarization by refraction. However,

*After its discoverer, David Brewster (1781–1868), a Scottish physicist.

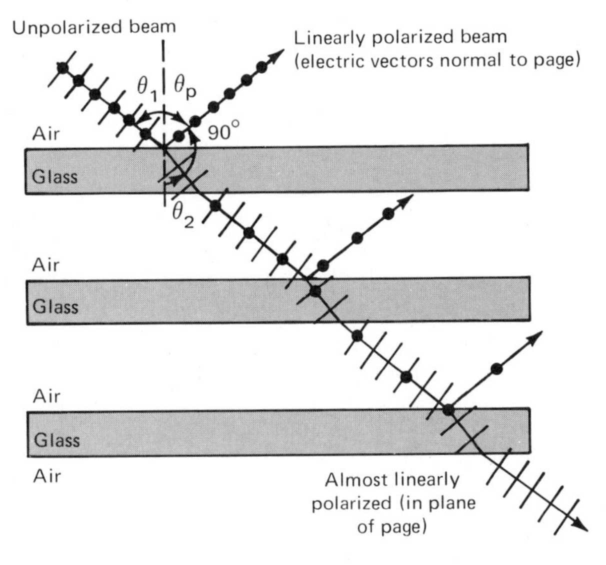

Fig. 47-4 Polarization by reflection. Maximum polarization occurs for a particular polarization angle θ_p which depends on the index of refraction of the material. Notice that the transmitted beam is partially polarized.

this generally refers to the *double refraction* exhibited by some crystals. In an optically isotropic medium, such as glass, light travels with the same speed in all directions. As a result, the material is characterized by a single index of refraction. In certain anisotropic crystals, however, the speed of light is not the same in all directions. The crystal calcite ($CaCO_3$, Iceland spar), for example, exhibits double refraction or birefringence, and is characterized by two indices of refraction.

When an unpolarized beam of light enters a calcite crystal, it splits into two polarized rays with polarizations in mutually perpendicular directions (Fig. 47-5). One beam is called the ordinary (O) ray, and the other, the extraordinary (E) ray. Because of this property, when something (e.g., typed print) is viewed through a calcite crystal, a double image is seen, corresponding to the two emergent rays.

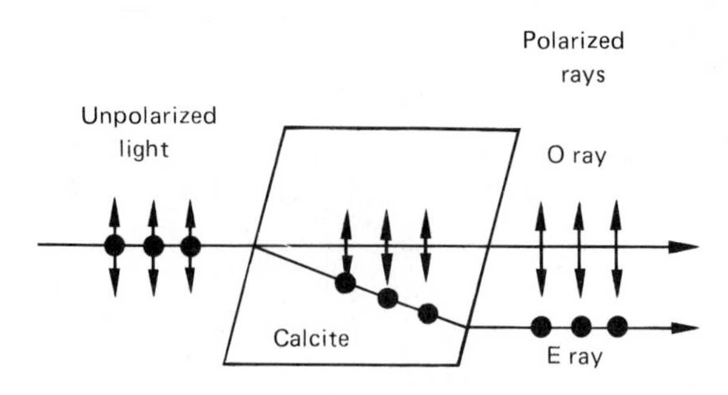

Fig. 47-5 Polarization by double refraction or birefringence. An unpolarized beam entering the crystal is split into two polarized beams.

D. *Polarization by Scattering*

Scattering is the process of a medium absorbing light, then reradiating it. For example, if light is incident on a gas, the electrons in the gas atoms or molecules can absorb and reradiate (scatter) part of the light. In effect, the electrons absorb the light by responding to the electric field of the light wave. Considering an electron as a small antenna, it radiates light in all directions, *except* along its axis of vibration. Hence, the scattered light is partially polarized.

Such scattering of sunlight occurs in the atmosphere and is known as Rayleigh scattering.* The condition for interaction and scattering is that the size d (diameter) of the molecules be much less than the wavelength λ of the light, $d \ll \lambda$. The scattering then varies as $1/\lambda^4$. This condition is satisfied for O_2 (oxygen) and N_2 (nitrogen) molecules in the atmosphere.

Sunlight incident on the atmosphere is white light with a spectrum of wavelength components (or colors). As a result of the $1/\lambda^4$ scattering relationship, the blue end (shorter wavelength) of the spectrum is scattered more than the red end (longer wavelength). The blue light is scattered and rescattered. When looking at the sky, we see this scattered light, and as a result, the sky appears to be blue. (Similarly, blue light is scattered from smoke, and the smoke, e.g., from a cigarette, appears to be blue.)

Hence, the blue "sky light" we see is partially polarized, even though we can't visually detect it (without a little help).

*After Lord Rayleigh (1842–1919), a British physicist who described the effect.

Optical Activity

Certain substances have the property of being able to rotate the plane of polarization of a beam of polarized light. This is called optical activity and is exhibited by crystalline mica, quartz, some sugars, and many long-chain molecular polymers. The principle is illustrated for an optically active crystal in Fig. 47-6. Notice that the crystal essentially changes the direction of one of the vector components of the polarized light along one of its optical axes. The (vector) resultant is then rotated 90° to the polarization plane of the incoming light and will be transmitted through the "crossed" analyzer, which would not ordinarily be the case without rotation.

In traveling through the crystal, the components travel at difference speeds. Suppose the thickness of the crystal is such that the vertical component gains (or falls behind) one-half wavelength compared with the horizontal component. The effect is a reversed vertical component and a 90° rotation of the plane of polarization as shown in the figure. Only for this particular wavelength of light and this particular crystal thickness (or appropriate multiple thereof) will a 90° rotation occur, and only this wavelength (color) of light will be transmitted. As a result of nonuniform crystal thickness, a colored pattern is observed when the crystal is viewed through crossed polarizers.

Optical Stress Analysis

An important application of polarized light is optical stress analysis of transparent materials. Glasses and plastics are usually optically isotropic. If polarized light is transmitted through an isotropic material and viewed with a crossed polarizer, the transmitted

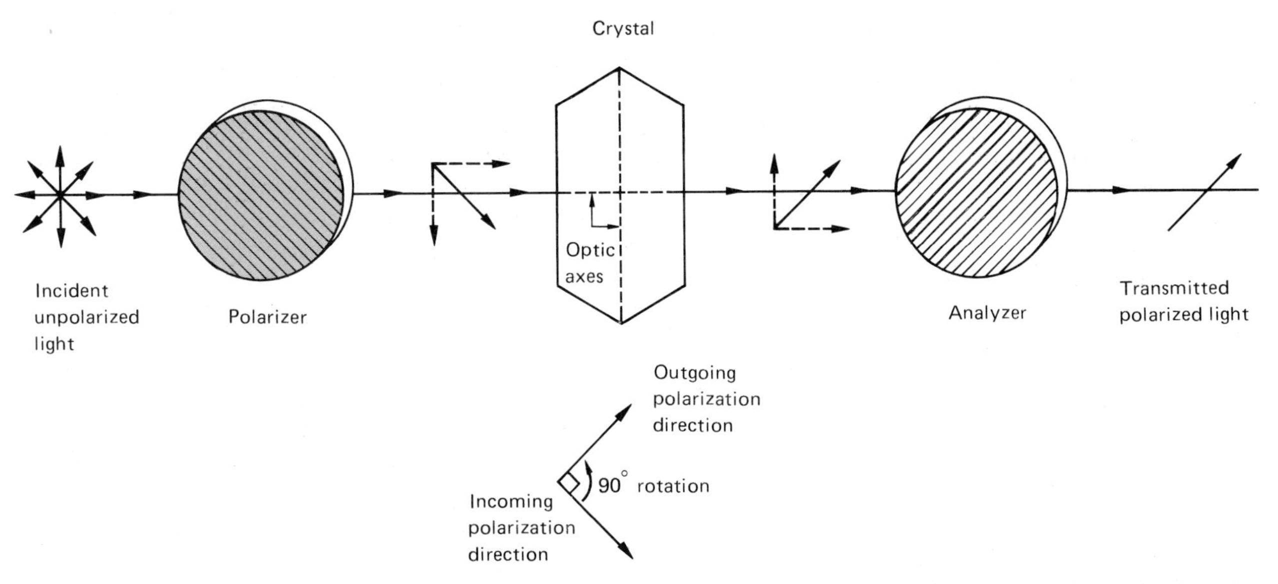

Fig. 47-6 Optical activity. Certain substances have the property of being able to rotate the plane of polarization of a beam of polarized light through 90°.

light intensity is minimal. However, when many materials are mechanically stressed, they become optically anisotropic (indices of refraction vary with direction), and the polarization of the transmitted polarized light is affected. Areas of strain may then be identified and studied through an analyzer (Fig. 47-7). For example, improperly annealed glass may have internal stresses and strains that may later give rise to cracks.

Liquid Crystal Displays (LCD's)

LCD's or liquid crystal displays are now commonly used in wrist watches, hand calculators, and even gas pumps. A liquid crystal is a liquid in which the molecules have some order or crystalline nature. Liquid crystals used in LCD's have the ability to rotate or "twist" the plane of polarization of polarized light. In a so-called twisted nematic display, a liquid crystal is sandwiched between crossed polarizing sheets and backed by a mirror (Fig. 47-8).

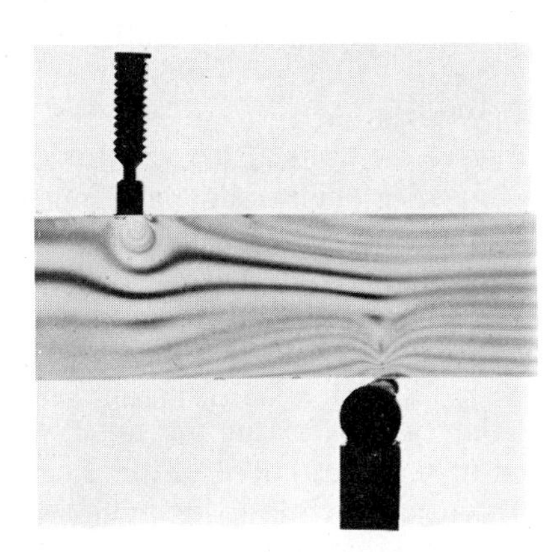

Fig. 47-7 Optical stress analysis with polarized light. A transparent beam reveals stress concentrations near the support and loading points. (Photograph supplied by Ealing Electro-Optics, Inc., 22 Pleasant Street, S. Natick, MA 01760.)

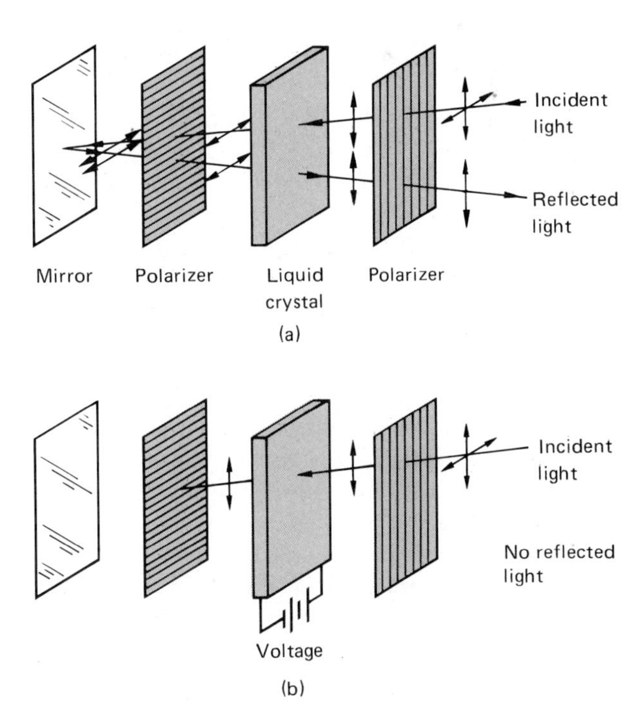

Fig. 47-8 LCD (liquid crystal display). (a) A liquid crystal has the property of being able to "twist" or rotate the plane of polarized light by 90°. (b) When a voltage is applied to the crystal, this property is lost. With no light re-emerging, the crystal appears dark.

Light falling on the surface of an LCD is polarized, twisted, reflected, twisted again, and then leaves the display. Hence, the display appears light when illuminated with polarized light coming from it. However, when a voltage is applied to the crystal, the incident polarized light on the crystals is not twisted in this area and the crystal appears dark. Such dark regions of the crystal are used to form numbers and letters in the display.

IV. EXPERIMENTAL PROCEDURE

Plane of Polarization (Transmission)

1. Inspect your polarizers to see if the planes of polarization or transmission are indicated on them. If not, these planes need to be determined. (Polarizer and analyzer directions will be needed later.) One method is to use a pair of polarizing sunglasses that has a vertical plane of polarization. Determine the planes of polarization for your polarizers by observing the orientations of maximum and minimum transmissions through a polarizer and sunglass lens. Mark the planes of polarization of the polarizers by some means (e.g., a wax pencil or pieces of tape).

2. Investigate the transmission through two polarizers as a function of angle θ between the planes of polarization (Fig. 47-3). At what angles is the transmission
(a) a maximum,
(b) reduced by one-half, and
(c) a minimum?
Record these angles in the Laboratory Report and show the theoretical prediction (using Eq. 47-1) of the angle for one-half transmission (i.e., $I/I_o = 0.5$).
(Optional) If a light exposure meter is available, set up an arrangement so as to measure the intensity through the polarizers as a function of the angle. Plot the intensity (or photocurrent) versus θ on Cartesian graph paper.

3. Orient two polarizers in a crossed position for minimum intensity. Place the third polarizer between these two. Viewing through the polarizers, rotate the middle one (keeping the outer two in a crossed orientation), and observe any intensity changes. As the middle polarizer is rotated, you should observe variations in the transmitted light intensity.
Note the orientation of the plane of polarization of the middle polarizer (relative to the planes of the outer two) for maximum transmission. Make a sketch of the polarizers in the Laboratory Report and indicate the planes of polarization for the polarizers for this condition. (*Hint:* Draw three polarizers of different sizes and label the outer two as 1 and 2, and the middle polarizer as 3.) Is there more than one orientation of the middle polarizer for maximum transmission? (Rotate through 360°.) If so, are the angles between the polarization planes different?
Explain in the Laboratory Report why this transmission through the outer crossed polarizers is observed.

Polarization by Reflection and Refraction

4. Using a general value of the index of refraction of glass to be $n = 1.6$, compute the reflection polarization angle θ_p. (Show work in Laboratory Report.)

5. (a) With a single glass microscope slide on the tripod stand positioned near the edge of the table, shine light on the slide at an angle of incidence equal to the computed θ_p. Observe the reflected light through an analyzer at the angle of reflection and examine for polarization. Note the axis of polarization of the reflected light. Observe the reflected light through polarizing sunglasses and comment.
Observe the light transmitted through the glass slide with an analyzer for any evidence of polarization.
(b) Observe the transmitted light for an increasing number of glass slides and report and explain any observable differences.

Polarization by Crystal Double Refraction

6. Place the calcite crystal on some written or printed material and observe the double image. (The images may appear slightly fuzzy owing to small defects in the crystal. Lay the crystal on a side that gives the clearest images.) Notice that when the crystal is rotated the images move, one more than the other. Examine the images with an analyzer.

7. With a pencil or pen, make a linear series of small, heavy dots on a piece of paper. The line of dots should be long enough to extend beyond the dimensions of the crystal.

   Placing the crystal on the line of dots, rotate the crystal. Notice that as the crystal is rotated, one of the dots of a double image remains relatively stationary and the second dot rotates about the first. The image of the nearly stationary dot is formed by the ordinary (O) ray and the rotating dot by the extraordinary (E) ray.

   Examine a set of dots with an analyzer and record the polarization direction of each ray.

Polarization by Scattering

8. If it's a sunny day, go outside (with the instructor's permission) and observe the sky light from different portions of the sky with an analyzer. Look in directions away from the sun and at angles of 90°. Once you find a region from which the light shows appreciable polarization, rotate your analyzer to see if there is any preferential direction of polarization. (Should it not be a sunny day, try this with your own polarizing sunglasses some fine day.)

Optical Activity

9. View a mica sheet between crossed polarizers. Rotate the analyzer and note the change. What changes, the general pattern shape or its colors?

10. Form a pattern or symbol by sticking various layers of cellophane tape on a glass plate or slide. For example, try a "V" or wedge shape with one, two, three, etc., layers of tape in different parts of the "V." Observe the tape symbol between crossed polarizers. Rotate the analyzer. (You may wish to make letters or symbols with pieces of tape cut with a sharp knife or razor blade, for example, your school letters. Could you give the letters your school colors?)

11. Observe the various-shaped pieces of plastic between crossed polarizers. Stress the pieces by pulling or pushing on them (but not so hard as to break them). Can you explain what is observed?

12. View an LCD through an analyzer. Rotate the analyzer. Is the light coming from the lighted portion of the display polarized? (*Note:* You can do this at home using polarizing sunglasses.)

Name .... Section .... Date ....

Lab Partner(s) ....

EXPERIMENT 47 *Polarized Light*

LABORATORY REPORT

Plane of Polarization

1. Transmission

	Angle θ between polarizer planes
Maximum intensity	....
One-half intensity	....
Minimum intensity	....

Calculation of angle for one-half intensity

2. *(Optional)* Exposure meter readings and angles

3. Three polarizers

Explanation:

Sketch

Name .... Section .... Date ....

Lab Partner(s) ....

EXPERIMENT 47 *Polarized Light*

Polarization by Reflection and Refraction

4. Polarization angle calculation

5. Observations on reflected light

6. Observations on transmitted refracted light

Polarization by Crystal Double Refraction

Observations

(continued)

QUESTIONS

1. In the polarization of light by reflection with a transparent material, would it be possible to have an optimum polarization angle (θ_p) less than 45°? Justify your answer. (*Hint:* Consider the definition of the index of refraction.)

2. At a sale of polarizing sunglasses, there is a half-price special on some glasses with lenses having a horizontal polarization direction. Would you buy the special? If not, why?

3. In the procedure using microscope slides, why was the polarization of the transmitted light more observable with an increasing number of slides?

4. Once polarized, can light be unpolarized? If so, suggest a method to do this.

Name .... Section .... Date ....

Lab Partner(s) ....

EXPERIMENT 48 *The Prism Spectrometer: Dispersion and Index of Refraction*

ADVANCE STUDY ASSIGNMENT

Read the experiment and answer the following questions.

1. What is meant by "dispersion"?

2. Is the index of refraction of a dispersive medium the same for all wavelengths? Explain.

3. What is meant by the "angle of deviation" or a prism?

4. What is the condition for the angle of minimum deviation? In particular, what is the relation of the transmitted ray to the base of the prism?

5. What are the four major components of the prism spectrometer and their functions?

EXPERIMENT **48**

The Prism Spectrometer: Dispersion and Index of Refraction

I. INTRODUCTION

In vacuum, the speed of light c is the same for all wavelengths or colors of light. However, when a beam of white light falls obliquely on the surface of and passes through a glass prism, it is spread out or dispersed into a spectrum of colors. This phenomenon led Newton to believe that light was a mixture of component colors. The dispersion arises in the prism because the wave velocity is slightly different for different wavelengths.

A **spectrometer** is an optical device used to observe and measure the angular deviations of incident light due to refraction and dispersion. Using Snell's law of refraction and geometry, the index of refraction of the prism glass for a specific wavelength or color can be easily determined.

II. EQUIPMENT NEEDED

- Prism spectrometer
- Incandescent light source and support stand

III. THEORY

A monochromatic (single color or wavelength) light beam obliquely incident on the surface of a transparent medium is refracted and deviated from its original direction in accordance with Snell's law (see Experiment 44),

$$n = \frac{c}{c_m} = \frac{\sin \theta_1}{\sin \theta_2} \qquad \textbf{(48-1)}$$

where n is the index of refraction, c the speed of light in vacuum (air), c_m the speed of light in the medium, and θ_1 and θ_2 the angles of incidence and refraction, respectively.

If the incident beam is not monochromatic, each component wavelength (color) is refracted differently because the wave speed is slightly different for different wavelengths in a material medium (e.g., a glass prism) (Fig. 48-1). We say that the substance exhibits **dispersion.**

The frequencies of the waves are unchanged, so the greater the wavelength λ_m, the greater the speed of the wave in the material ($c_m = \lambda_m f$). Consequently, the index of refraction is slightly different for different wavelengths ($n = c/c_m = c/\lambda_m f$).*

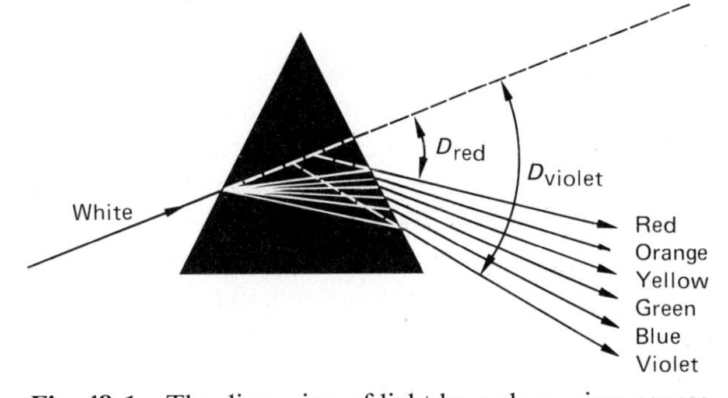

Fig. 48-1 The dispersion of light by a glass prism causes white light to be spread out into a spectrum of colors. The angle between the original direction of the beam and the emergent components of the beam is called the angle of deviation D for a particular component.

The dispersion of a beam of white light spreads the transmitted emergent beam into a spectrum of colors, red through violet (see Fig. 48-1). The red

*The wavelength of the light in a material λ_m differs from the wavelength of light in vacuum λ. $n = c/c_m = \dfrac{\lambda f}{\lambda_m f} = \dfrac{\lambda}{\lambda_m}$.

component has the longest wavelength, so it is deviated least. The angle between the original direction of the beam and an emergent component of the beam is called the **angle of deviation** D, and is different for each color or wavelength.

As the angle of incidence is varied, e.g., decreased from a large value, the angle of deviation of the component colors decreases, then increases, and hence goes through a respective angle of minimum deviation D_m. The angle of minimum deviation occurs for a particular component when the component ray passes through the prism symmetrically, i.e., parallel to the base of the prism (Fig. 48-2).

The angle of minimum deviation and the prism angle A are related to the index of refraction of the prism glass (for a particular color component) through Snell's law by the relationship

$$n = \frac{\sin [(A + D_m)/2]}{\sin (A/2)} \qquad \textbf{(48-2)}$$

The derivation of this equation can be seen from the geometry of Fig. 48-2. Note from the top triangle

$$2(90° - \theta_2) + A = 180°$$

and

$$\theta_2 = \frac{A}{2} \qquad \textbf{(48-3)}$$

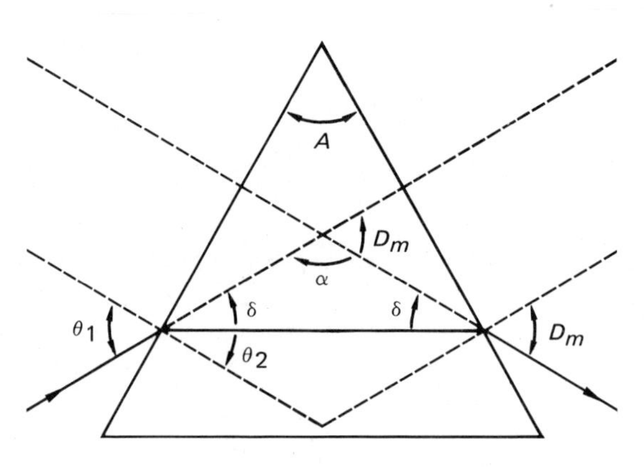

Fig. 48-2 The minimum angle of deviation D_m for a light ray.

Also, for the symmetric case, it can be seen that $D_m = 2\delta$ or

$$\delta = \frac{D_m}{2} \qquad \textbf{(48-4)}$$

(Note the interior triangle, $2\delta + \alpha = 180° = \alpha + D_m$.) Then

$$\theta_1 = \theta_2 + \delta = \frac{A}{2} + \frac{D_m}{2} = \frac{A + D_m}{2} \qquad \textbf{(48-5)}$$

Substituting Eqs. 48-3 and 48-5 into Snell's law (Eq. 48-1) yields Eq. 48-2.

IV. EXPERIMENTAL PROCEDURE*

1. Two types of prism spectrometers are shown in Fig. 48-3, one of which is an adapted force table (see Experiment 6). The four basic parts of a spectrometer are the (a) collimator, (b) prism, (c) telescope, and (d) divided circle.

   The **collimator** is a tube with a slit of adjustable width at one end and a converging lens at the other. Light from a light source enters the collimator. The collimator tube length is made equal to the focal length of the lens so as to make the rays of the emerging light beam parallel.

   The **prism** deviates and disperses the beam into a spectrum. The objective of the **telescope** converges the beam and produces an image of the slit which is viewed through the telescope eyepiece (see Experiment 46). The eyepiece is fitted with cross-hairs, which may be fixed on a particular spectral color. The **divided circle** allows the angle(s) of deviation to be measured.

2. After being given instructions by the instructor, study the various clamps and adjustment screws of your spectrometer. In particular, study the divided circle scale. Some spectrometers are equipped with vernier scales that permit readings to 1 min of arc. Care should be taken as the adjustments and alignments of the spectrometer are critical and it can be time consuming to restore proper adjustment.

3. *For spectrometer with telescope.* With the prism not on the spectrometer table, sight the telescope on some distant object and adjust the eyepiece until the cross-hairs are in good focus. Then mount the lamp near the collimator slit (adjusted to a small slit width). Move the telescope into the line of sight and adjust so that a sharp image of the illuminated slit is seen focused on the cross-hairs.

4. *Measurement of the prism angle A.* Mount the prism in the center of the spectrometer table and orient it as shown in Fig. 48-4. With the unaided eye, locate the white image of the slit reflected from a face of the prism on either side of the prism angle A. The prism may have to be ad-

*Some procedures may not apply to force table apparatus.

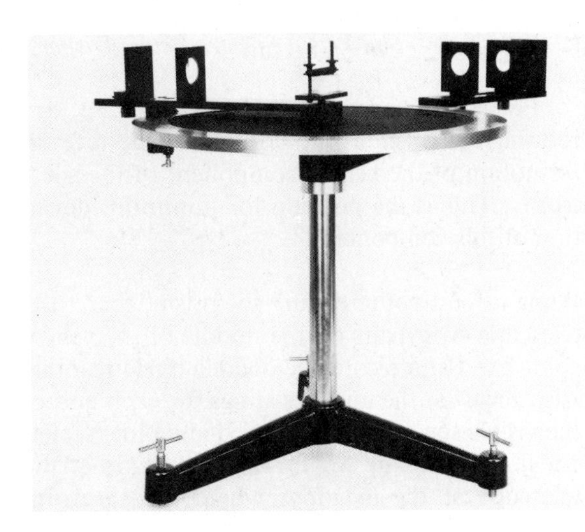

Fig. 48-3 Two types of prism spectrometers. (a) Light from the light source goes through the collimator, the prism, and the telescope to the viewer. Magnifying glasses for reading the divided circle are seen on the prism pedestal. (b) A force table adapted as a prism spectrometer. (Photo courtesy of Central Scientific Co., Inc.)

justed slightly. (You may also note the color spectrum in the prism face opposite A.)

Move the telescope in front of the eye and adjust the cross-hairs on the center of the slit image (with the fine-adjustment screw, if available). Make the slit as narrow as possible so that the best setting can be made. On a force table apparatus, adjustment is not required. Read the angle from the divided circle and record in the Laboratory Report.

Repeat this procedure for the other face of the prism. As shown in Fig. 48-4, the angle between the positions is equal to $2A$. Compute the angle A from the circle readings.

5. *Measurement of the angle of minimum deviation.* Rotate the prism to a position as shown in Fig. 48-5, and with the unaided eye, locate the emergent spectrum of colors. Move the telescope or slit holder in front of the eye and examine the spectrum. (Change the slit width if applicable and note any change.) Record the sequence of colors, beginning with red, in the Laboratory Report.

6. With the slit set as narrowly as possible, rotate the prism back and forth slightly and note the reversal of the direction of motion of the spectrum when the prism is rotated in one direction. Stop

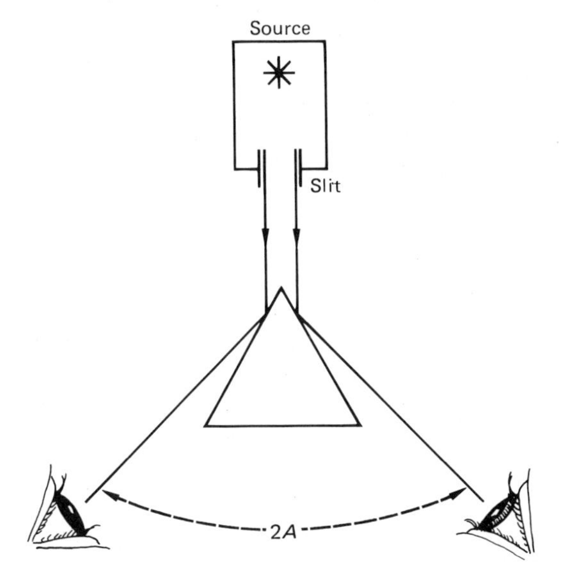

Fig. 48-4 The prism orientation for the experimental procedure to determine the prism angle A.

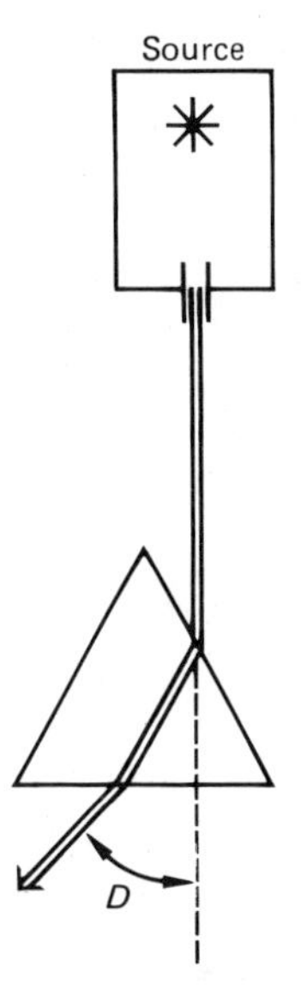

Fig. 48-5 The prism orientation for the experimental procedure to determine the angle of minimum deviation.

rotating the prism at the position of the reversal of motion of the yellow component of the spectrum. This is the position for minimum deviation of this component.

7. Being careful not to disturb the prism, center the telescope cross-hairs on the middle of the yellow color band and record the divided circle reading. Also measure the angle readings for each end of the visible spectrum [i.e., the red and blue (violet) ends]. Do this by setting the cross-hairs of the telescope at the locations where the spectrum ends are just visible (not at the center of the extreme bands).

8. Remove the prism and move the telescope into the line of sight of the slit and adjust so that a sharp image of the illuminated slit is seen on the cross hairs. Note and record the reading of the divided circle.

9. Compute the index of refraction for yellow light using Eq. 48-2.

LABORATORY REPORT

Measurement of Prism Angle A

Calculations (show work)

Circle readings for reflected images ....

Computation of $2A$ ....

Prism angle A ....

Measurement of Angle of Minimum Deviation

Spectrum (sequence) of colors

	Circle reading	Minimum deviation
Yellow	....	....
Red end	....	....
Blue end	....	....
Line of sight	....	

Calculations (show work)

Index of Refraction for Yellow Light

$n =$ ....

Name .... Section .... Date ....

Lab Partner(s) ....

EXPERIMENT 48 *The Prism Spectrometer: Dispersion and Index of Refraction*

QUESTIONS

1. Show that the angle between the two telescope settings of the reflected slit images is equal to $2A$ (procedure 4).

2. What is the speed of yellow light in the prism based on your experimental results?

3. What is the range of dispersion of the prism in terms of indices of refraction for visible light?

Name .... Section .... Date ....

Lab Partner(s) ....

EXPERIMENT 49 *Line Spectra and the Rydberg Constant*

ADVANCE STUDY ASSIGNMENT

Read the experiment and answer the following questions.

1. Distinguish between continuous and line spectra, and describe their causes.

2. Why does a gas discharge tube (e.g., a neon light) have a certain color?

3. What is (a) the Balmer series, and (b) the Rydberg constant?

4. Explain briefly how the prism spectrometer is calibrated.

5. Explain briefly how the Rydberg constant is determined experimentally.

EXPERIMENT **49**

Line Spectra and the Rydberg Constant

I. INTRODUCTION

In spectroscopic analysis, two types of spectra are observed: continuous spectra and line or discrete spectra. The spectrum of visible light from an incandescent source is found to consist of a **continuous spectrum** or band of merging colors and contains all the wavelengths of the visible spectrum. However, when the light from a gas discharge tube (e.g., mercury or helium) is observed through a spectroscope, only a few colors or wavelengths are observed. The colored images of the spectroscope slit appear as bright lines separated by dark regions: hence the name **line** or **discrete spectra.**

Each gas emits a particular set of spectral lines or has a characteristic spectrum. Thus, spectroscopy (the study of spectra) provides a method of element identification. The discrete lines of a given spectrum depend on the atomic structure of the atoms and are due to electron transitions. The line spectrum of hydrogen was explained by Bohr's theory of the hydrogen atom. However before this, the line spectrum of hydrogen was described by an empirical formula involving the Rydberg constant. In this experiment, line spectra will be observed and the relationship of the empirical Rydberg constant to the theoretical quantities of the Bohr theory will be investigated.

II. EQUIPMENT NEEDED

- Prism spectrometer
- Incandescent light source
- Mercury or helium discharge tube
- Hydrogen discharge tube
- Discharge-tube power supply
- 2 sheets of Cartesian graph paper

III THEORY

The electrons in an incandescent light source undergo thermal agitation and emit electromagnetic radiation (light) of many different wavelengths, hence producing a continuous spectrum. However, when light emitted from excited gases or vaporized liquids or solids is analyzed, line spectra such as those illustrated in Fig. 49-1 are observed.

Modern theory explains spectra in terms of photons of light of discrete wavelengths being emitted as the result of electron transitions between atomic energy levels. Different substances have characteristic spectra—that is, they have a characteristic set of lines at specific wavelengths. In a manner of speak-

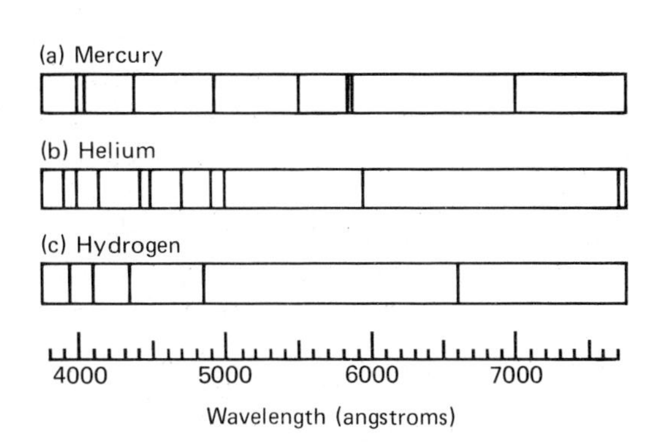

Fig. 49-1 Visible line spectra for (a) mercury, (b) helium, and (c) hydrogen.

ing, the spectrum of a substance acts as a "fingerprint" by which the substance can be identified.

The characteristic color of light from a gas discharge tube is indicative of the most intense spectral line(s) in the visible region. For example, light from a hydrogen discharge tube has a characteristic red glow resulting from an intense emission line with a wavelength of 6561 Å. (Wavelengths are commonly measured in angstrom units: 1 Å = 10^{-10} m = 10^{-8} cm.) When table salt (sodium chloride) is vaporized in a flame, yellow light is observed because of intense emission lines in the region of the sodium spectrum.

The systematic spacing of the spectral lines in the hydrogen spectrum was empirically described by spectroscopists in the late 1800s. For example, the wavelengths of spectral lines in the visible region, called the Balmer series, were found to fit the formula

$$\frac{1}{\lambda} = R\left(\frac{1}{2^2} - \frac{1}{n^2}\right) \qquad n = 3, 4, 5, \ldots \qquad \textbf{(49-1)}$$

where R is the **Rydberg constant,** with a value of 1.097×10^{-3} Å^{-1}.

The hydrogen spectrum is of particular theoretical interest because hydrogen, having only one proton and one electron, is the simplest of all atoms. Niels Bohr (1885–1962) developed a theory for the hydrogen atom that explains the spectral lines as resulting from electron transitions between energy levels or discrete electron orbits (Fig. 49-2), with the wavelengths of the spectral lines being given by the theoretical equation

$$\lambda = \frac{hc}{\Delta E} \qquad \textbf{(49-2)}$$

where $\Delta E = 13.6\ [(1/n_f^2) - (1/n_i^2)]$ eV is the energy difference between the initial and final states, n_i and n_f ($n = 1, 2, 3, 4, \ldots$, and are called the **principal quantum numbers**), $h = 6.63 \times 10^{-34}$ J-s = 4.14×10^{-15} eV-s (Planck's constant), and $c = 3 \times 10^8$ m/s (speed of light).

For spectral lines in the visible region, the final state is $n_f = 2$, and

$$\lambda = \frac{hc}{\Delta E} = \frac{hc}{13.6[(1/2^2) - 1/n^2]} \qquad n = 3, 4, 5, \ldots$$

or

$$\frac{1}{\lambda} = \frac{13.6}{hc}\left(\frac{1}{2^2} - \frac{1}{n_i^2}\right) \qquad n = 3, 4, 5, \ldots \qquad \textbf{(49-3)}$$

Comparing this theoretical equation with the empirical equation 49-1, it is seen that the forms are identical with the prediction that $R = 13.6/hc$.

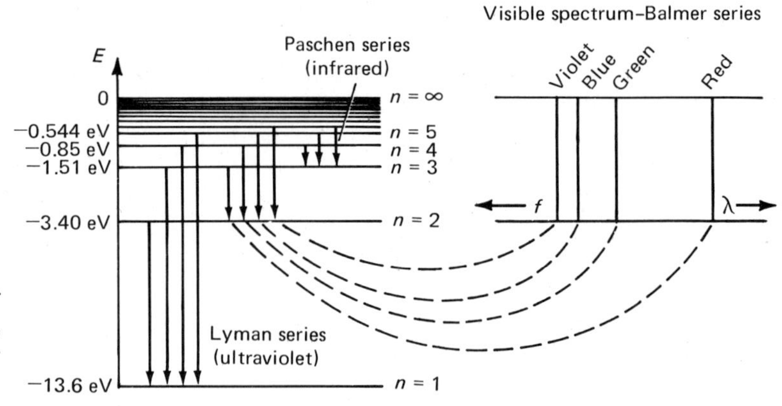

Fig. 49-2 The energy-level transitions for the hydrogen atom. The Balmer series, $n_f = 2$, produces a line spectrum in the visible region.

IV. EXPERIMENTAL PROCEDURE

1. A prism spectrometer will be used to analyze and study spectra in this experiment. The prism spectrometer is illustrated and its use described in Experiment 48. Review the operation of this instrument. Place the incandescent source in front of the collimator slit and observe the continuous spectrum that results from the prism dispersion (Experiment 48). List the colors of the spectrum in the Laboratory Report, beginning with red.

2. A convenient type of discharge tube and power supply is shown in Fig. 49-3. *Caution* should be observed, as the discharge tube operates at high voltage and you could receive an electrical shock.

   Mount a mercury (or helium) discharge tube in the power supply holder and place in front of the collimator slit. (If a larger mercury source is used, it should be properly shielded because of the ultraviolet radiation that may be emitted.

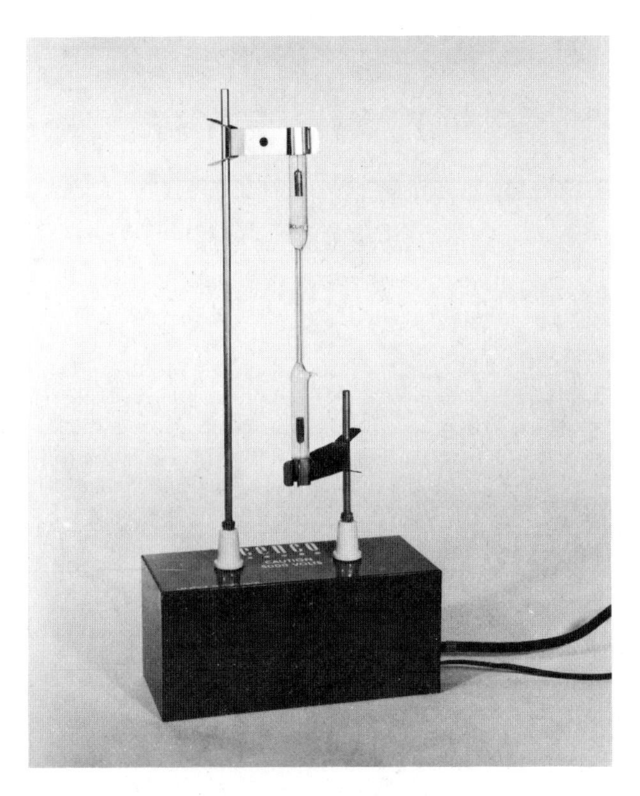

Fig. 49-3 Gas discharge tube apparatus and power supply. (Photo courtesy of Central Scientific Co., Inc)

Consult your instructor.) Turn on the power supply and observe the mercury spectrum through the telescope and note its line nature.

With the slit as narrow as possible, rotate the prism slightly back and forth and notice the reversal of the direction of motion of the spectrum when the prism is rotated in one direction. Focusing on the yellow line, stop rotating the prism at the position of the reversal of motion of this line. This sets the prism for minimum deviation for the yellow line (Experiment 48), which will be taken as an average for the spectrum. (The other lines have slightly different minimum deviations.)

3. (a) Without disturbing the prism, starting at the red end of the spectrum, set the cross-hairs of the telescope on the extreme red line and record the color and divided circle reading in the Laboratory Report. Repeat this procedure for each spectral line in order. (Turn off the discharge tube as soon as possible to conserve the life of the tube.)
   (b) Find the wavelengths of the spectral lines for the discharge tube gas in Appendix A, Table A8, and match to the line readings.
   (c) Using these data, plot the wavelength λ versus the divided circle reading θ. This calibrates the spectrometer, and unknown wavelengths can be determined from divided circle readings from the calibration curve.

4. With the discharge-tube power supply off, replace the mercury discharge tube with a hydrogen discharge tube. Turn on the power supply, and starting with the red line of the hydrogen spectrum, determine the divided circle reading for each spectral line with the cross-hairs of the telescope positioned on the center of the line. Record in the Laboratory Report. The red line is referred to as H_α in spectroscopic notation and the other sequential lines H_β, etc., with subscripts in Greek alphabetical order.

5. Determine the wavelengths of the hydrogen lines from the calibration curve, and plot the reciprocal of the wavelength $1/\lambda$ versus $1/n^2$. (Begin the abscissa scale with zero.) Draw the best straight line that fits the data points and determine the slope of the line.

   Note that Eq. 49-1, $1/\lambda = R(1/2^2 - 1/n^2) = R/4 - R/n^2$, has the form of a straight line, $y = ax + b$, with the negative slope equal to the Rydberg constant. Compare the slope of the line with the accepted value of the Rydberg constant by computing the percent error.

Name .... Section .... Date ....

Lab Partner(s) ....

EXPERIMENT 49 *Line Spectra and the Rydberg Constant*

LABORATORY REPORT

Colors of the Continuous Spectrum

MERCURY SPECTRUM

Color	Divided-circle reading	Wavelength (Å) (from Table A8)

HYDROGEN SPECTRUM

Line	Color	Divided-circle reading	Wavelength (Å)	$1/\lambda$	$1/n^2$
H_α, $n = 3$					
H_β, $n = 4$					
H_γ, $n = 5$					
H_δ, $n = 6$					

Name .... Section .... Date ....

Lab Partner(s) ....

EXPERIMENT 49 *Line Spectra and the Rydberg Constant*

Calculations (Slope of Graph)
(show work)

R (experimental) ....

Accepted value ....

Percent error ....

QUESTIONS

1. Compute the value of the Rydberg constant from the Bohr theory and compare it with the accepted empirical value.

2. Why are only four lines seen in the Balmer series? (Transitions for $n_i > 6$ also exist.) Justify your answer mathematically.

3. As n becomes very large, the wavelengths of the Balmer (and other) series approaches a minimum wavelength or a series limit (Eq. 49-1). What is the wavelength of the series limit for the Balmer series?

Name .... Section .... Date ....

Lab Partner(s) ....

EXPERIMENT 50 *The Transmission Diffraction Grating: Measuring the Wavelengths of Light*

ADVANCE STUDY ASSIGNMENT

Read the experiment and answer the following questions.

1. What is a diffraction grating? Distinguish between the two types of gratings.

2. What is the grating constant? What would be the grating constant for a grating with 12,000 lines/in.? (Express the constant in centimeters.)

3. Explain why there is a spectrum for each diffraction order when multicolored light is analyzed.

4. Is the red or violet end of the first-order spectrum nearest the central maximum? Justify your answer.

5. It will be observed that the second-order spectrum is "spread out" more than the first-order spectrum. Why?

EXPERIMENT **50**

The Transmission Diffraction Grating: Measuring the Wavelengths of Light

I. INTRODUCTION

In Experiment 48 the prism spectrometer had to be calibrated in terms of known wavelengths before we were able to experimentally determine unknown wavelengths of light. How, then, are the wavelengths of spectral lines or colors initially determined? This is most commonly done with a diffraction grating, which is a simple device that allows for the study of spectra and the measurement of wavelengths.

By replacing the prism with a diffraction grating, a prism spectrometer (Experiments 48 and 49) becomes a grating spectrometer. In the case of the diffraction grating, the angle(s) at which the incident beam is defracted is directly related to the wavelength(s) of the light. In this experiment the properties of a transmission grating will be investigated and the wavelengths of several spectral lines will be determined.

II. EQUIPMENT NEEDED

A. *Spectrometer Method*

- Spectrometer
- Diffraction grating and holder
- Mercury discharge tube
- Power supply for discharge tube
- Incandescent light source

B. *Optical Bench Method*

- Optical bench (commercial or meter stick with supports)
- Slit with attached scale and holder
- Diffraction grating and holder
- Sodium vapor lamp (or mercury discharge tube and power supply)
- Incandescent light source

III. THEORY

A **diffraction grating** consists of a piece of metal or glass with a very large number of evenly spaced parallel lines, usually 10,000 to 20,000 lines/in. There are two types of gratings: reflection gratings and transmission gratings.

Reflection gratings are ruled on polished metal surfaces and light is reflected from the unruled areas which act as a row of "slits." **Transmission gratings** are ruled on glass and the unruled slit areas transmit incident light. (The transmission type is used in this experiment.)

Diffraction refers to the "bending" or deviation of waves around sharp edges or corners. The slits of a grating give rise to diffraction and the diffracted light interferes so as to set up interference patterns (Fig. 50-1). Complete constructive interference of the waves occurs when the phase or path difference is equal to one wavelength, and the first-order maximum occurs for

$$d \sin \theta_1 = \lambda \quad \textbf{(50-1)}$$

where d is the grating constant or the distance between the grating lines, θ_1 the angle the rays are diffracted from the incident direction, and $d \sin \theta_1$ the path difference between adjacent rays. The grating constant is given by

$$d = 1/N \quad \textbf{(50-2)}$$

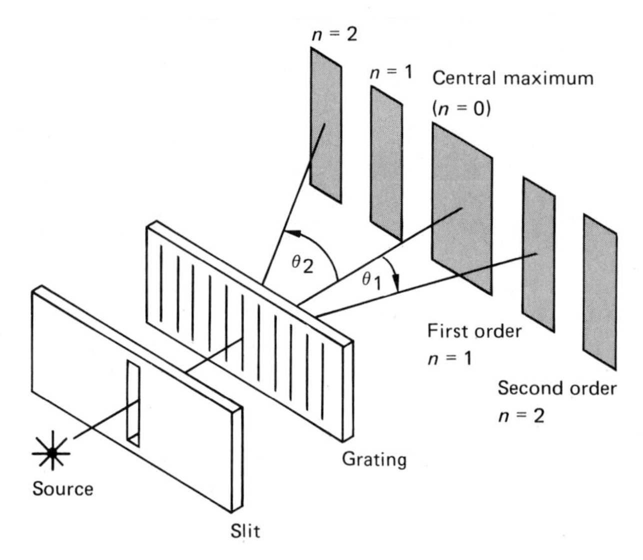

Fig. 50-1 The diffraction pattern (two orders) produced by a diffraction grating.

where N is the number of lines per length (usually per mm or per in.) of the grating.

A second-order maximum occurs for $d \sin \theta_2 = 2\lambda$, and so on, such that in general we may write

$$d \sin \theta_n = n\lambda \quad n = 1, 2, 3, \ldots \qquad \textbf{(50-3)}$$

where n is the order of the image maximum. The interference is symmetric on either side of an undeviated and undiffracted central maximum of the slit image, so the angle between symmetric image orders is $2\theta_n$ (Fig. 50-2).

In practice, only the first few orders are easily observed, with the number of orders depending on the grating constant. If the incident light is other than monochromatic, each order corresponds to a spectrum. That is, the grating spreads the light out into a spectrum. As can be seen from Eq. 50-1, since d is constant, each wavelength (color) is deviated by a slightly different angle, so that the component wavelengths are separated into a spectrum. Each diffraction order in this case corresponds to a spectrum order.

☐ **Example 50.1** In an experiment using a diffraction grating with 10,000 lines/in., the angle between the corresponding lines of a particular component of the first-order spectrum on either side of the incident beam is 27.4°. What is the wavelength of the spectral line?

Solution Given $2\theta_1 = 27.4°$ or $\theta_1 = 13.7°$, and with a grating ruling of 10^4 lines/in., the grating constant d is

$$d = \frac{1}{10^4} = 10^{-4} \text{ in. } (2.54 \text{ cm/in.})$$

$$= 2.54 \times 10^{-4} \text{ cm}$$

Then for first-order ($n = 1$) interference,

$$d \sin \theta_1 = \lambda$$

or

$$\lambda = d \sin \theta_1 = (2.54 \times 10^{-4} \text{ cm})(\sin 13.7°)$$

$$= (2.54 \times 10^{-4} \text{ cm})(0.237)$$

$$= 6.02 \times 10^{-5} \text{ cm } (1 \text{Å}/10^{-8} \text{ cm})$$

☐ $$= 6020 \text{ Å (angstroms)}$$

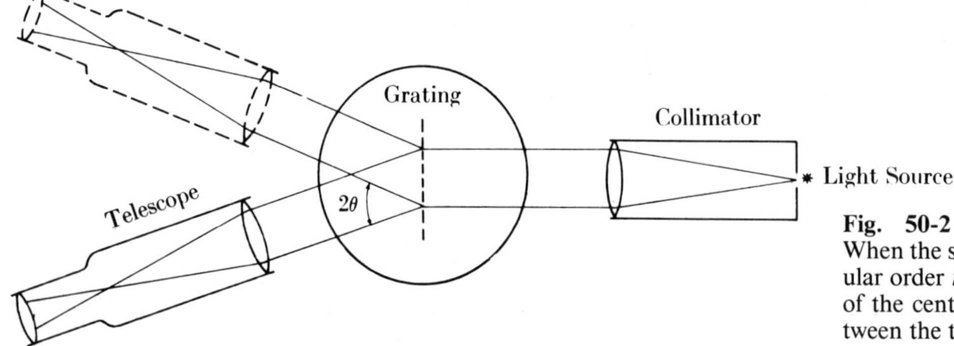

Fig. 50-2 A grating spectrometer. When the symmetric images of a particular order n are viewed from both sides of the central maximum, the angle between the two viewing positions is $2\theta_n$.

IV. EXPERIMENTAL PROCEDURE

A. *Spectrometer Method*

1. Review the operation of a spectrometer if necessary (Experiment 48). Record the number of lines per unit length of your diffraction grating in the Laboratory Report. Mount the grating on the spectrometer table with the grating ruling parallel to the collimator slit and the plane of the grating perpendicular to the collimator axis.

DETERMINATION OF THE WAVELENGTH RANGE OF THE VISIBLE SPECTRUM

2. Mount an incandescent light source in front of the collimator slit. Move the spectrometer telescope into the line of the slit of the collimator and focus the cross-hairs on the central slit image.

   Notice that this central maximum or "zeroth"-order image does not depend on the wavelength of light, so that a white image is observed. Then move the telescope to either side of the incident beam and observe the first- and second-order spectra. Note which is spread out more.

3. (a) Focus the cross-hairs on the blue (violet) end of the first-order spectrum at the position where you judge the spectrum just becomes visible. Record the divided circle reading (to the nearest minute of arc) in Data Table 1.
   (b) Move the telescope to the other (red) end of the spectrum and record the divided circle reading of its visible limit.
   (c) Repeat this procedure for the first-order spectrum on the opposite side of the central maximum. The angular difference between the respective readings corresponds to an angle of 2θ (Fig. 50-2).

4. Compute the grating constant d, and with the experimentally measured θ's, compute the range of the wavelengths of the visible spectrum in centimeters and angstrom units.

DETERMINATION OF THE WAVELENGTHS OF SPECTRAL LINES

5. Mount the mercury discharge tube in its power supply holder and place in front of the collimator slit. (*Caution* should be observed, as the discharge tube operates at high voltage and you could receive an electrical shock. If a large mercury source is used, it should be properly shielded because of the ultraviolet radiation that may be emitted. Consult with your instructor.) Turn on the power supply and observe the first- and second-order mercury line spectra on both sides of the central image.

6. Because some of the lines are brighter than others and the weaker lines are difficult to observe in the second-order spectra, the wavelengths of only the brightest lines will be determined. Find the listing of the mercury spectral lines in Appendix A, Table A8, and record the color and wavelength (in Å) in Data Table 2.

   Then, beginning with either first-order spectra, set the telescope cross-hairs on each of the four brightest spectral lines and record the divided circle readings (read to the nearest minute of arc). Repeat the readings for the first-order spectrum on the opposite side of the central image.

7. Repeat the measurement procedure for the four lines in the second-order spectra and, using Eq. 50-2, compute the wavelength of each of the lines for both order spectra. Compare with the accepted values by computing the percent error of your measurements in each case. (*Note:* In the second-order spectra, two yellow lines—a doublet—may be observed. Make certain that you choose the appropriate line.) (*Hint:* See the wavelengths of the yellow lines in Appendix A, Table A8. Which is closer to the red end of the spectrum?)

B. *Optical Bench Method*

8. Record the number of lines per unit length of your grating in the Laboratory Report. Mount the grating and slit scale on an optical bench as shown in Fig. 50-3. The planes of the slit scale and grating should be parallel.

   Position an incandescent light source behind the slit and observe the diffraction orders of the continuous spectrum superimposed on the scale with the distance s between the slit and the grating at 60, 80, and 100 cm.

   Note the difference between the pattern positions x_1 and x_2 for the first two orders in each case. The images of the slit for a given order should appear at equal distances from the center line. If they do not, rotate the grating slightly until they do.

9. Replace the incandescent light with the sodium vapor lamp (or mercury discharge tube) and observe the line spectra. (*Caution* should be observed, as the discharge tube operates at high voltage and you could receive an electric shock. If a large source is used, it should be properly shielded. Consult with your instructor.) You should see two yellow lines plus other colored lines in each image order. [If a grating having only a few hundred lines per inch (e.g., 600 lines/in.) is used, only one yellow line will be seen, as the grating cannot resolve the two lines.]

10. Choosing the two image orders farthest from the central line that can be clearly seen (for a grating

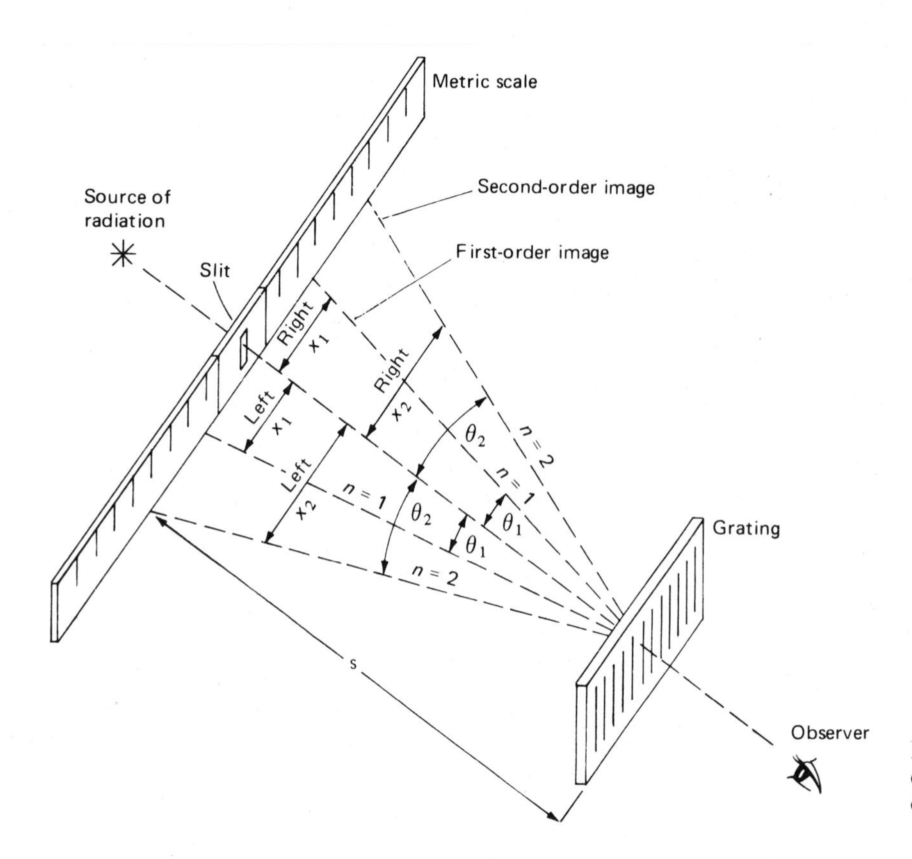

Fig. 50-3 The arrangement for the experimental procedure for the optical bench method.

with a small grating constant or large number of lines per inch, only two orders may be seen), measure the apparent displacement of the yellow lines of the sodium spectrum on each side of the slit with s equal to 60, 80, and 100 cm for each order, and record in Data Table 3. (If a mercury tube is used, choose one of the brightest lines in the spectrum.)

11. From Fig. 50-3, it can be seen that the sin θ_i for a give order by the Pythagorean theorem is equal to

$$\sin \theta_i = \frac{\text{side opposite } \theta}{\text{hypotenuse}} = \frac{x_i}{s^2 + x_i^2}$$

Compute sin θ_i for each measured order for each distance s and find the average value of sin θ_i for each order.

12. Compute the grating constant d, and using Eq. 50-3, find the average wavelength of the yellow lines of sodium from each of the order measurements.

Compare these experimental values with the accepted value of the average wavelength of the yellow sodium lines. (See Appendix A, Table A8, for the accepted wavelengths of the yellow lines.) Compute the percent error for the more accurate experimental value.

Name .... Section .... Date ....

Lab Partner(s) ....

EXPERIMENT 50 *The Transmission Diffraction Grating: Measuring the Wavelengths of Light*

LABORATORY REPORT

Number of lines per unit length on grating .... Grating constant, *d* (cm) ....

A. *Spectrometer Method*

Determination of the Wavelength Range of the Visible Spectrum

DATA TABLE 1

Spectrum limit	Divided-circle reading		2θ	θ	$\sin\theta$	Computed wavelength	
	Right	Left				cm	Å
Violet end							
Red end							

Calculations (show work)

(continued)

Determination of the Wavelengths of Spectral Lines

DATA TABLE 2

Mercury lines		Divided-circle reading		2θ	θ	$\sin\theta$	Computed λ (Å)	Percent error
Color	Wavelength	Right	Left					
First-order Spectrum								
Second-order Spectrum								

Calculations (show work)

Lab Partner(s) ....

EXPERIMENT 50 *The Transmission Diffraction Grating: Measuring the Wavelengths of Light*

B. *Optical Bench Method*

Number of lines per unit length on grating .... Grating constant, d (cm) ....

DATA TABLE 3

s Distance between grating and slit (cm)	Image order: n = ...		Average *displacement*	sin θ_i
	x_i left	x_i right		
			Average	

s Distance between grating and slit (cm)	Image order: n = ...		Average displacement	sin θ_i
	x_i left	x_i right		
			Average	

	n = ...	n = ...	Accepted λ
Average λ of yellow lines (Å)	....	....	....
Percent error of more accurate experimental value	....		

Calculations (show work)

(continued)

QUESTIONS

A. *Spectrometer Method*

Answer all of the questions below.

B. *Optical Bench Method*

Answer Questions 1, 3, and 4.

1. If a grating with more lines per unit length were used, how would the observed angles or spread of the spectra be affected?

2. Was there any difference in the accuracy of the determination of the wavelengths of the mercury lines for the different order spectra? If so, give an explanation.

3. Is it possible for the first-order spectrum to overlap the second-order spectrum? Explain and assume a continuous spectrum.

4. Is there a theoretical limit to the order of the spectrum one would be able to observe? Justify your answer mathematically.

Name .... Section .... Date ....

Lab Partner(s) ....

EXPERIMENT 51 *The Mass of an Electron: e/m Measurement*

ADVANCE STUDY ASSIGNMENT

Read the experiment and answer the following questions.

1. Why does an electron traveling perpendicular to a uniform magnetic field describe a circular path?

2. What causes the fan-shaped pattern in the vacuum tube?

3. How is the radius of curvature of the electron path measured in the experiment?

4. How does changing the current in the solenoid change the curvature of the electron path?

EXPERIMENT 51

The Mass of an Electron: e/m Measurement

I. INTRODUCTION

The mass of an electron, 9.1×10^{-31} kg, is much too small to measure directly. However, using electromagnetic methods involving the charge of the electron in electric and magnetic fields, the electron mass can be determined indirectly. In several situations, the ratio of the electron charge to mass (e/m) can be determined in terms of directly measurable parameters. Then, knowing the value of the electron charge ($e = 1.6 \times 10^{-19}$ C), for example from Millikan's oil-drop experiment, the mass of an electron can be computed. This will be the procedure used in this experiment.

II. EQUIPMENT NEEDED

- Tuning-eye vacuum tube (6AF6)*
- Air-core solenoid*
- Variable dc power supply (250 V dc)
- ac power supply (6.3 V ac)
- dc power supply (12 V dc)
- Rheostat
- dc ammeter (0 to 5 A)
- Connecting wires
- Vernier calipers and metric ruler
- Wooden dowels of different diameters

III. THEORY

An electron traveling with a speed v perpendicularly to a uniform magnetic field B will experience a force F with a magnitude of

$$F = evB \tag{51-1}$$

where e is the charge of the electron.

The direction of the force ($\mathbf{F} = e\mathbf{v} \times \mathbf{B}$) is perpendicular to the plane defined by $\mathbf{v}$ and $\mathbf{B}$. Thus, the force is always at right angles to the direction of the electron motion. This causes the electron to describe a circular path of radius r, with the magnetic force supplying the required centripetal force F_c. That is,

$$F = evB = F_c = \frac{mv^2}{r} \tag{51-2}$$

*Distributed by Eduquip-Macalaster Co.

where m is the mass of the electron. Solving for v, we have

$$v = \frac{eBr}{m} \tag{51-3}$$

The initial electron speed v is normally acquired by accelerating the electron through an electric potential V, and by the work–energy theorem,

$$W = eV = \tfrac{1}{2}mv^2 \tag{51-4}$$

Then, squaring Eq. 51-3 and substituting for v^2, we have, after rearranging,

$$\frac{e}{m} = \frac{2V}{B^2r^2} \tag{51-5}$$

Hence, by knowing or measuring e, V, B, and r, the mass of an electron can be computed.

In this experiment, the electron(s) will be produced, accelerated, and deflected in a commercial

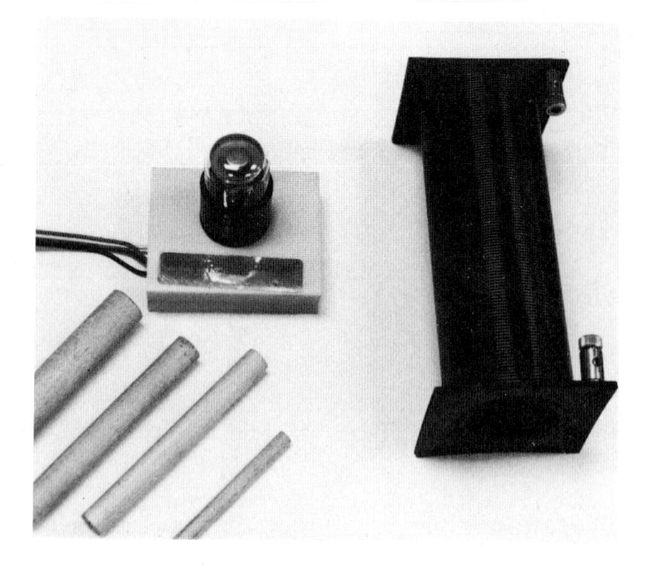

Fig. 51-1 The "tuning eye" vacuum tube, air-core solenoid, and dowels used in measuring e/m.

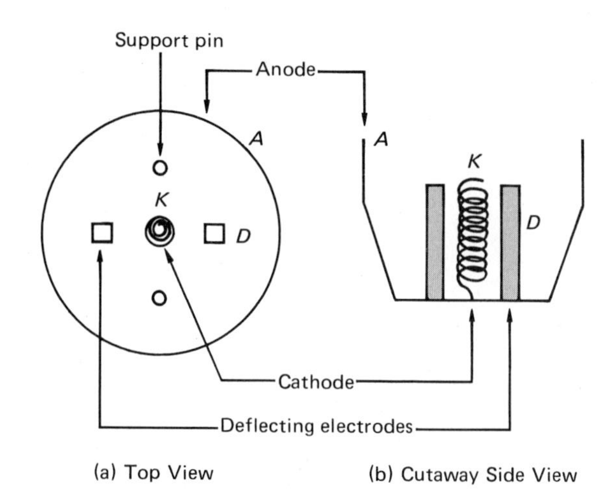

Fig. 51-2 The elements of the tuning-eye tube. See the text for a description.

"tuning-eye" vacuum tube (Fig. 51-1). Such tubes were once commonly used in tuning radios. The magnetic field will be supplied by a solenoid. In the tube, the electrons emitted by the cathode are accelerated by the potential difference applied between the center cathode and outer conical anode (Fig. 51-2).

The electrons move radially outward from the cathode coil, nearly reaching their maximum speed by the time they emerge from beneath the black metal cap covering the center of the tube. The conical anode is coated with a fluorescent material (e.g., zinc sulfide), which emits light when the electrons strike it. Hence, we can visually trace the beam (Fig. 51-3).

The beam pattern in the tube appears fan-shaped owing to two deflecting electrodes in the tube (Fig. 51-2). These electrodes are connected to the cathode, and therefore are negatively charged and repel electrons moving toward them from the cathode, forming wedge-shaped shadows.

When a uniform magnetic field is applied parallel to the tube, the electrons are deflected in an almost circular path. This is indicated by the edge of the shadow, from which the radius of curvature can be measured (Fig. 51-3). In the experiment, the magnetic field is produced by the solenoid, which conveniently fits over the tube. The magnetic field at the center of the solenoid coil carrying a current I is given by

$$B = \mu_o nI \tag{51-6}$$

where $n = N/L$ is the turn density (number of turns N per coil length L in meters).

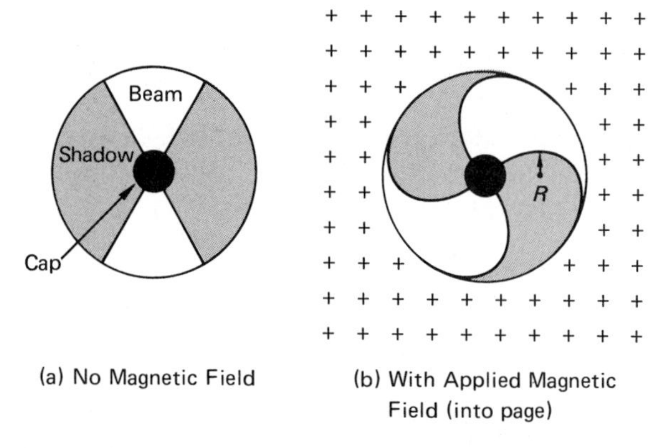

Fig. 51-3 The observed patterns of the tuning-eye tube.

IV. EXPERIMENTAL PROCEDURE

1. Set up the circuits as shown in the schematic diagrams in Fig. 51-4, and place the coil over the tube. Connect the tube leads to the ac and dc power sources and adjust the anode potential to about 150 V dc. Note the fan-shaped tuning-eye pattern. Then connect the solenoid to the 12-V dc power supply and observe how the electron beam is deflected into a curved path as the coil current is increased from zero to a maximum of about 5 A.

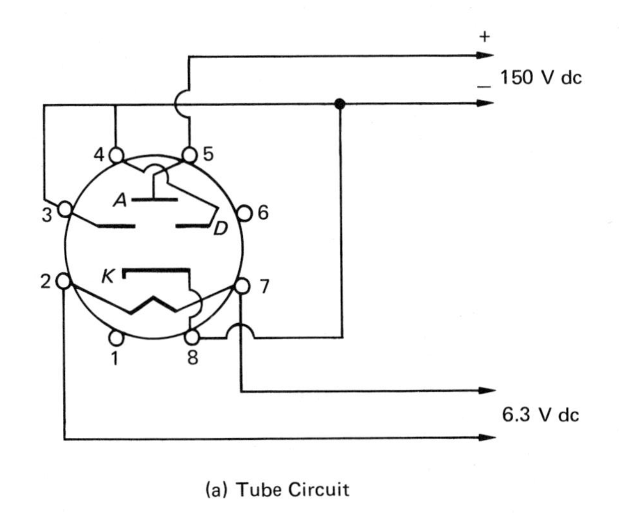

(a) Tube Circuit

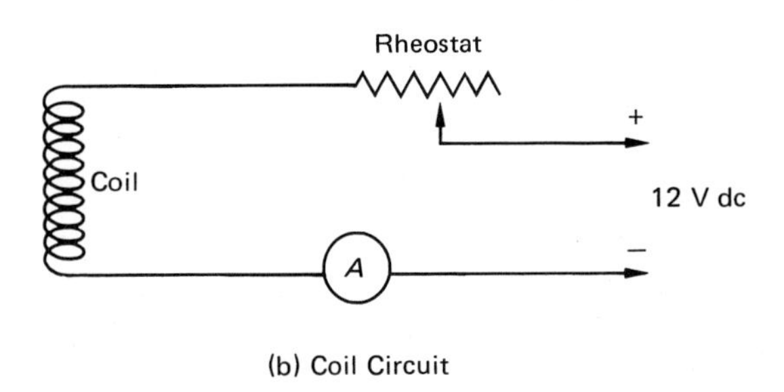

(b) Coil Circuit

Fig. 51-4 Circuit diagrams for the e/m apparatus.

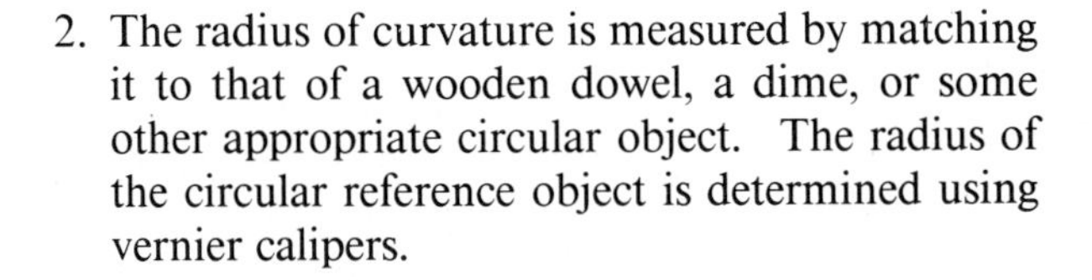

2. The radius of curvature is measured by matching it to that of a wooden dowel, a dime, or some other appropriate circular object. The radius of the circular reference object is determined using vernier calipers.

3. With the anode voltage at 150 V dc, adjust the coil current until the curvature of the shadow matches the reference object. Record the ammeter reading in the Laboratory Report. Determine the current value for two more trials so as to obtain an average. Do not watch the ammeter while matching the radii.

4. Repeat procedure 3 for anode voltages of 200 and 250 V.

5. Disconnect the apparatus from the power supplies. Measure the length L of the solenoid and count the number of turns N, so as to determine the turn density n if this is not given by the instructor. (The coil usually has 540 turns.)

6. Using the average current in each voltage case, compute the magnetic field B of the solenoid for each set of data.

7. Compute the mass of an electron for each set of data using $e = 1.6 \times 10^{-19}$ C. Average the results and compare the average value to the accepted value of the mass of an electron by computing the percent error.

Name .... Section .... Date ....

Lab Partner(s) ....

EXPERIMENT 51 *The Mass of an Electron: e/m Measurement*

LABORATORY REPORT

DATA TABLE

Anode voltage				
Radius, r				
Current (A)	Trial 1			
	2			
	3			
Average current				
Magnetic field, B				

Calculations (show work)

Turn density

N ....

L ....

n ....

Mass of electron

....

....

....

Average value ....

Percent error ....

Name .... Section .... Date ....

Lab Partner(s) ....

EXPERIMENT 51 *The Mass of an Electron: e/m Measurement*

QUESTIONS

1. What are the major sources of error in the experiment?

2. Suppose that protons were emitted in the vacuum tube instead of electrons. How would this affect the experiment?

3. Suppose that only a constant solenoid current were available. Could the mass of an electron still be determined? Explain.

Name .... Section .... Date ....

Lab Partner(s) ....

EXPERIMENT 52 *Exponential Functions*

ADVANCE STUDY ASSIGNMENT

Read the experiment and answer the following questions.

1. How does a time-dependent exponential or logarithmic function vary with time?

2. What is (a) a growth constant, (b) a decay constant, and (c) a time constant?

3. What is meant by (a) half-life, and (b) "doubling life" or doubling time?

4. When an exponential function is plotted on semilog graph paper (y versus x), what is actually being plotted?

5. Does the y intercept occur at $x = 0$ on (a) semilog graph paper, and (b) log-log graph paper? Explain.

EXPERIMENT **52**

Exponential Functions

I. INTRODUCTION

An exponential function is one characterized by a rate of growth (or decay) proportional to the present value of the process. Such processes occur in classical physics as well as modern physics. For example, the charging rate of a capacitor in an *RC* circuit (Experiment 37) and the cooling rate of a thermometer (Experiment 24) are exponential processes: growth and decay, respectively.

However, an in-depth look at the mathematics of exponential functions has been reserved until this section, so as to provide a better understanding of nuclear decay processes. The presentation is general, and as will be learned in the experiment, exponential functions describe the changes that occur in many processes.

II. EQUIPMENT NEEDED

- Pencil and straightedge
- Logarithm tables (Appendix B, Tables B7 and B8) or calculator with logarithmic functions
- 2 sheets of Cartesian graph paper
- 4 sheets of semilog graph paper (1-cycle, 2-cycle, and two 3-cycle)
- 1 sheet of log graph paper (2 × 2-cycle)

III. THEORY

In an exponential process, a variable quantity grows or decays at a rate proportional to the quantity's present value. Let N be the current value of the variable, and dN/dt the rate of change of N with time t (i.e., the time rate of change). By definition, a process is exponential if dN/dt is proportional to N (i.e., $dN/dt \propto N$), and we may write

$$\frac{dN}{dt} = \lambda N \quad \text{or} \quad \frac{dN}{dt} = -\lambda N \qquad \textbf{(52-1)}$$

growth process (N increasing) — decay process (N decreasing)

where λ is a constant of proportionality. The negative sign in the second equation indicates N decreases with time.

(The following derivation requires integral calculus. If you are unfamiliar with calculus, consider only the results of the following integration.)

The equation(s) may be integrated as follows:

$$\frac{dN}{dt} = \lambda N \quad \text{can be written} \quad \frac{dN}{N} = \lambda\, dt$$

and

$$\int_{N_0}^{N} \frac{dN}{N} = \int_0^t \lambda\, dt \qquad \textbf{(52-2)}$$

where the limits of the integration are N_0 at $t = 0$ and N at time t. Performing the integration, we obtain

$$\ln N \Big]_{N_0}^{N} = \lambda t \Big]_0^t$$

and

$$\ln N - \ln N_0 = \lambda(t - 0)$$

or

$$\ln \frac{N}{N_0} = \lambda t \qquad \textbf{(52-3)}$$

where ln is the natural (base e) logarithm, $e = 2.718$ (log is commonly used for the common or base 10 logarithm). Recalling that the equation $\ln_e y = x$ can be written in exponential form as $y = e^x$, Eq. 52-3 becomes

$$\frac{N}{N_0} = e^{\lambda t}$$

or

$$N = N_0 e^{\lambda t} \qquad \textbf{(52-4)}$$

growth process
(N increasing)

Similarly, for a decreasing N,

$$N = N_0 e^{-\lambda t} \qquad \textbf{(52-5)}$$

decay process
(N decreasing)

Because of the equivalency of the exponential and logarithmic forms of the equations, an exponential process is also referred to as a **logarithmic process.**

In the preceding equations, λ is called the **growth** or **decay constant.** Another constant, $\tau = 1/\lambda$, called the **time constant,** is often defined and the equations become

$$N = N_0 e^{t/\tau} \text{ and } N = N_0 e^{-t/\tau} \qquad \textbf{(52-6)}$$

growth process — decay process

In this form, it is convenient to speak of time in terms of time constants [e.g., $t = \tau$ (one time constant), $t = 2\tau$ (two time constants), etc.].

Then, for the various times, $t = \tau$, $t = 2\tau$, and so on,

Growth process

(a) $t = \tau$:

$$\frac{N}{N_0} = e^{t/\tau} = e^1 = 2.718$$

or

$$N = 2.718 N_0$$

(N is 2.718 times the original value N_0)

(b) $t = 2\tau$:

$$\frac{N}{N_0} = e^{t/\tau} = e^2 = 7.389$$

(N is 7.389 times N_0)

Decay process

(a) $t = \tau$:

$$\frac{N}{N_0} e^{-t/\tau} = e^{-1} = \frac{1}{e} = 0.368$$

or

$$N = 0.368\, N_0$$

(N is 0.368 times N_0 or has decayed to 36.8% of the original value N_0)

(b) $t = 2\tau$:

$$\frac{N}{N_0} = e^{-t/\tau} = e^{-2} = 0.135$$

(N is 13.5% of N_0)

A term used to describe decay processes is **half-life** ($t_{1/2}$), which is the time for a quantity to decay to one-half of its original value:

$$\frac{N}{N_0} = \frac{1}{2} = e^{-t_{1/2}/\tau}$$

Since

$$e^{-0.693} = \frac{1}{2}$$

we have, by comparison,

$$0.693 = \frac{t_{1/2}}{\tau}$$

or

$$t_{1/2} = 0.693\tau \qquad \textbf{(52-7)}$$

Hence, if the time constant of a process is known, the half-life can be computed.

☐ **Example 52.1** What fraction of the original value of a quantity (N/N_0) is present in an exponential decay process at the end of two half-lives?

Solution For

$$t = 2t_{1/2} = 2(0.693\tau) = 1.368\tau$$

we have

$$\frac{N}{N_0} = e^{-t/\tau} = e^{-1.386\tau/\tau} = e^{-1.386} = 0.25 = \frac{1}{4}$$

or

$$\frac{N}{N_0} = \frac{1}{4} \qquad \textbf{(52-8)}$$

Hence, after one half-life, the quantity present is one-half of the original value; and after another half-life, the quantity is again decreased by one-half, or is one-fourth of the original value ($\frac{1}{2} \times \frac{1}{2} = \frac{1}{4}$). ☐

The exponential variable is not always time. In some instances, the variable may be distance. For example, the absorption of nuclear radiations by materials decays or "falls off" exponentially with distance or material thickness. The intensity of the radiation is given by

$$I = I_0 e^{-\mu x} \quad \textbf{(52-9)}$$

where I_0 is the original intensity and the constant μ is the linear absorption coefficient.

GRAPHING EXPONENTIAL FUNCTIONS

In previous experiments, exponential functions of the form

$$N = N_0 e^{\lambda t} \quad \textbf{(52-10)}$$

$$(\text{or } y = Ae^{ax})$$

were plotted on Cartesian coordinates in linear form by first taking the natural or Naperian logarithm (base e) of both sides of the equation, e.g., for $N = N_0 e^{\lambda t}$,

$$\ln N = \ln (N_0 e^{\lambda t}) = \ln N_0 + \ln e^{\lambda t} = \ln N_0 + \lambda t$$

or

$$\ln N = \lambda t + \ln N_0 \quad \textbf{(52-11)}$$

This equation has the general form of a straight line, $y = ax + b$. By plotting ln N versus t on Cartesian coordinates, the slope of the line is λ and the intercept is ln N_0. The value of N_0 is obtained by taking the antilog of the intercept value ln N_0. (For a decaying exponential, $N = N_0 e^{-\lambda t}$, the slope would be negative.)

On a Cartesian graph, the ln N must be found for each value of N before it can be plotted. Because this type of function occurs quite often, special graph paper is printed with graduations along the y or ordinate axis that are logarithmically spaced rather than linearly spaced. With the x or abscissa axis linearly graduated, the graph paper is referred to as *semilog graph paper.* (Look at a sheet of semilog graph paper.)

When the variable N is plotted as the ordinate (y axis) value on semilog graph paper, the logarithmic graduations automatically take the logarithm of N, making the separate calculation of ln N unnecessary. Plotting N versus t on semilog paper then actually gives the graph of ln N versus t.

The logarithmic ordinate scale is called "one-cycle," "two-cycle," and so on, depending on the number of powers of 10 covered on the axis. The beginnings of the cycles are consecutively labeled in multiples of 10 (e.g., 0.1, 1.0, 10, or 1.0, 10, 100, etc.), depending on the range (cycles) of the function. (Common logarithms can also be plotted on semilog paper.)

Care must be taken in determining the slope of the line on a semilog plot. On an ordinary Cartesian graph, the slope of a line is given by $\Delta y/\Delta x = (y_2 - y_1)/(x_2 - x_1)$. However, on a semilog graph, the slope of a line is given by

$$\text{slope} = \frac{\Delta \ln N}{\Delta t} \quad \left(\text{or } \frac{\Delta \ln y}{\Delta x}\right) \quad \textbf{(52-12)}$$

On a semilog plot, the listed ordinate values are N, *not* ln N. Hence, one must explicitly take the logs of the ordinate values of the end points of the slope interval, N_2 and N_1, or the log of their ratio:

$$\text{slope} = \frac{\Delta \ln N}{\Delta t} = \frac{\ln N_2 - \ln N_1}{t_2 - t_1}$$

$$= \frac{\ln N_2/N_1}{t_2 - t_1} \quad \textbf{(52-13)}$$

The value of N_0 can be read directly from the y intercept of the graph.

Another common equation form in physics is

$$y = ax^n \quad \textbf{(52-14)}$$

For example, the electric field, $E = kq/r^2 = kqr^{-2}$, is of this form, with $a = kq$ and $n = -2$. By plotting y versus x^n on Cartesian graph paper, a straight line is obtained with a slope of a. However, in an experiment the measured values are usually y and x, so computation of the x^n's is required. But in some instances the exponent n may not be known. This constant, along with the constant a, may be found by plotting y versus x on log graph paper. (This is commonly called *log-log paper* because of the logarithmic graduations on both axes. Look at a sheet of log-log graph paper.)

The logarithmic graduations on axes again automatically take the logarithms of x and y. Working with common logarithms (base 10) in this instance, the log-log plot of y versus x yields a straight line, as can be seen by taking the (common) log of both sides of Eq. 52-14.

$$\log y = \log (ax^n) = \log a + \log x^n$$

$$= \log a + n \log x$$

or

$$\log y = n \log x + \log a \quad \textbf{(52-15)}$$

which has the general form of a straight line with a slope of n and an intercept of log a. For the electric field example, this would be

$$E = \frac{kq}{r^2} = kqr^{-2}$$

$$\log E = -2 \log r + \log kq$$

Again, care must be taken in determining the slope of a straight line on a log-log graph. In this case,

$$\text{slope} = \frac{\Delta \log y}{\Delta \log x} = \frac{\log y_2 - \log y_1}{\log x_2 - \log x_1} = \frac{\log y_2/y_1}{\log x_2/x_1} \qquad \textbf{(52-16)}$$

and the logs of the end points of the slope interval or their ratio must be found explicitly. (The ordinate and abscissa values on the log-log plot are y and x, *not* $\log y$ and $\log x$.)

As in the case of a semilog plot, the value of a in $y = ax^n$ can be read directly from the y intercept of the graph. However, in this case, the intercept is not at $x = 0$, *but at* $x = 1$, since the intercept $\log y = \log a$ requires that $\log x = 0$ and $\log 1 = 0$. (Natural logarithms can also be plotted on log paper.)

IV. EXPERIMENTAL PROCEDURE

A. *Cartesian and Semilog Comparison*

1. First, let's investigate the graphing of the exponential functions

$$\underset{\text{(growth)}}{y = Ae^{ax}} \quad \text{and} \quad \underset{\text{(decay)}}{y = Ae^{-ax}}$$

using the specific examples

$$\text{(a) } y = 0.3e^{2x} \quad \text{and} \quad \text{(b) } y = 90e^{-2x}$$

Various values of y for different values of x for these functions are given in Data Table 1.

2. On the same sheet of Cartesian graph paper, plot y versus x for both functions. Label as Graph I in the upper right-hand corner. (Review the proper graphical presentation in Experiment 1, if necessary.)

3. Taking the natural logs of both sides of the equations gives the equivalent forms (see Eq. 52-11)

$$\text{(a) } \ln y = 2x + \ln 0.3$$

and

$$\text{(b) } \ln y = -2x + \ln 90$$

On another sheet of Cartesian graph paper, plot $\ln y$ versus x for both functions. Label as Graph II. (Find the $\ln y$'s from Appendix B, Table B8, or from a calculator if one is available with a natural log function.)

4. On the same sheet of three-cycle semilog graph paper, plot y versus x for both functions (ordinate values are y, *not* $\ln y$). Label as Graph III.

From Graphs II and III, it should be evident that the logarithmic graph paper automatically takes the log of y. Determine the slope of each curve on Graph III, and find the growth constant and the decay constant of the exponential functions for each respective case. Show your calculations in detail in the Laboratory Report. Also, read and record the intercept values.

B. *Simple Interest—Compounded Annually*

5. An example of exponential growth is simple interest. Suppose that you deposited \$100 in a savings account with an annual interest rate of 6 or 12 percent. Your money "grows" each year as shown in Data Table 2.

Note that each year the growth is proportional to or depends on the amount present at the beginning of each year (end of the previous year). You not only get interest on the original principal each year, but also interest on the interest. Hence, the amount of money grows exponentially.

6. On the same sheet of one-cycle semilog graph paper, plot the amount of money present at the end of each year N versus time t in years for both interest rates. Label as Graph IV. Write the equations of the curves in exponential form in the Laboratory Report.

7. In this growth case, it is convenient to talk about "doubling life" or doubling time. This is the time it takes for the original or current amount of money to double. Determine the approximate doubling time in years for each interest rate from the graph. Extrapolate the curves if necessary. Record the doubling times in the Laboratory Report.

C. *Half-life of a Water Column**

8. When water in a relatively large diameter vertical glass tube is allowed to flow through a long horizontal capillary tube attached to the base of the vertical tube, the height of the water column decreases with time. It can be shown that the height h of the water column is given theoretically by

$$h = h_0 e^{-\lambda t} \qquad \textbf{(52-17)}$$

where h_0 is the height of the column at $t = 0$.

9. Typical experimental data are given in Data Table 3. On semilog graph paper, plot h versus t. Determine the decay constant from the graph and find the half-life of the water column. Show your work in the Laboratory Report.

D. *Gauge Numbers and Wire Diameters*

10. Wire sizes are designated by gauge numbers. One such common designation is the American Wire Gauge (AWG), the gauge (sometimes spelled gage) numbers and wire diameters for which are listed in Appendix A, Table A7. The gauge numbers increase linearly (after the large diameter multiple zero gauges), while the wire diameters decrease about 11 percent for each consecutive gauge number.

This leads to a change in diameter with gauge number that is proportional to a particular (present) diameter d_0 and gauge number n_0, and hence is an exponential function.

11. On a sheet of two-cycle semilog graph paper, plot the wire diameter d (in cm) versus gauge number n. Start with Gauge 0 (as n_0) and plot only multiple values of five up to Gauge 30. Label as Graph V.

12. Determine the decay constant and write an equation for the relationship between diameter and gauge number. Show your work in the Laboratory Report.

E. *Log-log Graphing*

13. A particle experiences a force F, whose magnitude is inversely proportional to some power of the distance r from the force source ($F \propto 1/r^n$). Experimental measurements yield data as given in Data Table 4. Plot F versus r on log-log paper (Graph VI) and determine the equation of the force. (Do not forget units of the constant of proportionality.)

*J. R. Smithson and E. R. Pinkston, "Half-life of a Water Column as a Laboratory Exercise in Exponential Decay," *American Journal of Physics,* Vol. 28, No. 8 (1960).

Name .... Section .... Date ....

Lab Partner(s) ....

EXPERIMENT 52 *Exponential Function*

LABORATORY REPORT

A. *Cartesian and Semilog Comparison*

DATA TABLE 1

$y = 0.3e^{2x}$			$y = 90e^{-2x} = 90/e^{2x}$		
y	$\ln y$	x	y	$\ln y$	x
0.37	....	0.1	74	....	0.1
0.45	....	0.2	60	....	0.2
0.55	....	0.3	49	....	0.3
0.82	....	0.5	33	....	0.5
1.0	....	0.6	27	....	0.6
1.5	....	0.8	18	....	0.8
2.2	....	1.0	12	....	1.0
6.0	....	1.5	4.5	....	1.5
16	....	2.0	1.6	....	2.0
45	....	2.5	0.61	....	2.5
			0.22	....	3.0

Calculations: Slopes and Intercepts (show work)

B. *Simple Interest—Compounded Annually*

DATA TABLE 2 Simple Interest—Compounded Annually (Beginning with $100)

Year	Amount at end of year	
	6% interest	12% interest
1	$106.00	$112.00
2	112.36	125.44
3	119.10	140.49
4	126.25	157.35
5	133.83	176.23
6	141.86	197.38
7	150.37	221.06
8	159.39	247.59

Calculations: Equations and Doubling Times (show work)

C. *Half-life of a Water Column*

DATA TABLE 3
Water Column Data

Time, t (min)	Column height, h (cm)
0.0	120.0
1.0	88.1
2.0	64.6
3.0	46.8
4.0	33.9
5.0	24.5
6.0	17.7
7.0	12.9
8.0	9.3

Calculations: Decay Constant and Half-life (show work)

D. *Gauge Numbers and Wire Diameters*

Calculations: Equation Relationship (show work)

After deriving equation, show its applicability by computing d for some arbitrary n not used in the graph data (e.g., $n = 40$).

E. *Log-log Graphing*

DATA TABLE 4 Force versus Distance

F (newtons)	r (meters)
11	2.0
5.8	2.5
3.3	3.0
2.1	3.5
1.4	4.0
0.72	5.0
0.42	6.0
0.26	7.0
0.18	8.0
0.12	9.0

Calculations: Force Equation (show work)

(continued)

QUESTIONS

1. The curve of $y = Ae^{-ax}$ on Cartesian graph paper is said to approach the x axis asymptotically. Describe what asymptotic means. (*Hint:* Does the curve ever intercept the x axis?)

2. How do (a) the growth constant and (b) the decay constant affect the growth rate and decay rate, respectively, of an exponential function? That is, what would occur if these constants were varied (e.g., increased)?

3. Do the doubling time and half-life of exponential functions apply to the quantities present at any given time? Justify your answer with examples from the function data in the experiment.

4. The population of a country increases exponentially. The yearly growth rate of the United States is about 1 percent. (Growth rate is analogous to monetary interest rate.) Assuming this growth rate to be constant, in how many years will the population of the United States double? What would be our approximate population at that time? (Show work)

5. After how many time constants is a decaying exponential process essentially complete (i.e., less than 1 percent of the original quantity remains)? How many half-lives is this? (Show work)

6. As shown in Section D, the wire diameter decreases exponentially with increasing gauge number. As a result, the electrical resistance per unit length of wire increases exponentially with increasing gauge number (cf. *Handbook of Chemistry and Physics*). Recall that resistance is given by (Experiment 31) $R = \rho L/A$, where ρ is the resistivity, L the length of the wire, and A the cross-sectional area. Then, the resistance per unit length is $R/L = \rho/A$. Would the growth constant for the resistance/length be the same as the decay constant for the diameter? If not, what would it be? Explain and justify your answer mathematically.

Name .... Section .... Date ....

Lab Partner(s) ....

EXPERIMENT 53 *The Chart of Nuclides*

ADVANCE STUDY ASSIGNMENT

Read the experiment and answer the following questions.

1. What is a nuclide?

2. What is the chart of nuclides?

3. How are natural stable isotopes and natuıal long-lived radioactive isotopes distinguished on the chart?

4. How are artificially produced isotopes and natural short-lived radioactive isotopes distinguished on the chart?

5. What are nuclear isomers, and how are they distinguished on the chart?

6. What is the difference between the mass number and the atomic weight of a nuclide?

7. List the change(s) in the proton number (Z) and/or the neutron number (N) for the following nuclear processes and give the relative location of the daughter nucleus to that of the parent nucleus on the chart:

   (a) alpha decay

   (b) beta minus (β^-) decay

   (c) beta plus (β^+) decay

   (d) gamma decay

EXPERIMENT 53*

The Chart of Nuclides

I. INTRODUCTION

The term **nuclide** refers to an atom or nucleus as characterized by the number of protons (Z) and neutrons (N) that the nucleus contains. By listing the known nuclei, both stable and radioactive, in an array on a graph of Z versus N, a chart of nuclides is formed.

The chart of nuclides is somewhat analogous to the periodic chart of elements. Pertinent nuclear information, such as stability, half-life, and decay mode(s), is indicated for each nuclide. Thus, the chart of nuclides gives the nuclear physicist information about nuclei and nuclear reactions, much the same as the periodic chart of elements gives the chemist information about chemicals and chemical reactions.

This experiment is an exercise in the use of the chart of nuclides. It will be explained how information is displayed on the chart; then questions will be asked about various radioactive properties, decay schemes, and nuclear reactions. The answers to the questions may be easily read from the chart once you are familiar with it.

II. EQUIPMENT NEEDED

- Chart of nuclides (colored charts in booklet form may be purchased for a reasonable price from: Manager, Order Service General Electric Co., Nuclear Energy Marketing Dept., 175 Curtiner Avenue, San Jose, California 95125)
- (Optional) 4 sheets of Cartesian graph paper

III. THEORY

The **chart of nuclides** is a display of the properties of the known nuclei at the positions (Z, N) on a graphical plot of Z versus N. Because of the large number of nuclides, the chart is usually shown in consecutive segments for convenience of handling.

The chemical element (symbol) is given at the extreme left of each horizontal row in a heavy bordered square next to the proton (atomic) number Z (Fig. 53-1). Below the chemical symbol is the atomic weight of the element. Below the atomic weight is the thermal neutron absorption cross section σ in barns (1 barn $= 10^{-24}$ cm^2). Thermal or slow neutrons have an average speed on the order of 0.025 eV. (This is the average energy of neutrons in a thermal nuclear reactor where fission neutrons are slowed down by collisions with the moderator atoms until they are in thermal equilibrium with the moderator.)

Each horizontal row in the chart represents the isotopes of a particular element (same Z, different N). Similarly, a vertical column on the chart represents the nuclei of different elements with the same number of neutrons. The neutron number N is given at the bottom of the column.

At the top of each nuclide square or space, the symbol of the nuclide is listed, followed by its mass

*After an experiment developed by and courtesy of Robert Neff, Suffern High School, Suffern, NY.

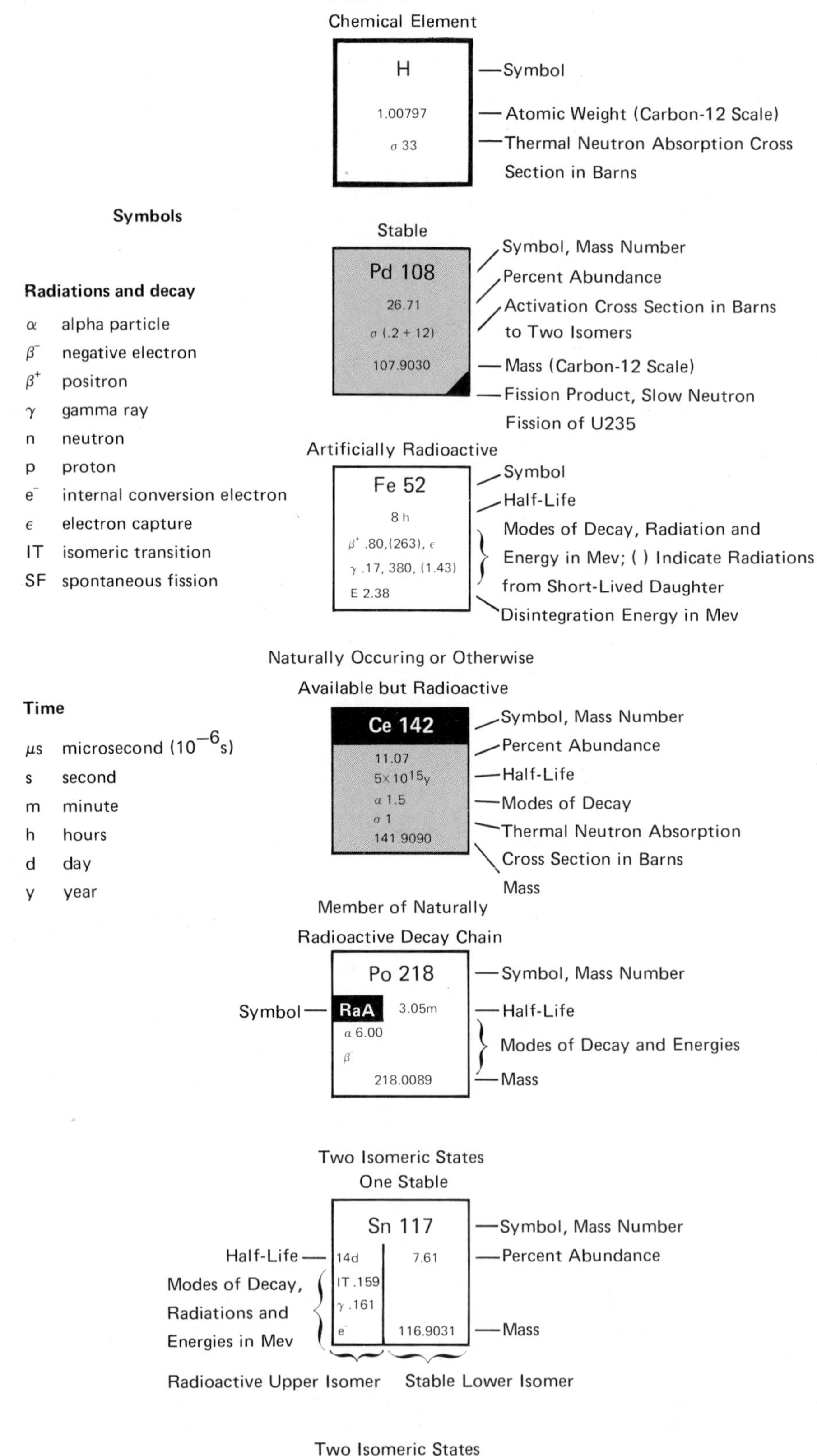

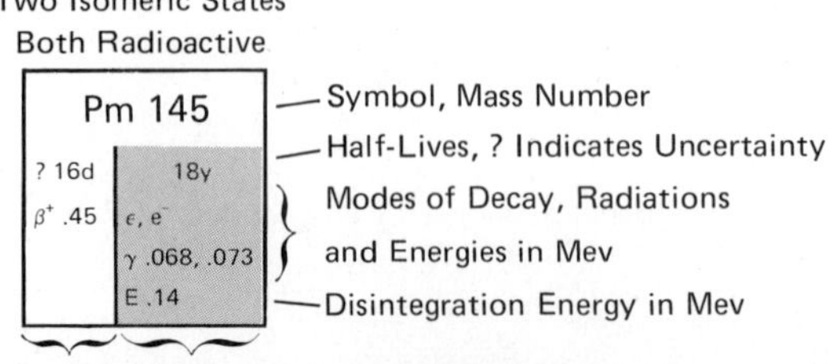

Fig. 53-1 Examples of the data display in the chart of nuclides.

number A ($A = Z + N$). For example, carbon-12 ($Z = 6$, $N = 6$).

There is a great deal of information listed on the chart. Some of this information is beyond the scope of the experiment. We will be concerned only with basic items. General information on nuclide stability may be obtained at a glance from the shading or color of the nuclide squares (see Fig. 53-1):

1. *Gray-shaded squares*—isotopes that occur in nature and are generally considered stable.
2. *White or "color" squares*—artificially produced radioactive isotopes. (Some charts of nuclides have a colored coding concerning the range of half-lives and neutron absorption properties. See the specific chart for information.)
3. *Black rectangles across the top of square*
   (a) On gray-shaded squares—radioactive nuclides found in nature with very long half-lives [e.g., Ce-142 ($t_{1/2} = 5 \times 10^{16}$ years)]. Because of the long half-lives, these nuclides are considered to be generally stable.
   (b) On white squares—radioactive nuclides found in nature with relatively short half-lives [e.g., C-14 ($t_{1/2} = 5730$ years)].
4. *Smaller black rectangle near top of square*—nuclide is a member of a natural radioactive decay chain. The historic symbol is inserted in the black area, for example, Ra A for Po-238 and UX_1 for Th-234.
5. *Black triangle at bottom corner of square*—indicates nuclide is formed by fission of U-235 or Pu-239, for example, Xe-140 and Sr-94 in the induced fission reaction

$$^{235}_{92}U + ^{1}_{0}n \rightarrow ^{140}_{54}Xe + ^{94}_{38}Sr + 2(^{1}_{0}n)$$

   Xe-140 and Sr-94 are radioactive. In other instances, the resulting nuclides may be stable, e.g., Pd-108.
6. *Vertically divided square*—indicates a nuclide with isomeric or metastable states. The nuclide possesses states of different radioactive properties. The nuclei of the different states of particular nuclide are called *nuclear isomers* (same Z and N numbers, but different radioactive properties). The isomeric states have different energies or energy levels. (The lower energy state is commonly referred to as the *ground state* and higher energy levels as *isomeric states.*) Frequently, the ground state is a stable nuclide (e.g., Sn-117).

   If two isomers exist, the higher-energy state is shown on the left. If three isomers exist, the higher-energy state is shown on the left, with the lower state below it or to the right of it, and the ground state to the right of both or below them.

Data are displayed on each nuclide square. Pertinent data for this experiment on the nuclide squares include (see Fig. 53-1):

1. *Gray-shaded squares* (stable nuclides)
   (a) First line—symbol and mass number
   (b) Second line—atom percent abundance, which is the percent of the natural element this isotope represents
   (c) Third line—thermal neutron cross-section
   (d) Fourth line—atomic weight of nuclide atom (mass of nucleus and surrounding electrons)
2. *Black-topped gray-shaded squares* (long-lived radioactive nuclides found in nature)
   Second line—atom percent abundance, followed by half-life and mode(s) of decay with decay particle energy in MeV. When more than one mode of decay occurs, the most prominent mode appears first (above or to the left of the other modes).
3. *White squares* (artificially produced radioactive nuclides) Similar to black-topped gray squares with atom percent abundance absent

Isomeric states have similar displays. Some charts of nuclides have color coding for the range of half-lives and neutron absorption cross-sections. The half-life coding generally appears in the upper half of the nuclide square and the cross-section coding in the bottom half. The color codes are defined on the particular chart.

The arrangement of the chart of nuclides allows one to quickly determine the nuclide change when a nuclear process occurs, either naturally or artificially induced. This is done in accordance with the scheme illustrated in Fig. 53-2.

1. *Induced reactions.* For an induced reaction, knowing the original nucleus and the incident or "in" particle, the *compound nucleus* of the reaction is given in the square of the "in" particle. For example, for the induced reaction of Be-9 being bombarded by an alpha particle,

$$^{9}_{4}Be + \alpha \rightarrow$$

   the compound nucleus is two squares diagonally upward to the right (refer to first diagram in Fig. 53-2),

$$^{9}_{4}Be + ^{4}_{2}He \rightarrow (^{13}_{6}C^{*})$$

   A neutron is the "out" particle in this reaction; hence, the product nucleus is one square to the left of the compound nucleus C-13, which is the square of C-12:

$$(^{13}_{6}C^{*}) \rightarrow ^{12}_{6}C + ^{1}_{0}n$$

Fig. 53-2 Diagrams showing the relative locations of the products of various nuclear processes and displacements caused by nuclear reactions.

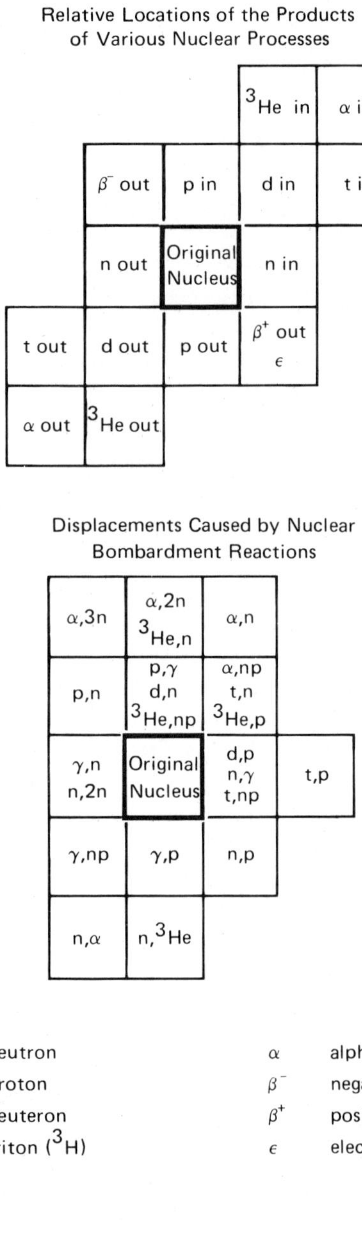

Relative Locations of the Products of Various Nuclear Processes

			^{3}He in	α in
	β^- out	p in	d in	t in
	n out	**Original Nucleus**	n in	
t out	d out	p out	β^+ out ϵ	
α out	^{3}He out			

Displacements Caused by Nuclear Bombardment Reactions

α,3n	α,2n ^{3}He,n	α,n	
p,n	p,γ d,n ^{3}He,np	α,np t,n ^{3}He,p	
γ,n n,2n	**Original Nucleus**	d,p n,γ t,np	t,p
γ,np	γ,p	n,p	
n,α	n,^{3}He		

n	neutron	α	alpha particle
p	proton	β^-	negative electron
d	deuteron	β^+	positron
t	triton (^{3}H)	ϵ	electron capture

In equation form the total reaction is

$$^{9}_{4}\text{Be} + ^{4}_{2}\text{He} \rightarrow (^{13}_{6}\text{C}^*) \rightarrow ^{12}_{6}\text{C} + ^{1}_{0}n$$

or in abbreviated notation,

$$^{9}\text{Be}(\alpha,n)^{12}\text{C}$$

The second diagram in Fig. 53-2 combines this process and the product nucleus is given directly by the location of the reaction particles.

2. *Radioactive decay.* In this case, we are concerned with only an "out" particle, and the location of the daughter nucleus is obtained from the first diagram. For example, C-14 (the parent nucleus) beta-decays (β^-), so the daughter nucleus is one square diagonally upward to the left (i.e., N-14).

Similarly, F-18 undergoes positron decay (β^+, a positively charged "electron") and the daughter nucleus O-18 is one square diagonally downward to the right.

IV. EXPERIMENTAL PROCEDURE

Answer the questions in the Laboratory Report using a chart of nuclides.

Name .... Section .... Date ....

Lab Partner(s) ....

EXPERIMENT 53 *The Chart of Nuclides*

LABORATORY REPORT

1. How are the isotopes of an element arranged on the chart?

2. Nuclides with the same number of neutrons are called *isotones.* How are they arranged on the chart?

3. Nuclides with the same mass number are called *isobars.* What would be the orientation of a line connecting an isobaric series?

4. List all of the naturally occurring generally stable nuclides* that have an odd number of protons and an odd number of neutrons. Also, comment on the ratio of neutrons to protons for increasingly heavier nuclides.

5. List the percent abundances of the naturally occurring nuclides of (a) oxygen; (b) uranium. Do they add up to 100 percent? If not, explain.

*Stable or with extremely long half-lives (> 10^9 years). List the stable nuclides and the long-life radioactive nuclides separately.

(continued)

6. List the elements that have only one stable isotope.

7. Which element(s) has the greatest number of stable isotopes? Give the number of isotopes.

8. Which element(s) has the greatest number of radioactive isotopes? Give the number of isotopes.

9. List an example radioactive nuclide and its half-life for each of the following half-life ranges:

   (a) μs (microseconds, 10^{-6} s)

   (b) s

   (c) min

   (d) h

   (e) days

   (f) years

10. Which nuclide on the chart has (a) the shortest half-life; (b) the longest half-life? List their half-lives.

11. How many nuclides are the products of slow neutron fission of U-235 or Pu-239, and how many of these nuclides are stable?

12. (Do the following exercise on a separate sheet of paper and attach to Laboratory Report.) Beginning with the following radioactive parent nuclei, trace their decay processes and depict the mode and direction of each decay process on the chart. For example,

$$^{14}N \xleftarrow{\beta^-} {}^{14}C$$

(a) O-20, (b) Fe-52, (c) Po-197, (d) Dy-150 (list the energies of the emitted alpha particles in this decay process beside the directional arrows), (e) Ho-162.

13. (Do the following exercise on separate sheets of paper or graph paper if available and attach to Laboratory Report. Graph paper is convenient for Z versus N plots.) Beginning with the following radioactive parent nuclei, trace the decay "chains." List the modes and direction of the process on the chart as in part 12. The first three decay chains are found in nature, but the fourth chain is not. When you have completed its decay processes, explain why the fourth chain is not found in nature. (a) U-238 (uranium series), (b) U-235 (actinium series), (c) Pu-244 (plutonium series—originally called the thorium series beginning with Th-232 before Pu-244 was discovered), (d) Np-237 (neptunium series). Why is this series not found in nature?

14. Using the chart of nuclides, supply the product nucleus of each of the following reactions. Also give the compound nucleus of each reaction.

Reaction	Product	Compound nucleus
(a) $^{10}B(n,\alpha)$	....	....
(b) $^{16}O(n,p)$	....	....
(c) $^{7}Li(p,\gamma)$	....	....
(d) $^{17}O(\gamma,np)$	....	....
(e) $^{32}S(n,p)$	....	....
(f) $^{3}H(d,n)$	....	....
(g) $^{2}H(t,n)$	....	....

Name .... Section .... Date ....

Lab Partner(s) ....

EXPERIMENT 54 *Detection of Nuclear Radiation: The Geiger Counter*

ADVANCE STUDY ASSIGNMENT

Read the experiment and answer the following questions.

1. What is the principle of the Geiger tube?

2. Define each of the following: (a) threshold voltage; (b) cumulative ionization; (c) plateau; and (d) dead time.

3. Are all radiations counted when the tube voltage is below the threshold voltage? Explain.

4. Approximately how may volts above the threshold voltage is the normal operating voltage of the Geiger tube, and why is this voltage selected?

5. What is background radiation?

6. How does the count rate vary with distance from a point source?

EXPERIMENT **54**

Detection of Nuclear Radiation: The Geiger Counter

I. INTRODUCTION

Nuclear radiations (alpha, beta, and gamma rays or particles) cannot be detected directly by our senses. Hence, we must use some observable detection method employing the interaction of nuclear decay particles with matter. There are several methods, but the most common is the **Geiger tube.** In a Geiger tube, the particles from radioactive decay ionize gas molecules, giving rise to electrical pulses that can be amplified and counted. The total instrument is referred to as a Geiger counter.

In this experiment, the characteristics of a Geiger tube and the inverse-square relationship for nuclear radiations will be investigated.

II. EQUIPMENT NEEDED

- Geiger counter (rate meter or scaler type)
- Radioactive source [e.g., Cs-137 (beta–gamma)]
- Laboratory timer or stopwatch
- Calibrated mounting board or meter stick
- 1 sheet of Cartesian graph paper
- 1 sheet of log (log-log) graph paper (3-cycle)

III. THEORY

The three types of radiation—alpha, beta, and gamma—are all capable of ionizing a gas. The degree of ionization depends on the energy of the particles and the amount of radiation absorbed by the gas. The ionization of gas molecules by nuclear radiations is the principle of the Geiger tube (also called a Geiger-Müller or G-M tube).

A **Geiger tube** consists of a fine wire running axially along a metal cylinder which is filled with a gas, usually argon, at a pressure of about 0.1 atm (Fig. 54-1). A potential difference or voltage is maintained between the central wire and the cylinder, the central wire being at a positive potential (+) with respect to the cylinder (−).

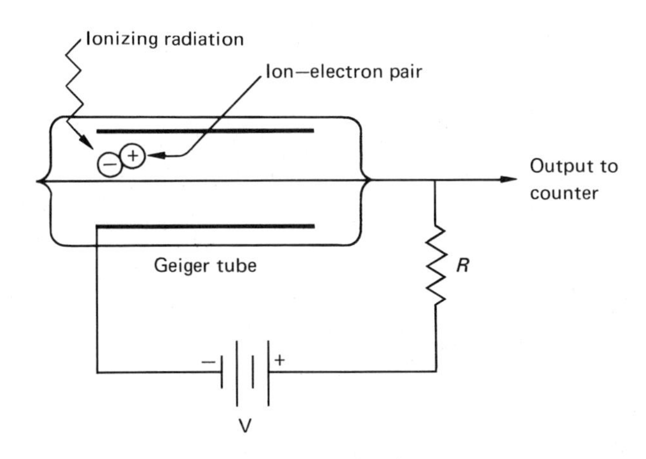

Fig. 54-1 A Geiger tube and circuit.

Energetic nuclear particles (ionizing radiation) passing through and entering the tube ionize the gas molecules. The freed electrons are attracted toward the wire and the positive ions toward the cylinder. If the voltage between the wire and cylinder is great enough, the accelerated electrons acquire enough energy to ionize other gas molecules on their way to the positive wire. The electrons from the secondary ionizations produce additional ionizations. This process is called *cumulative ionization.* As a result, an "avalanche" discharge sets in and a current is produced in the resistor. This reduces the potential difference between the wire and cylinder to the point where cumulative ionization does not occur. After the momentary current pulse, which lasts on the order of microseconds, the potential difference between the wire and the cylinder resumes its original value.

A finite time is required for the discharge to be cleared from the tube. During this time, the voltage of the tube is less than that required to detect other radiation that might arrive. This recovery time is referred to as the *dead time* of the tube. If a large amount of radiation arrives at the tube, the counting rate (counts per minute or cpm) as indicated on the counting equipment will be less than the true value.

There are two common types of Geiger tubes—a "normal" or side-window tube and an "end-window" tube. The side-window tube has a relatively thick wall that may not be penetrated by less penetrating radiation such as alpha particles (Fig. 54-2). The end-window tube has a thin end window, usually of mica, thin enough to be penetrated by most radiations.

The brief change in the potential that occurs when a discharge takes place in the tube produces a voltage pulse that can be detected and counted by appropriate instrumentation. Common instruments used for counting are scalers and count-rate meters.

A *scaler* displays the cumulative number of counts on a lighted panel. By using a separate timer, the number of counts per minute (cpm) can be obtained. Some scalers have internal timers that stop the counting after a preset time interval.

A *rate meter* displays the average counting rate directly by means of a dial needle (Fig. 54-2). The needle reading fluctuates back and forth. This is due to the electronic averaging of the number of counts received during a short period of time. A scaler–timer is usually preferred over a rate meter because of this effect.

A. *Tube Voltage and Count Rate*

When a Geiger tube is in the vicinity of a constant source of radiation and there is no voltage on the tube, no counts are observed on the counter. (Counters usually have a loudspeaker circuit so that the counts may also be heard as audible "clicks.") If the tube voltage is slowly increased from zero, at some applied voltage counts will be observed. The lowest applied voltage that will produce a count in the instrument is called the *starting* or **threshold voltage** (Fig. 54-3).

As the tube voltage is increased above the threshold voltage, the number of counts per minute increase rapidly. In this region (about 50 V wide beginning at about 600 to 700 V, depending on the tube), the count rate is almost linearly proportional to the voltage. This is because as the voltage increases, more of the less energetic particles are counted. Hence, in this region the tube discriminates between radiations of different energies. At a given voltage, only particles above a certain energy are detected. The tube then acts as a proportional counter—the voltage being proportional to the energies of the incident particles.

Eventually, as the voltage is increased, the number of counts per minute becomes almost independent

Fig. 54-2 Apparatus for radioactive experiments. The standard side-window Geiger tube probe on the mounting board is connected to a count-rate meter. The sheets of material (right) are used in absorption experiments (Experiment 56).

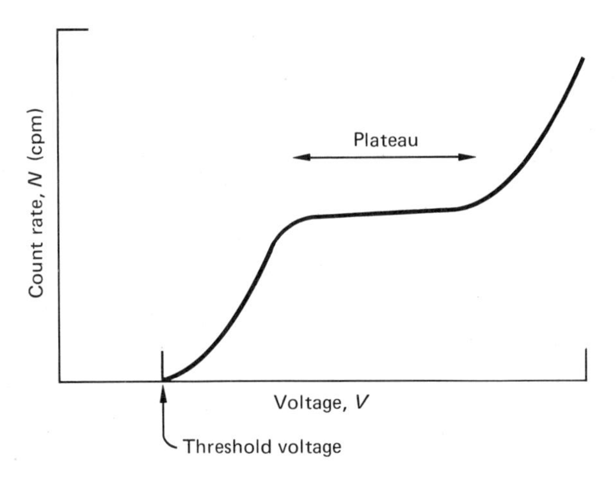

Fig. 54-3 Typical graph of count rate versus voltage for a Geiger tube. Normal operation is in the plateau region.

of the applied voltage (level region in Fig. 54-3). This region (about 200 V wide) is called the **plateau** of the tube. A change in voltage has little effect on the number of counts detected, Normally, the Geiger tube is operated at a voltage in about the middle of the plateau. Fluctuations in the applied voltage from the power supply will then have little effect on the counting rate.

The tube voltage should never be raised to a value far above that of the end of the plateau. At such high voltages, a continuous discharge sets in, and this may destroy the tube if allowed to persist.

B. *Inverse-Square Relationship*

In normal operation, the count rate depends on the number of particles per unit time that are entering the Geiger tube. Hence, the count rate depends on the distance of the tube from the source. For a point source emitting a total of N_0 particles/min, the particles are emitted in all directions. The number of particles/min N' passing through a unit area of a sphere of radius r is

$$N' = \frac{N_0}{A} = \frac{N_0}{4\pi r^2} \quad \text{(counts/min/area)} \quad \textbf{(54-1)}$$

where $A = 4\pi r^2$ is the area of the sphere.

A Geiger tube with a window area A' at a distance r from a point source then intercepts or receives N counts/min, given by

$$N = N'A' = \frac{N_0 A'}{4\pi r^2} \quad \textbf{(54-2)}$$

Although the effective area A' of the Geiger tube is usually not known, the equation shows that the count rate is inversely proportional to r^2 (inverse-square-law form):

$$N \propto \frac{1}{r^2} \quad \textbf{(54-3)}$$

Hence, for a point source, the count rate "falls off" as $1/r^2$ with the distance from the source.

IV. EXPERIMENTAL PROCEDURE*

1. Connect the Geiger tube probe to the counter by means of the coaxial cable. Before plugging in the counter to an ac outlet, familiarize yourself with the controls, particularly the high-voltage control.

   Scaler: Set the high-voltage control to the minimum setting.

   Rate meter: Set the high-voltage control to the minimum setting. The off-on switch is commonly on the high-voltage control. A selector switch is labeled with volts and counts per minute multiplier positions ($\times 1$, $\times 10$, etc.). When the Geiger tube voltage is adjusted by means of the high-voltage control, the selector switch should *always* be set on "volts." The selector switch is then turned to the appropriate count multiplier range for counting. The meter display scale usually had dual calibrations in volts and counts per minute.

2. Plug in and turn on the counter. Place the radioactive source near the Geiger tube, with the source facing the probe opening as in Fig. 54-2. (A tube mount may be available for an end-window-type tube. The source is placed at the bottom of the tube mount in this case.)

   Slowly increase the tube voltage by means of the high-voltage control until the first indication of counting is observed. Then, increase the voltage to about 75 to 100 V above this value.

3. Set the counter to the counting mode and adjust the distance of the source from the tube (or add aluminum sheets to the end-window tube mount) so that the count rate is several thou-

*Review the radiation safety procedures in Section B (General Laboratory Procedures) of the Introduction.

sand (3000 to 5000) counts per minute. The Geiger tube is then operating normally, and the dead time will not cause serious error.

A. *Tube Voltage and Count Rate*

4. Lower the high-voltage control to the minimum setting. Then slowly raise the voltage until the first indication of counting is observed (rate meter selector on "volts"). Record this threshold voltage in Data Table 1.

5. Increase the voltage to 25 V above the threshold voltage and record the tube voltage. Measure and record the number of counts per minute at this voltage setting. (A rate meter is switched to a counting position. Because the meter needle fluctuates, it is best to watch the meter for 30 s and note the highest and lowest meter readings. The count rate is then taken as the mean or average of these readings.)

6. Continue to repeat procedure 5, increasing the voltage by 25 V each time. Record the voltage and the corresponding count rate for each voltage setting. You will notice that the count rate first increases rapidly with voltage, then "levels off." This is the plateau region of the Geiger tube. Eventually, with a particular voltage step, a sharp increase in the count rate will be observed. *Do not increase the voltage above this value.* Quickly lower the tube voltage to the minimum setting after this reading to avoid damaging the tube.

7. Plot the count rate N (counts/min) versus voltage V on Cartesian graph paper. Include the threshold voltage. Draw a smooth curve that best fits the data.

B. *Background Radiation*

8. Remove the source several meters (across the room) from the Geiger tube and apply the mid-plateau voltage to the tube as determined from the graph. (If using an end-window tube with tube mount, remove the tube from the mount and lay the tube on the table. You will observe an occasional count on the counter. This is due to background radiation arising from cosmic rays and radioactive elements in the environment (e.g., in building materials). Time the number of counts for 4 to 5 min and determine the background count rate in counts per minute and record. If the background count rate is small compared to the source count, it may be considered negligible.

C. *Inverse-Square Relationship*

9. Bring the source toward the Geiger tube and locate the source a distance from the tube where the counting rate is 300 to 400 cpm. Record the distance r and the count rate N in Data Table 2. Record this r as the farthest distance.

10. Then bring the source relatively close to the tube and determine the distance from the source that gives a count rate of about 8000 cpm. Record the count rate and distance (closest).

11. Divide the length between the two measured distances into eight intervals or steps. Measure and record the count rate and source distance from the tube for each step as the source is moved away from the tube.

12. Plot N versus r on log graph paper and draw a straight line that best fits the data. Determine the slope of the line and compare it to the theoretical value by finding the percent error.

Name .... Section .... Date ....

Lab Partner(s) ....

EXPERIMENT 54 *Detection of Nuclear Radiation: The Geiger Counter*

LABORATORY REPORT

A. *Tube Voltage and Count Rate*

DATA TABLE 1

	Tube voltage	Count rate (cpm)
Threshold voltage		X

B. *Background Radiation*

Number of counts ....

Counting time (min) ....

Counts/min ....

C. *Inverse-Square Relationship*

DATA TABLE 2

	Source-counter distance (cm)	Count rate (cpm)
Closest distance		
Farthest distance		

Calculations (show work)

Slope of graph ....

Theoretical value ....

Percent error ....

Name .... Section .... Date ....

Lab Partner(s) ....

EXPERIMENT 54 *Detection of Nuclear Radiation: The Geiger Counter*

QUESTIONS

1. What is the average percent increase in the count rate over the voltge range of the Geiger tube plateau? (Obtain from a graph of the data.)

2. If dead time and background radiation corrections were taken into account, how would this affect the graph of N versus V?

3. Give possible reasons why the experimental result of N versus r is not exactly an inverse-square relationship.

4. A count rate of 8000 cpm is recorded at a distance of 5 cm from a point source. What would be the observed count rate at a distance of 20 cm?

Name .... Section .... Date ....

Lab Partner(s) ....

EXPERIMENT 55 *Radioactive Half-life*

ADVANCE STUDY ASSIGNMENT

Read the experiment and answer the following questions.

1. What is the half-life of a radioactive isotope in terms of (a) the amount of sample or number of nuclei; and (b) the activity of the sample?

2. How is the half-life related to the decay constant of a radioactive process?

3. Is the activity of a sample of radioisotope greater after one half-life or one time constant? Explain.

4. What is meant by "milking a cow"? Give the technical terms for "milking" and "cow."

5. Ba-137m is a nuclear isomer of Ba-137. Explain what this means.

EXPERIMENT 55

Radioactive Half-life

I. INTRODUCTION

The decrease in the activity of a radioactive isotope is characterized by its half-life. This is the time required for one-half of the nuclei of a sample to decay. Of course, the nuclei of a sample cannot be counted directly, but when one-half of the sample has decayed, the activity or the rate of emission of nuclear radiation has also decreased by one-half. Thus, by monitoring the sample with a Geiger counter, when the count rate (counts per minute, cpm) has decreased by one-half, one half-life has elapsed.

In this experiment, the half-life of a radioactive isotope will be determined.

II. EQUIPMENT NEEDED

- Geiger counter (rate meter with clip mount or scaler type with tube mount)
- Cesium-137/Barium-137m Minigenerator with 0.04 HCl-saline solution
- Laboratory timer or stopwatch
- Disposable plachet (small metal cuplike container to hold radioactive sample)
- 1 sheet of semilog graph paper (3-cycle)
- 1 sheet of Cartesian graph paper

III. THEORY

Since the activity of a radioactive isotope is proportional to the quantity of isotope present, the radioactive decay process is described by an exponential function,

$$N = N_0 e^{-\lambda t} = N_0 e^{-t/\tau} \qquad \textbf{(55-1)}$$

where N is the number of nuclei present at time t, N_o the original number of nuclei present at $t = 0$, λ the decay constant of the process, and $\tau = 1/\lambda$ the time constant. (See Experiment 52 for a discussion of exponential functions.) The variable N can also represent the activity (cpm) of an isotope sample.

The half-life $t_{1/2}$ is the time it takes for the number of nuclei present or the activity to decrease by one-half ($N = N_0/2$). Hence,

$$\frac{N}{N_0} = \frac{1}{2} = e^{-t_{1/2}/\tau}$$

Since

$$e^{-0.693} = \tfrac{1}{2}$$

by comparison

$$0.693 = \frac{t_{1/2}}{\tau}$$

and

$$t_{1/2} = 0.693\tau = \frac{0.693}{\lambda} \qquad \textbf{(55-2)}$$

Thus, the half-life can be computed if the time constant or the decay constant is known.

☐ **Example 55.1** A radioactive sample has an activity of 4000 cpm. What is the observed activity after three half-lives?

Solution After one half-life, the activity decreases by $\frac{1}{2}$, and after another half-life by another $\frac{1}{2}$, and so on. Hence, after three half-lives, the initial activity decreases by a factor of $\frac{1}{2} \times \frac{1}{2} \times \frac{1}{2} = \frac{1}{8}$. With $N_0 = 4000$ cpm,

☐ $$N = \tfrac{1}{8} N_0 = \tfrac{1}{8}(4000) = 500 \text{ cpm}$$

Theory of Minigenerator

The Cesium-137/Barium-137m Minigenerator† (Fig. 55-1) is an eluting system in which a short-lived daughter radioactive isotope is eluted (separated by washing) from a long-lived parent isotope. A small "generator" contains radioactive Cs-137, which has a half-life of 30 years. Cs-137 beta-decays into Ba-137m, which is an isomeric state of the stable nucleus Ba-137 (see Experiment 53). The excited isomer Ba-137m (or Ba-137*) has a relative short half-life and gamma-decays into Ba-137.

The Ba-137m isotope is washed or eluted from the generator by passing a hydrochloric acid-saline solution through the generator. Because of this process, the generator is commonly referred to as a "cow," with the Ba-137m being "milked" from the cow. The generator "cow" may be milked many times, but like an actual cow, a time interval must elapse between milkings.

Eluting removes the Ba-137m from the generator, and time is required for the "regeneration" of Ba-137m from the decay of Cs-137. Normally, the parent and daughter isotopes exist in equilibrium with equal activities. After eluting or milking of the cow, it takes about 12 min for the Ba-137m to build up and again reach equilibrium with the Cs-137.

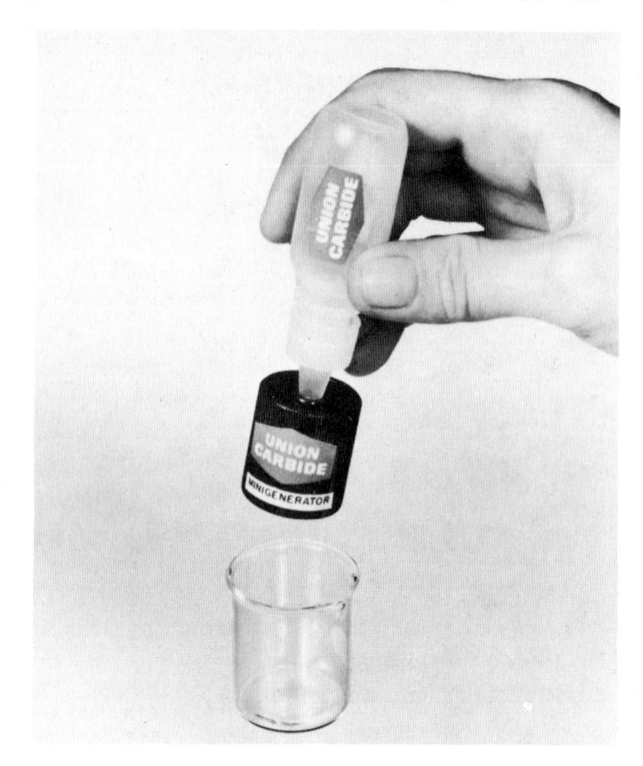

Fig. 55-1 The Cesium-137/Barium-137m Minigenerator system. (Photo courtesy of Sargent-Welch Scientific Company.)

IV. EXPERIMENTAL PROCEDURE‡

1. Before the Minigenerator is eluted, turn on the Geiger counter. Apply the appropriate tube voltage for normal operation (see Experiment 54). Over a period of 4 to 5 min measure and record the count of the background radiation as was done in Experiment 54.

2. Mount the Geiger probe so that a planchet with the radioactive Ba-137m sample can be quickly and carefully placed below and near the probe opening at a fixed distance.

3. The counting procedure is as follows. When given the sample and it is in place, the laboratory timer is started ($t = 0$) and allowed to run continuously. Simultaneously with the starting of the timer, the activity is measured on the Geiger counter for 15 s, and the count rate (cpm), together with the time elapsed on the timer, are recorded in the data table. [*Note:* If using a rate meter, the average of the high and low meter readings over the 15-s interval is taken as the count rate. If a scaler is used (with or without an internal timer), the count rate in cpm must be computed. Suppose that 500 counts are observed for the 15-s ($\frac{1}{4}$ min) interval. The count rate is then 500 counts/$\frac{1}{4}$ min $= 500 \times 4 =$ 2000 cpm.]

   After 1 min has elapsed on the timer, take another 15-s count of the activity. Repeat the 15-s count at the beginning of each elapsed time for 10 to 12 min. A dry run of the counting procedure is helpful.

4. The instructor will "milk the cow" or supervise you in doing so. Only a few (2 to 3) drops of the eluate (milk) are needed. Care should be taken in handling the sample. The milking should be done over a paper that can be discarded in case of a spill, and *if you should come in contact with the sample, immediately wash your hands.* The instructor may wish to give you a sample for a trial run of the counting procedure.

5. When given the actual data sample, carry out the counting procedure as described above.

†Registered trademark, Union Carbide.

‡Review the radiation safety procedures in Section B (General Laboratory Procedures) of the Introduction.

6. Correcting for the background radiation if necessary, plot the sample activity (N) in cpm versus the elapsed time (t) in minutes on both Cartesian and semilog graph paper. Determine the half-life of Ba-137m from the semilog plot.

7. Look up the half-life of Ba-137m in Appendix A, Table A9, and compare the experimental value with this accepted value by finding the percent error.

LABORATORY REPORT

Background count (cpm) ....

DATA TABLE

t Elapsed time (min)	N Observed activity (cpm)	Corrected for background radiation (cpm)

Calculations (show work)

Half-life of Ba-137m

Experimental ....

Accepted ....

Percent error ....

(continued)

QUESTIONS

1. In the experiment, if the Ba-137m sample were placed closer to the Geiger tube, the measured activity would be greater (inverse-square relationship). Would this affect the result of the half-life? Explain how this would affect the N versus t graph.

2. A cobalt-60 source has a measured activity of 12,000 cpm. After how long would the observed activity be 750 cpm? (The half-life of Co-60 is 5.27 years.)

3. An instructor buys a 10-microcurie Cs-137 source for laboratory experiments. After 5 years, what is (a) the strength of the source in μCi; (b) the activity of the source in disintegrations per second? [1 curie (Ci) = 3.70×10^{10} disintegrations/s.] (c) What is the strength of the source in becquerels (Bq)? (The becquerel is the official SI unit, 1 Ci = 3.7×10^{10} Bq.)

Name .... Section .... Date ....

Lab Partner(s) ....

EXPERIMENT 56 *The Absorption of Nuclear Radiations*

ADVANCE STUDY ASSIGNMENT

Read the experiment and answer the following questions.

1. On what parameters does the absorption of nuclear radiation depend?

2. Do the three types of nuclear radiations have definite ranges of penetration in materials? Explain. What is meant by half-thickness?

3. What is the mass absorption coefficient, and what are its units and advantages?

4. Explain how a single beta–gamma source emits radiations of both types.

5. Why is a beta–gamma source that is shielded with a relatively thin sheet of aluminum effectively a gamma source?

EXPERIMENT **56**

The Absorption of Nuclear Radiations

I. INTRODUCTION

The observed activity of a radioactive source of a given strength depends on several factors, for example, the distance of the counter from the source. As discussed in Experiment 54 for a point source, the observed activity varies inversely as the distance from the source (inverse-square relationship). This decrease is primarily due to the geometrical spreading of the emitted nuclear radiation outward from the source.

If a Geiger probe is a fixed distance from a long-lived source, the observed activity is relatively constant. However, if a sheet of material is placed between the source and the counter, a decrease in the activity may be observed. That is, the nuclear radiation is absorbed by the material. The amount of absorption depends on the nature of the radiation (type and energy), the material, and its density.

The absorption or degree of penetration of nuclear radiations is an important consideration in applications such as medical radioisotope treatment and nuclear shielding (e.g., around a nuclear reactor). Also, in industrial manufacturing processes, the absorption of nuclear radiations is used to automatically monitor and control the thickness of metal and plastic sheets and films.

In this experiment, the absorption properties of various materials for different nuclear radiations will be investigated.

II. EQUIPMENT NEEDED

- Geiger counter (rate meter or scaler type)
- Calibrated mounting board (or meter stick)
- Beta–gamma source, Cs-137 suggested
- Set of cardboard, aluminum, and lead sheets (about 1 mm thick, 10 sheets of each)
- Laboratory timer or stopwatch
- Micrometer caliper
- 1 sheet of Cartesian graph paper
- 1 sheet of semilog graph paper (3-cycle)

III. THEORY

The three types of nuclear radiations (alpha, beta, and gamma) are absorbed quite differently by different materials. The electrically charged alpha and beta particles interact with the material and produce ionizations along their paths. The greater the charge and slower the particle, the greater is the linear energy transfer (LET) and ionization along the path. The absorption or degree of penetration of the radiation also depends on the density of the material.

Alpha particles are easily absorbed. A few centimeters of air and even a sheet of paper will almost completely absorb them. Hence, alpha particles do not penetrate the walls or window of an ordinary Geiger tube and so are not counted by this method.

Beta particles can travel a few meters in air or a few millimeters in aluminum before being completely absorbed. Beta radiation, then, does penetrate a Geiger tube. Both alpha and beta particles of a given energy therefore have a definite range of penetration in a particular material. Fig. 56-1 illustrates the radiation intensity (in cpm) versus absorber thickness for a relatively low density absorber for radiation from a beta–gamma source. The

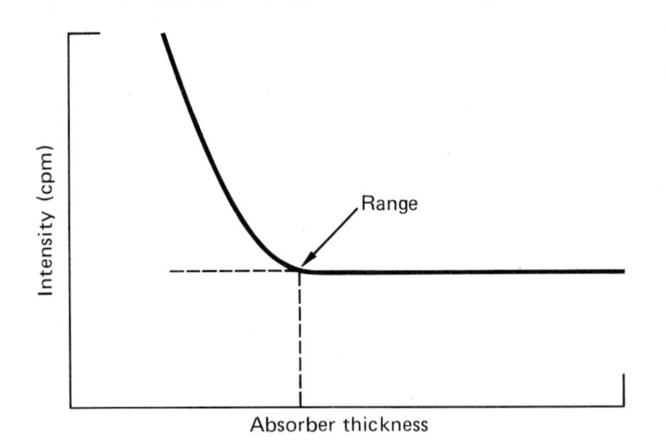

Fig. 56-1 A typical graph of radiation intensity versus absorber thickness for beta–gamma radiation by a low-density absorber. The range is that of the beta radiation.

"bend" in the curve indicates the range of the beta radiation. The penetration for thickness beyond this is due to gamma radiation.

Gamma rays, which consist of electromagnetic radiation of very short wavelength, are not readily absorbed. They can penetrate 1 cm or more of a dense material such as lead. In a given material, a beam of gamma rays is absorbed exponentially. The intensity I (in cpm) of the beam after passing through a certain thickness x of a material is given by

$$I = I_0 e^{-\mu x} \tag{56-1}$$

where I_0 is the original intensity (at $x = 0$) and the decay constant μ is called the **linear absorption coefficient.** The absorption coefficient is characteristic of the absorbing material (and the wavelength or energy of the gamma radiation). Notice that the unit of μ is inverse length, (e.g., 1/cm or cm^{-1}).

Since the gamma intensity decays exponentially, there is no definite penetrating or stopping range as in the case of beta radiation. Hence, it is convenient to speak in terms of a half-thickness $x_{1/2}$, which is the material thickness required to reduce the intensity by one-half (i.e., $I_{1/2} = I_0/2$ or $I_{1/2}/I_0 = \frac{1}{2}$). Then, by Eq. 56-1,

$$\frac{I_{1/2}}{I_0} = e^{-\mu x_{1/2}} = \frac{1}{2}$$

Taking the logarithm (base e) of both sides of the equation, we have

$$\ln (e^{-\mu x_{1/2}}) = \ln \tfrac{1}{2}$$

or

$$-\mu x_{1/2} = -\ln 2$$

and

$$x_{1/2} = \frac{\ln 2}{\mu} = \frac{0.693}{\mu} \tag{56-2}$$

Hence, knowing the absorption coefficient of a material, the half-thickness can be calculated.

The absorption of gamma radiation of a given wavelength or energy depends primarily on the density ρ of the material. Thus, it is convenient to define a *mass absorption coefficient* μ_m:

$$\mu_m = \frac{\mu}{\rho} \tag{56-3}$$

The mass absorption coefficient provides a "standardized" coefficient. Samples of a particular absorbing material may have different densities. Each sample would have a different linear absorption coefficient μ, but the mass absorption coefficient μ_m would have the same value for all the samples.

Notice from Eq. 56-3 that the units of μ_m with $\mu = 1/\text{cm}$ and $\rho = \text{g/cm}^3$ are cm^2/g:

$$\mu_m = \frac{\mu\,(1/\text{cm})}{\rho\,(\text{g/cm}^3)} = \frac{\mu}{\rho}\,(\text{cm}^2/\text{g})$$

If μ_m is used in Eq. 56-1 in place of μ, then

$$I = I_0 e^{-\mu x} = I_0 e^{-(\mu/\rho)(x\rho)} = e^{-\mu_m x'} \tag{56-4}$$

and the absorber thickness $x' = x\rho$ is in g/cm^2. Absorber thicknesses are frequently expressed in these units.

A beta–gamma source will be used to study the absorption of nuclear radiations. The decay scheme of the suggested Cs-137 source is illustrated in Fig. 56-2. The chief decay mode (94 percent) is beta decay to the excited (isomeric) state of Ba-137. This decays by gamma emission to the stable ground state of Ba-137. Only 6 percent of the Cs-137 beta decays directly to ground state Ba-137. Hence, for the most part, Cs-137 is beta–gamma source of 0.511-MeV beta particles and 0.662-MeV gamma rays.

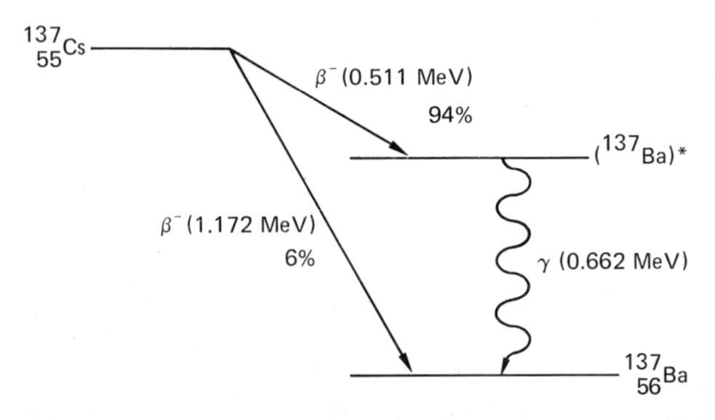

Fig. 56-2 The decay scheme of Cs-137.

IV. EXPERIMENTAL PROCEDURE*

1. First, measure the individual thickness of three different sheets of (a) cardboard, (b) aluminum, and (c) lead with the micrometer and determine the average sheet thickness of each. Record in Data Table 1.

2. Set up the Geiger counter with the probe on the mounting board (see Fig. 54-2). If an end window tube is used, lay the tube in the mounting board groove and tape down to immobilize (or tape to a meter stick). Turn on the counter. Place the radioactive source near the probe and adjust the tube voltage to the plateau operating voltage. (See Experiment 54 for a description of Geiger counter operation.)

A. *Beta Radiation Absorption*

3. Adjust the distance of the source from the probe so that the observed count rate is about 8000 cpm. (For a rate meter, the count rate is taken as the average of the high and low meter readings for 30-s time interval.) Record the count rate (I_0) in the cardboard column in Data Table 2.

   Place a sheet of cardboard between the source and the probe and measure and record the count rate. (Allow a rate meter to come to equilibrium before taking a 30-s reading.)

4. Add cardboard sheets between the source and the probe one at a time, measuring and recording the count rate after the addition of each sheet. Continue until the count rate is relatively constant with the addition of four successive sheets.

5. Remove the cardboard sheets and repeat the procedure with aluminum sheets.

6. Without recording data, repeat the procedure with lead sheets and mentally note the degree of beta absorption or penetration in lead.

7. Plot the intensity I (cpm) versus the number n of absorber sheets for both cardboard and aluminum on the same Cartesian graph. Dual-label the ordinate (y) axis so that the curve for each absorber occupies the majority of the graph paper.

   Determine the range of beta absorption for each absorber in sheet units from the graph and record. Multiply each range (in sheet units) by the respective average sheet thickness to determine the range in length units.

B. *Gamma Radiation Absorption*

8. Using the result of the range of beta absorption in aluminum from procedure 7, place the minimum number of sheets of aluminum in front of the probe that will completely absorb the beta radiation. Then move the source toward the probe until the intensity observed on the Geiger counter is 700 to 800 cpm. Record this intensity I_0 in Data Table 3. The observed intensity is then almost solely due to gamma radiation. Why?

9. Leaving the aluminum sheet(s) in place, insert lead sheets one at a time between the aluminum sheets and the source. Measure and record the count rate after each sheet is inserted. Be careful not to move the source. Insert a total of 10 sheets of lead. After inserting the sixth sheet, two sheets may be inserted at a time.

10. Remove all the sheets. Remove the source several meters (across the room) from the probe, and measure the background radiation intensity I_b over a 4- to 5-min interval. (See Experiment 54 for a procedure description, if necessary.)

11. Subtract the background count rate from each reading for the lead sheets to obtain the corrected intensities. Plot the corrected intensity I_c versus the number n of lead absorber sheets on semilog graph paper. Draw a straight line that best fits the data and determine its slope (see Experiment 52). When using the number of sheets as a variable, Eq. 56-1 has the form

$$I = I_0 e^{-\mu x} = I_0 e^{-\mu(n x_i)} = I_0 e^{-(\mu x_i) n}$$

   where x_i is the individual sheet thickness and the absorber thickness is $x = n x_i$. Hence, the magnitude of the slope of the line is μx_i.

   Divide the measured slope by x_i (average thickness of lead sheets) to determine the linear absorption coefficient (μ = slope/x_i) and record.

12. Compute the mass absorption coefficient μ_m for lead (ρ_{Pb} = 11.3 g/cm^3). Compare the experimental value to the accepted value of 0.1 cm^2/g for gamma rays of this energy by computing the percent error.

*Review the radiation safety procedures in Section B (General Laboratory Procedures) of the Introduction.

Name .... Section .... Date ....

Lab Partner(s) ....

EXPERIMENT 56 *The Absorption of Nuclear Radiations*

LABORATORY REPORT

DATE TABLE 1 SHEET THICKNESSES

	Cardboard	Aluminum	Lead
Average sheet thickness			

A. *Beta Radiation Absorption*

DATA TABLE 2

Cardboard		Aluminum	
Number of sheets n	Intensity I (cpm)	Number of sheets n	Intensity I (cpm)
0	(I_0)	0	(I_0)
Range of beta radiation (cm)		Range of beta radiation (cm)	

Lab Partner(s) ....

EXPERIMENT 56 *The Absorption of Nuclear Radiations*

B. *Gamma Radiation Absorption*

DATA TABLE 3

Number of lead sheets n	Intensity I (cpm)	Corrected intensity $I_c = I - I_b$
0	(I_0)	

Background radiation

Number of counts ....

Time interval ....

Intensity, I_b (cpm) ....

Calculations (show work)

Slope of line ....

Linear absorption coefficient, μ ....
(units)

Mass absorption coefficient, μ_m ....
(units)

Accepted μ_m ....

Percent error ....

(continued)

QUESTIONS

1. What is the half-thickness of 0.662-MeV gamma rays in lead?

2. Compute what percent of an incident beam of 0.662-MeV gamma rays is absorbed after passing through 2.5 mm of lead.

3. Would the semilog graph of I_c versus n (or x) be a straight line if the gamma radiation contained gamma rays of two different energies? Explain.

4. The mass absorption coefficient of iron is 0.058 for 1.24-MeV gamma rays. What percentage of the beam of such gamma rays is transmitted (if any) through an iron plate 3 cm thick? (ρ_{Fe} = 7.86 g/cm^3.)

Name .... Section .... Date ....

Lab Partner(s) ....

EXPERIMENT 57 *The Pi-Mu-e Decay Process*

ADVANCE STUDY ASSIGNMENT

Read the experiment and answer the following questions.

1. How may elementary particles be detected visually? Can all elementary particles be observed? Explain.

2. Explain how particle tracks indicate (a) the relative speeds of equally charged particles; (b) the kinetic energy of a particle that comes to rest; (c) the momentum of a particle.

3. Can the total momentum p of a particle be determined directly from a photograph of the particle track? Explain.

4. Given the average values of the positron radius of curvature $\overline{R}$ and the muon path length $\overline{L}_1$, explain how the rest masses of the muon and pion are determined by explicitly writing out the equations used.

EXPERIMENT 57

The Pi-Mu-e Decay Process

I. INTRODUCTION

The three fundamental building blocks of nuclei and atoms are electrons, protons, and neutrons. However, in the last half century many other "elementary" particles have been discovered. Some elementary particles are observed in cosmic rays and as the products of their interactions in the earth's upper atmosphere. But with the development of high-speed particle accelerators and sophisticated detection methods, the majority of the work in elementary particle physics is now done in accelerator laboratories around the world. Here, physicists study the properties of elementary particles (e.g., mass, charge, mean lifetimes, decay products, and interactions with other particles) in an effort to understand how they fit into the scheme of fundamental particles that make up matter.

In this experiment, a particular elementary particle decay process (π-μ-e) will be studied. By measuring the tracks of the particles in hydrogen-bubble-chamber photographs, the masses of the π (pion) and μ (muon) particles will be determined.

II. EQUIPMENT NEEDED*

- Four or five pi-mu-*e* photographs per student (to obtain good averages of the measurements, the data taken by all students will be pooled and the average values for the entire class used for calculations)
- Radius-of-curvature template for each student

III. THEORY

In the **pi-mu-*e* (π-μ-e) decay process,** a π^+ meson (commonly called a *pion*) decays with a mean lifetime of 2.5×10^{-8} s into a μ^+ meson (commonly called a *muon*) and a neutrino (a μ neutrino ν_μ):

$$\pi^+ \rightarrow \mu^+ + \nu_\mu$$

The μ^+ muon then decays with a mean lifetime of 2.2×10^{-6} s into a positron (e^+) and two neutrinos:

$$\mu^+ \rightarrow e^+ + \nu_e + \bar{\nu}_\mu$$

A positron is a particle with the mass of an electron but with a positive fundamental unit of charge—a "positive electron" or the antiparticle of the electron. The ν_e is an electron neutrino and $\bar{\nu}_\mu$ is a muon antineutrino.

*Photographs and materials are available from The Ealing Corporation, 22 Pleasant Street, South Natick, Massachusetts 01760.

Elementary particle decay processes or "events" are recorded and studied using visual detection techniques such as photographic emulsions, cloud chambers, and bubble chambers. All of these record the trails of ionization left in the wake of charged particles. For example, in a liquid hydrogen bubble chamber, small bubbles form along the trail of ionized molecules.

The arrangement of a bubble chamber and a photograph of π-μ-e decay events are shown in Fig. 57-1. An absorber is placed in the pion beam (produced by the bombardment of a copper target with accelerated protons), so that a sizable fraction of the pions come to rest in the liquid hydrogen. These decay into muons, which travel short distances before coming to rest and decay. The resulting long-lived positrons then travel in curved paths due to the applied magnetic field in the chamber. (The un-

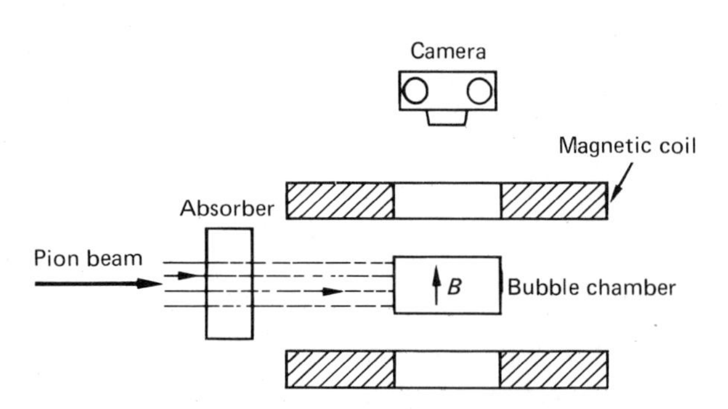

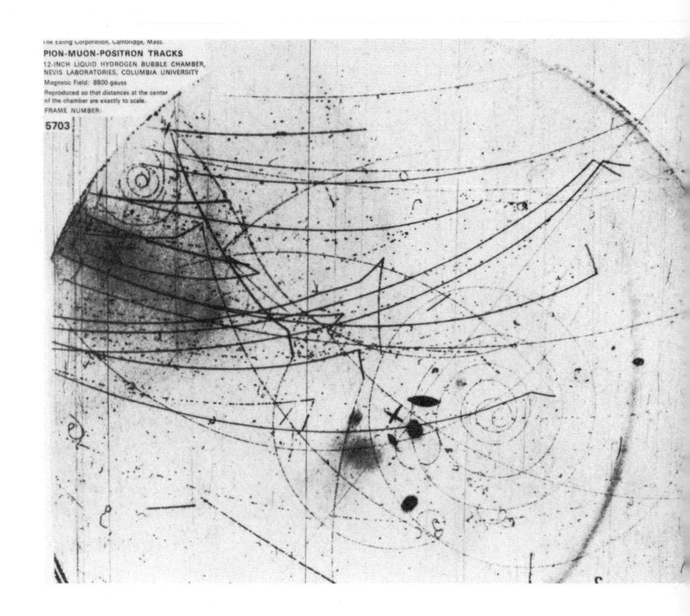

Fig. 57-1 Drawing of a bubble chamber and a bubble-chamber photograph. The dark tracks with a "hooked" signature near the top of the photo are π - μ - e events. The spiral track in the center of the photograph is produced by an electron.

charged neutrinos produce no ionizations and hence leave no tracks.)

Slower charged particles produce more ionizations and the density of bubbles along the tracks of such particles is heavier. If a particle comes to rest in the liquid, its kinetic energy can be determined from its range. The momentum of a circularly deflected particle can be determined from the curvature of its track.

Elementary particles travel at speeds that are appreciable fractions of the speed of light. Hence, relativistic mechanics must be used in analyzing the particle tracks. The total relativistic energy E of a particle is given by

$$E = K + m_0c^2 \qquad \textbf{(57-1)}$$

where K is the kinetic energy and m_oc^2 the rest energy. Also, in terms of the momentum p,

$$E^2 = (pc)^2 + (m_0c^2)^2 \qquad \textbf{(57-2)}$$

Eliminating E between these two equations, we have

$$(pc)^2 = K + 2K(m_0c^2) \qquad \textbf{(57-3)}$$

Notice that pc and m_0c^2 have the units of energy. In particle physics, this is commonly expressed in MeV. Thus, momentum p has the units MeV/c in this system.

When a charged particle moves with a velocity $v_\perp$ in the plane perpendicular to the magnetic field B, it experiences a force and describes a curved path with a radius of curvature R:

$$F = ev_\perp B = \frac{mv_\perp^2}{R}$$

or

$$R = \frac{mv_\perp}{eB} = \frac{p_\perp}{eB} \qquad \textbf{(57-4)}$$

where $p_\perp$ is the momentum in the plane perpendicular to B. Expressing $p_\perp$ in units of MeV/c, Eq. 57-4 becomes

$$p_\perp = (3 \times 10^{-4})BR \quad \text{MeV}/c \qquad \textbf{(57-5)}$$

where B is in gauss and R in centimeters. [1 gauss (G) = 10^{-4}. Wb/m^2 = 10^{-4} tesla (T).] A magnetic field of 8800 gauss was used in the hydrogen bubble chamber for the photographs of this experiment.

For the more general case of the momentum p of a particle being at an angle θ (rather than 90°) to **B** (Fig. 57-2), the particle travels in a helix, and

$$\frac{p_\perp}{p} = \sin\theta \quad \text{or} \quad p = \frac{p_\perp}{\sin\theta} \qquad \textbf{(57-6)}$$

In the photographs used in the experiment, only the projections of the tracks on the plane perpendicular to the magnetic field are obtained. By knowing the value of B and measuring R from a particle track on the photograph, $p_\perp$ may be found from Eq. 57-5.

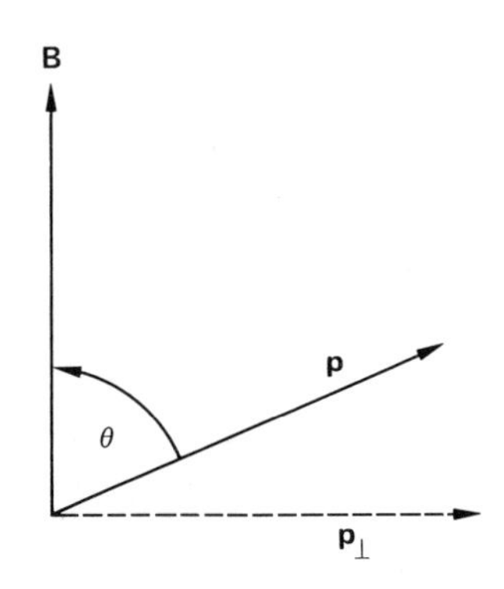

Fig. 57-2 The general case of the momentum p of a particle at an angle θ relative to the magnetic field B.

However, the angle θ cannot be determined from the photograph, and it is the value of p, rather than $p_\perp$, that is needed in the energy equations (Eqs. 57-2 and 57-3). An average value of p of a particle is obtained from photograph measurements by assuming that all directions are equally probable. Then, by measuring a large number of events, the average momentum $\overline{p}$ will be the average value of the $p_\perp$ measurements divided by the average value of sin θ (i.e., $\overline{p} = \overline{p}_\perp/\overline{\sin\theta}$). The average value of sin θ over a sphere can be shown to be equal to 0.785. Hence,

$$\overline{p} = \frac{\overline{p}_\perp}{0.785} \qquad \textbf{(57-7)}$$

By similar reasoning, the average length $\overline{L}$ of a track of a decay particle (e.g., μ^+) can be determined by measuring a large number of the perpendicular components $L_\perp$ on the photographs and dividing the average by sin $\theta = 0.785$:

$$\overline{L} = \frac{\overline{L}_\perp}{0.785} \qquad \textbf{(57-8)}$$

When a particle is created and also loses all its energy inside the bubble chamber, its measured track length L corresponds to its range in the liquid hydrogen. It can be shown that the kinetic energy of the particle is given by the expression

$$K = \sqrt{2.1(m_0c^2)\rho L} \qquad \textbf{(57-9)}$$

where ρ is the density of liquid hydrogen ($\rho = 0.07$ g/cm^3) and m_0c^2 is the rest energy of the particle in MeV.

Measurement of m_{0_μ}

When the μ^+ muon decays,

$$\mu^+ \rightarrow e^+ + \nu_e + \overline{\nu}_\mu$$

its rest energy $m_{0_\mu}c^2$ goes almost entirely into the kinetic energy of the decay products. Since the rest energy of the muon is over 200 times that of the positron (0.511 MeV) and the neutrinos are massless, we may write, to a good approximation,

$$\begin{aligned} E_\mu = m_{0_\mu}c^2 &\approx K_e + K_\nu + K_{\overline{\nu}} \\ &= E_e + E_\nu + E_{\overline{\nu}} \\ &\approx p_ec + p_\nu c + p_{\overline{\nu}}c \end{aligned} \qquad \textbf{(57-10)}$$

It has been shown that on the average for a large number of events, each of the decay products receives one-third of the total muon energy:

$$\overline{p}_ec \approx p_\nu c \approx p_{\overline{\nu}}c \approx \frac{E_\mu}{3} = \frac{m_{0_\mu}c^2}{3}$$

Hence, we write

$$\overline{p}_ec = \frac{m_{0_\mu}c^2}{3} \qquad \textbf{(57-11)}$$

Then, measuring $\overline{p}_e$ through Eq. 57-7, m_{0_μ} can be computed.

Measurement of $m_{0\pi}$

In the decay of a pion at rest,

$$\pi^+ \rightarrow \mu^+ + \nu_\mu$$

by the conservation of total energy,

$$E_{0\pi} = E_\mu + E_\nu$$

or

$$m_{0\pi}c^2 = \sqrt{(p_\mu c)^2 + (m_{0\mu}c^2)^2} + p_\nu c \qquad \textbf{(57-12)}$$

($E_\nu = p_\nu c$ since the neutrino is a massless particle.) But, by the conservation of momentum for the decay of a pion at rest,

$$0 = p_\mu - p_\nu$$

or

$$p_\mu = p_\nu$$

Hence,

$$m_{0\pi}c^2 = \sqrt{(p_\mu c)^2 + (m_{0\mu}c^2)^2} + p_\mu c \qquad \textbf{(57-13)}$$

By measuring the lengths of a number of muon tracks, $\overline{L}$ can be computed from Eq. 57-8. Using this value in Eq. 57-9, K_μ can be found. Then, using K_μ in Eq. 57-3, p_μ can be computed and substituted into Eq. 57-13 to find $m_{0\pi}$.

IV. EXPERIMENTAL PROCEDURE

1. Using the template, measure and record in Data Table 1 the lengths of muon tracks and the radii of curvature of the positron tracks on the bubble-chamber photographs. The curvatures of the positron tracks should be measured near the beginnings of the tracks. Do not mark on the photographs. Fig. 57-3 shows the appearance of π-μ-e decay tracks that should be measured. They have a characteristic "hooked" signature of double decay.

   Notice that one type of decay track series should be rejected. A pion may be scattered or

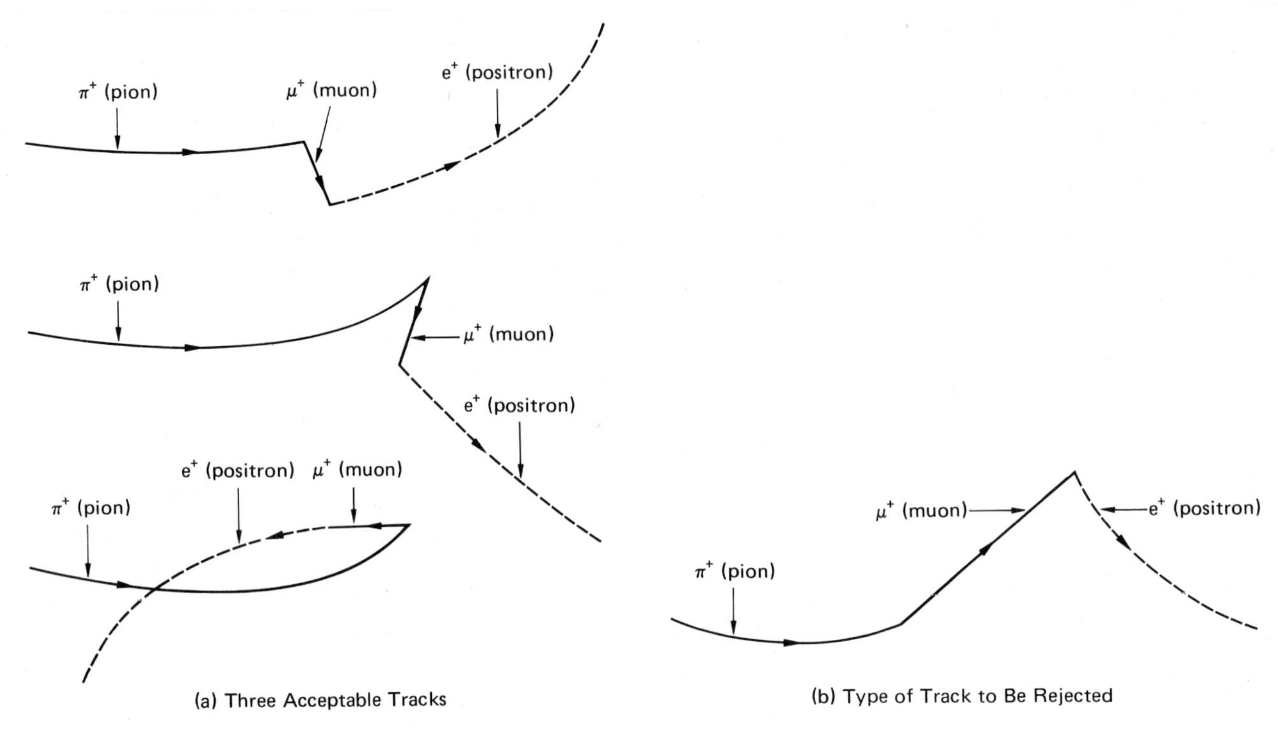

Fig. 57-3 π - μ - e decay tracks. (a) Acceptable tracks and (b) a type of track to be rejected because the muon track is too long and tends to be in the forward direction.

may decay in flight. In this case, the muon track will be much longer than normal and, in general, in the forward direction (direction of pion beam). Also, to provide good statistical measurements:

(a) Measure all appropriate muon tracks on the photographs, even if they are very short.
(b) The pion and muon decays are independent. Thus, it is all right to measure a muon track even if the positron track is obscured in the photograph, and vice versa.

2. Determine the average values of the measured $L_\perp$'s and R's, and give to the instructor, who will find the class average of $\overline{L}_\perp$ and $\overline{R}$. Record these values in Data Table 2 and use to compute the rest masses $m_{0\mu}$ and $m_{0\pi}$ of the μ meson (muon) and π meson (pion), respectively.

Name .... Section .... Date ....

Lab Partner(s) ....

EXPERIMENT 57 *The Pi-Mu-e Decay Process*

LABORATORY REPORT

DATA TABLE 1

Muon track length (cm)				Positron track curvature (cm)			
1		11		1		11	
2		12		2		12	
3		13		3		13	
4		14		4		14	
5		15		5		15	
6		16		6		16	
7		17		7		17	
8		18		8		18	
9		19		9		19	
10		20		10		20	

DATA TABLE 2

Class average of length of muon tracks, $\overline{L}_{\perp}$	....	cm
Class average of curvature of positron tracks, $\overline{R}$	....	cm
Strength of bubble-chamber field, B	8800	gauss
Density of liquid hydrogen, ρ	0.07	g/cm^3

(continued)

DATA TABLE 3

Average positron momentum, $\overline{p}_e$ (Eq. 57-5)	....	MeV/c
Average positron momentum, $\overline{p}_e$ (Eq. 57-6)	....	MeV/c
Rest energy of muon, $m_{0\mu}c^2$ (Eq. 57-11)	....	MeV
Muon rest mass, $m_{0\mu}$, in equivalent electron masses .... m_e	Percent error	
Accepted value ....207.............. m_e	....	
(m_ec^2 = 0.511 MeV)		
Kinetic energy of muon, K_μ (Eq. 57-9)	....	MeV
Momentum of muon, p_μ (Eq. 57-3) (Use accepted value of rest energy as determined from accepted rest mass.)	....	MeV/c
Rest energy of pion, $m_{0\pi}c^2$ (Eq. 57-12)	....	MeV
Pion rest mass, $m_{0\pi}$, in equivalent electron masses .... m_e	Percent error	
Accepted value ....273.............. m_e	....	
(m_ec^2 = 0.511 MeV)		

Calculations (show work)

Name .... Section .... Date ....

Lab Partner(s) ....

EXPERIMENT 57 *The Pi-Mu-e Decay Process*

QUESTIONS

1. In Eq. 57-10, why can the total energy of the positron be approximated by its kinetic energy?

2. Are the muon tracks actually straight lines? Explain.

3. Why should the curvature of the positron tracks be measured near the beginnings of the tracks?

4. What are the accepted values of the rest energies of the pion and muon in MeV?

APPENDIX **A**

Material Properties

TABLE A1 Densities of Materials (g/cm³)

Solids:	
Aluminum	2.7
Brass	8.4
Copper	8.9
Glass	
crown	2.5–2.7
flint	3.0–3.6
Gold	19.3
Iron and Steel (general)	7.88
Lead	11.3
Nickel	8.8
Silver	10.5
Wood	
oak	0.60–0.90
pine	0.35–0.50
Zinc	7.1
Liquids:	
Alcohol	
ethyl	0.79
methyl	0.81
Carbon tetrachloride	1.60
Gasoline	0.68–0.75
Glycerine	1.26
Mercury	13.6
Turpentine	0.87
Water	1.00
Gases (at STP):	
Air	0.001293
Carbon dioxide	0.001975
Helium	0.000179
Hydrogen	0.000089
Nitrogen	0.000125
Oxygen	0.00143

TABLE A2 Young's Modulus for Some Metals

	(N/m^2)	(dyn/cm^2)
Aluminum	6.5×10^{10}	6.5×10^{11}
Brass	9.0×10^{10}	9.0×10^{11}
Copper	12.0×10^{10}	12.0×10^{11}
Iron		
cast	9.0×10^{10}	9.0×10^{11}
wrought	19.0×10^{10}	19.0×10^{11}
Steel	19.2×10^{10}	19.2×10^{11}

TABLE A3 Coefficients of Linear Thermal Expansion (1/°C)

Aluminum	24.0×10^{-6}
Brass	18.8×10^{-6}
Copper	16.8×10^{-6}
Glass	
window	8.5×10^{-6}
Pyrex	3.3×10^{-6}
Iron	11.4×10^{-6}
Lead	29.4×10^{-6}
Nickel	12.8×10^{-6}
Silver	18.8×10^{-6}
Steel	13.4×10^{-6}
Tin	26.9×10^{-6}
Zinc	26.4×10^{-6}

TABLE A4 Specific Heats (cal/g-°C or kcal/kg-°C)

Aluminum	0.22
Brass	0.092
Copper	0.093
Glass	0.16
Iron	0.11
Lead	0.031
Mercury	0.033
Nickel	0.11
Silver	0.056
Steel	0.11
Tin	0.054
Water	1.00
Zinc	0.093

TABLE A5 Color Code for Resistors (Composition Type)

Bands A and B		Band C		Band D	
Color	*Significant figure*	*Color*	*Multiplier*	*Color*	*Resistance tolerance (percent)*
Black	0	Black	1	Silver	±10
Brown	1	Brown	10	Gold	±5
Red	2	Red	100		
Orange	3	Orange	1,000		
Yellow	4	Yellow	10,000		
Green	5	Green	100,000		
Blue	6	Blue	1,000,000		
Purple (violet)	7				
Gray	8	Silver	0.01		
White	9	Gold	0.1		

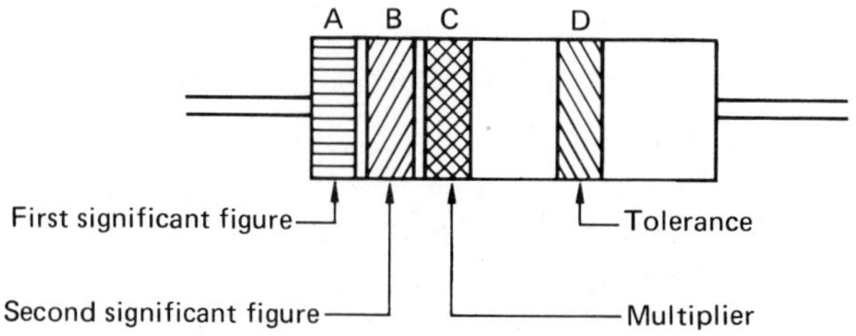

A B C D

First significant figure

Second significant figure

Tolerance

Multiplier

TABLE A6 Resistivities and Temperature Coefficients

Substance	*Resistivity, ρ* (Ω-cm)	*Temperature coefficient* (1/°C)
Aluminum	2.8×10^{-6}	0.0039
Brass	7×10^{-6}	0.002
Constantan	49×10^{-6}	0.00001
Copper	1.72×10^{-6}	0.00393
German silver (18% Ni)	33×10^{-6}	0.0004
Iron	10×10^{-6}	0.005
Manganin	44×10^{-6}	0.00001
Mercury	95.8×10^{-6}	0.00089
Nichrome	100×10^{-6}	0.0004
Nickel	7.8×10^{-6}	0.006
Silver	1.6×10^{-6}	0.0038
Tin	11.5×10^{-6}	0.0042

TABLE A7 Wire Sizes [American Wire Gauge (AWG)]

Gauge No.	*Diameter*	
	in.	*cm*
0000	0.4600	1.168
000	0.4096	1.040
00	0.3648	0.9266
0	0.3249	0.8252
1	0.2893	0.7348
2	0.2576	0.6543
3	0.2294	0.5827
4	0.2043	0.5189
5	0.1819	0.4620
6	0.1620	0.4115
7	0.1443	0.3665
8	0.1285	0.3264
9	0.1144	0.2906
10	0.1019	0.2588
11	0.09074	0.2305
12	0.08081	0.2053
13	0.07196	0.1828
14	0.06408	0.1628
15	0.05707	0.1450
16	0.05082	0.1291
17	0.04526	0.1150
18	0.04030	0.1024
19	0.03589	0.09116
20	0.03196	0.08118
21	0.02846	0.07229
22	0.02535	0.06439
23	0.02257	0.05733
24	0.02010	0.05105
25	0.01790	0.04547
26	0.01594	0.04049
27	0.01419	0.03604
28	0.01264	0.03211
29	0.01126	0.02860
30	0.01003	0.02548
31	0.008928	0.02268
32	0.007950	0.02019
33	0.007080	0.01798
34	0.006304	0.01601
35	0.005614	0.01426
36	0.005000	0.01270
37	0.004453	0.01131
38	0.003965	0.01007
39	0.003531	0.008969
40	0.003145	0.007988

TABLE A8 Major Visible Spectral Lines of Some Elements

Element	*Wavelength* (Å)	*Color*	*Relative intensity*
Helium	3889	Violet	1000
	3965	Violet	50
	4026	Violet	70
	4388	Blue-violet	30
	4471	Dark blue	100
	4713	Blue	40
	4922	Blue-green	50
	5015	Green	100
	5876	Yellow	1000
	6678	Red	100
	7065	Red	70
Mercury	4047	Violet	300
	4078	Violet	150
	4358	Blue	500
	4916	Blue-green	50
	5461	Green	2000
	5770	Yellow	200
	5790	Yellow	1000
	6907	Red	125
Sodium	4494	Blue	60
	4498	Blue	70
	4665	Blue	80
	4669	Blue	200
	4983	Green	200
	5149	Green	400
	5153	Green	600
	5670	Green	100
	5675	Green	150
	5683	Green	80
	5688	Green	300
	5890	Yellow-orange	9000
	5896	Yellow-orange	5000
	6154	Orange	500
	6161	Orange	500

Wavelengths of various colors (Å)

Color	*Representative*	*Limits*
Red	6500	6470–7000
Orange	6000	5840–6470
Yellow	5800	5750–5850
Green	5200	4912–5750
Blue	4700	4240–4912
Violet	4100	4000–4200

Visible spectrum ≃ 4000–7000

TABLE A9 Radioisotopes

Isotope	*Half-life*	*Principal Radiations (MeV)* Alpha	Beta	Gamma
Barium-133	10.4 years			0.356
Bismuth-210	5.01 days	4.654, 4.691	1.161	
Carbon-14	5730 years		0.156	
Cesium-137	30.1 years		0.512, 1.173	
Barium-137m	2.6 min			0.662
Cobalt-60	5.26 years		0.315	
Iodine-131	8.07 days		0.606	
Lead-210	22.3 years		0.017, 0.061	0.0465
Manganese-54	312.5 days			0.835
Phosphorus-32	14.3 days		1.710	
Polonium-210	138.4 days	5.305		
Potassium-42	12.4 hours		3.52 1.97	
Radium-226	1600 years	4.781 4.598		0.186
Sodium-22	2.60 years	0.545 1.82		1.275
Strontium-90	28.1 years		0.546	
Thallium-204	3.78 years		0.763	
Uranium-238	4.5×10^6 years	4.195		0.48
Yttrium-90	64.0 hours		2.27	
Zinc-65	243.6 days		0.329	1.116

TABLE A10 Elements: Atomic Numbers and Atomic Weights

The atomic weights are based on $^{12}C = 12.0000$. If the element does not occur naturally, the mass number of the most stable isotope is given in parentheses.

	Symbol	*Atomic number*	*Atomic weight*		*Symbol*	*Atomic number*	*Atomic weight*
Actinium	Ac	89	(227)	Californium	Cf	98	(251)
Aluminum	Al	13	26.9815	Carbon	C	6	12.011
Americium	Am	95	(243)	Cerium	Ce	58	140.12
Antimony	Sb	51	121.75	Cesium	Cs	55	132.9055
Argon	Ar	18	39.948	Chlorine	Cl	17	35.453
Arsenic	As	33	74.9216	Chromium	Cr	24	51.996
Astatine	At	85	(210)	Cobolt	Co	27	58.9332
Barium	Ba	56	137.34	Copper	Cu	29	63.545
Berkelium	Bk	97	(247)	Curium	Cm	96	(247)
Beryllium	Be	4	9.01218	Dysprosium	Dy	66	162.50
Bismuth	Bi	83	208.9806	Einsteinium	Es	99	(254)
Boron	B	5	10.81	Erbium	Er	68	167.26
Bromine	Br	35	79.90	Europium	Eu	63	151.96
Cadmium	Cd	48	112.40	Fermium	Fm	100	(253)
Calcium	Ca	20	40.08	Fluorine	F	9	18.9984

TABLE A10 (Continued)

The atomic weights are based on $^{12}C = 12.0000$. If the element does not occur naturally, the mass number of the most stable isotope is given in parentheses.

	Symbol	*Atomic number*	*Atomic weight*
Francium	Fr	87	(223)
Gadolinium	Gd	64	157.25
Gallium	Ga	31	69.72
Germanium	Ge	32	72.59
Gold	Au	79	196.967
Hafnium	Hf	72	178.49
Hahnium	Ha	105	(260)
Helium	He	2	4.00260
Holmium	Ho	67	164.9303
Hydrogen	H	1	1.0080
Indium	In	49	114.82
Iodine	I	53	126.9045
Iridium	Ir	77	192.22
Iron	Fe	26	55.847
Krypton	Kr	36	83.80
Lanthanum	La	57	138.9055
Lawrencium	Lr	103	(257)
Lead	Pb	82	207.12
Lithium	Li	3	6.941
Lutetium	Lu	71	174.97
Magnesium	Mg	12	24.305
Manganese	Mn	25	54.9380
Mendelevium	Md	101	(256)
Mercury	Hg	80	200.59
Molybdenum	Mo	42	95.94
Neodymium	Nd	60	144.24
Neon	Ne	10	20.179
Neptunium	Np	93	(237)
Nickel	Ni	28	58.71
Niobium	Nb	41	92.9064
Nitrogen	N	7	14.0067
Nobelium	No	102	(253)
Osmium	Os	76	190.2
Oxygen	O	8	15.9994
Palladium	Pd	46	106.4
Phosphorus	P	15	30.9738
Platinum	Pt	78	195.09
Plutonium	Pu	94	(224)
Polonium	Po	84	(209)
Potassium	K	19	39.102
Praseodymium	Pr	59	140.9077
Promethium	Pm	Pm	(145)
Protactinium	Pa	91	(231)
Radium	Ra	88	(226)
Radon	Rn	86	(222)
Rhenium	Re	75	186.2
Rhodium	Rh	45	102.9055
Rubidium	Rb	37	85.4678
Ruthenium	Ru	44	101.07
Rutherfordium	Rf	104	(257)
Samarium	Sm	62	150.4
Scandium	Sc	21	44.9559
Selenium	Se	34	78.96
Silicon	Si	14	28.086
Silver	Ag	47	107.868
Sodium	Na	11	22.9898
Strontium	Sr	38	87.62
Sulfur	S	16	32.06
Tantalum	Ta	73	180.9479
Technetium	Tc	43	(99)
Tellurium	Te	52	127.60
Terbium	Tb	65	158.9254
Thallium	Tl	81	204.37
Thorium	Th	90	232.0381
Thulium	Tm	69	168.9342
Tin	Sn	50	118.69
Titanium	Ti	22	47.90
Tungsten	W	74	183.85
Uranium	U	92	238.029
Vanadium	V	23	50.9414
Xenon	Xe	54	131.30
Ytterbium	Yb	70	173.04
Yttrium	Y	39	88.9059
Zinc	Zn	30	65.37
Zirconium	Zr	40	91.22

APPENDIX **B**

Mathematical and Physical Constants

TABLE B1 Metric Prefixes

Multiple		*Name*	*Abbreviation*
1,000,000,000,000,000,000	10^{18}	exa	E
1,000,000,000,000,000	10^{15}	peta	P
1,000,000,000,000	10^{12}	tera	T
1,000,000,000	10^{9}	giga	G
1,000,000	10^{6}	mega	M
1,000	10^{3}	kilo	k
100	10^{2}	hecto	h
10	10^{1}	deka	da
1	1	—	—
0.1	10^{-1}	deci	d
0.01	10^{-2}	centi	c
0.001	10^{-3}	milli	m
0.000001	10^{-6}	micro	μ
0.000000001	10^{-9}	nano	n
0.000000000001	10^{-12}	pico	p
0.000000000000001	10^{-15}	femto	f
0.000000000000000001	10^{-18}	atto	a

TABLE B2 Physical Constants

Quantity	Symbol	Value
Acceleration due to gravity	g	$9.8\ \text{m/s}^2 = 980\ \text{cm/s}^2 = 32.2\ \text{ft/s}^2$
Universal gravitational constant	G	$6.67 \times 10^{-11}\ \frac{\text{N-m}^2}{\text{kg}^2} = 6.67 \times 10^{-8}\ \frac{\text{dyne-cm}^2}{\text{g}^2}$
Electron charge	e	1.60×10^{-19} C
Speed of light	c	$3.0 \times 10^8\ \text{m/s} = 3.0 \times 10^{10}\ \text{cm/s} = 1.86 \times 10^5\ \text{mi/s}$
Boltzmann's constant	k	$1.38 \times 10^{-23}\ \text{J/K} = 1.38 \times 10^{-16}\ \text{erg/K}$
Planck's constant	h	$6.63 \times 10^{-34}\ \text{J-s} = 6.63 \times 10^{-27}\ \text{erg-s} = 4.14 \times 10^{-15}\ \text{eV-s}$
	$\hbar$	$h/2\pi = 1.05 \times 10^{-34}\ \text{J-s} = 1.05 \times 10^{-27}\ \text{erg-s} = 6.58 \times 10^{-16}\ \text{eV-s}$
Electron rest mass	m_e	$9.11 \times 10^{-31}\ \text{kg} = 5.49 \times 10^{-4}\ \text{amu} \leftrightarrow 0.511\ \text{MeV}$
Proton rest mass	m_p	$1.672 \times 10^{-27}\ \text{kg} = 1.00783\ \text{amu} \leftrightarrow 938.3\ \text{MeV}$
Neutron rest mass	m_n	$1.674 \times 10^{-27}\ \text{kg} = 1.00867\ \text{amu} \leftrightarrow 939.1\ \text{MeV}$
Coulomb's law constant	k	$1/4\pi\varepsilon_0 = 9.0 \times 10^9\ \text{N-m}^2/\text{C}^2$
Permittivity of free space	ϵ_0	$8.85 \times 10^{-12}\ \text{C}^2/\text{N-m}^2$
Permeability of free space	μ_0	$4\pi \times 10^{-7} = 1.26 \times 10^{-6}\ \text{Wb/A-m}$
Astronomical and earth data		
Radius of earth		
equatorial		$3963\ \text{mi} = 6.378 \times 10^6\ \text{m}$
polar		$3950\ \text{mi} = 6.357 \times 10^6\ \text{m}$
Mass of earth		6.0×10^{24} kg
Mass of moon		$7.4 \times 10^{22}\ \text{kg} = \frac{1}{81}$ mass of earth
Mass of sun		2.0×10^{30} kg
Average distance of earth from sun		$93 \times 10^6\ \text{mi} = 1.5 \times 10^8\ \text{km}$
Average distance of moon from earth		$2.4 \times 10^5\ \text{mi} = 3.8 \times 10^5\ \text{km}$
Diameter of moon		$2160\ \text{mi} \approx 3500\ \text{km}$
Diameter of sun		$864{,}000\ \text{mi} \approx 1.4 \times 10^6\ \text{km}$

TABLE B3 Conversion Factors

Quantity	Conversions	
Mass	$1\ \text{g} = 10^{-3}\ \text{kg} = 6.85 \times 10^{-5}$ slug	1 metric ton = 1000 kg
	$1\ \text{kg} = 10^3\ \text{g} = 6.85 \times 10^{-2}$ slug	
	$1\ \text{slug} = 1.46 \times 10^4\ \text{g} = 14.6\ \text{kg}$	
	$1\ \text{amu} = 1.66 \times 10^{-24}\ \text{g} = 1.66 \times 10^{-27}\ \text{kg}$	
Length	$1\ \text{cm} = 10^{-2}\ \text{m} = 0.394$ in.	$1\ \text{Å} = 10^{-10}\ \text{m} = 10^{-8}\ \text{cm}$
	$1\ \text{m} = 10^{-3}\ \text{km} = 3.28\ \text{ft} = 39.4$ in.	
	$1\ \text{km} = 10^3\ \text{m} = 0.621\ \text{mi}$	
	1 in. $= 2.54\ \text{cm} = 2.54 \times 10^{-2}\ \text{m}$	
	1 ft = 12 in. = 30.48 cm = 0.3048 m	
	1 mi = 5280 ft = 1609 m = 1.609 km	
Area	$1\ \text{cm}^2 = 10^{-4}\ \text{m}^2 = 0.1550\ \text{in.}^2 = 1.08 \times 10^{-3}\ \text{ft}^2$	
	$1\ \text{m}^2 = 10^4\ \text{cm}^2 = 10.76\ \text{ft}^2 = 1550\ \text{in.}^2$	
	$1\ \text{in.}^2 = 6.94 \times 10^{-3}\ \text{ft}^2 = 6.45\ \text{cm}^2 = 6.45 \times 10^{-4}\ \text{m}^2$	
	$1\ \text{ft}^2 = 144\ \text{in.}^2 = 9.29 \times 10^{-2}\ \text{m}^2 = 929\ \text{cm}^2$	

TABLE B3 Conversion Factors (continued)

Volume	1 cm^3 = 10^{-6} m^3 = 3.53 × 10^{-5} ft^3 = 6.10 × 10^{-2} $in.^3$ 1 m^3 = 10^6 cm^3 = 10^3 liters = 35.3 ft^3 = 6.10 × 10^4 $in.^3$ = 264 gal 1 liter = 10^3 cm^3 = 10^{-3} m^3 = 1.056 qt = 0.264 gal 1 $in.^3$ = 5.79 × 10^{-4} ft^3 = 16.4 cm^3 = 1.64 × 10^{-5} m^3 1 ft^3 = 1728 $in.^3$ = 7.48 gal = 0.0283 m^3 = 28.3 liters 1 qt = 2 pt = 946.5 cm^3 = 0.946 liter 1 gal = 4 qt = 231 $in.^3$ = 3.785 liters
Time	1 h = 60 min = 3600 s 1 day = 24 h = 1440 min = 8.64 × 10^4 s 1 year = 365 days = 8.76 × 10^3 h = 5.26 × 10^5 min = 3.16 × 10^7 s
Angle	360° = 2π rad 180° = π rad 1 rad = 57.3° 90° = $\pi/2$ rad 60° = $\pi/3$ rad 1° = 0.0175 rad 45° = $\pi/4$ rad 30° = $\pi/6$ rad
Speed	1 m/s = 3.6 km/h = 3.28 ft/s = 2.24 mi/h 1 km/h = 0.278 m/s = 0.621 mi/h = 0.911 ft/s 1 ft/s = 0.682 mi/h = 0.305 m/s = 1.10 km/h 1 mi/h = 1.467 ft/s = 1.609 km/h = 0.447 m/s 60 mi/h = 88 ft/s
Force	1 newton = 10^5 dynes = 0.225 lb 1 dyne = 10^{-5} N = 2.25 × 10^{-6} lb 1 lb = 4.45 × 10^5 dynes = 4.45 N Equivalent weight of 1-kg mass = 2.2 lb = 9.8 N
Pressure	1 Pa (N/m^2) = 1.45 × 10^{-4} $lb/in.^2$ = 7.5 × 10^{-3} torr (mmHg) = 10 $dynes/cm^2$ 1 torr (mm Hg) = 133 Pa (N/m^2) = 0.02 $lb/in.^2$ = 1333 $dynes/cm^2$ 1 atm = 14.7 $lb/in.^2$ = 1.013 × 10^5 N/m^2 = 1.013 × 10^6 $dynes/cm^2$ = 30 in. Hg = 76 cm Hg 1 bar = 10^6 $dynes/cm^2$ 1 millibar = 10^3 $dynes/cm^2$
Energy	1 J = 10^7 ergs = 0.738 ft-lb = 0.239 cal = 9.48 × 10^{-4} Btu = 6.24 × 10^{18} eV 1 kcal = 4186 J = 4.186 × 10^{10} ergs = 3.968 Btu 1 Btu = 1055 J = 1.055 × 10^{10} ergs = 778 ft-lb = 0.252 kcal 1 cal = 4.186 J = 3.97 × 10^{-3} Btu = 3.09 ft-lb 1 ft-lb = 1.356 J = 1.36 × 10^7 ergs = 1.29 × 10^{-3} Btu 1 eV = 1.60 × 10^{-19} J = 1.60 × 10^{-12} ergs
Power	1 W = 0.738 ft-lb/s = 1.34 × 10^{-3} hp = 3.41 Btu/h 1 ft-lb/s = 1.36 W = 1.82 × 10^{-3} hp 1 hp = 550 ft-lb/s = 745.7 W = 2545 Btu/h
Rest Mass–Energy Equivalents	1 amu = 1.66 × 10^{-27} kg ↔ 931 MeV 1 electron mass = 9.11 × 10^{-31} kg = 5.49 × 10^{-4} amu ↔ 0.511 MeV 1 proton mass = 1.672 × 10^{-27} kg = 1.00728 amu ↔ 938.3 MeV 1 neutron mass = 1.674 × 10^{-27} kg = 1.00867 amu ↔ 939.1 MeV

TABLE B4 Squares, Cubes, and Roots

N	N^2	$\sqrt{N}$	$\sqrt{10N}$	N^3	$\sqrt[3]{N}$
1	1	1.00 000	3.16 228	1	1.00 000
2	4	1.41 421	4.47 214	8	1.25 992
3	9	1.73 205	5.47 723	27	1.44 225
4	16	2.00 000	6.32 456	64	1.58 740
5	25	2.23 607	7.07 107	125	1.70 998
6	36	2.44 949	7.74 597	216	1.81 712
7	49	2.64 575	8.36 660	343	1.91 293
8	64	2.82 843	8.94 427	512	2.00 000
9	81	3.00 000	9.48 683	729	2.08 008
10	100	3.16 228	10.00 00	1 000	2.15 443
11	121	3.31 662	10.48 81	1 331	2.22 398
12	144	3.46 410	10.95 45	1 728	2.28 943
13	169	3.60 555	11.40 18	2 197	2.35 133
14	196	3.74 166	11.83 22	2 744	2.41 014
15	225	3.87 298	12.24 74	3 375	2.46 621
16	256	4.00 000	12.64 91	4 096	2.51 984
17	289	4.12 311	13.03 84	4 913	2.57 128
18	324	4.24 264	13.41 64	5 832	2.62 074
19	361	4.35 890	13.78 40	6 859	2.66 840
20	400	4.47 214	14.14 21	8 000	2.71 442
21	441	4.58 258	14.49 14	9 261	2.75 892
22	484	4.69 042	14.83 24	10 648	2.80 204
23	529	4.79 583	15.16 58	12 167	2.84 387
24	576	4.89 898	15.49 19	13 824	2.88 450
25	625	5.00 000	15.81 14	15 625	2.92 402
26	676	5.09 902	16.12 45	17 576	2.96 250
27	729	5.19 615	16.45 17	19 683	3.00 000
28	784	5.29 150	16.73 32	21 952	3.03 659
29	841	5.38 516	17.02 94	24 389	3.07 232
30	900	5.47 723	17.32 05	27 000	3.10 723
31	961	5.56 776	17.60 68	29 791	3.14 138
32	1 024	5.65 685	17.88 85	32 768	3.17 480
33	1 089	5.74 456	18.16 59	35 937	3.20 753
34	1 156	5.83 095	18.43 91	39 304	3.23 961
35	1 225	5.91 608	18.70 83	42 875	3.27 107
36	1 296	6.00 000	18.97 37	46 656	3.30 195
37	1 369	6.08 276	19.23 54	50 653	3.33 222
38	1 444	6.16 441	19.49 36	54 872	3.36 198
39	1 521	6.24 500	19.74 84	59 319	3.39 121
40	1 600	6.32 456	20.00 00	64 000	3.41 995
41	1 681	6.40 312	20.24 85	68 921	3.44 822
42	1 764	6.48 074	20.49 39	74 088	3.47 603
43	1 849	6.55 744	20.73 64	79 507	3.50 340
44	1 936	6.63 325	20.97 62	85 184	3.53 035
45	2 025	6.70 820	21.21 32	91 125	3.55 689
46	2 116	6.78 233	21.44 76	97 336	3.58 305
47	2 209	6.85 565	21.67 95	103 823	3.60 883
48	2 304	6.92 820	21.90 89	110 592	3.63 424
49	2 401	7.00 000	22.13 59	117 649	3.65 931
50	2 500	7.07 107	22.36 07	125 000	3.68 403
N	N^2	$\sqrt{N}$	$\sqrt{10N}$	N^3	$\sqrt[3]{N}$

N	N^2	$\sqrt{N}$	$\sqrt{10N}$	N^3	$\sqrt[3]{N}$
50	2 500	7.07 107	22.36 07	125 000	3.68 403
51	2 601	7.14 143	22.58 32	132 651	3.70 843
52	2 704	7.21 110	22.80 35	140 608	3.73 251
53	2 809	7.28 011	23.02 17	148 877	3.75 629
55	3 025	7.41 620	23.45 21	166 375	3.80 295
56	3 136	7.48 331	23.66 43	175 616	3.82 586
57	3 249	7.54 983	23.87 47	185 193	3.84 850
58	3 364	7.61 577	24.08 32	195 112	3.87 088
59	3 481	7.68 115	24.28 99	205 379	3.89 300
60	3 600	7.74 597	24.49 49	216 000	3.91 487
61	3 721	7.81 025	24.69 82	226 981	3.93 650
62	3 844	7.87 401	24.89 98	238 328	3.95 789
63	3 969	7.93 725	25.09 98	250 047	3.97 906
64	4 0.96	8.00 000	25.29 82	262 144	4.00 000
65	4 225	8.06 226	25.49 51	274 625	4.02 073
66	4 356	8.12 404	25.69 05	287 496	4.04 124
67	4 489	8.18 535	25.88 44	300 763	4.06 155
68	4 624	8.24 621	26.07 68	314 432	4.08 166
69	4 761	8.30 662	26.26 79	328 509	4.10 157
70	4 900	8.36 660	26.45 75	343 000	4.12 129
71	5 041	8.42 615	26.64 58	357 911	4.14 082
72	5 184	8.48 528	26.83 28	373 248	4.16 017
73	5 329	8.54 400	27.01 85	389 017	4.17 934
74	5 476	8.60 233	27.20 29	405 224	4.19 854
75	5 625	8.66 025	27.38 61	421 875	4.21 716
76	5 776	8.71 780	27.56 81	438 976	4.23 582
77	5 929	8.77 496	27.74 89	456 533	4.25 432
78	6 084	8.83 176	27.92 85	474 552	4.27 266
79	6 241	8.88 819	28.10 69	493 039	9.29 084
80	6 400	8.94 427	28.28 43	512 000	4.30 887
81	6 561	9.00 000	28.46 05	531 441	4.32 675
82	6 724	9.05 539	28.63 56	551 368	4.34 448
83	6 889	9.11 043	28.80 97	571 787	4.36 207
84	7 056	9.16 515	28.98 28	592 704	4.37 952
85	7 225	9.21 954	29.15 48	614 125	4.39 683
86	7 396	9.27 362	29.32 58	636 056	4.41 400
87	7 569	9.32 738	29.49 58	658 503	4.43 105
88	7 744	9.38 083	29.66 48	681 472	4.44 796
89	7 921	9.43 398	29.83 29	704 969	4.46 475
90	8 100	9.48 683	30.00 00	729 000	4.48 140
91	8 281	9.53 939	30.16 62	753 571	4.49 794
92	8 464	9.59 166	30.33 15	778 688	4.51 436
93	8 649	9.64 365	30.49 59	804 357	4.53 065
94	8 836	9.69 536	30.65 94	830. 584	4.54 684
95	9 025	9.74 679	30.82 21	857 375	4.56 290
96	9 216	9.79 796	30.98 39	884 736	4.57 886
97	9 409	9.84 886	31.14 48	912 673	4.59 470
98	9 604	9.89 949	31.30 50	941 192	4.61 044
99	9 801	9.94 987	31.46 43	970 299	4.62 607
100	10 000	10.00 000	31.62 28	1 000 000	4.64 159
N	N^2	$\sqrt{N}$	$\sqrt{10N}$	N^3	$\sqrt[3]{N}$

TABLE B4 Squares, Cubes, and Roots (Continued)

N	N^2	$\sqrt{N}$	$\sqrt{10N}$	N^3	$\sqrt[3]{N}$
100	10 000	10.00 00	31.62 28	1 000 000	4.64 159
101	10 201	10.04 99	31.78 05	1 030 301	4.65 701
102	10 404	10.09 95	31.93 74	1 061 208	4.67 233
103	10 609	10.14 89	32.09 36	1 092 727	4.68 755
104	10 816	10.19 80	32.24 90	1 124 864	4.70 267
105	11 025	10.24 70	32.40 37	1 157 625	4.71 769
106	11 236	10.29 56	32.55 76	1 191 016	4.73 262
107	11 449	10.34 41	32.71 09	1 225 043	4.74 746
108	11 664	10.39 23	32.86 34	1 259 712	4.76 220
109	11 881	10.44 03	33.01 51	1 295 029	4.77 686
110	12 100	10.48 81	33.16 62	1 331 000	4.79 142
111	12 321	10.53 57	33.31 67	1 367 631	4.80 590
112	12 544	10.58 30	33.46 64	1 404 928	4.82 028
113	12 769	10.63 01	33.61 55	1 442 897	4.83 459
114	12 996	10.67 71	33.76 39	1 481 544	4.84 881
115	13 225	10.72 38	33.91 16	1 520 875	4.86 294
116	13 456	10.77 03	34.05 88	1 560 896	4.87 700
117	13 689	10.81 67	34.20 53	1 601 613	4.89 097
118	13 924	10.86 28	34.35 11	1 643 032	4.90 487
119	14 161	10.90 87	34.49 64	1 685 159	4.91 868
120	14 400	10.95 45	34.64 10	1 728 000	4.93 242
121	14 641	11.00 00	34.78 51	1 771 561	4.94 609
122	14 884	11.04 54	34.92 85	1 815 848	4.95 968
123	15 129	11.09 05	35.07 14	1 860 867	4.97 319
124	15 376	11.13 55	35.21 36	1 906 624	4.98 663
125	15 625	11.18 03	35.35 53	1 953 125	5.00 000
126	15 876	11.22 50	35.49 65	2 000 376	5.01 330
127	16 129	11.26 94	35.63 71	2 048 383	5.02 653
128	16 384	11.31 37	35.77 71	2 097 152	5.03 968
129	16 641	11.35 78	35.91 66	2 146 689	5.05 277
130	16 900	11.40 18	36.05 55	2 197 000	5.06 580
131	17 161	11.44 55	36.19 39	2 248 091	5.07 875
132	17 424	11.48 91	36.33 18	2 299 968	5.09 164
133	17 689	11.53 26	36.46 92	2 352 637	5.10 447
134	17 956	11.57 58	36.60 60	2 406 104	5.11 723
135	18 225	11.61 90	36.74 23	2 460 375	5.12 993
136	18 496	11.66 19	36.87 82	2 515 456	5.14 256
137	18 769	11.70 47	37.01 35	2 571 353	5.15 514
138	19 044	11.74 73	37.14 84	2 628 072	5.16 765
139	19 321	11.78 98	37.28 27	2 685 619	5.18 010
140	19 600	11.83 22	37.41 66	2 744 000	5.19 249
141	19 881	11.87 43	37.55 00	2 803 221	5.20 483
142	20 164	11.91 64	37.68 29	2 863 288	5.21 710
143	20 449	11.95 83	37.81 53	2 924 207	5.22 932
144	20 736	12.00 00	37.94 73	2 985 984	5.24 148
145	21 025	12.04 16	38.07 89	3 048 625	5.25 359
146	21 316	12.08 30	38.20 99	3 112 136	5.26 564
147	21 609	12.12 44	38.34 06	3 176 523	5.27 763
148	21 904	12.16 55	38.47 08	3 241 792	5.28 957
149	22 201	12.20 66	38.60 05	3 307 949	5.30 146
150	22 500	12.24 74	38.72 98	3 375 000	5.31 329

N	N^2	$\sqrt{N}$	$\sqrt{10N}$	N^3	$\sqrt[3]{N}$
150	22 500	12.24 64	38.72 98	3 375 000	5.31 329
151	22 801	12.28 82	38.85 87	3 442 951	5.32 507
152	23 104	12.32 88	38.98 72	3 511 808	5.33 680
153	23 409	12.36 93	39.11 52	3 581 577	5.34 848
154	23 716	12.40 97	39.24 28	3 652 264	5.36 011
155	24 025	12.44 99	39.37 00	3 723 875	5.37 169
156	24 336	12.49 00	39.49 68	3 796 416	5.38 321
157	24 649	12.53 00	39.62 32	3 869 893	5.39 469
158	24 964	12.56 98	39.74 92	3 944 312	5.40 612
159	25 281	12.60 95	39.87 48	4 019 679	5.41 750
160	25 600	12.64 91	40.00 00	4 096 000	5.42 884
161	25 921	12.68 86	40.12 48	4 173 281	5.44 012
162	26 244	12.72 79	40.24 92	4 251 528	5.45 136
163	26 569	12.76 71	40.37 33	4 330 747	5.46 256
164	26 896	12.80 62	40.49 69	4 410 944	5.47 370
165	27 225	12.84 52	40.62 02	4 492 125	5.48 481
166	27 556	12.88 41	40.74 31	4 574 296	5.49 586
167	27 889	12.92 28	40.86 56	4 657 463	5.50 688
168	28 224	12.96 15	40.98 78	4 741 632	5.51 785
169	28 561	13.00 00	41.10 96	4 826 809	5.52 877
170	28 900	13.03 84	41.23 11	4 913 000	5.53 966
171	29 241	13.07 67	41.35 21	5 000 211	5.55 050
172	29 584	13.11 49	41.47 29	5 088 448	5.56 130
173	29 929	13.15 29	41.59 33	5 177 717	5.57 205
174	30 276	13.19 09	41.71 33	5 268 024	5.58 277
175	30 625	13.22 88	41.83 30	5 359 375	5.59 344
176	30 976	13.26 65	41.95 24	5 451 776	5.60 408
177	31 329	13.30 41	42.07 14	5 545 233	5.61 467
178	31 684	13.34 17	42.19 00	5 639 752	5.62 523
179	32 041	13.37 91	42.30 84	5 735 339	5.63 574
180	32 400	13.41 64	42.42 64	5 832 000	5.64 622
181	32 761	13.45 36	42.54 41	5 929 741	5.65 665
182	33 124	13.49 07	42.66 15	6 028 568	5.66 705
183	33 489	13.52 77	42.77 85	6 128 487	5.67 741
184	33 856	13.56 47	42.89 52	6 229 504	5.68 773
185	34 225	13.60 15	43.01 16	6 331 625	5.69 802
186	34 596	13.63 82	43.12 77	6 434 856	5.70 827
187	34 969	13.67 48	43.24 35	6 539 203	5.71 848
188	35 344	13.71 13	43.35 90	6 644 672	5.72 865
189	35 721	13.74 77	43.47 41	6 751 269	5.73 879
190	36 100	13.78 40	43.58 90	6 859 000	5.74 890
191	36 481	13.82 03	43.70 35	6 967 871	5.75 897
192	36 864	13.85 64	43.81 78	7 077 888	5.76 900
193	37 249	13.89 24	43.93 18	7 189 057	5.77 900
194	37 636	13.92 84	44.04 54	7 301 384	5.78 896
195	38 025	13.96 42	44.15 88	7 414 875	5.79 889
196	38 416	14.00 00	44.27 19	7 529 536	5.80 879
197	38 809	14.03 57	44.38 47	7 645 373	5.81 865
198	39 204	14.07 12	44.49 72	7 762 392	5.82 848
199	39 601	14.10 67	44.60 94	7 880 599	5.83 827
200	40 000	14.14 21	44.72 14	8 000 000	5.84 804

TABLE B4 Squares, Cubes, and Roots (Continued)

N	N^2	$\sqrt{N}$	$\sqrt{10N}$	N^3	$\sqrt[3]{N}$
200	40 000	14.14 21	44.72 14	8 000 000	5.84 804
201	40 401	14.17 74	44.83 30	8 120 601	5.85 777
202	40 804	14.21 27	44.94 44	8 242 408	5.86 746
203	41 209	14.24 78	45.05 55	8 365 427	5.87 713
204	41 616	14.28 29	45.16 64	8 489 664	5.88 677
205	42 025	14.31 78	45.27 69	8 615 125	5.89 637
206	42 436	14.35 27	45.38 72	8 741 816	5.90 594
207	42 849	14.38 75	45.49 73	8 869 743	5.91 548
208	43 264	14.42 22	45.60 70	8 998 912	5.92 499
209	43 681	14.45 68	45.71 65	9 129 329	5.93 447
210	44 100	14.49 14	45.82 58	9 261 000	5.94 392
211	44 521	14.52 58	45.93 47	9 393 931	5.95 334
212	44 944	14.56 02	46.04 35	9 528 128	5.96 273
213	45 369	14.59 45	46.15 19	9 663 597	5.97 209
214	45 796	14.62 87	46.26 01	9 800 344	5.98 142
215	46 225	14.66 29	46.36 81	9 938 375	5.99 073
216	46 656	14.69 69	46.47 58	10 077 696	6.00 000
217	47 089	14.73 09	46.58 33	10 218 313	6.00 925
218	47 524	14.76 48	46.69 05	10 360 232	6.01 846
219	47 961	14.79 86	46.79 74	10 503 459	6.02 765
220	48 400	14.83 24	46.90 42	10 648 000	6.03 681
221	48 841	14.86 61	47.01 06	10 793 861	6.04 594
222	49 284	14.89 97	47.11 69	10 941 048	6.05 505
223	49 729	14.93 32	47.22 29	11 089 567	6.06 413
224	50 176	14.96 66	47.32 86	11 239 424	6.07 318
225	50 625	15.00 00	47.43 42	11 390 625	6.08 220
226	51 076	15.03 33	47.53 95	11 543 176	6.09 120
227	51 529	15.06 65	47.64 45	11 697 083	6.10 017
228	51 984	15.09 97	47.74 93	11 852 352	6.10 911
229	52 441	15.13 27	47.85 39	12 008 989	6.11 803
230	52 900	15.16 58	47.95 83	12 167 000	6.12 693
231	53 361	15.19 87	48.06 25	12 326 391	6.13 579
232	53 824	15.23 15	48.16 64	12 487 168	6.14 463
233	54 289	15.26 43	48.27 01	12 649 337	6.15 345
234	54 756	15.29 71	48.37 35	12 812 904	6.16 224
235	55 225	15.32 97	48.47 68	12 977 875	6.17 101
236	55 696	15.36 23	48.57 98	13 144 256	6.17 975
237	56 169	15.39 48	48.68 26	13 312 053	6.18 846
238	56 644	15.42 72	48.78 52	13 481 272	6.19 715
239	57 121	15.45 96	48.88 76	13 651 919	6.20 582
240	57 600	15.49 19	48.98 98	13 824 000	6.21 447
241	58 081	15.52 42	49.09 18	13 997 521	6.22 308
242	58 564	15.55 63	49.19 35	14 172 488	6.23 168
243	59 049	15.58 85	49.29 50	14 348 907	6.24 025
244	59 536	15.62 05	49.39 64	14 526 784	6.24 880
245	60 025	15.65 25	49.49 75	14 706 125	6.25 732
246	60 516	15.68 44	49.59 84	14 886 936	6.26 583
247	61 009	15.71 62	49.69 91	15 069 223	6.27 431
248	61 504	15.74 80	49.79 96	15 252 992	6.28 276
249	62 001	15.77 97	49.89 99	15 438 249	6.29 119
250	62 500	15.81 14	50.00 00	15 625 000	6.29 961
N	N^2	$\sqrt{N}$	$\sqrt{10N}$	N^3	$\sqrt[3]{N}$

N	N^2	$\sqrt{N}$	$\sqrt{10N}$	N^3	$\sqrt[3]{N}$
250	62 500	15.81 14	50.00 00	15 625 000	6.29 961
251	63 001	15.84 30	50.09 99	15 813 251	6.30 799
252	63 504	15.87 45	50.19 96	16 003 008	6.31 636
253	64 009	15.90 60	50.29 91	16 194 277	6.32 470
254	64 516	15.93 74	50.39 84	16 387 064	6.33 303
255	65 025	15.96 87	50.49 75	16 581 375	6.34 133
256	65 536	16.00 00	50.59 64	16 777 216	6.34 960
257	66 049	16.03 12	50.69 52	16 974 593	6.35 786
258	66 564	16.06 24	50.79 37	17 173 512	6.36 610
259	67 081	16.09 35	50.89 20	17 373 979	6.37 431
260	67 600	16.12 45	50.99 02	17 576 000	6.38 250
261	68 121	16.15 55	51.08 82	17 779 581	6.39 068
262	68 644	16.18 64	51.18 59	17 984 728	6.39 883
263	69 169	16.21 73	51.28 35	18 191 447	6.40 696
264	69 696	16.24 81	51.38 09	18 399 744	6.41 507
265	70 225	16.27 88	51.47 82	18 609 625	6.42 316
266	70 756	16.30 95	51.57 52	18 821 096	6.43 123
267	71 289	16.34 01	51.67 20	19 034 163	6.43 928
268	71 824	16.37 07	51.76 87	19 248 832	6.44 731
269	72 361	16.40 12	51.86 52	19 465 109	6.45 531
270	72 900	16.43 17	51.96 15	19 683 000	6.46 330
271	73 441	16.46 21	52.05 77	19 902 511	6.47 127
272	73 984	16.49 24	52.15 36	20 123 648	6.47 922
273	74 529	16.52 27	52.24 94	20 346 417	6.48 715
274	75 076	16.55 29	52.34 50	20 570 824	6.49 507
275	75 625	16.58 31	52.44 04	20 796 875	6.50 296
276	76 176	16.61 32	52.53 57	21 024 576	6.51 083
277	76 729	16.64 33	52.63 08	21 253 933	6.51 868
278	77 284	16.67 33	52.72 57	21 484 952	6.52 652
279	77 841	16.70 33	52.82 05	21 717 639	6.53 434
280	78 400	16.73 32	52.91 50	21 952 000	6.54 213
281	78 961	16.76 31	53.00 94	22 188 041	6.54 991
282	79 524	16.79 29	53.10 37	22 425 768	6.55 767
283	80 089	16.82 26	53.19 77	22 665 187	6.56 541
284	80 656	16.85 23	53.29 17	22 906 304	6.57 314
285	81 225	16.88 19	53.38 54	23 149 125	6.58 084
286	81 796	16.91 15	53.47 90	23 393 656	6.58 853
287	82 369	16.94 11	53.57 24	23 639 903	6.59 620
288	82 944	16.97 06	53.66 56	23 887 872	6.60 385
289	83 521	17.00 00	53.75 87	24 137 569	6.61 149
290	84 100	17.02 94	53.85 16	24 389 000	6.61 911
291	84 681	17.05 87	53.94 44	24 642 171	6.62 671
292	85 264	17.08 80	54.03 70	24 897 088	6.63 429
293	85 849	17.11 72	54.12 95	25 153 757	6.64 185
294	86 436	17.14 64	54.22 18	25 412 184	6.64 940
295	87 025	17.17 56	54.31 39	25 672 375	6.65 693
296	87 616	17.20 47	54.40 59	25 934 336	6.66 444
297	88 209	17.23 37	54.49 77	26 198 073	6.67 194
298	88 804	17.26 27	54.48 94	26 463 592	6.67 942
299	89 401	17.29 16	54.68 09	26 730 899	6.68 688
300	90 000	17.32 05	54.77 23	27 000 000	6.69 433
N	N^2	$\sqrt{N}$	$\sqrt{10N}$	N^3	$\sqrt[3]{N}$

TABLE B5 Trigonometric Relationships

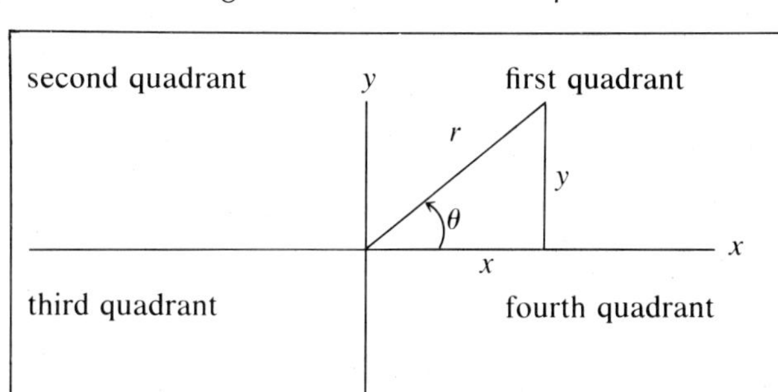

$\sin\theta = \frac{y}{r}$ $\cos\theta = \frac{x}{r}$ $\tan\theta = \frac{\sin\theta}{\cos\theta} = \frac{y}{x}$

$\theta°$ (rad)	$\sin\theta$	$\cos\theta$	$\tan\theta$
0° (0)	0	1	0
30° ($\pi/6$)	0.500	0.866	0.577
45° ($\pi/4$)	0.707	0.707	1.00
60° ($\pi/3$)	0.866	0.500	1.73
90° ($\pi/2$)	1	0	$\rightarrow\infty$

See Table 6 for other angles.

The sign of trigonometric functions depends on the quadrant, or sign of x and y, e.g., in the second quadrant $(-x,y)$, $-x/r = \cos\theta$ and $y/r = \sin\theta$, or by:

Reduction Formulas

	(θ in second quadrant)		(θ in third quadrant)		(θ in fourth quadrant)
$\sin\theta =$	$\cos(\theta - 90°)$	=	$-\sin(\theta - 180°)$	=	$-\cos(\theta - 270°)$
$\cos\theta =$	$-\sin(\theta - 90°)$	=	$-\cos(\theta - 180°)$	=	$\sin(\theta - 270°)$

Fundamental Identities

$\sin^2\theta + \cos^2\theta = 1$

$\sin 2\theta = 2\sin\theta\cos\theta$

$\cos 2\theta = \cos^2\theta - \sin^2\theta = 2\cos^2\theta - 1 = 1 - 2\sin^2\theta$

$\sin^2\theta = \frac{1}{2}(1 - \cos 2\theta)$

$\cos^2\theta = \frac{1}{2}(1 + \cos 2\theta)$

For half angle ($\theta/2$) identities, replace θ with $\theta/2$, e.g.,

$\sin^2\theta/2 = \frac{1}{2}(1 - \cos\theta)$ $\cos^2\theta/2 = \frac{1}{2}(1 + \cos\theta)$

$\sin(\alpha \pm \beta) = \sin\alpha\cos\beta \pm \cos\alpha\sin\beta$

$\cos(\alpha \pm \beta) = \cos\alpha\cos\beta \mp \sin\alpha\sin\beta$

For very small angles:

$\cos\theta \approx 1$

$\sin\theta \approx \theta$ (radians) $\tan\theta = \frac{\sin\theta}{\cos\theta} \approx \theta$

Law of sines:

$$\frac{a}{\sin\alpha} = \frac{b}{\sin\beta} = \frac{c}{\sin\gamma}$$

Law of cosines:

$a^2 = b^2 + c^2 - 2bc\cos\alpha$

$b^2 = a^2 + c^2 - 2ac\cos\beta$

$c^2 = a^2 + b^2 - 2ab\cos\gamma$

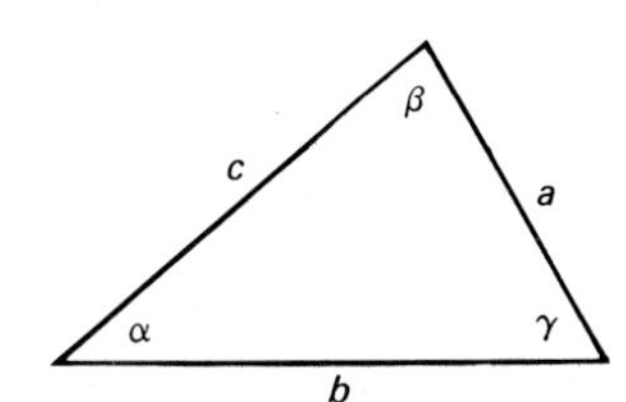

TABLE B6 Trigonometric Tables

Angle					*Angle*				
Degree	*Radian*	*Sine*	*Cosine*	*Tangent*	*Degree*	*Radian*	*Sine*	*Cosine*	*Tangent*
0°	.000	0.000	1.000	0.000					
1°	.017	.018	1.000	.018	46°	0.803	0.719	0.695	1.036
2°	.035	.035	0.999	.035	47°	.820	.731	.682	1.072
3°	.052	.052	.999	.052	48°	.838	.743	.669	1.111
4°	.070	.070	.998	.070	49°	.855	.755	.656	1.150
5°	.087	.087	.996	.088	50°	.873	.766	.643	1.192
6°	.105	.105	.995	.105	51°	.890	.777	.629	1.235
7°	.122	.122	.993	.123	52°	.908	.788	.616	1.280
8°	.140	.139	.990	.141	53°	.925	.799	.602	1.327
9°	.157	.156	.988	.158	54°	.942	.809	.588	1.376
10°	.175	.174	.985	.176	55°	.960	.819	.574	1.428
11°	.192	.191	.982	.194	56°	.977	.829	.559	1.483
12°	.209	.208	.978	.213	57°	.995	.839	.545	1.540
13°	.227	.225	.974	.231	58°	1.012	.848	.530	1.600
14°	.244	.242	.970	.249	59°	1.030	.857	.515	1.664
15°	.262	.259	.966	.268	60°	1.047	.866	.500	1.732
16°	.279	.276	.961	.287	61°	1.065	.875	.485	1.804
17°	.297	.292	.956	.306	62°	1.082	.883	.470	1.881
18°	.314	.309	.951	.325	63°	1.100	.891	.454	1.963
19°	.332	.326	.946	.344	64°	1.117	.899	.438	2.050
20°	.349	.342	.940	.364	65°	1.134	.906	.423	2.145
21°	.367	.358	.934	.384	66°	1.152	.914	.407	2.246
22°	.384	.375	.927	.404	67°	1.169	.921	.391	2.356
23°	.401	.391	.921	.425	68°	1.187	.927	.375	2.475
24°	.419	.407	.914	.445	69°	1.204	.934	.358	2.605
25°	.436	.423	.906	.466	70°	1.222	.940	.342	2.747
26°	.454	.438	.899	.488	71°	1.239	.946	.326	2.904
27°	.471	.454	.891	.510	72°	1.257	.951	.309	3.078
28°	.489	.470	.883	.532	73°	1.274	.956	.292	3.271
29°	.506	.485	.875	.554	74°	1.292	.961	.276	3.487
30°	.524	.500	.866	.577	75°	1.309	.966	.259	3.732
31°	.541	.515	.857	.601	76°	1.326	.970	.242	4.011
32°	.559	.530	.848	.625	77°	1.344	.974	.225	4.331
33°	.576	.545	.839	.649	78°	1.361	.978	.208	4.705
34°	.593	.559	.829	.675	79°	1.379	.982	.191	5.145
35°	.611	.574	.819	.700	80°	1.396	.985	.174	5.671
36°	.628	.588	.809	.727	81°	1.414	.988	.156	6.314
37°	.646	.602	.799	.754	82°	1.431	.990	.139	7.115
38°	.663	.616	.788	.781	83°	1.449	.993	.122	8.144
39°	.681	.629	.777	.810	84°	1.466	.995	.105	9.514
40°	.698	.643	.766	.839	85°	1.484	.996	.087	11.43
41°	.716	.658	.755	.869	86°	1.501	.998	.070	14.30
42°	.733	.669	.743	.900	87°	1.518	.999	.052	19.08
43°	.751	.682	.731	.933	88°	1.536	.999	.035	28.64
44°	.768	.695	.719	.966	89°	1.553	1.000	.018	57.29
45°	.785	.707	.707	1.000	90°	1.571	1.000	.000	∞

TABLE B7 Common Logarithms (base 10)

Examples: log 3.12 = 0.4942
log 256 = log (2.56×10^2) = 2.4082

No.	0	1	2	3	4	5	6	7	8	9	1	2	3	4	5	6	7	8	9
10	0000	0043	0086	0128	0170	0212	0253	0294	0334	0374	4	8	12	17	21	25	29	33	37
11	0414	0453	0492	0531	0569	0607	0645	0682	0719	0755	4	8	11	15	19	23	26	30	34
12	0792	0828	0864	0899	0934	0969	1004	1038	1072	1106	3	7	10	14	17	21	24	28	31
13	1139	1173	1206	1239	1271	1303	1335	1367	1399	1430	3	6	10	13	16	19	23	26	29
14	1461	1492	1523	1553	1584	1614	1644	1673	1703	1732	3	6	9	12	15	18	21	24	27
15	1761	1790	1818	1847	1875	1903	1931	1959	1987	2014	3	6	8	11	14	17	20	22	25
16	2041	2068	2095	2122	2148	2175	2201	2227	2253	2279	3	5	8	11	13	16	18	21	24
17	2304	2330	2355	2380	2405	2430	2455	2480	2504	2529	2	5	7	10	12	15	17	20	22
18	2553	2577	2601	2625	2648	2672	2695	2718	2742	2765	2	5	7	9	12	14	16	19	21
19	2788	2810	2833	2856	2878	2900	2923	2945	2967	2989	2	4	7	9	11	13	16	18	20
20	3010	3032	3054	3075	3096	3118	3139	3160	3181	3201	2	4	6	8	11	13	15	17	19
21	3222	3243	3263	3284	3304	3324	3345	3365	3385	3404	2	4	6	8	10	12	14	16	18
22	3424	3444	3464	3483	3502	3522	3541	3560	3579	3598	2	4	6	8	10	12	14	15	17
23	3617	3636	3655	3674	3692	3711	3729	3747	3766	3784	2	4	6	7	9	11	13	15	17
24	3802	3820	3838	3856	3874	3892	3909	3927	3945	3962	2	4	5	7	9	11	12	14	16
25	3979	3997	4014	4031	4048	4065	4082	4099	4116	4133	2	3	5	7	9	10	12	14	15
26	4150	4166	4183	4200	4216	4232	4249	4265	4281	4298	2	3	5	7	8	10	11	13	15
27	4314	4330	4346	4362	4378	4393	4409	4425	4440	4456	2	3	5	6	8	9	11	13	14
28	4472	4487	4502	4518	4533	4548	4564	4579	4594	4609	2	3	5	6	8	9	11	12	14
29	4624	4639	4654	4669	4683	4698	4713	4728	4742	4757	1	3	4	6	7	9	10	12	13
30	4771	4786	4800	4814	4829	4843	4857	4871	4886	4900	1	3	4	6	7	9	10	11	13
31	4914	4928	4942	4955	4969	4983	4997	5011	5024	5038	1	3	4	6	7	8	10	11	12
32	5051	5065	5079	5092	5105	5119	5132	5145	5159	5172	1	3	4	5	7	8	9	11	12
33	5185	5198	5211	5224	5237	5250	5263	5276	5289	5302	1	3	4	5	6	8	9	10	12
34	5313	5328	5340	5353	5366	5378	5391	5403	5416	5428	1	3	4	5	6	8	9	10	11
35	5441	5453	5465	5478	5490	5502	5514	5527	5539	5551	1	2	4	5	6	7	9	10	11
36	5563	5575	5587	5599	5611	5623	5635	5647	5658	5670	1	2	4	5	6	7	8	10	11
37	5682	5694	5705	5717	5729	5740	5752	5763	5775	5786	1	2	3	5	6	7	8	9	10
38	5798	5809	5821	5832	5843	5855	5866	5877	5888	5899	1	2	3	5	6	7	8	9	10
39	5911	5922	5933	5944	5955	5966	5977	5988	5999	6010	1	2	3	4	5	7	8	9	10
40	6021	6031	6042	6053	6064	6075	6085	6096	6107	6117	1	2	3	4	5	6	8	9	10
41	6128	6138	6149	6160	6170	6180	6191	6201	6212	6222	1	2	3	4	5	6	7	8	9
42	6232	6243	6253	6263	6274	6284	6294	6304	6314	6325	1	2	3	4	5	6	7	8	9
43	6335	6345	6355	6365	6375	6385	6395	6405	6415	6425	1	2	3	4	5	6	7	8	9
44	6435	6444	6454	6464	6474	6484	6493	6503	6513	6522	1	2	3	4	5	6	7	8	9
45	6532	6542	6551	6561	6571	6580	6590	6599	6609	6618	1	2	3	4	5	6	7	8	9
46	6628	6637	6646	6656	6665	6675	6684	6693	6702	6712	1	2	3	4	5	6	7	7	8
47	6721	6730	6739	6749	6758	6767	6776	6785	6794	6803	1	2	3	4	5	5	6	7	8
48	6812	6821	6830	6839	6848	6857	6866	6875	6884	6893	1	2	3	4	4	5	6	7	8
49	6902	6911	6920	6928	6937	6946	6955	6964	6972	6981	1	2	3	4	4	5	6	7	8
50	6990	6998	7007	7016	7024	7033	7042	7050	7059	7067	1	2	3	3	4	5	6	7	8
51	7076	7084	7093	7101	7110	7118	7126	7135	7143	7152	1	2	3	3	4	5	6	7	8
52	7160	7168	7177	7185	7193	7202	7210	7218	7226	7235	1	2	2	3	4	5	6	7	7
53	7243	7251	7259	7267	7275	7284	7292	7300	7308	7316	1	2	2	3	4	5	6	6	7
54	7324	7332	7340	7348	7356	7364	7372	7380	7388	7396	1	2	2	3	4	5	6	6	7
	0	1	2	3	4	5	6	7	8	9	1	2	3	4	5	6	7	8	9

TABLE B7 (Continued)

No.	0	1	2	3	4	5	6	7	8	9	1	2	3	4	5	6	7	8	9
55	7404	7412	7419	7427	7435	7443	7451	7459	7466	7474	1	2	2	3	4	5	5	6	7
56	7482	7490	7497	7505	7513	7520	7528	7536	7543	7551	1	2	2	3	4	5	5	6	7
57	7559	7566	7574	7582	7589	7597	7604	7612	7619	7627	1	2	2	3	4	5	5	6	7
58	7634	7642	7649	7657	7664	7672	7679	7686	7694	7701	1	1	2	3	4	4	5	6	7
59	7709	7716	7723	7731	7738	7745	7752	7760	7767	7774	1	1	2	3	4	4	5	6	7
60	7782	7789	7796	7803	7810	7818	7825	7832	7839	7846	1	1	2	3	4	4	5	6	6
61	7853	7860	7868	7875	7882	7889	7896	7903	7910	7917	1	1	2	3	4	4	5	6	6
62	7924	7931	7938	7945	7952	7959	7966	7973	7980	7987	1	1	2	3	3	4	5	6	6
63	7992	8000	8007	8014	8021	8028	8035	8041	8048	8055	1	1	2	3	3	4	5	5	6
64	8062	8069	8075	8082	8089	8096	8102	8109	8116	8122	1	1	2	3	3	4	5	5	6
65	8129	8136	8142	8149	8156	8162	8169	8176	8182	8189	1	1	2	3	3	4	5	5	6
66	8195	8202	8209	8215	8222	8228	8235	8241	8248	8254	1	1	2	3	3	4	5	5	6
67	8261	8267	8274	8280	8287	8293	8299	8306	8312	8319	1	1	2	3	3	4	5	5	6
68	8325	8331	8338	8344	8351	8357	8363	8370	8376	8382	1	1	2	3	3	4	4	5	6
69	8388	8395	8401	8407	8414	8420	8426	8432	8439	8445	1	1	2	2	3	4	4	5	6
70	8451	8457	8463	8470	8476	8482	8488	8494	8500	8506	1	1	2	2	3	4	4	5	6
71	8513	8519	8525	8531	8537	8543	8549	8555	8561	8567	1	1	2	2	3	4	4	5	5
72	8573	8579	8585	8591	8597	8603	8609	8615	8621	8627	1	1	2	2	3	4	4	5	5
73	8633	8639	8645	8651	8657	8663	8669	8675	8681	8686	1	1	2	2	3	4	4	5	5
74	8692	8698	8704	8710	8716	8722	8727	8733	8739	8745	1	1	2	2	3	4	4	5	5
75	8751	8756	8762	8768	8774	8779	8785	8791	8797	8802	1	1	2	2	3	3	4	5	5
76	8808	8814	8820	8825	8831	8837	8842	8848	8854	8859	1	1	2	2	3	3	4	5	5
77	8865	8871	8876	8882	8887	8893	8899	8904	8910	8915	1	1	2	2	3	3	4	4	5
78	8921	8927	8932	8938	8943	8949	8954	8960	8965	8971	1	1	2	2	3	3	4	4	5
79	8976	8982	8987	8993	8998	9004	9009	9015	9020	9025	1	1	2	2	3	3	4	4	5
80	9031	9036	9042	9047	9053	9058	9063	9069	9074	9079	1	1	2	2	3	3	4	4	5
81	9085	9090	9096	9101	9106	9112	9117	9122	9128	9133	1	1	2	2	3	3	4	4	5
82	9138	9143	9149	9154	9159	9165	9170	9175	9180	9186	1	1	2	2	3	3	4	4	5
83	9191	9196	9201	9206	9212	9217	9222	9227	9232	9238	1	1	2	2	3	3	4	4	5
84	9243	9248	9253	9258	9263	9269	9274	9279	9284	9289	1	1	2	2	3	3	4	4	5
85	9294	9299	9304	9309	9315	9320	9325	9330	9335	9340	1	1	2	2	3	3	4	4	5
86	9345	9350	9355	9360	9365	9370	9375	9380	9385	9390	1	1	2	2	3	3	4	4	5
87	9395	9400	9405	9410	9415	9420	9425	9430	9435	9440	0	1	1	2	2	3	3	4	4
88	9445	9450	9455	9460	9465	9469	9474	9479	9484	9489	0	1	1	2	2	3	3	4	4
89	9494	9499	9504	9509	9513	9518	9523	9528	9533	9538	0	1	1	2	2	3	3	4	4
90	9542	9547	9552	9557	9562	9566	9571	9576	9581	9586	0	1	1	2	2	3	3	4	4
91	9590	9595	9600	9605	9609	9614	9619	9624	9628	9633	0	1	1	2	2	3	3	4	4
92	9638	9643	9647	9652	9657	9661	9666	9671	9675	9680	0	1	1	2	2	3	3	4	4
93	9685	9689	9694	9699	9703	9708	9713	9717	9722	9727	0	1	1	2	2	3	3	4	4
94	9731	9736	9741	9745	9750	9754	9759	9763	9768	9773	0	1	1	2	2	3	3	4	4
95	9777	9782	9786	9791	9795	9800	9805	9809	9814	9818	0	1	1	2	2	3	3	4	4
96	9823	9827	9832	9836	9841	9845	9850	9854	9859	9863	0	1	1	2	2	3	3	4	4
97	9868	9872	9877	9881	9886	9890	9894	9899	9903	9908	0	1	1	2	2	3	3	4	4
98	9912	9917	9921	9926	9930	9934	9939	9943	9948	9952	0	1	1	2	2	3	3	4	4
99	9956	9961	9965	9969	9974	9978	9983	9987	9991	9996	0	1	1	2	2	3	3	3	4
	0	1	2	3	4	5	6	7	8	9	1	2	3	4	5	6	7	8	9

TABLE B8 Natural or Naperian Logarithms (base e)

To find the natural logarithm of a number that is $\frac{1}{10}, \frac{1}{100}, \frac{1}{1000}$, etc., of a number of whose logarithm is given, subtract from the given logarithm $\log_e 10$, $2 \log_e 10$, $3 \log_e 10$, etc.

To find the natural logarithm of a number that is 10, 100, 1000, etc., times a number whose logarithm is given, add to the given logarithm $\log_e 10$, $2 \log_e 10$, $3 \log_e 10$, etc.

$\log_e 10 =$ 2.30258 50930	$6 \log_e 10 =$ 13.81551 05580
$2 \log_e 10 =$ 4.60517 01860	$7 \log_e 10 =$ 16.11809 56510
$3 \log_e 10 =$ 6.90775 52790	$8 \log_e 10 =$ 18.42068 07440
$4 \log_e 10 =$ 9.21034 03720	$9 \log_e 10 =$ 20.72326 58369
$5 \log_e 10 =$ 11.51292 54650	$10 \log_e 10 =$ 23.02585 09299

1.00–4.99

N	0	1	2	3	4	5	6	7	8	9
1.0	0.00000	.00995	.01980	.02956	.03922	.04879	.05827	.06766	.07696	.08618
.1	.09531	.10436	.11333	.12222	.13103	.13976	.14842	.15700	.16551	.17395
.2	.18232	.19062	.19885	.20701	.21511	.22314	.23111	.23902	.24686	.25464
.3	.26236	.27003	.27763	.28518	.29267	.30010	.30748	.31481	.32208	.32930
.4	.33647	.34359	.35066	.35767	.36464	.37156	.37844	.38526	.39204	.39878
.5	.40547	.41211	.41871	.42527	.43178	.43825	.44469	.45108	.45742	.46373
.6	.47000	.47623	.48243	.48858	.49470	.50078	.50682	.51282	.51879	.52473
.7	.53063	.53649	.54232	.54812	.55389	.55962	.56531	.57098	.57661	.58222
.8	.58779	.59333	.59884	.60432	.60977	.61519	.62058	.62594	.63127	.63658
.9	.64185	.64710	.65233	.65752	.66269	.66783	.67294	.67803	68310	.68813
2.0	0.69315	.69813	.70310	.70804	.71295	.71784	.72271	.72755	.73237	.73716
.1	.74194	.74669	.75142	.75612	.76081	.76547	.77011	.77473	.77932	.78390
.2	.78846	.79299	.79751	.80200	.80648	.81093	.81536	.81978	.82418	.82855
.3	.83291	.83725	.84157	.84587	.85015	.85442	.85866	.86289	.86710	.87129
.4	.87547	.87963	.88377	.88789	.89200	.89609	.90016	.90422	.90826	.91228
.5	.91629	.92028	.92426	.92822	.93216	.93609	.94001	.94391	.94779	.95166
.6	.95551	.95935	.96317	.96698	.97078	.97456	.97833	.98208	.98582	.98954
.7	.99325	.99695	*.00063	*.00430	*.00796	*.01160	*.01523	*.01885	*.02245	*.02604
.8	1.02962	.03318	.03674	.04028	.04380	.04732	.05082	.05431	.05779	.06126
.9	.06471	.06815	.07158	.07500	.07841	.08181	.08519	.08856	.09192	.09527
3.0	1.09861	.10194	.10526	.10856	.11186	.11514	.11841	.12168	.12493	.12817
.1	.13140	.13462	.13783	.14103	.14422	.14740	.15057	.15373	.15688	.16002
.2	.16315	.16627	.16938	.17248	.17557	.17865	.18173	.18479	.18784	.19089
.3	.19392	.19695	.19996	.20297	.20597	.20896	.21194	.21491	.21788	.22083
.4	.22378	.22671	.22964	.23256	.23547	.23837	.24127	.24415	.24703	.24990
.5	.25276	.25562	.25846	.26130	.26413	.26695	.26976	.27257	.27536	.27815
.6	.28093	.28371	.28647	.28923	.29198	.29473	.29746	.30019	.30291	.30563
.7	.30833	.31103	.31372	.31641	.31909	.32176	.32442	.32708	.32972	.33237
.8	.33500	.33763	.34025	.34286	.34547	.34807	.35067	.35325	.35584	.35841
.9	.36098	.36354	.36609	.36864	.37118	.37372	.37624	.37877	.38128	.38379
4.0	1.38629	.38879	.39128	.39377	.39624	.39872	.40118	.40364	.40610	.40854
.1	.41099	.41342	.41585	.41828	.42070	.42311	.42552	.42792	.43031	.43270
.2	.43508	.43746	.43984	.44220	.44456	.44692	.44927	.45161	.45395	.45629
.3	.45862	.46094	.46326	.46557	.46787	.47018	.47247	.47476	.47705	.47933
.4	.48160	.48387	.48614	.48840	.49065	.49290	.49515	.49739	.49962	.50185
.5	.50408	.50630	.50851	.51072	.51293	.51513	.51732	.51951	.52170	.52388
.6	.52606	.52823	.53039	.53256	.53471	.53687	.53902	.54116	.54330	.54543
.7	.54756	.54969	.55181	.55393	.55604	.55814	.56025	.56235	.56444	.56653
.8	.56862	.57070	.57277	.57485	.57691	.57898	.58104	.58309	.58515	.58719
.9	.58924	.59127	.59331	.59534	.59737	.59939	.60141	.60342	.60543	.60744

*Starred decimals should be preceded by the integer on the next line.

TABLE B8 Natural or Naperian Logarithms (Continued)

5.00–9.99

N	0	1	2	3	4	5	6	7	8	9
5.0	1.60944	.61144	.61343	.61542	.61741	.61939	.62137	.62334	.62531	.62728
.1	.62924	.63120	.63315	.63511	.63705	.63900	.64094	.64287	.64481	.64673
.2	.64866	.65058	.65250	.65411	.65632	.65823	.66013	.66203	.66393	.66582
.3	.66771	.66959	.67147	.67335	.67523	.67710	.67896	.68083	.68269	.68455
.4	.68640	.68825	.69010	.69194	.69378	.69562	.69745	.69928	.70111	.70293
.5	.70475	.70656	.70838	.71019	.71199	.71380	.71560	.71740	.71919	.72098
.6	.72277	.72455	.72633	.72811	.72988	.73166	.73342	.73519	.73695	.73871
.7	.74047	.74222	.74397	.74572	.74746	.74920	.75094	.75267	.75440	.75613
.8	.75786	.75958	.76130	.76302	.76473	.76644	.76815	.76985	.77156	.77326
.9	.77495	.77665	.77834	.78002	.78171	.78339	.78507	.78675	.78842	.79009
6.0	1.79176	.79342	.79509	.79675	.79840	.80006	.80171	.80336	.80500	.80665
.1	.80829	.80993	.81156	.81319	.81482	.81645	.81808	.81970	.82132	.82294
.2	.82455	.82616	.82777	.82938	.83098	.83258	.83418	.83578	.83737	.83896
.3	.84055	.84214	.84372	.84530	.84688	.84845	.85003	.85160	.85317	.85473
.4	.85630	.85786	.85942	.86097	.86253	.86408	.86563	.86718	.86872	.87026
.5	.87180	.87334	.87487	.87641	.87794	.87947	.88099	.88251	.88403	.88555
.6	.88707	.88858	.89010	.89160	.89311	.89462	.89612	.89762	.89912	.90061
.7	.90211	.90360	.90509	.90658	.90806	.90954	.91102	.91250	.91398	.91545
.8	.91692	.91839	.91986	.92132	.92279	.92425	.92571	.92716	.92862	.93007
.9	.93152	.93297	.93442	.93586	.93730	.93874	.94018	.94162	.94305	.94448
7.0	1.94591	.94734	.94876	.95019	.95161	.95303	.95445	.95586	.95727	.95869
.1	.96009	.96150	.96291	.96431	.96571	.96711	.96851	.96991	.97130	.97269
.2	.97408	.97547	.97685	.97824	.97962	.98100	.98238	.98376	.98513	.98650
.3	.98787	.98924	.99061	.99198	.99334	.99470	.99606	.99742	.99877	*.00013
.4	2.00148	.00283	.00418	.00553	.00687	.00821	.00956	.01089	.01223	.01357
.5	.01490	.01624	.01757	.01890	.02022	.02155	.02287	.02419	.02551	.02683
.6	.02815	.02946	.03078	.03209	.03340	.03471	.03601	.03732	.03862	.03992
.7	.04122	.04252	.04381	.04511	.04640	.04769	.04898	.05027	.05156	.05284
.8	.05412	.05540	.05668	.05796	.05924	.06051	.06179	.06306	.06433	.06560
.9	.06686	.06813	.06939	.07065	.07191	.07317	.07443	.07568	.07694	.07819
8.0	2.07944	.08069	.08194	.08318	.08443	.08567	.08691	.08815	.08939	.09063
.1	.09186	.09310	.09433	.09556	.09679	.09802	.09924	.10047	.10169	.10291
.2	.10413	.10535	.10657	.10779	.10900	.11021	.11142	.11263	.11384	.11505
.3	.11626	.11746	.11866	.11986	.12106	.12226	.12346	.12465	.12585	.12704
.4	.12823	.12942	.13061	.13180	.13298	.13417	.13535	.13653	.13771	.13889
.5	.14007	.14124	.14242	.14359	.14476	.14593	.14710	.14827	.14943	.15060
.6	.15176	.15292	.15409	.15524	.15640	.15756	.15871	.15987	.16102	.16217
.7	.16332	.16447	.16562	.16677	.16791	.16905	.17020	.17134	.17248	.17361
.8	.17475	.17589	.17702	.17816	.17929	.18042	.18155	.18267	.18380	.18493
.9	.18605	.18717	.18830	.18942	.19054	.19165	.19277	.19389	.19500	.19611
9.0	2.19722	.19834	.19944	.20055	.20166	.20276	.20387	.20497	.20607	.20717
.1	.20827	.20937	.21047	.21157	.21266	.21375	.21485	.21594	.21703	.21812
.2	.21920	.22029	.22138	.22246	.22354	.22462	.22570	.22678	.22786	.22894
.3	.23001	.23109	.23216	.23324	.23431	.23538	.23645	.23751	.23858	.23965
.4	.24071	.24177	.24284	.24390	.24496	.24601	.24707	.24813	.24918	.25024
.5	.25129	.25234	.25339	.25444	.25549	.25654	.25759	.25863	.25968	.26072
.6	.26176	.26280	.26384	.26488	.26592	.26696	.26799	.26903	.27006	.27109
.7	.27213	.27316	.27419	.27521	.27624	.27727	.27829	.27932	.28034	.28136
.8	.28238	.28340	.28442	.28544	.28646	.28747	.28849	.28950	.29051	.29152
.9	.29253	.29354	.29455	.29556	.29657	.29757	.29858	.29958	.30058	.30158

1 2 3 4 5 6 7 8 9 0